# THE IMPACT OF SPACE EXPERIMENTS ON OUR KNOWLEDGE OF THE PHYSICS OF THE UNIVERSE

# THE IMPACT OF SPACE EXPERIMENTS ON OUR KNOWLEDGE OF THE PHYSICS OF THE UNIVERSE

*by*

FRANCO GIOVANNELLI
*Instituto di Astrofisica Spaziale e Fisica Cosmica, CNR, Roma, Italy*

and

LOLA SABAU-GRAZIATI
*Departamento de Ciencias del Espacia y Tecnologías Electronícas, INTA, Torrejón de Ardóz, Spain*

Reprinted from *Space Science Reviews*, Volume 112, Nos. 1–4, 2004

A.C.I.P. Catalogue record for this book is available from the Library of Congress

ISBN 978-90-481-6621-3 ISBN 978-94-017-3358-8 (eBook)
DOI 10.1007/978-94-017-3358-8

---

*Printed on acid-free paper*

TABLE OF CONTENTS

# THE IMPACT OF SPACE EXPERIMENTS ON OUR KNOWLEDGE OF THE PHYSICS OF THE UNIVERSE*

FRANCO GIOVANNELLI[1] and LOLA SABAU-GRAZIATI[2]
[1]*Istituto di Astrofisica Spaziale e Fisica Cosmica, CNR – Area di Ricerca di Roma–2*
*Via del Fosso del Cavaliere 100, I 00133, Roma, Italy*
[2]*Departamento de Ciencias del Espacio y Tecnologías Electrónicas, INTA*
*Ctra de Ajalvir Km 4, E 28850 Torrejón de Ardóz, Spain*

**Abstract.** With the advent of space experiments it was demonstrated that cosmic sources emit energy practically across all the electromagnetic spectrum via different physical processes. Several physical quantities give witness to these processes which usually are not stationary; those physical observable quantities are then generally variable. Therefore *simultaneous multifrequency observations* are strictly necessary in order to understand the actual behaviour of cosmic sources.

Space experiments have opened practically all the electromagnetic windows on the Universe. A discussion of the most important results coming from multifrequency *photonic astrophysics* experiments will provide new inputs for the advance of the knowledge of the physics, very often in its more extreme conditions.

A multitude of high quality data across practically the whole electromagnetic spectrum came at the scientific community's disposal a few years after the beginning of the Space Era. With these data we are attempting to explain the physics governing the Universe and, moreover, its origin, which has been and still is a matter of the greatest curiosity for humanity.

In this paper we will try to describe the last steps of the investigation born with the advent of space experiments, to note upon the most important results and open problems still existing, and to comment upon the perspectives we can reasonably expect.

Once the idea of this paper was well accepted by ourselves, we had the problem of how to plan the exposition. Indeed, the exposition of the results can be made in different ways, following several points of view, according to:

- a division in diffuse and discrete sources;
- different classes of cosmic sources;
- different spectral ranges, which implies in turn a sub-classification in accordance with different techniques of observations;
- different physical emission mechanisms of electromagnetic radiation;
- different vehicles used for launching the experiments (aircraft, balloons, rockets, satellites, observatories).

In order to exhaustively present *The Impact of Space Experiments on our Knowledge of the Physics of the Universe* it would then have been necessary to write a kind of Encyclopaedia of the Astronomical Space Research, which is not our desire. On the contrary, since our goal is to provide an useful tool for the reader who has not specialized in space astrophysics and for the students, we decided to write this paper in the form of a review, the length of which can be still considered reasonable, taking into account the complexity of the arguments discussed.

*Editor's Note: Due to the breadth and length of this paper and the mixture of topics in each chapter, the level of refereeing has been less stringent than usual.

*Space Science Reviews* **112:** 1–443, 2004.

Because of the impossibility of realizing a complete picture of the physics governing the Universe, we were obliged to select how to proceed, the subjects to be discussed the more or the less, or those to be rejected.

Because this work was born in the Ph.D. thesis of one of us (LSG) (Sabau-Graziati, 1990) we decided to follow the 'astronomical tradition' used there, namely: the spectral energy ranges. Although such energy ranges do not determine physical objects (even if in many cases such ranges are used to define the sources as: radio, infrared, optical, ultraviolet, X-ray, $\gamma$-ray emitters), they do determine the methods of study, and from the technical point of view they define the technology employed in the relative experiments. However, since then we have decided to avoid a deep description of the experiments, satellites, and observatories, simply to grant a preference to the physical results, rather than to technologies, however fundamental for obtaining those results.

The exposition, after an introduction (Section 1) and some crucial results from space astronomy (Section 2), has been focussed into three parts: the physics of the diffuse cosmic sources deduced from space experiments (Section 3), the physics of cosmic rays from ground- and space-based experiments (Section 4), and the physics of discrete cosmic sources deduced from space experiments (Section 5). In this first part of the paper we have used the logic of describing the main results obtained in different energy ranges, which in turn characterize the experiments on board space vehicles. Within each energy range we have discussed the contributions to the knowledge of various kind of cosmic sources coming from different experiments. And this part is mainly derived by the bulk of the introductory part of LSG's Ph.D. thesis.

In the second part of the paper, starting from Section 6, we have preferred to discuss several classes of cosmic sources independently of the energy ranges, mainly focussing the results from a multifrequency point of view, making a preference for the knowledge of the physics governing the whole class. This was decided also because of the multitude of new space experiments launched in the last fifteen years, which would have rendered almost impossible a discussion of the results divided into energy ranges without weakening the construction of the entire puzzle.

We do not pretend to cover every aspect of every subject considered under the heading of *the physics of the universe*. Instead a cross section of essays on historical, modern, and philosophical topics are offered and combined with personal views into tricks of the space astrophysics trade.

The reader is, then, invited to accept this paper even though it obviously lacks completeness and the arguments discussed are certainly biased by a selection effect owed essentially to our knowledge, and to it being of a reasonable length.

Some parts of it could seem, in certain sense, to belong to an older paper, in which the 'news' is not reported. But this is owed to our own choice, just in full accord with the goals of the text: we want to present those results which have, in our opinion, been really important, in the development of the science. These impacting results do not necessarily constitute the last news.

This text was formally closed just on the day of the launch of the INTEGRAL satellite: October 17, 2002. After that date only finishing touches have been added.

**Table of Contents**

## 1. Introduction

### 1.1. Historical background

The beginning of the Space Astrophysics Era is commonly located around the end of the fifties of the last century with the first space experiments, in the energy range 0.2–0.5 MeV, on board balloons. They were devoted to the detection of $\gamma$-rays generated in solar activity (Peterson & Winckler, 1958). But actually $\gamma$-ray astronomy was born in the last year of the XIXth century with the discoveries of penetrating gamma radiation (Villard, 1900), and the atmospheric ionization (Wilson, 1900). Wilson suggested that the extraterrestrial gamma radiation could be responsible for the atmospheric ionization. With balloon flights Hess (1912) demonstrated the extraterrestrial and extra-solar origin of the ionizing radiation, which was called *Cosmic Rays*. Until 1927 it was thought that cosmic rays consisted of $\gamma$-rays. Thanks to the discovery of the dependence of the cosmic ray flux on the geomagnetic latitude during a trip from Java to Genoa made by Clay (1927), it was possible to recognize that the composition of cosmic rays included charged particles. Later Hayakawa (1952) determined the contribution of $\gamma$-rays to the composition of cosmic rays as less than 1%.

The experiments outside the atmosphere started in 1946, soon after the end of the second world war, when the Naval Research Laboratory (NRL) launched a V2 rocket with a payload which observed the Sun's UV spectrum.

Since that time many space experiments were prepared and several fundamental results were reached, as summarized in Table 1 (updated from Bonnet, 1992). In our opinion the actual beginning of the Space Era for studying the Universe is the year 1962. An X-ray experiment – prepared by Giacconi, Gursky, Paolini & Rossi – launched on board an Aerobee rocket discovered a strong X-ray emission from an extra-solar object, namely Sco X-1 (Giacconi *et al.*, 1962). After this first historical experiment many others were launched on board rockets and later balloon and satellites. These experiments brought to our knowledge a really and until that time unknown, of an X-ray sky which started to give experimental proofs of the first theories of Baade & Zwicky (1934) about the possible existence of neutron stars. Space orbiting observatories with larger and more sophisticated experiments – from Uhuru launched in 1970 (Giacconi *et al.*, 1971) up to HEAO-1 launched in 1977 (Wood *et al.*, 1984) – discovered the most luminous galactic and extragalactic X-ray sources, such as pulsars, X-ray binaries, supernova remnants (SNRs), bursters, and active galactic nuclei (AGNs). But the qualitative jump in the observational capabilities was obtained with the HEAO-2 satellite (Einstein) in which the X-ray focussing optics of the instruments enhanced the sensitivity in the soft X-ray range by a factor of about 1000 with respect to the old generation of detectors. Also the angular resolution was improved up to ~2 arcsec. This allowed a re-definition of the positions of the already known X-ray sources and the discovery of a large number of weaker ones, such as the *normal galaxies* and *normal stars* spread on the entire HR diagram (Vaiana *et al.*, 1981).

The detected X-ray fluxes from these stars are definitively larger than those expected from theories of formation and heating of stellar coronae. This led to a revolution in the comprehension of the role of the star rotation and of the magnetic field in the turbulent transport of energy from the nucleus to the external parts of a star.

Thanks to the improving technologies experiments in other energy ranges, such as Gamma, UV, Optical and IR, were also launched on board rockets, balloons and satellites (see Table 1). The results from the whole set of space experiments improved the knowledge of the entire structure of the Universe and the *monochromatic optical old sky* became a *polychromatic new sky*, containing all the information necessary in understanding the physics governing the Universe itself.

Only 46 years passed from the first successful launch of an Earth satellite, 'Sputnik'. From then on astronomy has evolved so fast to be able to produce an enormous

TABLE 1

Major Past and Future Space Astronomy Missions (updated from Bonnet, 1992).

| Event or Mission | Year | Highlights |
|---|---|---|
| Balloon flight (Europe) | 1912 | Discovery of cosmic rays |
| NRL V2 rocket (USA) | 1946 | First observation of Sun's UV spectrum |
| NRL V2 rocket (USA) | 1949 | First observation of solar X-rays |
| Sputnik 1 (USSR) | 1957 | First artificial satellite |
| Explorer III (USA) | 1958 | Discovery of Earth's radiation belts |
| Balloon flight (USA) | 1958 | First observation of $\gamma$-rays from a solar flare |
| Luna 1 (USSR) | 1959 | Discovery of the solar wind |
| Luna 2 (USSR) | 1959 | First image of the Moon's hidden side |
| Aerobee rocket (USA) | 1962 | First observation of an X-ray star (Sco X1) |
| Luna 9 (USSR) | 1966 | First picture from the lunar surface |
| OAO 2 (USA) | 1968 | First orbiting astronomical observatory |
| Apollo 11 (USA) | 1969 | First human on the Moon |
| Uhuru (USA) | 1970 | First large X-ray survey from space |
| Copernicus (USA) | 1970 | First far-UV observatory |
| Skylab (USA) | 1973 | High resolution images of solar corona in X-rays |
| SAS 2 (USA) | 1973 | First $\gamma$-ray satellite |
| Mariner 10 (USA) | 1973–74 | First detailed picture of Mercury |
| COS B (ESA) | 1975 | First $\gamma$-ray map of Galaxy |
| Venera 9 (USSR) | 1975 | First picture of venusian surface |
| Viking 1, 2 (USA) | 1976 | First pictures taken on the martian surface |
| Voyager 1, 2 | 1977–79 | First images of Jupiter, Saturn, Uranus, and Neptune satellite and ring systems |
| Einstein (USA) | 1978 | First X-ray observatory |
| IUE (USA-UK-ESA) | 1978 | First international space UV observatory |
| Relict 1 (USSR) | 1983 | First large scale measurement of cosmic background anisotropy |
| IRAS (USA) | 1983 | First large IR survey from space |
| Vega 1 (USSR) | 1986 | First close encounter with Halley's comet |
| Giotto (ESA) | 1986 | First high resolution image of Halley's nucleus |
| TDRSS (USA) | 1987 | First VLBI measurement from space |
| MIR/KVANT (USSR) | 1987 | First space orbiter pad |
| Hipparcos (ESA) | 1989 | First space astrometry mission |
| Granat (USSR) | 1989 | X and $\gamma$-ray observatory |
| Galileo (USA) | 1989 | First orbiter and probe to Jupiter |
| COBE (USA) | 1989 | Cosmic background explorer |

TABLE 1
(Continued)

| Event or Mission | Year | Highlights |
|---|---|---|
| Hubble Space Telescope (USA-ESA) | 1990 | First large space observatory |
| ROSAT (FRG-UK-ESA) | 1990 | High sensitivity soft X-ray observatory |
| COMPTON-GRO (USA) | 1991 | Broad band Gamma-Ray Observatory |
| Ulysses (ESA-USA) | 1990–95 | First flight above solar poles |
| EUVE (NASA) | 1991 | Extreme UV astronomy explorer |
| ASCA (ASTRO D) (Japan) | 1993 | High sensitivity X-ray observatory |
| ISO (ESA) | 1995 | IR space observatory |
| RXTE | 1995 | Wide band X-ray Timing Explorer |
| Beppo-SAX (ASI, Italy) | 1996 | Wide band X-ray observatory |
| Cassini/Huygens (NASA-ESA) | 1997–2004 | Saturn orbiter and probe of Titan's atmosphere |
| AXAF-CHANDRA (NASA) | 1999 | Advanced X-ray astronomy facility |
| XMM (ESA) | 1999 | High throughput X-ray spectroscopy |
| Integral (ESA) | 2002 | High sensitivity wideband $\gamma$-ray observatory |
| Rosetta (ESA) | 2004 | Europe's Comet Chaser: First spacecraft to orbit a comet's nucleus |
| SPECTRUM X (CSI-USA-ESA) | 2004? | Wide band (EUV-hard X-ray) Observatory |
| AGILE (ASI, Italy) | 2004 | HE $\gamma$-ray and hard X-ray imager and X-ray spectroscopy |

amount of data. Indeed, thanks to the new windows opened onto the Universe by space experiments, starting roughly from the end of the 1970s almost all the electromagnetic spectrum was continuously surveyed. A large amount of data of excellent quality coming from space experiments has rendered the data, acquired during many centuries from the ground, only a small fraction of the total now available. Figure 1 (updated from Lena, 1988 and Giovannelli & Sabau-Graziati, 1995, 1997) shows schematically the amount of data in different wavelength regions versus time. The grey area is roughly proportional to the total amount of data. From these data a strong impetus was given to the knowledge of the physics of cosmic sources. In this paper we want to discuss the main results coming from measurements of different classes of cosmic sources made with space experiments following a general division in Diffuse Sources and Discrete Sources. However, the classical division into energy ranges is also used since this is one of the main criteria, which can be used in describing the advances in science thanks to space experiments. Such a criterion probably is not the best, but certainly the most convenient. Indeed, to a first approximation the classification of the sources as $\gamma$-ray

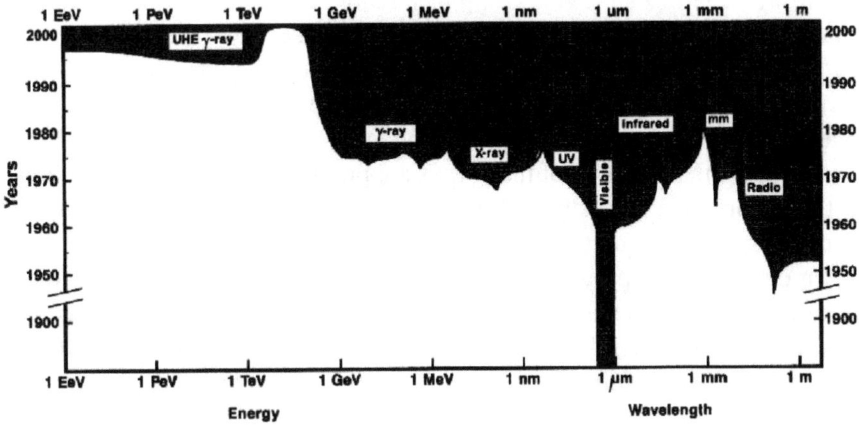


*Figure 1.* Amount of astronomical data acquired across the electromagnetic spectrum during last century (updated from Lena, 1988).

emitters, X-ray emitters, IR emitters, etc., is strictly correlated with the technology of detectors used in the experiments launched on board space-vehicles. Such experiments provide measurements in those selected energy ranges, and therefore characterize the whole space mission itself.

## 1.2. Some general remarks

We know for certain that the Universe has an absolute power limit $L_{\max} \sim \epsilon_{\rm pl}/t_{\rm pl} \sim c^5/G \sim 3.6 \times 10^{59}$ erg s$^{-1}$, where $\epsilon_{\rm pl}$ and $t_{\rm pl}$ are the Planck energy and Planck time, respectively, $c$ the light velocity and $G$ the gravitational constant. This amount of power is produced in different kind of cosmic sources, namely, the Early Universe (EU), Quasars (QSOs) and Active Galactic Nuclei (AGNs), Supernovae (SNe), Neutron Stars (NSs), and Black Holes (BHs), all types of Galaxies with their stars and Interstellar Medium (ISM), and Intergalactic Medium (IGM) (e.g., Lipunov, 1995). They radiate particles and photons at different levels of energy across the entire electromagnetic spectrum from their origins. However, the cosmic particle radiation arriving near Earth, with energies from $\sim 10^6$ to $\sim 10^{20}$ eV, is nearly isotropic, because of the galactic magnetic field, which cancels any particular directionality in the Galaxy. Such cosmic particle radiation apparently includes the nuclei of all known elements, as well as electrons, positrons, and antiprotons. So in spite of the cosmic rays being carriers of rich astrophysical information, it is very difficult to understand their message clearly. Indeed, although it is evident that they must originate in different sources, it is at the same time extremely difficult to separate the different contributions. Very reasonable attempts at searching for the origin and composition of cosmic rays have been made by many authors (e.g., Silberberg *et al.*, 1991; Biermann, 1995). In spite of this the problem of the origin of high energy cosmic rays is still far to be solved (e.g., Müller, 1993; Erlykin,

1997; Wolfendale & Zan, 1997). Help in solving the problem of the origin of cosmic rays can come from *photonic astrophysics*. Indeed, in the low energy region, below $\sim 10^{10}$ eV, $\gamma$-ray astronomy comes to the rescue by giving information about the average cosmic ray intensity across various lines of sight. The method is quite simple: the measured $\gamma$-ray intensity is divided by the column density of gas, the argument being that the bulk of the $\gamma$-rays are produced by cosmic rays interacting with the gas in the ISM. Although there are difficulties, associated with discrete sources, the actual gas components (mainly H, $H_2$, $H^+$, and He) and the division between electrons and protons (e.g., Wolfendale, 1995) it is generally agreed that the particles responsible for the $\gamma$-rays detected by satellite-borne detectors are produced in the Galaxy. The energies of the $\gamma$ rays range from $\sim 3\times 10^7$ eV to $\sim 3\times 10^{10}$ eV and the initiating particles are principally electrons up to $\sim 10^9$ eV and protons and heavier nuclei up to $\sim 10^{10}$ eV (Wolfendale, 1995). The Galactic origin identification comes from an inferred gradient of intensity, viz., a reduction in intensity with increasing Galactocentric distance (e.g., Strong *et al.*, 1978; Bhat, Mayer & Wolfendale, 1984; Bloemen, 1985).

An attempt at giving, via photonic astrophysics, the possible galactic sources of low energy cosmic rays has been discussed by Giovannelli (1997), who tentatively also quantifies the contributions of individual classes of galactic objects as sources of cosmic rays. Karakuła (1997) suggested that the highest energy cosmic ray showers are produced by Extreme High Energy (EHE) photons.

Thus photonic astrophysics appears to be the most promising and suitable probe in sounding the Universe.

With the last generation of detectors in different energy regions it has been largely demonstrated that all the cosmic sources are variable in their emission even if at very different levels in all the energy ranges. The sources considered constant (and therefore sometimes used as standard sources) appeared constant because of the limited resolutions in sensitivity (and in time) of the detectors (e.g., Giovannelli & Sabau-Graziati, 1995). So we can assume, with a high degree of confidence, that cosmic sources, and especially the high energy ones, are highly variable both in flux intensity and in the shape of the emitted spectra. These variations are the consequences of different processes occurring at different times because of different interactions of the *engines* – producing energy – with the surrounding matter and/or radiation and/or electromagnetic fields (e.g., Giovannelli & Sabau-Graziati, 1993).

Therefore multifrequency observations, possibly simultaneous, are necessary for the actual comprehension of the intriguing physics governing the behaviour of the high energy cosmic sources. These are the frontier objects between astrophysics and particle physics, and then the natural laboratories to test the most advanced theories of the modern physics (e.g., Giovannelli & Sabau-Graziati, 1994).

Exhaustive books on this topic, containing excellent reviews, have been published by the Italian Physical Society (Giovannelli & Mannocchi, 1989, 1991, 1993, 1995, 1997, 1999, 2001, 2003) and by the Italian Astronomical Society (Giovannelli & Sabau-Graziati, 1996a; 1999, 2002a, 2002b).

It is then possible to introduce the *new* hybrid field in science, namely **Multifrequency Astrophysics**, as proposed and discussed by Giovannelli & Sabau-Graziati (1996b). A pioneering book on this topic was that of the first 1984 Frascati Workshop (Giovannelli, 1985).

With passive physics experiments we observe our Universe, whilst with active physics experiments we try to reproduce some of the physical conditions and then processes occurring somewhere in the Universe. Both kinds of experiments converge to the knowledge of the physics governing the Universe.

In this paper we would like to emphasize the main results coming from multifrequency space experiments, and finally discuss the prospects for the next decades.

## 1.3. MULTIFREQUENCY ASTROPHYSICS

The philosophy of multifrequency observations of cosmic sources is now generally accepted by most experimental astrophysicists. As such, after pioneering works, begun in the 1970s (e.g., Giovannelli, 1985 and references therein) by a few people, it is very often defined *cum fide* as *researchers wasting their energies* or *mala fide* as *researchers without a clear line of investigation.* Now, for a better and faster development of science, at last this methodology is becoming the norm. Examples of it can be found in the books *Multifrequency Behaviour of High Energy Cosmic Sources* (Giovannelli & Sabau-Graziati, 1996a; 1999, 2002a, 2002b).

The importance of multifrequency astrophysics today was remarked upon by Giovannelli & Sabau-Graziati (1996b). They discussed this apparent trivial methodology in programming multifrequency measurements, and in obtaining and interpreting the experimental data. They also emphasized the extreme importance of studying physical processes independently of particular cosmic sources, in order to obtain good fits with experimental data from individual sources without making strong *ad hoc* hypotheses about their morphology and their surroundings. We will discuss this problem better in Section 6.

The present generation of experiments are sometimes *multifrequency experiments* themselves, such as for instance the BeppoSAX X-ray observatory active, until April 30, 2002, in the energy range 0.1–300 keV (e.g., Piro, 1996), and RXTE (0.2–200 keV) (Bradt, Rothschild & Swank, 1996).

A race parallel to that in rendering as wide as possible the energy ranges covered by an individual observatory is that in improving the sensitivities, angular and spectral resolutions of each instrument constituting sub-systems of that observatory (e.g., Giovannelli & Sabau-Graziati, 1994). In general each kind of improvement is the fruit of a compromise between technical performances, required by scientific goals one plans to obtain, and actual feasibility and cost.

Figure 2 shows the diameters (or effective areas) of different kind of detectors from radio to $\gamma$-ray energy ranges. The diameter $D$ of the detectors is correlated with wavelength roughly as $D \propto \lambda^{1/3}$.

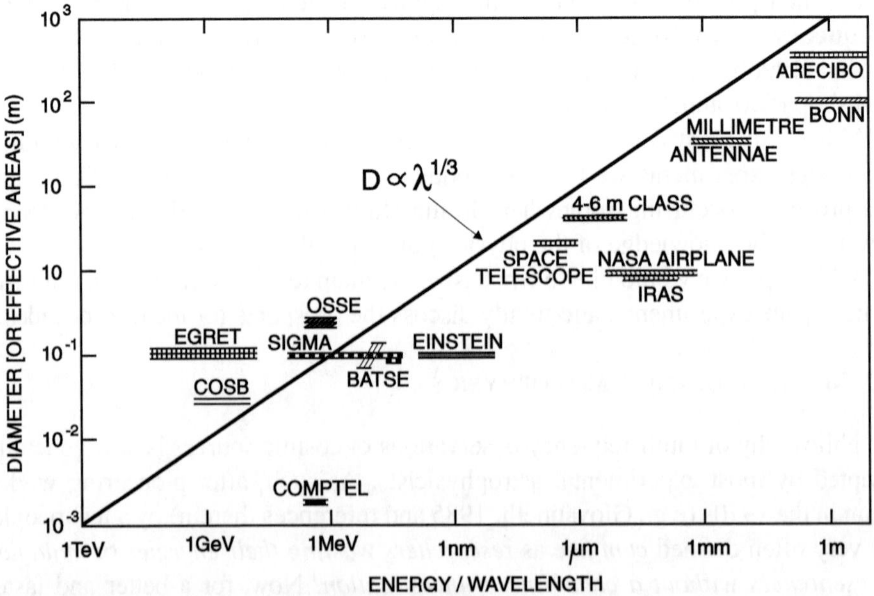


*Figure 2.* Diameters, or effective areas, of a sample of detectors from radio to $\gamma$-ray energy ranges. The curve $D \propto \lambda^{1/3}$ is superimposed (updated from Lena, 1988).

### 1.4. Active Space Observatories

Many space experiments, from IR to $\gamma$-ray energy ranges were successfully orbiting during the last ~15 years. Amongst them the IUE was definitively switched off, after 18 years of excellent behaviour, as well as the ISO, ROSAT, and CGRO (Batse, Osse, Comptel, Egret), and MIR-KVANT experiments (Art-P&S, Konus, Phebus, Sigma, Tournesol). Figure 3 shows the space observatories and some of the experiments particularly important for the study of high energy cosmic sources of roughly the last fifteen years.

They cover about twelve decades of the electromagnetic spectrum; therefore at present astrophysics is one of the most studied fields of science in which particular fruitful results are coming and rapidly improving the knowledge of the Universe. In the following section we will comment some of these crucial results.

### 1.5. Advantages of the Space Astronomy

Space Astronomy has undisputed advantages with respect to Ground Astronomy. They are:

- Absence of the restrictions owed to atmospheric absorption;
- Absence of atmospheric phenomena, such as flickering, ‘seeing’, etc.;
- Absence of restrictions owed to meteorological conditions;
- Absence of the day–night cycle;

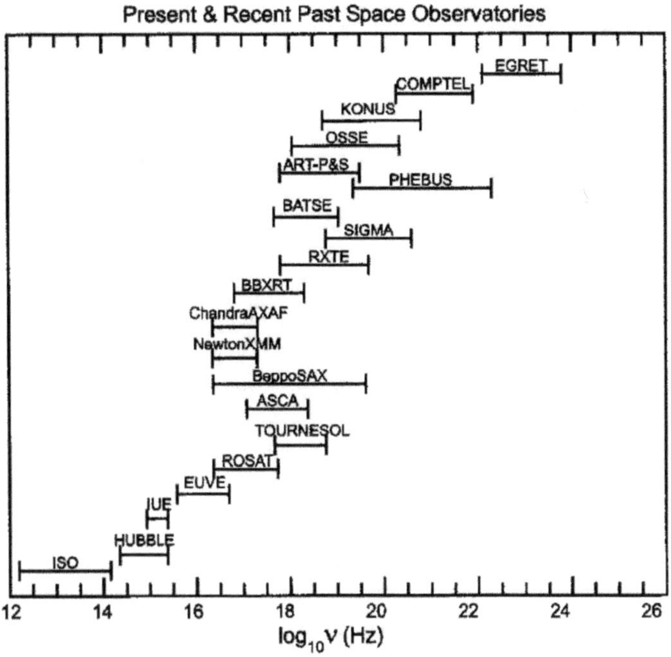


*Figure 3.* A selection of present and recent past Space Observatories from IR to $\gamma$-ray ranges.

– Access to the entire electromagnetic spectrum with only a few limitations, such as the absorption of the interstellar medium (ISM) in the extreme ultraviolet (EUV) range and the absorption of the ISM and interplanetary medium (IPM) in the radio range.

Indeed, the interstellar gas, essentially concentrated in the galactic disk and mainly comprise of neutral atomic hydrogen, imposes severe limitations on any kind of extra-solar observations in the EUV between 912 Å and 100–10 Å. The former limit corresponds to the photo-ionization threshold of hydrogen in the fundamental state (13.5 eV), the latter limit depends on the hydrogen columns density existing between the cosmic source and the observer. At long radio wavelengths (∼50–100 m) the ground based observations are affected by variable absorption and by the ionosphere's effects, and therefore they are not reliable. In contrast, space radioastronomy can extend the observations of extra-solar radio sources up to wavelengths of several thousand metres; beyond this limit, ISM becomes opaque because of free–free absorption and reflection owed to the same plasma at ∼100–300 km. This is the upper limit of wavelength for radio measurements.
Therefore, with the exception of the latter limitations, the spectral region within which it is possible to perform measurements from space ranges over about 23 orders of magnitude: $\sim 10^{-9}$–$10^{14}$ eV. This property renders the Space Astronomy one of the most powerful facilities of physics for the scientific investigation of the

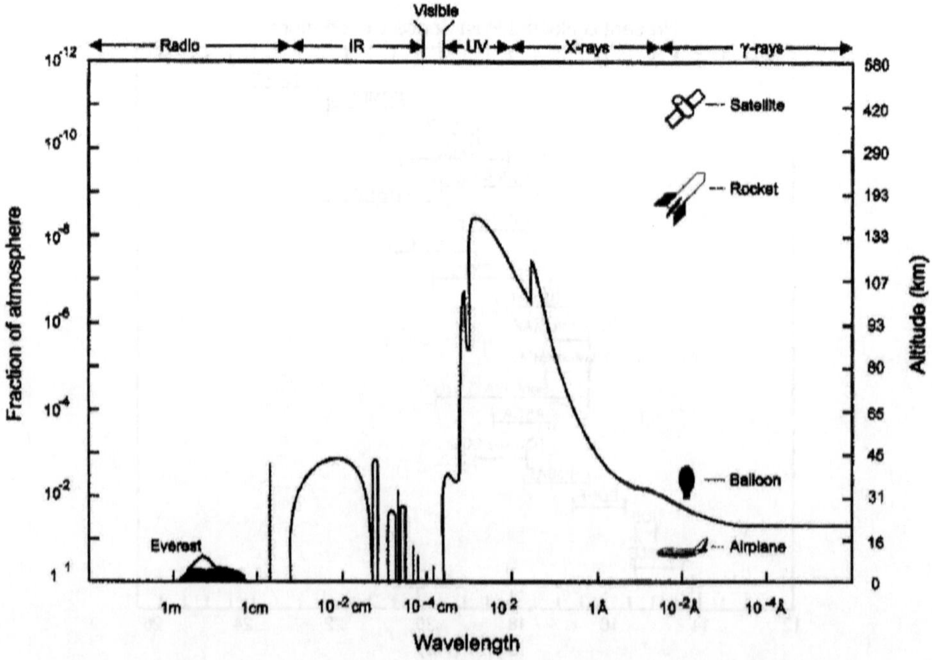


*Figure 4.* The altitude at which the Earth's atmosphere is 50% transparent to incoming electromagnetic radiation (after Giacconi *et al.*, 1968; Charles & Seward, 1995).

Universe. The spectral regions accessible for measurements entering the Earth's atmosphere become suddenly strongly limited. Figure 4 shows the altitude at which the Earth's atmosphere is 50% transparent to incoming electromagnetic radiation as a function of energy (Giacconi, Gursky & Van Speybroek, 1968; Charles & Seward, 1995).

## 2. Some Crucial Results from Space Astronomy

Only space experiments can observe most of the physical processes in cosmic sources – occurring from the Big Bang to the formation of the stars and galaxies, and until their extreme fate – which produce observable effects across the whole electromagnetic spectrum. From Table 1 it is already possible to infer the enormous impact of space experiments on science. However, a selection of some crucial results amongst the so many important ones obtained from space experiments is necessary in order to give a general idea of the impact of space experiments in accelerating the knowledge of the physics. For this reason we will briefly summarize the highlights obtained in different energy regions before discussing in more detail several selected examples.

## 2.1. A SWIFT CRUISE THROUGH COSMOLOGY

Modern physical cosmology has now converged on the Big Bang framework, although several documented facts could play an important role against the Big Bang picture (e.g., Ellis, 1984; Arp, Keys & Rudnicki, 1993).

The Big Bang theory is founded on four principal pillars:

- **Hubble expansion;**
- **Microwave background;**
- **Light element abundances;**
- **Inflation.**

The first pillar is a necessary condition for the Big Bang, but hardly sufficient since alternative cosmologies such as the steady state also include it. The second and third pillars do indeed tend to force us to an early universe which was hot and dense: this can be nicknamed a Big Bang universe (e.g., Schramm, 1993, 1998; Steigman, Hata & Felten, 1998; Rees, 1998; Walker, 1998). The fourth pillar is necessary for a flat universe (e.g., Kellerman, 1993).

The Hubble expansion can be argued by looking at the features in the spectra of redshifted galaxies. The existence of the microwave background tells us that the entire universe was at one time hot enough to be in thermal equilibrium, which required a temperature $T \geq 1$ eV. This temperature occurred – in the Big Bang model (e.g., Rees, 1989) – at a time of $\sim 10^5$ yr.

The light element abundances are a crucial argument in testing theories. Indeed, the Big Bang model predicts that when the universe was at a temperature of $\sim 10^{10}$ K ($\sim$1 second old), nuclear processes originated and would eventually yield certain very well specified abundances for the light isotopes (e.g., Boesgaard & Steigman, 1985; Rebolo, 1996; Schramm, 1998).

Giovannelli & Sabau-Graziati (1997, 1999b and the references therein) discussed on the four pillars of cosmology. The situation can be summarized as follows:

### 2.1.1. *The Hubble Expansion*

The initial phase of experimental cosmology ended three years after Hubble's death with the publication of the fundamental paper of Humason, Mayall & Sandage (1956). In this work, as in the successive ones, the discussion about the velocity–distance relationship is based on the Hubble diagram also known as the magnitude–redshift relationship. In principle the determination of the Hubble constant – whose value determines the scale and age of the present universe – is simple.

The absolute magnitude $M$ of the galaxies can be determined through the measurement of the apparent magnitude $m$ and redshift $z$. This allows the use of an equation between the modulus of the distance $m - M$ and redshift $z$, in which $H_0$ (the Hubble constant) and $\Omega_0$ (the present average mass density of the Universe) appear as parameters (Matting, 1958, 1959). However, the experimental determination of Hubble constant remains one of the fundamental problems in cosmology in spite of the numerous attempts also through the most powerful observatories such as the Hubble Space Telescope (HST).

Up to now the many determinations of $H_0$ with different techniques have given discrepant results for its value. Ground based observations of Cepheid variables in NGC 4571 give a value of $H_0 = 87 \pm 7$ km s$^{-1}$ Mpc$^{-1}$ (Pierce *et al.*, 1994); this result creates a potentially serious crisis in cosmology, since it conflicts with the ages of the oldest known stars. The distance determinations with the HST (Tammann *et al.*, 1996) give values of $H_0$ ranging between 55 and 80 km s$^{-1}$ Mpc$^{-1}$. By using the distances of all distant SNe Ia, for which the maximum of their light curves is known, and velocities ranging from 1,100 to 30,000 km s$^{-1}$, the large scale value of the Hubble constant is $H_0 = 55 \pm 10$ km s$^{-1}$ Mpc$^{-1}$. Introducing a correction for a monotonic correlation between the peak luminosity of SNIa and the decay rate after the peak (intrinsically brighter SN decay more slowly) the value of $H_0$ slightly increases to the range 60–64 km s$^{-1}$ Mpc$^{-1}$. A critical review on the determination of the Hubble constant gives a value $H_0 = 59 \pm 6$ km s$^{-1}$ Mpc$^{-1}$ (Panagia, 1999a).

Therefore $H_0$ seems to be, in the last few years, *the less constant constant* of physics.

The last unpublished paper *Reflections on the Hubble Constant* of Professor Livio Gratton (Gratton, 1990) deeply discusses all available information about the Hubble constant. A value of $52 > H_0 > 45$ km s$^{-1}$ Mpc$^{-1}$ is the result of such a discussion. Analyzing the papers on this topic already cited, **a critical evaluation of the Hubble constant is $H_0 = 56 \pm 6$ km s$^{-1}$ Mpc$^{-1}$** (Giovannelli & Sabau-Graziati, 1997).

Van den Bergh (1989) concluded that the most probable value of the Hubble constant is $H_0 = 67 \pm 8$ km s$^{-1}$ Mpc$^{-1}$. He defined his review as perhaps *the last large review of the scale of distances before the true value of $H_0$ is determined.*

How the value of the Hubble constant influences cosmology was largely discussed in the concluding remarks of the Vulcano Workshop 1996 by Lipari (1997).

Supposing the Universe closed to the critical density in light and ~30 eV neutrinos which decay radiatively with a lifetime of $\sim 10^{23}$ s, Sciama (1997) derived the parameters of the decaying hot dark matter cosmology rather precisely, namely: $m_\nu = 27.4 \pm 0.2$ eV $\longrightarrow \Omega_\nu h^2 = 0.293 \pm 0.003$, $\tau_\nu \cong$ (1–2) $\times 10^{23}$ s, $h = 0.548 \pm 0.003$, for $\Omega_\nu + \Omega_b = 1$, where $\Omega_\nu$ and $\Omega_b$ denote the fraction of the critical density in neutrinos and baryons, respectively, and $h$ the Hubble constant in units of 100 km s$^{-1}$ Mpc$^{-1}$. This shows that the Hubble constant must be ~55 km s$^{-1}$ Mpc$^{-1}$.

From such a universe, an expected UV line at $13.7 \pm 0.1$ eV should be detected (Sciama, 1998) by the EURD (Espectrógrafo Ultravioleta extremo para la Radiación Difusa) detector (Bowyer, Edelstein & Lampton, 1997) aboard the Spanish MINISAT-01, described by Giménez & Sabau-Graziati (1996). In contrast, the EURD data appear to be completely incompatible with Sciama's model of radiatively decaying massive neutrinos (Bowyer *et al.*, 1999).

### 2.1.2. *The Microwave Background*

The cosmic background radiation in the microwave band (CMBR) was first discovered by Penzias and Wilson (1965) and soon became one of the major cosmological discoveries of all time. This radiation fills the whole of space and is thought to be the remnant of the radiation of the early universe. Indeed, the universe appears to be isotropic and homogeneous at very large scales. Observationally it is found that at small scales the distribution of galaxies is inhomogeneous but become increasingly more homogeneous as the scale increases, with the greatest degree of isotropy in the universe being exhibited by the CMBR. The surface of last scattering of the CMBR is the Universe at redshift $z \sim 1,100$–1,000; this radiation contains detailed information about the history of the primeval Universe. However, the CMBR has been observed to be isotropic to an extremely high degree and may in fact be too isotropic for the good of the standard Big Bang cosmology. Without some fluctuations in the early universe it would be very hard to explain the formation of the structures in the universe we are now observing.

A small spatial anisotropy would indicate that matter was not homogeneously distributed when the CMBR originated, and small deviations with respect to the black body spectrum should indicate the presence of high energy sources in the primeval Universe. Sometimes this is noted with the analogy that *you cannot make an omelet without breaking eggs; i.e., you cannot make galaxies without disturbing the microwave background.*

The CMBR temperature in the Rayleigh–Jeans region was determined before COBE (COsmic Background Explorer) as $2.73 \pm 0.05$ K (Smoot *et al.*, 1985). This value is better than 2% with respect to that from COBE.

Measurements at 15 GHz, performed with the ground experiment in Tenerife, have established a fluctuation of $1.8 \times 10^{-5}$ on the background temperature at an angular scale of $\sim 5°$ (Watson *et al.*, 1992). This fluctuation is consistent with that found by the Differential Microwave Radiometers (DMR) of COBE. Indeed, in COBE's maps at 31, 53, and 90 GHz, a deviation of $\sim 1.1 \times 10^{-5}$ on the background temperature of $2.735 \pm 0.06$ (Mather *et al.*, 1991) – generally very stable in an angular scale of 10° – has been found (Smoot *et al.*, 1991, 1992).

Also the soviet spacecraft Relict-1 detected some anisotropy of the CMBR at 37 GHz. For an inflationary perturbation spectrum, the relative magnitude of the quadrupole component was estimated with a confidence level of 90% between $6 \times 10^{-6}$ and $3.3 \times 10^{-5}$. An anomaly of the microwave radiation has been found with a confidence level of 99%, in a region with area $\sim 1$ sr near the point $l = 150°$ and $b = -70°$. The magnitude of this anomaly is $-71 \pm 43$ $\mu$K, with 90% confidence level (Strukov *et al.*, 1992).

These measurements imply stringent limits on the energy release in the early Universe after $t \sim 1$ yr and redshift $z \sim 3 \times 10^6$.

These results are a milestone in our understanding of the large scale structure of the Universe if the temperature fluctuations are truly of cosmological origin.

How the former fluctuations are present at angular scales larger than $\sim 10°$ is still an open problem.

At the very large scale the universe is observed to be extremely isotropic and homogeneous; therefore any viable cosmological model must contain this exactly, or at least in the limit. However, at the scale of superclusters ($\cong$ 50 Mpc) observations show that the universe has some filamentary and bubble-like structures, making it generally inhomogeneous.

Excellent reviews explaining these large scale structures and cosmic evolution are those written by Longair (1991) and Alexander (1992).

The origin of the initial fluctuations is a fundamental problem in cosmology. The most likely explanation for the origin of the primordial spectrum comes from inflation, and it is stunning. The largest structures in the observed universe are owed to quantum mechanical fluctuations at microscopic scales which are stretched to astrophysical size by the tremendous expansion that occurred during inflation.

As noted by Lipari (1997) *it is hard to imagine a more remarkable connection between particle physics, astrophysics, and cosmology. Inflation is the bold idea that the universe very early on, perhaps after $10^{-32}$ seconds, underwent a burst of exponential expansion, driven by the potential energy of a scalar field locked in a 'false vacuum' state. This enormous growth could explain the flatness and smoothness of the universe, and provide together with the mechanism described above an explanation for the origin of the primordial fluctuations in the energy density of the universe. A very important prediction of the inflationary paradigm, is that the geometry of universe is, to a very good approximation, flat, or equivalently that $\Omega_{tot} = 1$.*

The growth of initial perturbations depends on the cosmological parameters $\Omega_{tot}$, $\Lambda$ and $H_0$ and on the nature of the matter composing the cosmological fluid (what fraction of baryons, neutrinos, supersymmetric particles...). $\Omega_{tot}$, $\Lambda$, and $H_0$ are the ratio of the present density to the critical density $\rho_c$ when the universe decoupled matter and radiation, the cosmological constant, and the Hubble constant, respectively.

The comparison of fluctuations at different cosmological epochs gives important constraints on both cosmological parameters and the properties of the matter composing the universe.

In order to explain the observed evolution of the density fluctuations it is necessary to invoke the existence of weakly interacting non-baryonic matter that can de-couple early from the radiation. The growth of the fluctuations at the galactic scale is significantly accelerated if the particle composing the non-baryonic matter is 'cold' when the structure is forming.

Light neutrinos (of a few eV) are historically the best candidates for the dominant component of the universe. But light neutrinos moving at relativistic velocities (then 'hot') stream out of the smaller scale (galaxy sized) inhomogeneities, clustering mostly at larger (supercluster) scales, in contrast with the observations.

We do not know of any particle with the properties required by this structure formation scenario: a particle, with a mass greater than several GeV, which interacts only weakly. We are then led to an extraordinary prediction (a theoretical and experimental challenge): most of the energy of the universe is owed to a yet unobserved particle. It is remarkable that, independently, theoretical developments in elementary particle physics have led to predictions of the existence of new particles that can have naturally the mass and interaction properties required by these cosmological considerations.

A detailed comparison of the data of the CMBR with the data of the large scale distribution of galaxies seems to indicate that a pure *cold dark matter model* requires modifications. One possibility is that there is energy density in the vacuum (a non-zero cosmological constant), another is that there is a mixture of cold and hot dark matter ($\Omega_{\rm cold} \sim 0.7$ and $\Omega_{\rm hot} \sim 0.2$) enhancing the 'power' in the very large scale structures. Therefore neutrinos, after all, could be one of the fundamental components of the energy of the universe (Lipari, 1997).

Colafrancesco (1997) discussed details of the cosmology that can be derived by multifrequency observations at different cosmic epochs of clusters of galaxies.

### 2.1.3. *The Abundances of Light Elements*

The Big Bang Nucleosynthesis (BBN) provides quantitative experimental tests of the standard and non-standard cosmological models. Nuclear reactions took place in the primordial plasma during the first minutes after the Big Bang, starting at $t \sim 1$ s, when the temperature was $\sim 10^{10}$ K.

The predicted abundances of light elements depend mainly on the universal baryonic density, which may be considered the only free parameter in the nucleosynthesis calculations.

Measurements of the primordial abundances of the light elements test the consistency of primordial nucleosynthesis and in turn set very valuable constraints on the baryonic density $\eta = n_{\rm b}/n_{\gamma}$. $\eta$ has remained constant since the positron–electron annihilation epoch to the present day. This topic has been extensively reviewed (e.g., Rebolo, 1996; Schramm, 1998; Walker, 1998).

The abundances of these light elements have all been accurately determined to be in impressive agreement with the Big Bang predictions to within the accuracy of the measurements. Furthermore, the Big Bang model predicts that the abundances would fit well only if there were no more than three families of neutrinos, and this was exactly what was observed at LEP (Large Electron Proton collider). So the light elements with abundances ranging from 76% for H to $10^{-10}$ for Li all fit with the cosmological predictions with the single adjustable parameter being the baryon density $\Omega_{\rm b} \sim 0.05 \pm 0.03$. So if one *prefers* a universe with $\Omega_{\rm tot} = 1$ one must also demand that the bulk of the matter in the universe ($\geq 90\%$) be something other than baryons (Schramm, 1993).

Figure 5 shows the BBN abundance versus baryonic density and baryon/photon ratio (Wagoner, 1992). The final abundances of nuclei produced by the range of

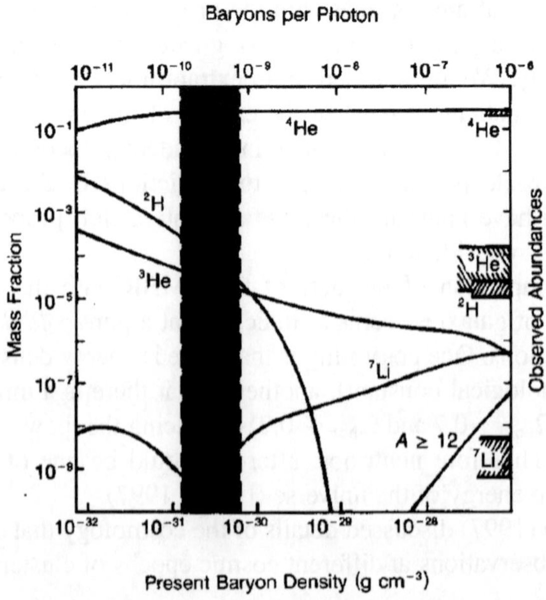


*Figure 5.* The Big Bang nucleosynthesis abundance versus baryonic density and baryon/photon ratio (Wagoner, 1992).

the SBBN model (Standard BBN) are compared with their observed abundances. The band indicates those models which are consistent with the allowed values of primordial abundances, which can be deduced from the observational data.

Figure 6 shows the density parameter $\Omega$ versus the scale at which it is determined. The baryon density is consistent with galactic halo densities but exceeds visible matter densities. Note also that only at the largest scales is there any hint that $\Omega$ exceeds $\Omega_b$.

The mass fraction of the hot gas in clusters of galaxies has been obtained through the measurements performed with ROSAT and ASCA satellites. This can provide the baryonic content, and then a test for the BBN (or alternatively an indirect determination of the Hubble constant) assuming that the baryonic content is in agreement with the BBN.

ASCA and earlier data show that the masses of clusters of galaxies are dominated by dark matter, but the hot intra-cluster gas substantially contributes (~20–25%) to the total mass in rich clusters, within a radius of 1 Mpc (e.g., Schindler, 1996; Ohashi *et al.*, 1996). This may be common to many clusters and could well indicate the primordial ratio between dark matter and baryons, considering the large spatial extent of clusters. Moreover, the fraction of the total mass which is owed to hot intra-cluster gas rises rapidly at large radii.

These results pose a number of problems for the standard cosmological models. First, the fraction of the mass in clusters which is baryonic is much larger than would be allowed by cosmic nucleosynthesis models with a critical density $\Omega_{tot} =$

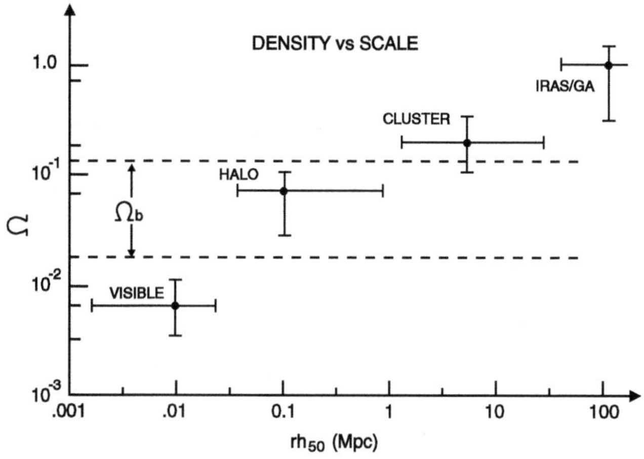


*Figure 6.* The density parameter Ω versus the scale on which it is determined (Schramm, 1993).

1. Secondly, the mass to light ratio in clusters is much smaller than would be need to close the universe (i.e. make $\Omega_{\rm tot} = 1$). Finally, the degree of central concentration of the dark matter in clusters may conflict with the idea of dissipationless dark matter. On the other hand, the galaxy distributions are poorly determined, and the gas may have been inflated by heating (Sarazin, 1996).

However, observations with the ROSAT satellite have shown rich clusters of galaxies to have a mass fraction ($\sim 0.3 \times h_{50}^{-3/2}$) of hot X-ray gas (Mushotzky, 1992). Since $\Omega_{\rm cluster} \sim 0.2$, this yields $\Omega_{\rm b} \sim 0.06$ for these clusters in reasonable agreement with BBN, but only if $h_{50} = H_0/50 \sim 1$.

This result strongly favors, once more, a value of the Hubble constant $H_0 = 50$ km s$^{-1}$ Mpc$^{-1}$, which is in agreement with the determinations already discussed.

### 2.1.4. *Inflation*

It is well known that inflation predicts a flat universe and produces Gaussian density fluctuations with a power spectrum

$$\left(\frac{\partial \rho}{\rho}\right)_K^2 \sim K^n$$

with $n \sim 1$, $K$ the wave number of the fluctuation ($K = 2\pi/L$, where $L$ is the length scale). Standard inflation yields the flat ($n = 1$) Harrison–Zeldovich spectrum with equal power at all scales.

The COBE satellite has given results consistent with a Gaussian spectrum and yields $n = 1.2^{+0.5}_{-0.6}$ (Smoot *et al.*, 1992). Fluctuations detected by COBE are at angular scales $\theta \geq 7°$, which are greater than the causal horizon ($\theta \sim 2°$) at the time of radiation de-coupling; the observed fluctuations then support the need for inflation. In fact, minimal fluctuations, existing at the scales of the observed galaxies and structures, would roughly incoherently add all together at larger scales and give rise to a power spectrum $\propto K^4$. COBE results clearly contradict this property.

The primordial spectrum is then not a simple superposition of the fluctuations produced by galaxies, clusters, and other observable structures. In other words, this means that some primordial fluctuations at larger scales must exist. It is still necessary to check other eventual additional seeds, perhaps non-Gaussian, but in any case **COBE results are consistent with inflation.**

However, this consistency is necessary but not sufficient. Indeed, the idea of a flat spectrum ($n = 1$) existed prior to the inflation hypothesis. Then, although obtaining such a spectrum, the unicity of inflation is not proved.

Another prediction of inflation is a flat universe ($\Omega_{\rm tot} = 1$). This is supported by the angular size versus redshift measurements of compact radio jets measured with a very long baseline interferometer (Kellerman, 1993).

A problem for inflation could be the value of the Hubble constant. Indeed, the age of the Universe is $t = 2/(3H_0)$ for a value $\Omega_{\rm tot} = 1$, a matter dominated universe. But the age of globular clusters is $t_{\rm GC} = 15 \pm 3$ Gyr (Schramm, 1990), which can be consistent with an $\Omega_{tot} = 1$ universe only if $H_0 \leq 60$ km s$^{-1}$ Mpc$^{-1}$.

Then, once more a value of the Hubble constant consistent with those previously reported and discussed comes indirectly out under the hypothesis that the inflation is valid. Vice versa, since we can now affirm that $H_0 = 56 \pm 6$ km s$^{-1}$ Mpc$^{-1}$ the inflation idea is proved.

### 2.2. CONVERGING PROOFS FOR THE STANDARD MODEL

Gravitational forces compressed the primordial plasma until resistance from photon pressure reversed the motion, leading to acoustic oscillations. Because compression raises the temperature, this results in *hot spots* and *cold spots* which are visible in the microwave sky today. Flat models ($\Omega_{\rm m} + \Omega_\Lambda = 1$) produce an acoustic peak at the Legendre multipole $l = 200$ ($\sim 1°$ on the sky). Open models have a peak that is shifted to smaller scales (larger $l$ values). The height of the peak depends on additional parameters, including $\Omega_{\rm m}$, $\Omega_\Lambda$, $\Omega_{\rm b}$, $H_0$, and tilt.

Before the BOOMERanG (Balloon Observations Of Millimetric Extragalactic Radiation and Geomagnetism) experiment (de Bernardis *et al.*, 2000), the experimental evidence favored a flat and lightweight (low dark matter density) universe model. Such a model has an initial spectrum of density fluctuations with constant power across all length scales (Bahcall *et al.*, 1999). This standard model is strongly inconsistent with the observed lack of prominent second peak in the power spectrum of the CMB. The possible explanations for the missing peak are:

- Initial density fluctuations could actually increase with length scale, thereby suppressing small scale fluctuations. This is the so-called *tilted* model. This solution would have important implications for the particle physics of inflation and observations of gravitational waves;

- The baryon density is as much as 50% less than the value implied by the abundance of light elements in the Universe as the nucleosynthesis theory predicts. Any extra baryons cannot be in the stars we see today. If this were the solution the question of where most of the baryons are today becomes even more puzzling (Fukugita, Hogan & Peebles, 1998). The standard value can also be made to work by lowering the predicted height of the peaks relative to the COBE measurements at the 10° scale by any one of several effects. A testable consequence of either baryon density solution is that the third peak should be higher in amplitude than the second;
- The final, and perhaps most speculative, solution is if the formation of atomic hydrogen were to be delayed until the Universe was nearly 30% older, either by an unknown source of energy or through a change in our understanding of atomic physics at early times. This would increase the time available for the acoustic oscillations to dissipate and hence suppress the smaller peaks.

A combination of some or all of these solutions may also provide the answer and perhaps avoid any extreme departures from the Standard Model (Hu & Peebles, 2000).

### 2.2.1. *CMBR Anisotropy after BOOMERanG*

BOOMERanG data (de Bernardis *et al.*, 2000) place a lower limit on the baryon density that is comparable with the nucleosynthesis estimate. A value for the dark matter density higher than the standard one third of the critical density also helps fit the power spectrum better (three times as much dark matter and 50% more baryons (Bahcall *et al.*, 1999), at the expense of agreement with other cosmological data.

The first images at four frequencies of resolved structure in the CMB anisotropies clearly distinguish the CMB from foreground emission. The angular power spectrum of the CMB gives a peak at $l_{\text{peak}} = 197 \pm 6$, with an amplitude $\Delta T_{200} = 69 \pm 8\ \mu\text{K}$. This is consistent with that expected for cold dark matter models in a flat Euclidean Universe, as favored by the standard inflationary model.

### 2.2.2. *CP Violation*

Looking at the known elementary particles one discovers that a symmetry of *existence* is present (e.g., electron and positron, proton and antiproton, etc.). These particles of matter and antimatter, and in particular their constituents (quarks), are copiously produced by means of accelerators. This perfect symmetry, encountered in the laboratory, is not found in nature. Indeed, in the observed universe there are no signs of the presence of antimatter.

Something *strange* occurred at the beginning in order to have produced the disappearance of antimatter after the Big Bang.

A fundamental theorem of field theory affirms that *all the forces are invariant under CPT (Charge, Parity, Time)*. In other words, if we apply the three operations $C$, $P$, and $T$ in any succession to a system of interacting particles (and then also to macroscopic objects), the system obtained can be realized in laboratory.

Matter and antimatter can exist transforming into each other (in a sort of oscillation) via weak forces. This phenomenon has been discovered through the neutral $K^0$ mesons, whose mass is roughly half of the proton mass, and represent the lightest particles containing the quark strange ($q_s$). When in a p–p collision a $K^0$ is created it rapidly goes away and transforms itself in an anti-$K^0$. Such a matter–antimatter oscillation continues until the disintegration of the particle into a $\pi$-meson. This oscillating system is a very sensitive probe in studying eventual differences in the behavior between matter and antimatter. Indeed, it has been discovered that the weak forces producing such transformations do not satisfy invariance under time. *It is more probable for antimatter to become matter rather than vice versa.* In fact, the experimental distribution of the decay time of $K^0$ mesons and anti-$K^0$ mesons are different. Therefore the forces regulating the mixing of neutral $K^0$ mesons violate the $CP$, and then $T$ if $CPT$ is preserved.

The current theoretical models explain this fact introducing an *ad hoc* number in the CKM (Cabibbo–Kobayashi–Maskawa) matrix, which regulates the intensity of transformations amongst quarks. However, the reality could be different: e.g., the special numbers in the CKM matrix might not be present and the $CP$ violation could be provoked by the existence of another *very weak* force, which are able to directly transform $K^0$ mesons into anti-$K^0$ mesons. Only experiments can demonstrate which one of the theories is correct.

The theoretical model of the CKM matrix suggests a very small variation in the number of times in which $K$ mesons decay into two charged–pions and two neutral–pions; on the contrary, the *super-weak* model does not predict such an effect. Recent first results (from KTeV and NA48 experiments, together with those of 10 year old results from NA31 experiment at CERN) show a difference between *short* and *long* decays. They are then against the *super-weak* model (e.g., Calvetti, 2000 and the references therein).

Two questions still survive:

– Is it possible that the arrow of time, the difference between matter and antimatter, and then the survival of the former and the disappearance of the latter are consequence of a small number in the CKM matrix, as the standard model predicts? Or are other reasons responsible for that?
– Why did antiprotons and antineutrons with their baryonic numbers disappear at the beginning of the evolution of our Universe?

**Which open problems in cosmology still survive?**

In spite of the enormous progress in cosmology obtained in the last few years, a number of open problems still survive, namely:

– The Hubble constant is still uncertain, although the most recent determination from HST (Panagia, 1999a) gives a value of $H_0 = 59 \pm 6$ km s$^{-1}$ Mpc$^{-1}$;
– If the fluctuations of $\sim 1.1 \times 10^{-5}$ on the background temperature, very stable at 2.735 K, at a scale of 10° (COBE results: Smoot *et al.*, 1991; 1992) are truly of cosmological origin, they represent a milestone in understanding the large scale structure of the universe. BOOMERanG data show the background

temperature fluctuations in a scale of 1° and put a lower limit on the baryon density, which is comparable with the nucleosynthesis estimate (de Bernardis *et al.*, 2000). However, the problem with the height of the second peak – which depends on physical parameters, including $\Omega_m$, $\Omega_\Lambda$, $\Omega_b$, $H_0$, and tilt – in the experimental power spectrum must be solved.

– Light element abundances fit cosmological predictions with a baryon density $\Omega_b = 0.05 \pm 0.03$. So a Universe with $\Omega_{tot} = 1$ must have the bulk of the matter ($\geq$ 90%) formed by something other than baryons: dark matter (Schramm, 1993). BOOMERanG data imply a flat Euclidean universe with $0.88 < \Omega_0 = \Omega_m + \Omega_\Lambda < 1.12$. From the 2dF Galaxy Redshift Survey (2dFGRS) $\Omega_0 = 1.11 \pm 0.07$ (Peacock *et al.*: 2001).

## 2.3. GRAVITATIONAL LENSES

The Universe appears distorted because of gravitational lenses: light rays from distant galaxies travel a tortuous path through a Universe filled with clustering dark matter. Every bend in the path of a bundle of light from a distant galaxy stretches its apparent image. The orientation of the resulting elliptical images of galaxies contains information about the size and mass of the gravitational lenses distributed over the path of the light. Weak gravitational lensing by large scale mass structure can occur: distant galaxy orientation is correlated at scales characteristic of the lensing dark matter structures. Light bundles from two distant galaxies which are projected closely together in the sky follow similar paths and undergo similar gravitational deflections by intervening dark matter concentrations. Apparent orientations of distant galaxies are thus correlated at angular scales of less than a few degrees. *The larger the mass in the gravitational deflectors, the larger the faint galaxy ellipticity correlations of a given angular scale*. These ellipticity correlations of distant galaxies reveal the statistics of the large scale dark matter distribution in the intervening universe – a central diagnostic of the underlying cosmology (Wittman *et al.*, 2000).

By the end of 1999 more than 50 gravitational lens systems were known, thanks to the observations of the Hubble Space Telescope (HST), in which one single galaxy acts as a lens creating multiple images of quasars and radio sources located behind it. Almost all lenses are massive elliptical galaxies. They represent a special group, inasmuch as they are selected on the basis of their masses, rather than in the usual way of according to their luminosities. There is a big project, named CASTLES (CfA Arizona Space Telescope Lens Survey, cfa-www.harvard.edu/castles), whose goals are:

– to find as yet undiscovered lens galaxies;
– to measure the exact positions of the images and the lens;
– to determine the redshift of the lens by photometry;
– to prove the existence of lensed galaxies (host galaxies) around active galactic cores;

*Figure 7.* The galaxy cluster CL0024+1654, in which one background galaxy is imaged seven times (blue) (Photo NASA/ESA) (Staude & Rix, 2001a).

– to determine the structure and evolution of the galaxies which act as lenses;
– to measure the extinction in the galaxies acting as lenses;
– to detect dark matter;
– to determine the Hubble constant.

This list underscores the far reaching possibilities offered by the gravitational lens effect.

Figure 7 shows the image of the galaxy cluster CL0024+1654, in which one background galaxy is imaged seven times. In this case the gravitational potential of a whole galaxy cluster is acting as the lens (Staude & Rix, 2001a).

## 2.4. A SWIFT CRUISE THROUGH SPACE ASTROPHYSICS

Independently of the origin of the universe we are living now in our Universe as it is. Photonic astrophysics is a powerful probe in sounding the Universe. We will briefly summarize the main results coming from space experiments in different energy regions.

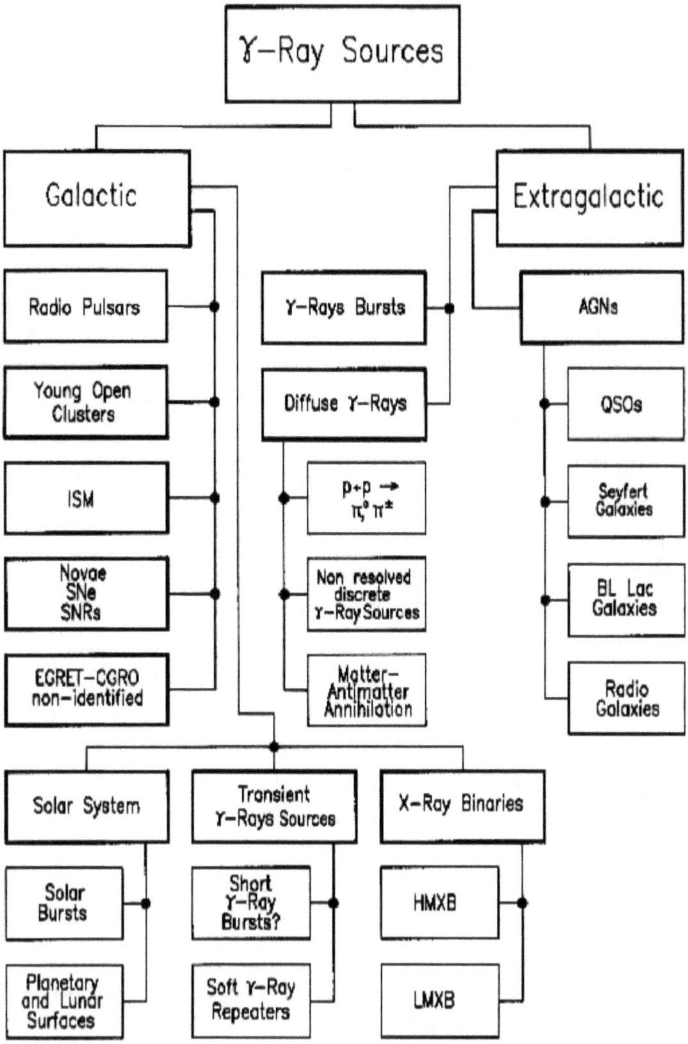


*Figure 8.* Classification of $\gamma$-ray sources (updated from Giovannelli & Sabau-Graziati, 1993).

2.4.1. *Gamma Ray Astronomy*

In spite of the relatively low fluxes coming from cosmic sources in the $\gamma$-ray energy range, the results obtained in $\gamma$-ray astronomy are extremely important in understanding the physics governing the collapsed objects and their interactions with the surrounding matter and/or radiation. These results can be summarized as follows:

– Direct relationship amongst $\gamma$-rays, high energy nuclear reactions and high energy physical phenomena, such as radiative de-excitation of nuclei and baryons, decay of mesons, and matter–antimatter annihilation;

- Direct relationship between $\gamma$-rays and particles of cosmic rays (electrons and protons) via collisional bremsstrahlung radiation, synchrotron radiation, inverse Compton scattering;
- Large penetration of $\gamma$-rays into galactic and extragalactic matter;
- Discovery of discrete and diffuse $\gamma$-ray sources with SAS II, COS B and CGRO satellites;
- Discovery of the diffuse $\gamma$-ray background which is formed by extragalactic and galactic components. The $\gamma$-ray continuum spectrum is coming from the Galactic Center disk and point sources (e.g., Gehrels & Tueller, 1992; Schönfelder, 1995; Strong *et al.*, 1996; Purcell *et al.*, 1996). The extragalactic component seems to come mostly by the sum of AGN contributions (e.g., Stecker, Salamon & Malkan, 1993; Hasinger, 1996). The main contribution is from optically violent variable quasars (OVV) and BL Lac objects (Setti and Woltjer, 1994), although these measurements can be affected by the interactions of energetic cosmic rays with the material surrounding the instruments (Kniffen *et al.*, 1996; Kappadath *et al.*, 1996);
- In the field of $\gamma$-ray line spectroscopy COMPTEL-CGRO has achieved a major break-through with: i) establishment of the first map of the galactic plane in the light of the 1.8 MeV line from radioactive $^{26}$Al (e.g., Diehl, 1996); ii) the discovery of the 1.15 MeV $^{44}$Ti-line from Cas A; and iii) the first detection of $\gamma$-ray emission from the Orion complex in the 3 to 7 MeV range that firstly was attributed to the 4.44 MeV and 6.13 MeV nuclear de-excitation lines of $^{12}C^*$ and $^{16}O^*$, respectively (Bloemen *et al.*, 1994a). Later (Bloemen *et al.*, 1999b), re-analyzing the data from Orion complex, stated that the $\gamma$-ray emission measured is only an excess with respect to the background and then it is not possible to attribute such an excess to the latter nuclear de-excitation lines. However, this is a promising measure for a possible detection of such nuclear lines from Orion complex and other sites, such as the young open cluster Berkeley 87 and similar clusters, with the forthcoming INTEGRAL satellite (Giovannelli *et al.*, 1999; Giovannelli, 2002).

For a tutorial review of $\gamma$-ray sources see the paper by Giovannelli & Sabau-Graziati (1993). Such sources can be schematically grouped as in Figure 8 (updated from the latter paper).

### 2.4.2. *X-Ray Astronomy*

With the many surveys and selected observations at different X-ray energy ranges several important results have been obtained, namely:

- Distribution of X-ray sources: around the galactic plane most sources are galactic, whilst at high galactic latitude most sources are extragalactic. Figure 9 shows the historical first Uhuru Catalog, in which the positions of sources are in galactic coordinates and the size of circles proportional to the sources intensities (Forman *et al.*, 1978).

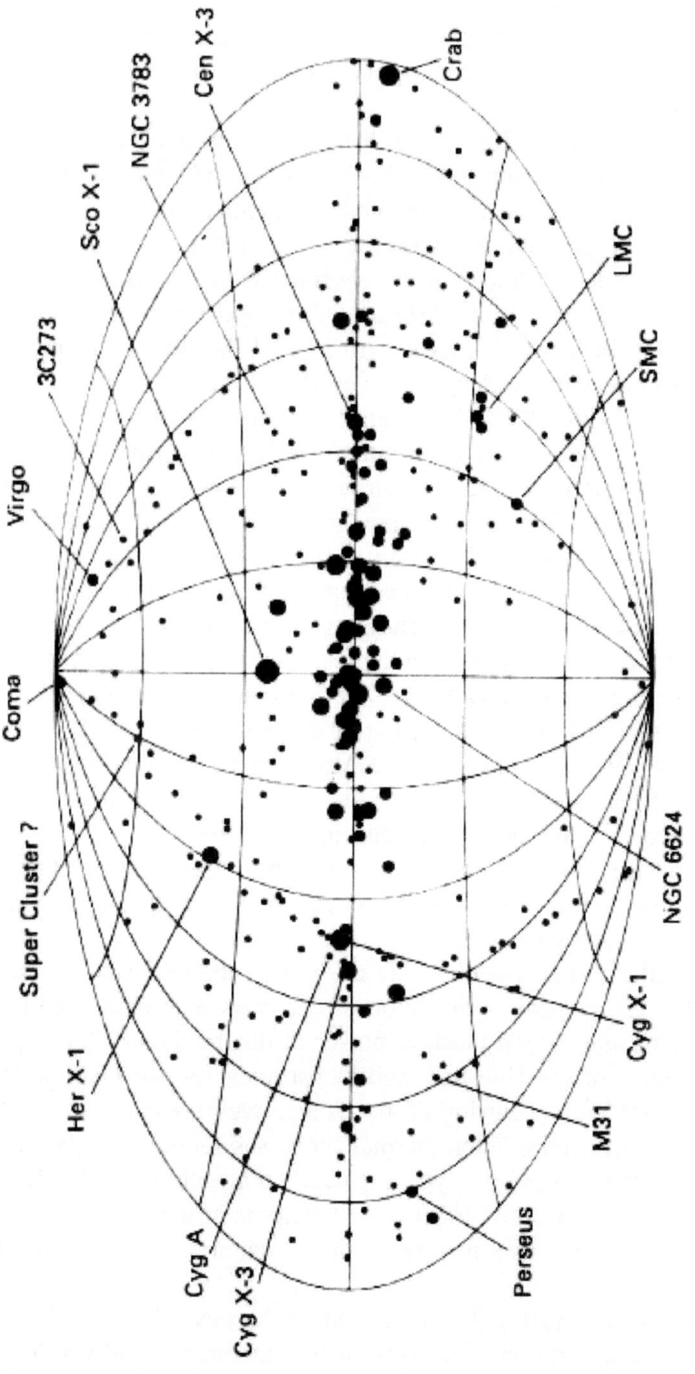


*Figure 9.* The historical fourth Uhuru Catalog. Circles' sizes are proportional to the sources' intensities (Forman *et al.*, 1978).

- Identification of point-like X-ray sources with objects already known in optical and radio ranges.
- Nature of the identified X-ray sources as normal stars, X-ray binary systems, SNRs, pulsars, normal galaxies, active galaxies, quasars and clusters of galaxies.
- Discovery of the diffuse X-ray background, galactic and extragalactic, partially coming from unresolved X-ray sources and from processes in the galactic disk, ISM and IGM originated by interactions between matter and radiation. The problem of extragalactic X-ray background is now practically resolved; indeed, almost all the observed background, in most of the X-ray domain, is coming from QSOs, Seyferts, and other active galaxies. In particular, in the range 1–2 keV, virtually all the flux is accounted for by resolved sources in the deepest surveyed ROSAT fields (Hasinger *et al.*, 1993; Hasinger, 1994, 1996). For higher energies (2–100 keV) the X-ray background spectrum seems to be fitted in terms of thermal Comptonization and reflection of X-rays from cool surfaces of AGNs (disks?) (Zdziarski, Zycki & Krolik, 1993). Matsuoka & Cappi (1996) discussed the contribution of AGNs to the X-ray background. Parmar *et al.* (1999) discussed the low energy (0.1–7.0 keV) X-ray background spectrum observed with the BeppoSAX satellite at galactic latitudes $> |25|^\circ$. They suggest that the observed spectral complexity may originate primarily in the local hot bubble (soft thermal component) and in more distant absorbed region (hard thermal component).

  Reviews of cosmic X-ray background are those by Boldt (1987), Fabian & Barcons (1992), Hasinger (1995), Matt (1995), Barcons (1996).

  Figure 10 shows the X and $\gamma$-ray background; the different contributors are indicated (Schönfelder & Weidenspointner, 1999).

For a tutorial review of X-ray sources see the paper by Giovannelli & Sabau-Graziati (1993). Such sources can be schematically grouped as in Figures 11 (updated from the latter paper).

Undoubtedly X-ray astronomy is the most suitable branch of space science both for the relative easier realization of detectors with respect to those in other energy regions and for the intrinsic power of this method of investigation on the hot components of the Universe, and in particular of our Galaxy. Thermal processes ($\sim 10^6$-$10^8$ K), occurring in most stars, manifest in the softer X-ray region (0.1–10 keV). In contrast, non-thermal processes require 50–500 GeV electrons for synchrotron radiation in a typical galactic magnetic field of $\sim 10^{-5}$ G (Hillas, 1984), 5–50 MeV electrons for inverse Compton scattering of starlight photons, and 0.3–3 GeV electrons for inverse Compton scattering on microwave background photons.

The process dominating the interactions of X-rays below $\sim 10$ keV with matter is photoelectric absorption. The cross section increases rapidly at lower energies, scaling approximately as $E^{-3}$. Figure 12 shows the mean free path of X-rays as a function of energy assuming an average interstellar density of 1 H atom $cm^{-3}$ that

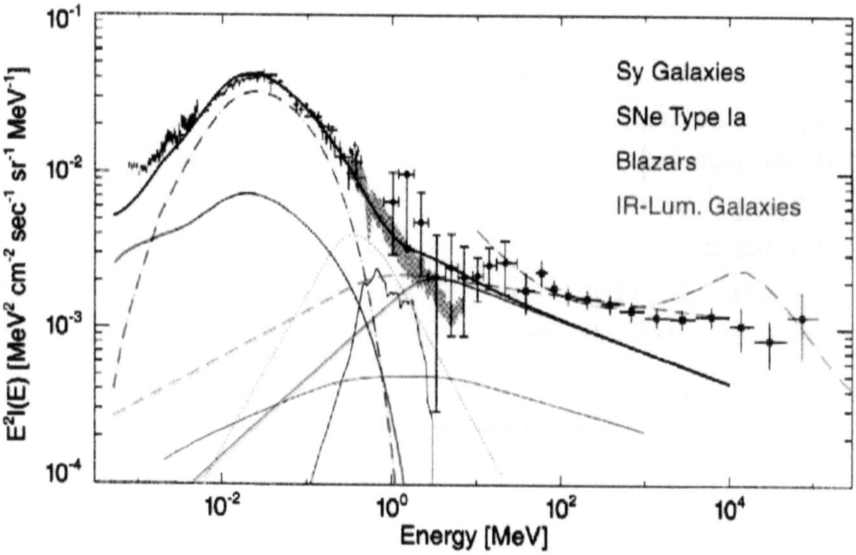


*Figure 10.* X and Gamma ray background with the different contributors (Shönfelder & Weidenspointner, 1999).

encompasses the entire range of galactic distance scales (McCammon & Sanders, 1990).

A very rough interpretation is that 0.1 keV X-rays can reach us from neighboring stars, although the use of an average density is here definitively not correct, whilst at 10 keV the entire diameter of the disk is transparent. Interstellar scattering can be ignored over the whole range. The sharp jumps in the mean free path are owed to the atomic absorption edges of the more abundant heavy elements. These provide a possibility of direct measurement of interstellar column densities of particular species.

### 2.4.3. *UV Astronomy*

In 1946 an UV experiment launched on board a V2 rocket observed the Sun's UV spectrum for the first time (Malina, 1946). This experiment opened the route to UV astronomy, which has been mostly developed thanks to the Copernicus, IUE, and EUVE satellites. The main results are:

- Hot stars radiate most of their energy in the UV range, in which many important spectral characteristics are demonstrated, such as low excitation levels, absorption and emission lines. Through these results it was possible to evaluate the element abundances.
- Emission lines of highly ionized elements in UV and EUV ranges belong to hot plasmas, such as stellar chromospheres and coronas or expanding envelopes from stars.

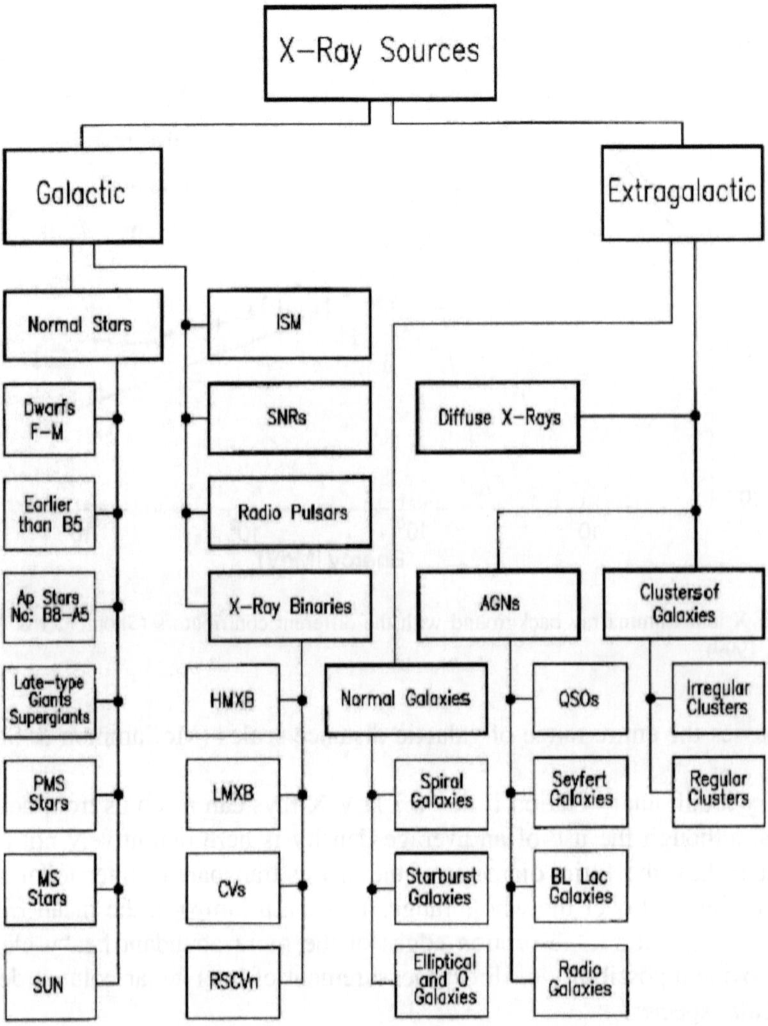


*Figure 11.* Classification of X-ray sources (updated from Giovannelli & Sabau-Graziati, 1993).

– The existence of absorption lines of neutral interstellar atoms and sometimes ionized atoms below 3000 Å.
– Large differences between the scattering characteristics of interstellar grains of different natures.
– Molecular bands owed to molecular dissociation. They are interesting in the investigation of planetary atmospheres, cometary tails, and ISM.

In our opinion the highlight of UV astronomy was the discovery of the galactic halo through the detection of OVI noting the existence of a hot gas emitting high energy radiation via thermal processes (e.g., Jenkins, 1978a,b). An independent

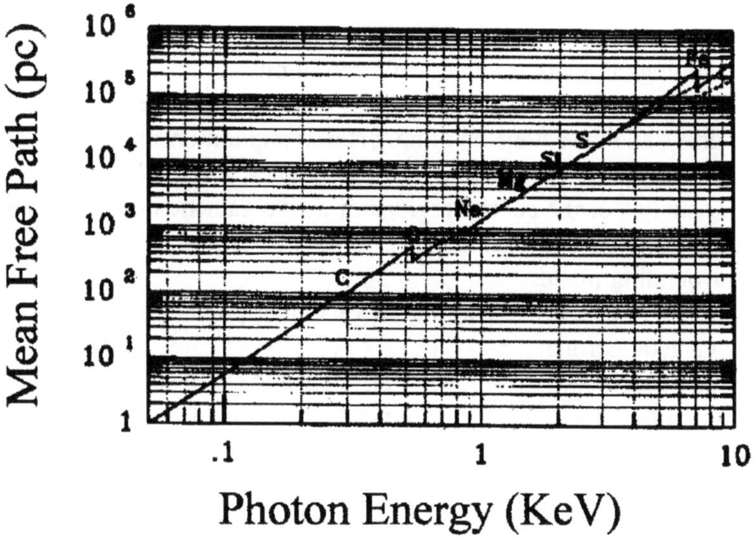


*Figure 12.* The mean free path of X-rays assuming an average interstellar density of 1 H atom $cm^{-3}$ that encompasses the entire range of galactic distance scales (McCammon & Sanders, 1990).

proof of the thermal origin of this radiation was the detection of an OVII line at 0.57 keV made by the Hakucho satellite in a spectrum best fitted by a thermal emission of a plasma at a temperature of $1.4 \times 10^6$ K (Inoue *et al.*, 1979).

The ORFEUS (Orbiting and Retrievable Far and Extreme Ultraviolet Spectrometers) telescope, aboard the space platform Astro-SPAS during the mission of the space shuttle *Discovery* flown in September 1993, provided high resolution observations of the Small Magellanic Cloud (SMC) star NGC 346 No.1 near the OVI resonance doublet (1031.9, 1037.6 Å). These observations support the cooling fountain models. They provided a value of the OVI column density of $\sim 1.4 \times 10^{14}$ $cm^{-2}$. This allows one to estimate a scale height of $\sim$1.2 kpc (Hurwitz & Bowyer, 1995), in agreement with the predictions (Shull & Slavin, 1994), and a factor 4 greater than the old predictions made by Jenkins (1978a,b) based on Copernicus' observations.

### 2.4.4. *Optical Astronomy*

In the optical astronomy the Hubble Space Telescope (HST) is providing one of the most important contributions to knowledge of the Universe. We can summarize then as follows:

- A fundamental contribution to the determination of the distance scale and age of the Universe, by extending by a factor 10 the distance over which it was possible to measure the properties of standard candles, used in the logical ladder leading to the determination of the Hubble constant;

- Improvements in the study of clusters of galaxies and individual galaxies at higher redshifts than possible from the ground, and then it is providing tests for cosmological models and cosmic evolution;
- An extension of morphological classification of galaxies up to redshifts of order unity, and the extension of faint stellar population studies, which provide important inputs about the formation, dynamics, and evolution of galaxies;
- Deeper studies of the central regions of active galactic nuclei, which provide information about the central powerful engine and about its interaction with the surrounding matter;
- The studies in the Milky Way of the very faint population of stars belonging to the galactic corona;
- Improvements in the knowledge of physical and chemical composition of ISM with direct consequences for evolutionary models of the outer atmosphere of stars, stellar winds, coronal winds, circumstellar shells, planetary nebulae, and SNRs;
- A crucial contribution in knowing the birth and evolution of single stars, binary stars, and globular clusters;
- Exceptional results in the knowledge of the solar system.

Although it is difficult to select highlights from the HST, because the impact of this telescope is really great in many fields of astrophysics; in our opinion, two of the most exciting results are those connected with the galactic fountain and with the SN 1987A:

- According to Bregman (1980) supernovae occurring in the galactic disk generate an upward and outward expansion of hot ($\sim 10^6$ K) gas which constitutes the corona. The coronal gas, cooling radiatively, becomes thermally unstable, and so clouds condense and then fall ballistically back to the disk. The hot wind from supernovae represents the rising fluid of the galactic fountain, and the high velocity hydrogen clouds, detected in the 21 cm hydrogen line, are the return flow. Spitzer & Fitzpatrick (1992) investigated highly ionized atoms across the line of sight to HD 93521, an apparent population-I O9.5 star at the unlikely height of $\sim$1.5 kpc above the galactic plane. In their study of the interstellar features in the spectrum of HD 93521 they noted a 50 km $s^{-1}$ FWHM line width of CIV and SiIV. These broad lines are attributed to mass motions, and the large negative central velocities of the CIV and SiIV ($-67$ and $-60$ km $s^{-1}$, respectively) are produced by matter falling toward the galactic plane. So these observations seem consistent with the presence of a Galactic Fountain;
- Splendid FOC (Faint Object Camera) repeated images (separated by $\sim$500 days) of the SN 1987A revealed the surrounding ring nebula, and the SNR remained in free expansion with the diameter growing linearly with time (Jakobsen *et al.*, 1991; Panagia *et al.*, 1991; Jakobsen, Macchetto & Panagia, 1993).

Maran & Kinney (1993) reviewed results from HST. Results from HST can be found in the book *Science with the Hubble Space Telescope* (Benvenuti, Macchetto & Schreier, 1996). Panagia (1999b, 2002a,b) discussed highlights from HST.

2.4.5. *IR Astronomy*

Space IR astronomy has been mainly developed by the IRAS satellite in the range 12–100 $\mu$m. More than two hundred thousand sources have been discovered and catalogued. The main results are (Beichman, 1987):

- The distribution of the IR sources shows a marked concentration in the galactic plane;
- The identification of IR point-like sources with objects already known in the optical and radio ranges. K and M giant stars, which can present (or not) an IR excess owed to mass loss processes, are the most common stars in the IRAS catalog;
- The discovery of the nature of the identified IR sources as protostars, evolved stars, T Tauri type stars, cold circumstellar shells, circumstellar dust shells, diffuse galactic objects, interstellar grains, Galactic Center (invisible in the optical range), extragalactic objects;
- The identification of the diffuse IR radiation background. This radiation is best fitted to a black body radiation at 2.7 K.

Later the Infrared Space Observatory (ISO) (2.5–240 $\mu$m) has provided undoubtedly important advances in knowledge of the Galaxy, the Solar System and extragalactic objects. A short review of ISO results is that by Ulla *et al.* (1999). A special issue of *Astronomy and Astrophysics* of November 1996 was dedicated to the ISO mission.

The map at 15 $\mu$m of several cuts of the galactic plane around 12° squared, in the inner region of the Galaxy at $|b| < 0.5°$ and $|l| < 30°$ has been obtained. With these results it has been possible to trace the galactic structure of the inner obscured galactic structures (Omont & Blommaert, 1997). One of the most outstanding results in the field of dusty interstellar and circumstellar material is the rich emission spectrum between 30 and 45 $\mu$m observed from a number of stars with an O-rich shell. It appears that many of these features coincide with laboratory features of crystalline silicates. One of them at 43 $\mu$m coincides with that of crystalline $H_2O$ ice. Since formation of such silicates requires high densities and temperatures followed by relatively long periods of slow cooling, their detection provides very interesting clues about the formation of dust during stellar evolution. Observations towards three evolved C-rich stars (one post AGB (Asymptotic Giant Branch) and two planetary nebulae) show the presence of Polycyclic Aromatic Hydrocarbons (PAHs). PAHs have been detected also from compact HII regions.

These facts pushed interstellar chemistry studies a lot. Extremely interesting results show that a large variety of gas phase molecules are present in the ISM and that many of them are organic, and that they are found in dense molecular clouds in which stars are forming. Recent reviews of interstellar organic chemistry are those by Charnley (2001) and Ehrenfreund *et al.* (2001). Beckman (2001) discussed the conditions for the formation of pre-biological molecules in the ISM.

The ISO mission also provided spectroscopic studies (ISO-SWS: 2.5–45 $\mu$m, and ISO-LWS: 45–196 $\mu$m) of solar system objects, such as Jupiter, whose atmosphere has been deeply sounded, and Saturn, Neptune, and Uranus. Spectra have also been obtained for several asteroids. Thermal emission of $CH_4$ at 3.3 $\mu$m have been detected in Jupiter, as well as $NH_3$ ice, associated with the Jovian clouds. These, together with the detection of gaseous $H_2O$ in the upper atmospheres of Jupiter, Saturn, Uranus, and Neptune, and $CO_2$ in Saturn and Neptune, provided a major contribution to the progress in knowledge of the Solar System (e.g., Ulla *et al.*, 1999).

The infrared camera (ISOCAM) (2.5–17 $\mu$m) provided important information about external galaxies, such as the Whirpool Galaxy (Sauvage *et al.*, 1996), where the images clearly show regions of star formation across the spiral arms. Maps of IR-luminous galaxies, such as the Antennae galaxies, allowed the investigation of the sources of the immense amount of radiated energy (Vigroux *et al.*, 1996).

The imaging photo-polarimeter (ISOPHOT) (2.5–240 $\mu$m) in the Serendipity Survey near 200 $\mu$m allowed the exploration of $\sim$15% of the sky during the otherwise unused slew time between the ISO's fine pointing. Thousands of galaxies brighter than the detection limit of 1 Jy have been detected (Bogun *et al.*, 1996).

A Conference on *The Universe As Seen By The ISO* was held in Paris in 1998 and the proceedings appeared one year later (Cox & Kessler, 1999).

## 3. Physics of the Diffuse Cosmic Sources from Space Experiments

It is very difficult to concentrate in a single review a homogeneous presentation of the known physics of cosmic sources, simply because this means writing an updated encyclopaedia of physics. For this reason we are obliged to select a line of thinking which is, in our opinion, the most convenient for noting the fundamental advances in the knowledge of the physics governing the Universe brought by space experiments.

Space experiments are a means of approaching the natural cosmic laboratories where processes in extreme physical conditions occur. These are the same processes we are trying to reproduce on the Earth with enormous efforts both intellectual and technological. And in spite of this we are not able to reproduce all physical processes simply because we do not have the immense amount of energy necessary for reproducing most of them. Therefore in order to understand why our Universe is as it is observed, the modern physics has the obligation to intersect the two fields of Astrophysics and Particle Physics, which must be enlarged until they overlap in a new one, namely *Astroparticle Physics*. A few symptoms that physics was moving toward this direction began to be clear about twenty five years ago, when the most audacious projects, such as the Space Telescope, and Big Accelerators and Collision Rings, started to be developed. We then address our tracing of the physics of cosmic sources along the main line of this new branch of science, *Astroparticle Physics*. The high energy processes we are trying to reproduce in Earth's laboratories developed into the primeval Universe. Similar processes also originate in several astrophysical objects, such as the large scale structures of the Universe, active galaxies, neutron stars, and black holes, where the temperatures can reach values greater than $10^{10}$ K, which render these objects the most suitable natural high energy laboratories accessible from space. Therefore our tracing will follow the logic to scour, in a certain sense, the outstanding stages of the formation of the Universe: from the Big Bang to nowadays.

For obvious reasons we will discuss only a few examples, but sufficient to demonstrate the impact of space experiments on the advances in knowledge of the physics governing the Universe, avoiding the chronicle of space experiments, on the other hand already synthesized in Table 1.

Nowadays current theories of Cosmology are based on the hypothesis that space and time were generated at the moment of the Big Bang; after that the Universe started to expand with a fast cooling, and then to form the first structures such as stars and galaxies. *Stellar physics* is the result of the combination of gravitational and nuclear forces, and *galactic physics* is the result of the gravitational force, considering the galaxies as accumulation of stars, neglecting the microscopic effects.

All these processes, from the Big Bang to the formation of the stars and galaxies, and up to their extreme fate, produce effects giving rise to observable quantities, which can be observed in practice across the whole electromagnetic spectrum. Therefore experiments in space are the unique possibility of detecting those effects.

### 3.1. EXTRAGALACTIC DIFFUSE BACKGROUND

After the Big Bang the Universe started to expand with a fast cooling. The cosmic radiation observed now is probably a melting of different components which had their origin in different stages of the evolution as the results of different processes. This is the Diffuse Extragalactic Background Radiation (DEBRA), which, if observed in different energy ranges, allows the study of many astrophysical, cosmological, and particle physics phenomena (Ressel & Turner, 1990).

The first component of this radiation was detected in the microwave region. Its spectrum is best fitted to that of a black body at 2.7 K. The temperature is uniform in all the sky to a precision of $\sim 10^{-4}$. This is the residual radiation, predicted by the theory of the Big Bang, in a phase in which the entire Universe was warm, dense, and opaque. The Universe became transparent $\sim 10^5$–$10^6$ years later (cosmological redshift $\sim$1100–1000) when the temperature decreased below 3,000 K and hydrogen recombined. Some theories about the formation of complex structures predicted the existence of a background radiation also in the IR and optical spectral regions originating during the formation of the first stars in the first galaxies, at a cosmological redshift of ∼4–5 (Partridge & Pebles, 1967a,b). Also in the X-ray and $\gamma$-ray ranges a diffuse background radiation exists, but it is not yet completely understood, although the new generations of detectors of ROSAT and CGRO (recently taken out of service), ASCA, AXAF–Chandra, BeppoSAX, and XMM–Newton nowadays in flight, are improving knowledge of it. Also each class of residual particles, with the relative radiative decay, should contribute to the cosmic diffuse background. Therefore it is possible to use measurements of the diffuse radiation in order to evaluate the masses (or at least some limits) and couplings of each class of particles.

The survey of the whole sky made with the COBE satellite is playing an important role in studying the DEBRA, thanks to the high sensitivity at 3.3, 5.7 and 9.5 mm. In Section 3.1.2 we will comment upon some important results coming from this satellite.

It is possible to consider the DEBRA as a radiation produced by a cosmic source: the whole Universe. We will summarize the knowledge about the DEBRA, dividing the discussion into energy bands.

#### 3.1.1. *Radio Band* ($10^5$–$10^2$ *cm*)

Data from $10^5$ to $10^2$ cm are available in the radio band. For wavelengths greater than $10^5$ cm the opacity of the ISM is very high because of free–free absorption of electrons.

It is believed that the diffuse radio radiation is formed by three components:

- Synchrotron radiation coming from the galactic disk;
- Synchrotron radiation coming from the galactic halo;
- Diffuse extragalactic background radiation owed to the integrated emission of all non-resolved extragalactic sources. It is difficult to separate the three components from one another, therefore it is better to consider the diffuse background radiation in the radio band as owed to a combination of galactic and extragalactic components (Ressel & Turner, 1990).

#### 3.1.2. *Microwave and Submillimeter Wave Bands* ($10^2$–$10^{-2}$ *cm*)

The cosmic background radiation in the microwave band (CMBR) is intensively studied because the residual radiation coming from the Big Bang dominates this spectral region. Since the surface of last scattering for the CMBR is the Universe at redshift $z \sim 1100$–$1000$, this radiation contains detailed information on the history of the primeval Universe. The CMBR being the best known amongst the components of the DEBRA, the measure of its parameters constitutes a fruitful experimental field to test cosmological models.

The current researches are essentially of two kinds:

- Studies of the spatial anisotropy (Partridge, 1987; Wilkinson, 1987; Strukov *et al.*, 1992; Strukov, Skulachev & Brjukhanov, 1995; White, Scott & Silk, 1994 and references therein).

- Studies of the spectral shape (Richards, 1988; Smoot *et al.*, 1992; Wright *et al.*, 1994).

Indeed, small spatial anisotropies could indicate that the matter was not homogeneously distributed when CMBR originated, and small deviations with respect to the black body spectrum should indicate the presence of high energy sources in the primeval Universe. Sometimes this is noted with the analogy that *you cannot make an omelet without breaking eggs; i.e., you cannot make galaxies without disturbing the microwave background.*

As is well known, the scattering of cosmic background photons by energetic electrons would produce a variation in the black body spectrum. This perturbation would reveal the presence of a hot ionized gas produced by an injection of energy somewhat after the de-coupling between radiation and matter. This heating could have been produced by the formation of galaxies and stars.

Matsumoto *et al.* (1988a) found an excess of $\sim$10% from a black body spectrum with a temperature of 2.74 K in their submillimeter measurements in the range $10^{-2}$–$10^{-1}$ cm made with a payload on board a rocket. Because of the low flux on the Wien side of the spectrum, these results must be taken with caution.

If the distortion is real the enormous energy budget suggests that the explanation involves fundamental physics, such as the decay of some relic species.
We would like to remark how, sometimes, news from experiments activates theoretical works that are not able to justify those results because of either weakness of models or wrong measurements. So our warning is that to avoid a long chain of theoretical works before verifying the strength of the experimental data – in this case the Matsumoto *et al.* (1988a) results.

Indeed, for instance, after those results, Dar, Loeb & Nussinov (1989) discussed how unstable relic particles could distort the CMBR.

The amount of energy corresponding to the reported excess is so large ($\Delta\epsilon\gamma \approx 3 \times 10^{-5}$ keV cm$^{-3}$, or about 300 keV per baryon) that no presently known astrophysical heat source can account for it. A decaying weakly interacting relic particle with cosmological abundance ($n_x \cong n_\gamma$) could, in principle, supply such energies. In general three classes of scenarios can be envisioned:

- Weakly interacting relic particles which decayed radiatively ($x' \longrightarrow x + \gamma$) slightly before recombination (at redshift of $1 + z_r \approx 1300$) and produced photons which avoided complete thermalization (Kawasaki & Sato, 1986; Silk & Stebbins, 1983; Stebbins & Silk, 1986) could provide the reported excess in the Wien region. In particular, this possibility requires a particles mass $m_x \approx 2(1+z_r)\times(2.70\,T_0') \approx 3$ eV, where $T_0' \approx 4.9$ K is the approximate temperature of the excess photons (Daly, 1988) and $2.70\,T_0'$ is the average energy of black body photons with a temperature $T_0'$.
  These scenarios do not require any new particle. The $x'$ and $x$ particles could be massive $\nu_\tau$, $\nu_\mu$, or $\nu_e$ with radiative decays.
- More massive particles with $m_x = 1$–$10$ keV could decay after recombination, mainly into invisible massless decay products, but also radiatively ($x' \longrightarrow$

$x+\gamma$), with a small branching ratio $B_\gamma$. The resulting photons could re-ionize the universe (Fukugita, 1988), and distort the CMBR via Comptonization (Zel'dovich & Sunyaev, 1969).

This scenario can involve decays of neutrinos $\nu' \longrightarrow \nu + \Phi^{(0)}$, $\Phi^{(0)}$ being massless majoron or familon particles (Fukugita & Yanagita, 1985), which have been predicted but have not yet been experimentally verified.

– A presently radiatively decaying light particle with a mass $m_x \approx 2 \times (2.70\, T_0')$ $\approx 2 \times 10^{-3}$ eV could generate directly the reported spectral excess by Doppler broadening.

This scenario involves completely novel particles invoked exclusively for explaining the spectral distortion measured by Matsumoto *et al.* (1988a).

After a thorough discussion of the three possible scenarios, Dar, Loeb & Nussinov (1989) concluded that the shape of the measured distortion could be best fitted by a greybody spectrum (Daly, 1988). In spite of this, general cosmological and particle physics considerations indicate severe difficulties in accounting via such mechanisms for the CMBR distortion of magnitude $\sim 3 \times 10^{-5}$ keV cm$^{-3}$, or even substantially small values.

Indeed, as confirmation of Matsumoto *et al.*'s measurements which are hard to believe, the first results in the range $10^{-0.5}$ mm from the FIRAS (Far Infrared Absolute Spectrophotometer) of COBE show a spectrum well fitted by a black body with a temperature of $2.735 \pm 0.06$ K and do not show any evidence for the submillimeter excess previously reported by Matsumoto *et al.* (1988a) in the CMBR (Mather *et al.*, 1990).

In contrast, measurements at 15 GHz, performed with the ground experiment in Tenerife, have established a fluctuation of $1.8 \times 10^{-5}$ on the background temperature in an angular scale of $\sim 5°$ (Watson *et al.*, 1992). This fluctuation is consistent with that found by the Differential Microwave Radiometers (DMR) of COBE. Indeed, in the COBE maps at 31, 53, and 90 GHz a deviation of $\sim 1.1 \times 10^{-5}$ on the background temperature, generally very stable at 2.735 K, in an angular scale of 10° has been found (Smoot *et al.*, 1991; 1992).

Also Relict 1 spaceborne soviet experiment detected some anisotropy of the CMBR at 37 GHz. For an inflationary perturbation spectrum, the relative magnitude of the quadrupole component was estimated with a confidence level of 90% between $6 \times 10^{-6}$ and $3.3 \times 10^{-5}$. An anomaly of the microwave radiation has been found with a confidence level of 99% in a region with area $\sim 1$ sr near the point $l = 150°$ and $b = -70°$. The magnitude of this anomaly is $-71 \pm 43\ \mu$K, with a confidence level of 90% (Strukov *et al.*, 1992).

These results are a milestone in our understanding of the large scale structure of the Universe if the temperature fluctuations are truly of cosmological origin.

Indeed, Bennett *et al.* (1992) and Bennett, Stebbins & Bouchet (1992) have constructed preliminary models of microwave emission from our Galaxy based on COBE and other data in order to distinguish cosmic and galactic signals. When galactic emission is removed from the DMR data, the residual fluctuations are virtually unaffected, and therefore they are not dominated by any known galactic emission components. Exotic sources of emission could produce the fluctuations seen in the DMR maps and mimic cosmic anisotropy.

A clumpy population of cold dust needles (Wright, 1991) in the galactic halo would be difficult to distinguish from CMBR anisotropies.

Hoyle & Burbidge (1992) attempted an explanation of the results of Smoot *et al.* (1991, 1992), considering a few possibilities, namely: i) a primordial effect taking place at a look back redshift $z \sim 1000$; ii) a primordial effect associated with little big bangs taking place at a more modest redshift $z \sim 4$ (Arp *et al.*, 1992); iii) a conversion of galactic infrared radiation into microwaves by iron whiskers taking place at still smaller redshifts, say $z \sim 0.3$; iv) a highly local effect owed to clouds of iron particles receding from the halo of our Galaxy.

It is worth mentioning other interesting results amongst the many coming from ground based experiments realized in this fruitful microwave energy range.

With ground based experiments De Amici *et al.* (1991) have measured the low frequency spectrum of the CMBR at 7.9 cm (3.8 GHz), during the austral spring of 1989, from the South Pole. They obtained a brightness temperature of $2.64 \pm 0.07$ K (confidence level of 68%), in agreement with that obtained in 1986–88, at the same frequency, from a very different site: the White Mountain Research Station (WMRS) in Eastern California. From the same site near the geographical South Pole Levin *et al.* (1992) have obtained a brightness temperature of $2.70 \pm 0.08$ K at 4 cm (7.5 GHz) (confidence level of 68%). Their combined results of the 1988 and 1989 measurements give a temperature of $2.64 \pm 0.06$ K. A temperature of $2.70 \pm 0.07$ K was obtained at 6.3 cm (4.75 GHz) by Mandolesi *et al.* (1986) with a radiometer placed at WMRS.

The weighted average of the measurements of all radiometers operating at WMRS is $2.73 \pm 0.05$ K (Smoot *et al.*, 1985).

This means that the CMBR temperature in the Rayleigh Jeans region was determined to better than 2% before COBE.

A measurement of the large scale CMBR anisotropy at 1.8, 1.1, 0.63, and 0.44 mm was performed with an experiment launched on board a balloon from Fort Summer (New Mexico, USA) in 1989. It placed the most stringent constraints on fluctuations in the CMBR on angular scales greater than $\sim 4°$ at these wavelengths. The analysis of the sky map at 1.8 mm gives an upper limit on fluctuations of the CMBR of $\sim 1.6 \times 10^{-5}$, at a confidence level of 95%, for Gaussian fluctuations at a correlation angle of 13° (Meyer, Cheng & Page, 1991).

Whatever the cause of this fluctuation, the latter upper limit strongly constrains CMBR anisotropy at levels comparable to those of COBE.

However, COBE results from FIRAS instrument give for the CMBR a best fit to a black body spectrum within $3.4 \times 10^{-8}$ erg cm$^{-2}$ s$^{-1}$ sr$^{-1}$ over the range 5–0.5 mm (2–20 cm$^{-1}$) (Bennett *et al.*, 1994).

These measurements imply stringent limits on energy release in the early Universe after $t \sim 1$ yr and redshift $z \sim 3 \times 10^6$. The deviations are less than 0.03%

of the peak brightness. The temperature of the CMBR is $2.726 \pm 0.010$ K at a confidence level of 95% (Mather *et al.*, 1994).

### 3.1.3. *Infrared* (100–1 $\mu m$)

The IR range of the DEBRA is particularly important in cosmology, but it is one of the critical regions in which the measurements are extremely difficult to perform because of the high opacity of the Earth's atmosphere and localized dusts.

Many cosmological and astrophysical events giving rise to the diffuse IR radiation occur in the range of redshift between $\sim$1000 and a few units. These contributions are coming from population III stars, galaxy formation, dust filled galaxies, decaying relic particles, etc. (e.g., Bond, Carr & Hogan, 1986; Carr, 1988). In this energy range it is very difficult to compare theoretical predictions with the observations because of the presence of dust, which contaminates every observation; therefore the contribution of this contamination can be removed from the total flux only entering in models estimating the dust emission. Ressel & Turner (1990) attempted a correction of most of the data they used to draw the Grand Unified Photon Spectrum (GUPS) in the critical range 100–1 $\mu$m, in order to remove the dust emission, always model dependent. Around 100 $\mu$m only a few observations are available: the first one obtained by Matsumoto *et al.* (1988a) in three narrow IR bands centered at 262, 137, and 102 $\mu$m, but uncertain. These data are best fitted by interstellar dust at a temperature of $\sim$20 K; the second data set is from the IRAS observations (Hauser *et al.*, 1984; Boulanger & Perault, 1988).

With a particular dust emission model, Boulanger & Perault (1988) deduced the residual DEBRA component, which is substantially lower than that of Matsumoto *et al.* (1988a). The third observation has been recently made in order to obtain the surface brightness of the Far Infrared (FIR) sky at 134, 154, and 186 $\mu$m at high galactic latitude by using a liquid helium cooled telescope launched on board the Japanese sounding rocket S-520-15 (Kawada *et al.*, 1994). The measured brightness is well correlated with the 100 $\mu$m brightness measured by the IRAS and, in regions excluding molecular clouds, with HI column density. The spectrum of emission from the interstellar dust (ISD) is similar in all regions observed ($5^\circ \times 20^\circ$), including low column density CO clouds. The temperature is $T = 16.4^{+2.3}_{-1.8}$ K, assuming emissivity $\propto \lambda^{-2}$. Assuming a constant FIR dust emissivity per hydrogen nucleus, the ratio of the $H_2$ column density to the velocity integrated CO intensity in the high latitude molecular clouds is $N(H_2)/W_{CO} = (1.6 \pm 0.3) \times 10^{20}$ cm$^{-2}$ (K km s$^{-1}$)$^{-1}$. This result is lower than that for the Galactic Plane, $2.8 \times 10^{20}$ cm$^{-2}$ (K km s$^{-1}$)$^{-1}$ (Bloemen *et al.*, 1986), but is higher than that of de Vries, Heithausen & Thaddeus (1987), $(0.5 \pm 0.3) \times 10^{20}$ cm$^{-2}$ (K km s$^{-1}$)$^{-1}$, over a larger region of the same CO cloud. The difference between the results of de Vries, Heithausen & Thaddeus (1987) and Kawada *et al.* (1994) could be owed to the small size of the CO regions sampled by the latter authors. The residual brightness obtained by Kawada *et al.* (1994), after subtracting the emission correlated with HI column density, is $\lambda I_\lambda(154\ \mu\text{m}) = (1.4 \pm 0.6) \times 10^{-12}$

W cm$^{-2}$ sr$^{-1}$, yielding an upper limit to the FIR extragalactic background radiation of $\lambda I_\lambda(154\ \mu\text{m}) \leq 2.6 \times 10^{-12}$ W cm$^{-2}$ sr$^{-1}$, which excludes some high evolution models of IR galaxies.

In the range 50–10 $\mu$m only total IR flux measurements exist and up to date nobody has attempted to extract the DEBRA component. Ressel & Turner (1990) gave upper limits deduced by measurements of Hauser *et al.* (1984) at galactic poles and earlier measurements of Soifer, Houck & Harwit (1971).

In the Near Infrared (NIR) range (10–1 $\mu$m) absolute measurements of the DEBRA can be performed, although dust emission is still a problem. The measurements in this region are very interesting because it is possible that the DEBRA includes a component from the redshifted light associated with the initial epoch of galaxy formation. Most of the data – corrected for dust emission – used by Ressell & Turner (1990) to produce the GUPS are coming from the rocket flight of Matsumoto, Akiba & Murakami (1988b). Even if this experiment had problems with contamination from rocket exhaust, the data at 2.2 $\mu$m are considered real by the authors. In any case, these data must be used with extreme caution because the amount of dust emission to be removed is a factor of ~10 higher than the diffuse flux; therefore the model's dependence on removing the dust's contribution is crucial.

Matsuura *et al.* (1994) have flown a liquid helium cooled spectrometer on board the already mentioned Japanese rocket S-520-15 in order to search for extragalactic background radiation at high galactic latitude at NIR range 4–1.4 $\mu$m. The observations clearly show time-dependent components of terrestrial environmental origin. The observed residual brightness was slightly brighter than that of previous rocket borne experiments and the recent result of COBE–DIRBE. They set upper limits on the extragalactic continuum intensity of $\lambda I_\lambda$ (154 $\mu$m) $\leq 5 \times 10^{-11}$ W cm$^{-2}$ sr$^{-1}$ at 2.5 $\mu$m and $\lambda I_\lambda$ (154 $\mu$m) $\leq 2.5 \times 10^{-11}$ W cm$^{-2}$ sr$^{-1}$ at 4.0 $\mu$m and on the extragalactic line intensity of $I \leq 5 \times 10^{-13}$ W cm$^{-2}$ sr$^{-1}$ at 1.7–2.5 $\mu$m.
In the last ten years we have seen a *crescendo* of discoveries which have revolutionized our understanding of the process by which the tiny graininess seen in the CMB resulted in the formation of galaxies. Recently, after a brief historical excursion on the galaxy evolution models, De Zotti *et al.* (2002) discussed over the intensity of the Cosmic FIR Background and the strong evolution of galaxies in the FIR to mm energy ranges, which demonstrate that the bulk of star light emitted during the early phases of galaxy evolution was reprocessed by dust. Therefore the optical view of the galaxy formation process is highly incomplete and biased, and must be complemented with FIR/sub-mm observations. In particular, they reviewed the impact of sub-mm surveys on our understanding on the evolutionary history of spheroidal galaxies, and discussed a model in which the inter-relationships between formation and early evolution of spheroidal galaxies and quasars is taken into account.

3.1.4. *Optical and Ultraviolet* ($10^4$–10 Å)

Following the usual split up of the energy ranges employed in astronomy, the range $10^4$–10 Å includes the optical, UV, FUV, EUV, and ultra soft X-ray bands.

Several measurements have been performed in order to search for the optical component of the DEBRA, but no detection has been claimed. Only upper limits to the DEBRA have been reported in the optical, such as those at 6500 Å (Boughn & Kuhn, 1986), at 5115 Å (Dube, Wickes & Wilkinson, 1979), at 4400 Å (Toller, 1983), and at 4000 Å (Spinrad & Stone, 1978). These upper limits, converted to cgs units by Ressell & Turner (1990), have been used to construct the GUPS, reported in Figure 15. From the end of the optical (~3000 Å) to the $L_\alpha$ limit (912 Å) observations are especially exciting because neutrinos with cosmologically interesting masses could decay and produce a diffuse background (de Rujula & Glashow, 1980). Many measurements in the latter range have been performed with space based experiments.

Historically the study of the FUV background was approached in the hope of detecting redshifted recombination radiation from a dense intergalactic medium (Weymann, 1967; Davidsen, *et al.*, 1975). This possible detection has remained unsolved up to date together with other exciting cosmological possibilities.

A major step in knowledge of the FUV background came from a telescope flown on the Apollo–Soyuz test project to perform a photometric survey at 1400 Å in ~300 directions. The conclusion obtained was different from that of all previous work. The intensity varied from 300 to 2000 photons cm$^{-2}$ s$^{-1}$ Å$^{-1}$ sr$^{-1}$ with no smooth dependence on galactic coordinates (Paresce & Jakobsen, 1980). The intensity showed a moderate correlation with the HI column. A confirmation of this result came from a re-analysis of data obtained with a spectrophotometer flown on the DB 2-Aura satellite (Maucherat-Joubert, Deharveng & Cruvellier, 1980). However, the possibility of residual airglow sensitivity in both experiments led to doubts about systematic uncertainties. Jakobsen *et al.* (1984) eliminated this difficulty with their flight on board a rocket, confirming the results of Paresce & Jakobsen (1980): the diffuse background shows significant variations over scales of several degrees and is correlated with $N_H$. The same conclusions have been obtained by Hurwitz & Bowyer (1988) from the Berkeley EUV/FUV Shuttle telescope. Consistent results were obtained also by Murthy *et al.* (1989), although they interpreted their data as evidence of a uniform background.

The sum of these measurements summarized in the paper by Martin & Bowyer (1989) is convincing evidence that the majority of the FUV background originates in the Milky Way. The minimum intensities observed by experiments with small stellar and instrumental subtractions are: ~300 photons cm$^{-2}$ s$^{-1}$ Å$^{-1}$ sr$^{-1}$ (Paresce & Jakobsen, 1980), 400 photons cm$^{-2}$ s$^{-1}$ Å$^{-1}$ sr$^{-1}$ (Joubert *et al.*, 1983; Jakobsen *et al.*, 1984). The minimum predicted intensity in directions with $6 \times 10^{19} < N_H < 1.5 \times 10^{20}$ cm$^{-2}$ is 30–150 photons cm$^{-2}$ s$^{-1}$ Å$^{-1}$ sr$^{-1}$. Therefore a spatially uniform component with $I_0 < 300$ photons cm$^{-2}$ s$^{-1}$ Å$^{-1}$ sr$^{-1}$ is consistent with the survey data. This isotropic FUV background is indeed cosmological, and has an intensity of ~30 photons $cm^{-2}$ $s^{-1}$ $Å^{-1}$ $sr^{-1}$ at Earth (Martin & Bowyer, 1989). One possible source is the integrated light from normal galaxies. Whilst galaxies have long been considered as a source of visible diffuse background, Tinsley (1972a,b) realized that an extragalactic UV background is produced in proportion to the hot star content of spirals, and is therefore a sensitive indicator of the mean star formation rate in the Universe. Fluctuations in the background intensity could be owed to the clustering of galaxies in space. With a rocket borne imaging camera Martin & Bowyer (1989) searched for small scale fluctuations in the FUV background that would be the signature of an integrated background contribution from normal spiral galaxies. In the range 1400–1800 Å, they found the evidence that these fluctuations, at a level of ~4% at $6'$–$12'$ scales, are the integrated light of distant galaxies. Using this result as an upper limit to the integrated light from galaxies, one obtains the mean UV luminosity density, in the redshift range $0.1 < z < 0.6$, of $L_{gal} < 7 \times 10^7$ $L_\odot$ $Mpc^{-3}$. When compared to estimates of the present UV luminosity density ($L_{gal} = 4.5 \times 10^7$ $L_\odot$ $Mpc^{-3}$ for $H_0 = 100$ km $s^{-1}$ $Mpc^{-1}$), this places a strong constraint on the FUV evolution of galaxies: very little evolution has occurred since $z = 0.6$. This constraint can be interpreted as a limit on the average past star formation rate relative to the present $\dot{M}/\dot{M}_\odot < 1.5$ if dust obscuration was not significantly greater in the past. With the latter assumption, Martin & Bowyer (1989) derived an upper limit to the average star formation rate in the redshift range $0.1 < z < 0.6$ of $\dot{M} < 0.033$ $M_\odot$ $yr^{-1}$ $Mpc^{-3}$ $h^{-2}$.

The origin of the remaining isotropic flux is unknown:

- It could be formed in part by scattering from a smooth component of dust for which there is evidence in the IRAS survey data and this would imply the existence of anomalous dust to gas ratios;
- It could arise from a tepid intergalactic medium radiating at moderate ($z \sim 4$) redshifts (Weymann, 1967; Sherman & Silk, 1979; Jakobsen, 1980). The power spectrum and other observations put many constraints on the characteristics of such a medium, but do not rule out its existence (Martin, 1986);
- Finally, the isotropic background may originate from a process unknown heretofore.

In order to derive information on these possible contributors, it is important to mention the work of Martin, Hurwitz & Bowyer (1991) performed with the Berkeley FUV spectrometer (600–1900 Å), which flew on Shuttle mission 61-C during the week of January 12–20, 1986. The long wavelength data (~1400–1900 Å) obtained during two pointed observations allowed the properties of the extragalactic FUV background to be constrained. The mean continuum level $I_{EG} = 280 \pm 35$ photons $cm^{-2}$ $s^{-1}$ $Å^{-1}$ $sr^{-1}$ was obtained in a direction ($l = 155°$, $b = 58°$) with very low HI column density. This represents a firm upper limit to any extragalactic background in the 1400–1900 Å band. They find no evidence in the spectrum of line or continuum features that would signify additional extragalactic components. They find also that the spectrum of a nearby target with higher $N_{HI}$, when cor-

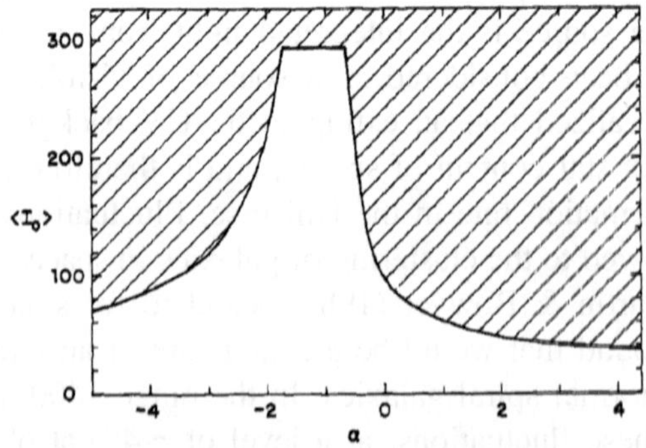


*Figure 13.* Upper limit of 90% confidence of the unknown UV Extragalactic Background component – modelled as $\lambda^{\alpha}$ – viz., $\alpha$ (Martin, Hurwitz & Bowyer, 1991).

rected for $H_2$ fluorescence, is very similar to the spectrum obtained in the low HI view direction. The observed spectrum can be fitted by a power law with an index $\alpha = -1.4 \pm 0.15$. Since they know that one observation is dominated by dust scattered starlight, they assume the same also for the other observation. Then the measured spectrum, represented by $a \sim \lambda^{-1.4}$ power law, is mainly representative of dust scattered starlight. The unknown extragalactic component is modelled as a power law $\lambda^{\alpha}$. So the confidence upper 90% limit to $\overline{I_{EG}}$ as a function of the assumed power law index is shown in Figure 13.

For example, an extragalactic component with $\alpha = 1$ could contribute no more than $\overline{I_{EG}} = 60$ photons cm$^{-2}$ s$^{-1}$ Å$^{-1}$ sr$^{-1}$. Taking into account the observed intensity in low $N_{HI}$ directions, Martin, Hurwitz & Bowyer (1991) then derived the limits on possible contributors to the FUV background, as reported in Table 2. In order to construct the GUSP (see Figure 17), all the former data, together with those used by Ressel & Turner (1990) [at 3300 and 2980 Å (Paresce & Jakobsen, 1980), at 2200 and 1690 Å (Joubert *et al.*, 1983), at 1715 and 1440 Å (Anderson *et al.*, 1979; Feldman, Brune & Henry, 1981), at 1360 Å (Weller, 1983), with upper limits in the region 1200–500 Å (Holberg, 1986), with data in the regions 1700–1200 Å (Murthy *et al.*, 1989), and 2850–1700 Å (Tennyson *et al.*, 1988)] have been used.

Results both observational and theoretical have identified a variety of additional sources of the FUV background which are of fundamental importance in their own right (Bowyer, 1991; Henry, 1991). Such additional sources have been discussed by Henry (1999a). They are:

– Within the Galaxy diffuse UV radiation is produced by several components of the ISM, from the cold neutral phase (Holberg, 1990) to the hot highly ionized phase;
– Additional emission from the cold neutral phase is expected from $H_2$ fluorescence (Witt *et al.*, 1989);

TABLE 2

FUV background in low $N_{\rm HI}$ directions: Limits on possible contributors (adapted from Martin, Hurwitz & Bowyer (1991).

| Component | I (1600 Å) ph cm$^{-1}$ s$^{-1}$ Å$^{-1}$ sr$^{-1}$ | References |
|---|---|---|
| Observed Intensity | 280 ± 50 | Martin *et al.* (1991); Paresce *et al.* (1980) |
| Airglow | <10 | Martin *et al.* (1991) |
| Zodiacal Light | <10 | Martin *et al.* (1991); Martin & Bowyer (1989); Hurwitz *et al.* (1990) |
| H Fluorescence | <12 | Martin *et al.* (1990) |
| HII Two-Photons | 55 | Martin *et al.* (1991); Hurwitz et al (1990) |
| Dust | 80–200 | Martin *et al.* (1991); Hurwitz *et al.* (1990) |
| CIV Gas | 10–15 | Martin & Bowyer (1990) |
| Galaxies | 30 | Martin & Bowyer (1989) |
| QSO/AGN | <10 | Martin & Bowyer (1989) |
| Redshifted Line* | <50 | Martin *et al.* (1991) |
| IGM | <20 | Martin *et al.* (1991) |
| Unexplained | 0–110 | Martin *et al.* (1991) |

*Produced in matter with significant large-scale structure.

- FUV line radiation from collisionally ionized gas at temperatures of up $10^{5.5}$ K, which is observable in emission at no other wavelengths, except for the highly obscured region shortward of 912 Å. The presence of such a gas in the ISM has been inferred by the detection of absorption lines such as the OVI by the Copernicus satellite (Jenkins, 1978b);
- There is evidence of an extragalactic component. Indeed, the observed correlation between the diffuse UV background and the hydrogen column density is not found to account for the entire background signal. Then a residual signal is inferred. Such a residual high galactic latitude background longward of Lyman–$\alpha$ (1216 Å) is ~300 photons units. The contribution from galactic processes cannot account for that residual, so an extragalactic component is implied (Henry, 2002).

The spectrum of the visible, ultraviolet, and soft X-ray portions of the diffuse background radiation is reported in Figure 14 (Henry, 2002).

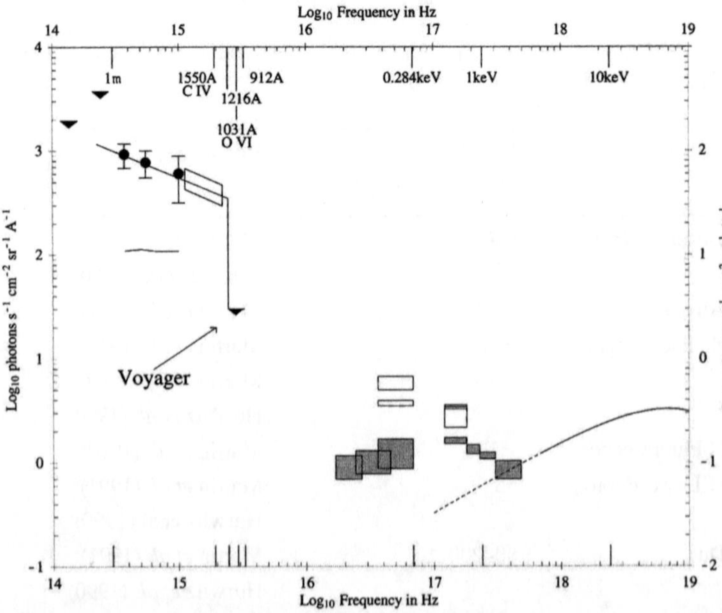


*Figure 14.* The spectrum of the visible, ultraviolet, and soft X-ray portions of the diffuse background radiation. The wavelengths of two important interstellar medium emission lines, OVI (1031 Å) and CIV (1550 Å), are identified, and the Carbon edge at 0.284 keV is also identified (Henry, 2002).

### 3.1.5. *X-Ray and Gamma Ray* (1 *keV* - 4 *GeV*)

This energy range is particularly important for the DEBRA since apart from the very bright CMBR and highly uncertain optical and UV backgrounds, the extragalactic sky is dominated by the X-ray background (XRB) at ~40 keV.

The diffuse extragalactic background radiation in the X and $\gamma$-ray ranges between ~0.1 keV to 100 MeV (~$10^{-6}$ cm – $10^{-12}$ cm) has been well measured since the beginning of the X-ray astronomy era. With the results from BBXRT (Broad Band X-Ray Telescope) on board NASA's ASTRO-1 Shuttle mission, ROSAT and ASCA (ASTRO D), BeppoSAX, AXAF–Chandra, and XMM–Newton satellites the softer part of this range is deeply explored and then some ideas on the origin of the softer XRB have now been verified. With the exception of the CMBR, the XRB is the best studied amongst the extragalactic background radiation. Indeed, the XRB was discovered during the first successful rocket flight, the same that discovered Sco X-1 (Giacconi *et al.*, 1962). The diffuse emission was later established to be highly isotropic for energies $\geq$ 3 keV within a few per cent on large scales, with the exception of a weak Galactic anisotropy observed in the energy range 2–10 keV (Warwick, Pye & Fabian, 1980) and a dipole anisotropy (Shafer & Fabian, 1983). The direction of the dipole is consistent with that of the dipolar component of the CMB; this can be explained as a consequence of the motion of the Milky Way in the Hubble flow. The XRB does not show any correlation with the galactic latitude outside the Galactic Plane. This fact, together with the

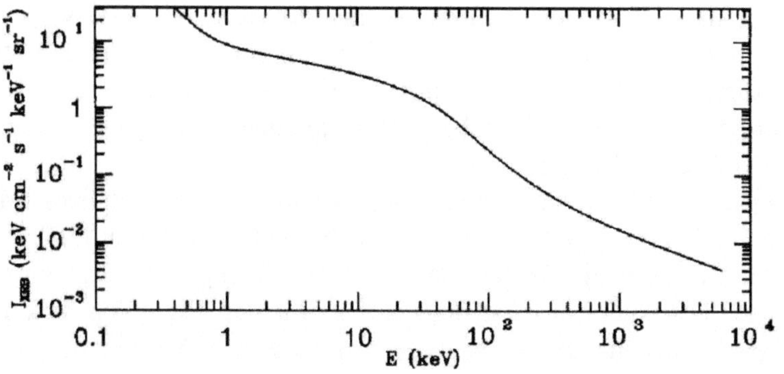


*Figure 15.* The spectrum of the extragalactic X-ray and $\gamma$-ray background (Matt, 1995, after Gruber, 1992, Gendreau, 1995, and Gendreau *et al.*, 1995).

dipolar anisotropy strongly favor the extragalactic origin of the XRB for energies $\geq$ 3 keV. For energies below 3 keV the extragalactic contribution is uncertain but less than 10% of the observed intensity below ~0.25 keV (e.g., McCammon & Sanders, 1990). The uncertainty is essentially owed to a strong Galactic component dominating the XRB below ~0.5 keV and to the attenuation of extragalactic X-rays by photoelectric absorption in our Galaxy's ISM at lower energies (e.g., Fabian & Barcons, 1992). Several reviews on XRB have been published recently in the literature, namely those of McCammon & Sanders (1990) in the range 0.1–10 keV, Fabian & Barcons (1992) in the range 3–300 keV, and Matt (1995) in the range 0.1 keV–6 MeV. Barcons (1996) reviewed in general the XRB.

It is now generally thought that the excess below 0.25 keV is owed to thermal radiation from a $10^6$ K component of the interstellar gas in our Galaxy, and that most of it originates within a few hundred parsecs of the Sun. In contrast, in the range ~0.5–1 keV it is not clear how much of the observed radiation is galactic or extragalactic.

Several fits have been proposed in order to explain the available experimental data. In the range 1–3 keV, the XRB intensity is best fitted by a power law (Schwartz, 1979): $I_E = 7.7(E/1\ \text{keV})^{-0.4}$ erg cm$^{-2}$ s$^{-1}$ erg$^{-1}$ sr$^{-1}$.

In the range 3 keV–6 MeV the XRB intensity is best fitted by the following laws (Gruber, 1992): $I_E = 7.877 \times E^{-0.29} \times e^{-E/41.13}$ for $E \leq 60$ keV, and $I_E = 1652 \times E^{-2} + 1.754 \times E^{-0.7}$ for $E > 60$ keV.

The spectrum of the XRB is reported in Figure 15. This spectrum is featureless. No absorption or emission edges or lines have been unambiguously revealed in that range. However, this is doubtless of extragalactic origin. Therefore the spectra of the contributing sources are themselves featureless or the features are smeared out by the superposition of sources at different redshifts (Matt, 1995).

Inoue *et al.* (1997) reported the ASCA sky surveys in the energy range 2–10 keV. A comparison with the ROSAT survey in the range 0.5–2 keV suggests the presence of a significant number of harder sources than the usual AGNs at the flux level around the ASCA detection limit. In fact, the average 2–10 keV spectral index of sources in the flux range of (1–25) $\times 10^{-13}$ erg cm$^{-2}$ s$^{-1}$ detected in the ASCA large sky survey is $0.5 \pm 0.2$. This is closer to the spectral index of the cosmic X-ray background than that obtained in the GINGA fluctuation analysis for sources in the flux range above $10^{-12}$ erg cm$^{-2}$ s$^{-1}$.

The presence of a number of hard sources with the energy indices smaller than 0.7 in the 2–10 keV range with fluxes near the ASCA detection limit could mean that they are responsible for the Cosmic X-ray Background Radiation (CXBR) in this band. The nature of these objects is still under discussion. One possibility is that they are distant QSOs. However, from the ASCA observations of distant QSOs the measured energy spectral indices are still in a rather narrow range of values around 0.7, although their cosmological redshifts, $z$, exceed unity. Hence, the new detected hard sources would not be distant type-1 QSOs evolved from those observed in the flux range of $10^{-12}$–$\sim 10^{-11}$ erg cm$^{-2}$ s$^{-1}$.

An optical follow up observation of the ASCA deep sky surveys and large sky survey, discovered a type-2 QSO at $z \cong 0.9$ (Ohta *et al.*, 1996). This supports the idea that the new detected hard sources would be highly obscured ($N_H \approx 10^{24}$ H atoms cm$^{-2}$ (Madau, Ghisellini & Fabian, 1994) AGNs at cosmological distances.

If these sources are responsible for the CXBR, the total number of highly obscured AGNs should be larger than that of unobscured AGNs. Otherwise the spectral index of the CXBR should have been unchanged from the average of bright Seyfert galaxies, which is around 0.7–0.8.

Such highly obscured AGNs are usually observed as Seyfert-2 galaxies when the sources are fairly close to us (Inoue *et al.*, 1997; Ueda *et al.*, 1999, 2000).

Before the advent of CGRO, only a few measurements at energies greater than 100 keV were available, namely:

- 100 keV–1 Mev (Gruber *et al.*, 1985);
- 0.3–10 MeV (Trombka *et al.*, 1977);
- Two upper limits in the range 10–35 MeV (Trombka *et al.*, 1977);
- Data from the SAS II satellite in the range 35–100 MeV (Fichtel, Simpson & Thompson, 1978);
- Measurements from COS B satellite at 70 MeV (Sacher & Schönfelder, 1983). All these data were used to construct the GUSP reported by Ressel & Turner (1990) (Figure 17).

Now, with the results of CGRO, especially those from COMPTEL and EGRET, together with those from COS B, Strong (1995) has analyzed the problem of the diffuse galactic $\gamma$-ray continuum emission. The key issue includes the role of emission from cosmic ray interactions with molecular hydrogen and its energy dependence, emissivity gradients and their interpretation, the cosmic ray electron spectrum, and the effect of discrete sources.

The relative contribution of the various emission processes at low and high galactic latitudes is estimated. The contribution of unresolved sources to the diffuse

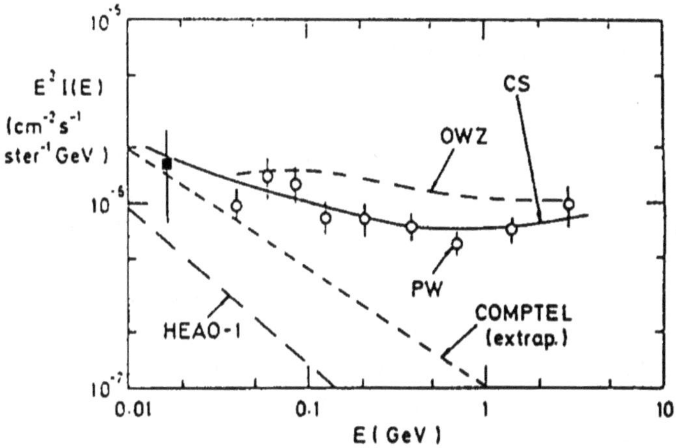


*Figure 16.* Energy spectrum of the diffuse isotropic background, plotted as $E^2I(E)$ vs $E$ (Smialkowski, Wolfendale & Zhang, 1997).

emission is unknown but probably lies in the range 10–20%. At high latitudes the galactic emission is intense enough to significantly complicate the identification of the extragalactic component. In particular, the inverse Compton emission from a halo of a few kpc in extent can account for much of the high latitude galactic emission.

It is conventional to cut off the Galactic Halo at a few kpc. Smialkowski, Wolfendale & Zhang (1997) advocate a value of $z_h = 2.5$ kpc for the electron component. However, this does not preclude a 'weak' Halo extending to many tens of kpc.

If there is no extended Galactic Halo, the curve CS – the composite spectral line drawn to connect the COMPTEL intensities with the PW point (Smialkowski, Wolfendale & Zhang, 1997) – is shown in Figure 16. CS represents the best estimate of the diffuse extragalactic energy spectrum. This new estimate of the diffuse $\gamma$-ray background then yields intensities lower than those estimated by Osborne, Wolfendale & Worral (1994).

About the extended Galactic Halo it is important to mention the studies of the Magellanic Stream by Moore & Davies (1994) and Weiner & Williams (1996). They make the case for an extended dilute halo of diffuse ionized gas surrounding the Galaxy. This gas is at $T \cong 2.2 \times 10^6$ K and, at a distance of 65 kpc, the density is $\sim 5 \times 10^{-5}$ atom cm$^{-3}$. The authors invoke ram pressure of this gas on the LMC and on the Magellanic Stream to explain some of their features. Some 40% of the observed diffuse background radiation in the X-ray M band (0.5–1 keV) would come from the gas. The authors make the case for the ionized gas extending to at least 100 kpc.

The implication for the diffuse $\gamma$-ray background can be very significant. If the Halo exists it is unlikely that galactic electrons will escape from the Halo and the Inverse Compton process will have a near isotropic term (particularly in the Outer Galaxy). Thus it leads to an inferred *true Extragalactic Intensity* appreciably lower than that given by Smialkowski, Wolfendale & Zhang (1997). The result is clearly model dependent. However, magnetic energy density and proton cosmic ray intensity are coupled and fall with $z_h$ from the usual Galactic Plane value to the necessary value of $\cong 3 \times 10^{-5}$ eV cm$^{-3}$ at $z_h \cong 100$ kpc. An approximate analysis which takes that into account then yields a reduction of ~40% of the extragalactic intensity.

An independent method, using the work of Chi & Wolfendale (1989) on $\gamma$-rays from electrons escaping from galaxies, gives a similar result.

Therefore the possibility of a Giant Galactic Halo leads to the conclusions that the true extragalactic isotropic diffuse intensity may be even lower than that quoted (by some tens of percent) and in turn models for explanation in terms of AGN will need revision.

The detection of $\gamma$-ray emission from the LMC (Sreekumar *et al.*, 1992) and the upper limit from the SMC (Sreekumar *et al.*, 1993) based on EGRET data are of great importance and relevant to the interpretation of the galactic diffuse emission. Indeed, they found that the flux from the LMC appears quite consistent with the cosmic ray intensity similar to that in the Galaxy. The non-detection of the SMC provides the long awaited proof that cosmic rays cannot be of universal origin. The upper limit (for $E > 100$ MeV) is only 20% of the expected flux based on the galactic emissivity (Strong, 1995). The detection (or not) of high energy $\gamma$-rays from other galaxies could give the final answer to the problem of the universal origin of cosmic rays. Unfortunately, with this generation of detectors the prospects for observing other spiral galaxies appear poor. Indeed, Pohl (1994) estimated the flux ($E > 100$ MeV) from M31 and M82 less than $2 \times 10^{-8}$ photon cm$^{-2}$ s$^{-1}$, below the EGRET sensitivity.

The INTEGRAL and AGILE (Morselli, 2002) missions could provide useful data from spiral galaxies. The description of the four experiments aboard the INTEGRAL observatory can be found e.g.: Optical Monitor Camera (OMC) (Giménez, 1999); Joint European Monitor for X-rays (JEM-X) (Lundt, Westergaard & Budtz-Jørgensen, 1999); Spectrometer for INTEGRAL (SPI) (Vedrenne *et al.*, 1999); Imager on Board the INTEGRAL Satellite (IBIS) (Lebrun *et al.*, 1999b).

Kappadath *et al.* (1996) have reported the preliminary cosmic diffuse $\gamma$-ray spectrum from 0.8 to 30 MeV measured with CGRO-COMPTEL.

We want to note the difficulties in making these measurements; indeed, the prompt background is produced primarily by the interactions of protons and neutrons with the spacecraft. Kappadath *et al.* (1996) constructed the spectrum by measuring the count rate of $\gamma$-rays from high galactic latitudes and first extrapolating the rate to a zero cosmic ray flux environment to eliminate the prompt background. Second, this rate was then corrected for the effects of delayed internal radioactivity to arrive at $\gamma$-ray count rate due to CDG (Cosmic Diffuse $\gamma$-ray).

The preliminary measurements of the CDG flux in the 2–10 MeV range is 5–10 times lower than earlier reports. The 2–30 MeV flux is consistent with power law extrapolation from lower and higher energies. The preliminary measurements show no evidence of a *MeV bump* above 2 MeV. The bump, if present, is restricted to energies less than $\sim$2 MeV. The CDG spectrum in the 60 keV to 10 GeV range, except for perhaps the range 0.8–2 MeV, can be well described by a broken power law that evolves from soft to hard with increasing energy. The break energy seems to occur in the 1–10 MeV range.

The all sky survey in high energy $\gamma$-rays ($E \geq 30$ MeV) carried out with the CGRO-EGRET (Kniffen *et al.*, 1996) has allowed the first study of the extragalactic diffuse $\gamma$-ray emission since the SAS-2 experiment in 1972–73 (Fichtel, Simpson & Thompson, 1978).

The all sky survey shows a strong galactic emission that extends well away from the galactic plane making the selection of the region, used for analysis, critical. Kniffen *et al.* (1996) restricted their analysis to regions at latitudes greater than 30° (absolute). The spectrum obtained is consistent with a power law distribution in energy with a slope of $-2.07 \pm 0.03$ and an integrated intensity above 100 MeV of $1.5 \times 10^{-5}$ cm$^{-2}$ s$^{-1}$ sr$^{-1}$.

This result is in excellent agreement with that reported by Fichtel, Simpson & Thompson (1978) by using the SAS-2 data. The first spectrum of the diffuse radiation, in the energy range $\sim$35–150 MeV, was characterized with a power law distribution in energy with a slope of $-2.34 \pm 0.4$ and an integrated intensity above 100 MeV of $1.5 \times 10^{-5}$ cm$^{-2}$ s$^{-1}$ sr$^{-1}$.

The SAS-2 and EGRET data are consistent with continuation of the results of Kappadath *et al.* (1996), already commented, to higher energies.

As remarked by Kniffen *et al.* (1996) there is much work to be done to gain confidence that there is no residual contamination from the extension of the galactic diffuse emission to high latitudes. There is surely a contribution from unresolved extragalactic sources. Only those with strong EGRET detections have been removed.

It is premature to draw conclusions based on models that all of the intensity results from unresolved sources, but nevertheless a credible case can be made as shown by Chiang *et al.* (1995) and Salamon & Stecker (1994). For instance, following the unification model for AGNs (e.g., Antonucci, 1993 for a review), Matt & Fabian (1994) developed a model based on Seyfert-1 plus Seyfert-2 galaxies, with which they best fitted the XRB from 0.1 to 100 keV reported by Gruber (1992). Furthermore, it remains to explore the scale sizes of spatial fluctuations to look for evidence of structures beyond the size of point sources.

Hasinger (1996) reviewed both X-ray and $\gamma$-ray extragalactic background, and Erlykin *et al.* (1996) attempted a theoretical interpretation. Comastri *et al.* (1995) discussed the contribution of various classes of AGNs (Seyfert galaxies and quasars) to the extragalactic XRB, under the hypothesis of the validity of the unification schemes of AGNs. The integrated emission from AGNs can provide a good fit to the XRB from several to ~100 keV if data are folded with an appropriate cosmological evolution law. Such an emission contributes only ~74% of the ROSAT soft XRB.

### 3.1.6. *HE, VHE, UHE $\gamma$-rays and Cosmic Rays ($10^8$–$10^{20}$ eV)*

The spectral region between $10^8$ and $10^{20}$ eV is very interesting, but unfortunately the diffuse flux within this range is extremely difficult to measure. At these energies there are no claimed measurements of a diffuse extragalactic component. There are several detections of the diffuse component that can be used as upper limits to any extragalactic component. However, many parts of this range are still unexplored for technical reasons. For instance, the range 4–400 GeV is completely unknown although its exploration should be fundamental in obtaining information about the dark matter decay and annihilation in the halo (Primack, Sadoulet & Seckel, 1988; Silk, 1989; Saggion & Bastieri, 2002). Results in the range 1 GeV – 1 TeV could come from proposed experiments on board the Freedom Space Station. Ressell & Turner (1990) used cosmic ray measurements in order to provide absolute upper limits. At these energies photon and hadron induced showers are very similar, with the exception of the muon content and shower shape. When both $\gamma$-ray and cosmic ray data are available it is possible to see that the $\gamma$-ray flux is 3–4 orders of magnitude smaller. Important reliable data (in the wedge $\pm 30°$ from the galactic plane) are those from COS B in the range 70 MeV – 3.3 GeV (Sacher & Schönfelder, 1983) since they are quite free of galactic contamination. For this reason they provide a reasonable good upper limit to any extragalactic flux. It is remarkable to note that the ratio between the flux in the direction of the galactic center and that in the direction of the galactic anti-center is about 5.

In the era of CGRO effort has been devoted to investigating the low energy diffuse cosmic $\gamma$-ray spectrum owed to the cosmological photino annihilation. Gao, Stecker & Cline (1991), using a corrected assumption for the annihilation cross section, studied the annihilation of a general type of the best motivated cold dark matter (CDM) candidates, the lightest supersymmetric particle (LSP). They concluded that the cosmological LSP type of annihilation cannot be a major contributor to the observed extragalactic $\gamma$-ray background (EGB), since the LSP type annihilation fluxes are almost negligible compared with the observed EGB in the energy range from a few MeV to ~1 GeV.

In the energy range 30 GeV – 1 TeV data are very scarce; however, Nishimura *et al.* (1980) during a balloon flight for measuring the electron flux of cosmic rays obtained a measurement of atmospheric $\gamma$-ray flux with a spectrum best fitted by:

$$I_E = 1.2 \times 10^{-6} \times (100\,\mathrm{GeV}/E)^{1.75}\,\mathrm{erg\,cm^{-2}\,s^{-1}\,erg^{-1}\,sr^{-1}}. \qquad (3.1)$$

They attributed the entire flux to secondary $\gamma$-rays from the decay of $\pi^0$, which are produced by the interaction of primary hadronic cosmic rays in the atmosphere. These data are very important since they provide an absolute upper limit for any diffuse $\gamma$-ray flux, being the atmospheric depth of the experiment of 4 g cm$^{-2}$. The interaction length for $\gamma$-rays in the atmosphere is 47 g cm$^{-2}$ (e.g., Longair, 1981).

Since $\gamma$-rays with energies greater than 100 GeV behave like hadronic cosmic rays when they interact with matter, they should have been detected in any experiment measuring the total cosmic ray flux. For this reason the total cosmic ray flux above 100 GeV is a definitive upper limit to the $\gamma$-ray flux at the same energies. Then these data coming from Hillas (1984), including data from PROTON 4 satellite and Akeno, Haverah Park, Tien Shan and Yakutsk experiments have been used by Ressell & Turner (1990) for constructing their GUPS.

At greater energies ($10^{17}$–$10^{20}$ eV) cosmic ray data are coming from the Fly's Eye detectors (Baltrusaitis *et al.*, 1985) as well as from Haverah Park and Yakutsk experiments. Analysis of the Fly's Eye data shows a cutoff at $\sim 7 \times 10^{19}$ eV, which is owed to the threshold energy of the reaction producing $\pi$ mesons via interaction of protons with the cosmic-background photons: $p + \gamma_{3K} \longrightarrow \pi$.

The other important feature in the cosmic ray data is the very well known *knee* located between $10^{15}$ and $10^{16}$ eV which traces to the large gyro radius for cosmic rays above this energy which permits them an easier escape from magnetic confinement within the Galaxy.

Gamma-ray astronomy is objectively a difficult field of astrophysics, since the small residual flux of primary $\gamma$-rays is very difficult to separate from the large hadronic cosmic ray flux that constitutes, in this case, a *terrible* background. Indeed, the expected $\gamma$-ray flux is about $10^{-3}$–$10^{-4}$ of the cosmic ray flux, which in turn is falling rapidly with energy (e.g. Wdowczyk & Wolfendale, 1983). Moreover, it is necessary to differentiate between $\gamma$ and hadron induced showers, and the differences are subtle in shower development and shape and in muon content. Photon induced showers are narrower and muon poor. It is obvious that whilst cosmic rays are influenced by the galactic magnetic field $\gamma$-rays maintain their directionality; this fact strongly favors the research of discrete cosmic $\gamma$-ray sources. Therefore in this kind of investigation, very well described by Weekes (1988, 1996, and 1997), the hadronic cosmic ray background must be removed. This removal is obtained by means of measurements obtained pointing the detectors toward the source for a time (on source) and then, for an equal time, toward a zone not containing the source (off source). In this way it is quite easy to eliminate the background under the hypothesis that it is constant; but at the same time the DEBRA component is eliminated too. However, several measurements of a diffuse component associated with the galactic disk and in some directions on the sky, which have been largely discussed in the fundamental review paper of Ressell & Turner (1990), will be summarized in the following.

Starting from the La Jolla International Cosmic Ray Conference in 1985, detections of several discrete $\gamma$-ray sources have been reported (e.g., the review paper of Nagle, Gaisser & Protheroe, 1988), although the claimed detections were later subject to reconsideration. In spite of this, some of those sources have been confirmed as $\gamma$-ray emitters, such as for instance Cyg X-3 and Her X-1. These sources are placed at several kpc distant from the Earth; therefore the primary particles must be neutral in order to maintain the directionality and to be long lived. Both characteristics force it to be admitted that these primary particles must be photons. But if these observations are correct the primary particles are not photons or photon induced showers at energies $E \geq 1$ TeV are not muon poor.

The threshold of the reaction of $\gamma$-rays with the cosmic microwave background photons producing electron–positron pairs $\gamma + \gamma_{3K} \longrightarrow e^- + e^+$ is between $10^{14}$ and $10^{15}$ eV. The mean free path of primary $\gamma$-rays at such energies is then short ($\sim$10 s of kpc).

The mean free path of photons at higher energies slowly increases, but even for energies of $\sim 10^{20}$ eV this path is far from the order of cosmological distances. Therefore the possibility of measuring an actual diffuse extragalactic background at these energies is completely helpless.

Individual measurements of the diffuse $\gamma$-ray background have been used by Ressell & Turner (1990) for energies between 0.5 TeV up to $\sim$800 TeV in order to construct the GUPS. Weekes (1988) argues that Tien Shan EAS (Extensive Air Shower) gives the most reliable data at energies of $\sim$1 PeV and above. Indeed, this experiment has a good particle discrimination, and with a selection of shower events with very few muons and hadrons it was possible to derive an all sky diffuse $\gamma$-ray flux (Stamenov *et al.*, 1983). Their flux corresponds to 5 $\gamma$-rays in 12,900 hours of observations.

So a remark on the objective difficulties in making this kind of experiments is as obvious as essential.

Diffuse $\gamma$-ray flux in the galactic plane, at energies greater than $2 \times 10^{16}$ eV, has been measured by Dzikowski *et al.* (1985); this value ($\sim 10^{14}$ erg cm$^{-2}$ s$^{-1}$ erg$^{-1}$ sr$^{-1}$) has then been used by Ressell & Turner (1990) in the GUPS as an upper limit to the diffuse extragalactic $\gamma$-ray flux. The last point at $E \geq 10^{17}$ eV (a few units in $10^{-18}$ erg cm$^{-2}$ s$^{-1}$ erg$^{-1}$ sr$^{-1}$) is that from the Yakutsk EAS array, based on muon poor showers (Glushkov *et al.*, 1985).

It is important to note that all the claimed detections of a diffuse $\gamma$-ray flux correspond to $\sim$0.1% of the cosmic ray flux, like predicted by theories. However, it is interesting to remark that no definitive measurements of diffuse $\gamma$-ray flux at energies above $\sim$10 GeV exist and that there is not yet the answer to the question if $\gamma$-ray induced showers are muon poor.

In order to make a comparison, Ressell & Turner (1990) included in their graph of the GUPS the cosmic ray electron flux (plus positron flux), the atmospheric muon flux and the atmospheric neutrino flux. The atmospheric neutrino spectrum is the theoretical spectrum based on known nuclear and particle physics and the measured flux of muons at sea level.

Therefore, in conclusion, the review of the DEBRA leads to the Grand Unified Photon Spectrum (GUPS), covering 29 orders of magnitude of the electromagnetic spectrum, from $10^{-9}$ to $10^{20}$ eV, as shown in Figure 17.

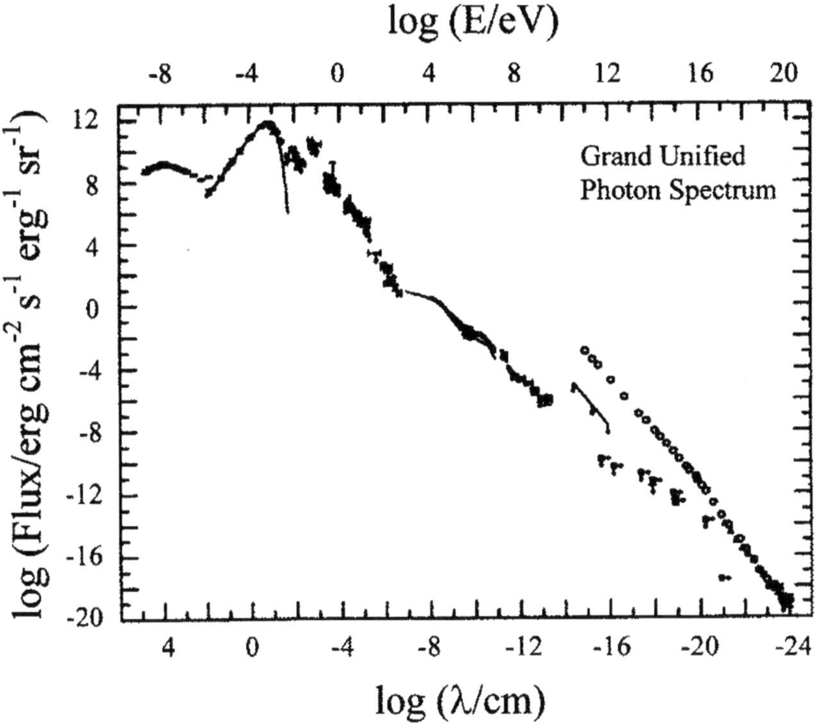


*Figure 17.* The Grand Unified Photon Spectrum of the Diffuse Extragalactic Background Radiation (Ressel & Turner, 1990).

Henry (1999b, 2002) thoroughly discussed the updated experimental situation of the cosmic background. Henry (2002) presented his paper as a supplement to that published in ApJ Letters (Henry, 1999b). The first such supplement was to present the discussion of units for use in the display of spectra *without* the compression that appears in ApJ Letters, so that the material is much more readable:

If one is interested in *energy content* the most meaningful units in which to display the spectrum of diffuse radiation are, remarkably enough, photons $s^{-1}$ $cm^{-2}$ $sr^{-1}$ $nm^{-1}$. If there is an *equal* amount of energy present in every logarithmic interval of frequency, then these units have the virtue of assuming *constant* value, as demonstrated in Henry (2002).

The results are shown in Figure 18, where is reported the spectrum of the diffuse background radiation:

1) Radio background; 2) Microwave background (Fixsen *et al.* 1994); 3) Firas excess (Fixsen *et al.* 1998); 4) DIRBE observations and upper limits (Hauser *et al.* 1998); 5) Visible background (Bernstein 1998); 6) Background from *Voyager* (Murthy *et al.* 1999); 7) Interstellar photo-ionization cross section (use the right hand scale for this); 8) Soft X-ray background; and 9) Hard X-ray and $\gamma$-ray back-

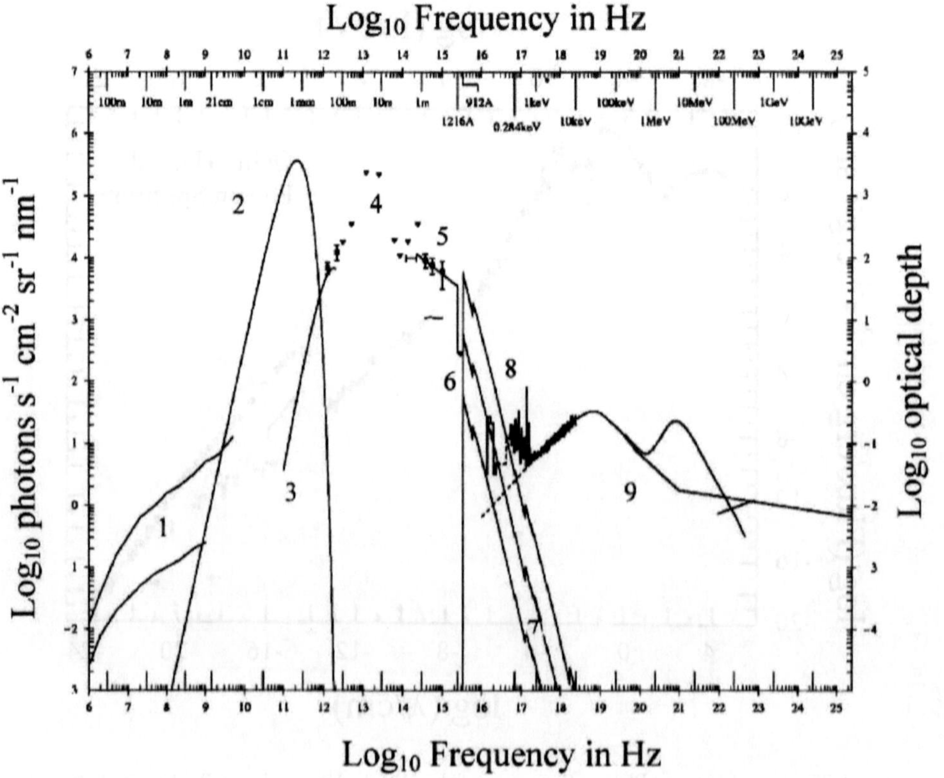


*Figure 18.* The spectrum of the diffuse background radiation (Henry, 2002).

ground. This figure differs from the figure in Henry (1999b) in several respects. Only one FIRAS spectrum is shown (instead of observed and $\pm 1\sigma$). Instead of actual observations of the soft X-ray background, what is shown here is a simulation of what might be expected from David Burrows' CUBIC experiment (Burrows 1996). And finally, in addition to the new $\gamma$-ray background, he showed the earlier spectrum that was reported by Fichtel, Simpson & Thompson (1978). The short straight line near $10^{22}$ Hz is the estimate of a galactic component that was provided by the latter authors. The large MeV bump proved to be an artifact (Sreekumar *et al.*, 1998); for a detailed independent description of the *Comptel* data analysis that led to this important advance, see Weidenspointner (1999).

Another version of this same figure appears in Henry (1999c); that version includes *both* the Burrows simulation *and* the boxes, for comparison purposes.

## 3.2. INTERSTELLAR MEDIUM

### 3.2.1. *General Background*

The interstellar medium (ISM) is the cauldron in which the Universe we are observing was generated. At one time all galaxies, or at least the parts we can observe, consisted entirely of interstellar gas; therefore understanding the ISM is literally fundamental in understanding galaxy formation, galaxy evolution and cosmology. Furthermore, the physics of ISM is classical physics. The physical processes producing the radiation we observe are well understood, therefore the observations of the ISM are a very powerful probe into the physical conditions in a galaxy. Galaxies are made of stars and ISM, with a variety of properties which correlate remarkably with the forms of galaxies as they appear in the sky. The evolution of the stars and gas in galaxies has been thoroughly discussed in a review paper by Beatrice M. Tinsley (1980).

The chemical composition of galaxies in general shows much less diversity than their stellar populations. There are few examples of interstellar gas having heavy elements underabundant or overabundant by more than a factor of four relative to the Sun. Although stellar metallicities range over more than two orders of magnitude, they are usually within a factor of three of solar metallicity *(we remind the reader that metals are in this context all elements heavier than helium)*. Many of the differences amongst chemical compositions in galaxies are systematic: metallicities of both stars and gas tend to decrease outward from the centers of galaxies, and average metallicities tend to increase with galaxy luminosity. Models for the chemical evolution of galaxies aim to account for their composition in terms of production of elements by stars (mainly) and the mixing of stellar ejecta with interstellar gas. Gas flows often play important roles in chemical evolution, diluting the products of nucleosynthesis with non-enriched material from outside the galaxy, and carrying metals from one part of the galaxy to another. In general, composition differences amongst the old stars of galaxies serve as frozen-in tracers of their early chemical evolution, whilst abundances in young stars and the ISM indicate the progress of continuing enrichment (Tinsley, 1980).

In the conventional cosmological picture the primeval gas emerging from the Big Bang, some 10–15 Gyr ago, consisted almost entirely of hydrogen and helium (e.g., Boesgaard & Steigman, 1985). In order to allow this gas for some regions to be locally bound and collapse, despite the expansion of the Universe, there must have been large enough perturbations of the gas density. The existence of galaxies and bound clusters with a $10^{10}$ fold range of masses shows that there was a very wide spread of density perturbations. A favored mass scale is $\sim 10^6$ $M_\odot$, which is the Jeans mass in the universe after re-combination of the primeval plasma about $10^6$ yr after the Big Bang. Maybe star formation first began in lumps of this mass, which could themselves have been part of large scale perturbations destined to collapse eventually into a massive galaxy (Tinsley, 1980).

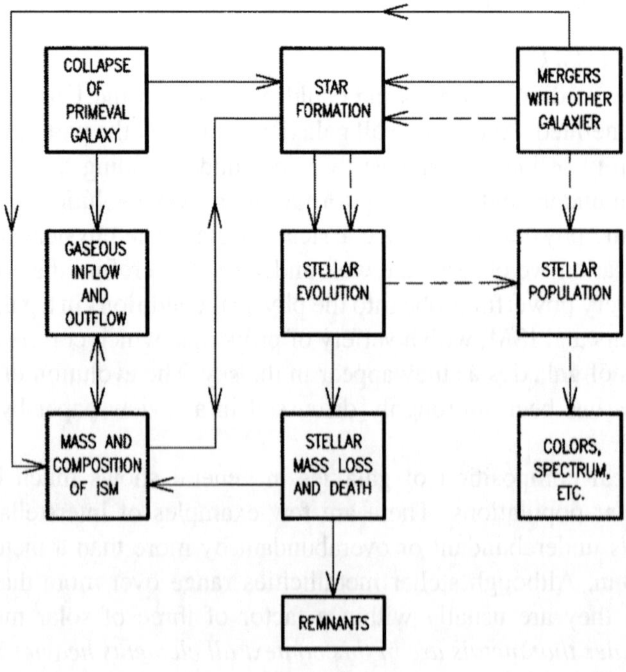


*Figure 19.* A schematic diagram of relations among some of the most important factors in galactic evolution (Tinsley, 1980).

Figure 19 shows a schematic diagram of relations among some of the most important factors in galactic evolution. It serves to summarize the steps in the evolution and the main sets of processes and constituent:

i) Chemical evolution depends mainly on the areas connected by solid line, which indicate that stars form from the ISM and return some of their mass into it via stellar winds. Moreover, new products of nucleosynthesis are injected into the ISM at the star's death (e.g., the review of Mazurek & Wheeler (1980), and Isern *et al.* (1997)). Recently a book on *The Chemical Evolution of the Milky Way* discusses the updated observational and theoretical arguments on this topic (Matteucci & Giovannelli, 2000).

The composition of the ISM also depends on gas flow to and from the region under consideration.

ii) Photometric properties depend mainly on the areas connected by dashed lines, indicating that the history of star formation and the evolution of individual stars determine the population at any time (Tinsley, 1980).

So, summarizing, the knowledge of the mass and composition of the ISM has importance in all the evolution of stars and galaxies. The bulk of the mass is apparently in the form of HI and $H_2$; the other components of the ISM, at least in the Galaxy, contain insignificant fractions of the mass but are of primary importance for the energy transport and cycling of the ISM. Most of the heavy atoms are in particular form as interstellar dust, whose IR emission is a most important diagnostic tool of both dust content and star formation. It appears as though, with the advent of molecular line (e.g., Wilson, 1970) and X-ray (Forman, Jones & Tucker, 1985) observations, all the ISM components in galaxies of all types have been discovered and are susceptible to detailed observational investigation, both as to their physical condition and to their distribution within a galaxy. As well as the physical properties of the ISM, the energy input needs to be investigated.

The energy input has three main contributors:

a) The microwave background;

b) The stellar radiation – starlight, supernovae, and cosmic rays;

c) Mechanical (collisional) energy.

The effect of b) is to dump large amounts of energy into a relatively small fraction of the gaseous mass. This causes regions of much higher than average temperature (the HII regions with $T_e \sim 10^4$ K, and the coronal gas, heated by supernova shocks, with $T > 10^5$ K). The relative effect of the mechanical energy input is a strong function of galaxy properties, i.e., of the relative amounts of stellar and gaseous mass and of ordered and random motion. Collisions in the disks of spiral galaxies take place between clumps of cold interstellar gas at relative speeds of a few km $s^{-1}$ with energy input comparable to the input from other processes. In elliptical galaxies collisions take place at relative speeds of hundreds of km $s^{-1}$ and dominate the energy input to the ISM, with the result that most of the interstellar gas is at high temperature and emits at X-ray energies (Knapp, 1990).

Several important discoveries in the 1960s have provided important connections between *astrophysics* and *cosmic ray physics*. Indeed, primary electrons and positrons were confirmed in the cosmic rays. These components provide a powerful trace of the matter and magnetic field distributions (e.g., the review of Verschuur, 1980) encountered by the cosmic rays in the Galaxy. At about the same time extra-terrestrial X-rays and $\gamma$-rays were discovered near the top of the Earth's atmosphere. So from our present vantage point we might list the possible components of these rays from the outer space as:

- Electrically charged particles, both positive and negative;
- Photons;
- Electrically neutral particles;
- Dust particles.

As is well known, each of these components has been found above the Earth's protective atmosphere and all have been involved in the new branch of Physics named *Astroparticle Physics*.

Cosmic dust, for instance, has been studied on board high flying crafts, balloons, and space missions and shown to consist of some grains of interplanetary origin and others, possibly, of interstellar origin (e.g., Brownlee, 1985).

The study of the morphology and the element/isotope patterns in these grains will provide new insights into the formation of our solar system and processes in the ISM. Therefore from this point of view it is possible to illustrate in Figure 20

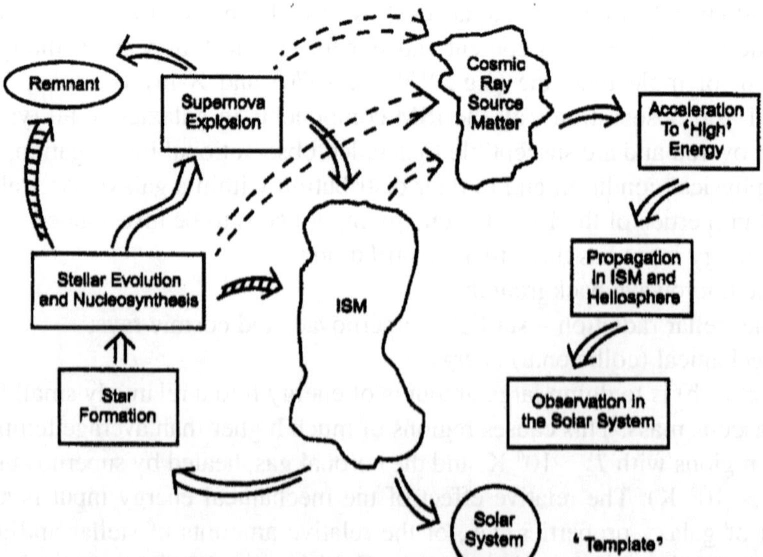


*Figure 20.* The relationships between cosmic ray processes and astrophysical processes which may contribute matter to the cosmic radiation (Wefel, 1988).

the relationships between cosmic ray processes and astrophysical processes which may contribute matter to the cosmic radiation (Wefel, 1988).

The origin of cosmic ray particles from the matter is shown on the right. Such a matter is to become cosmic rays through acceleration to high energy. Subsequently cosmic rays propagate in the Galaxy and in the heliosphere (the region of space around the sun controlled by the outflowing solar wind) before being observed at the Earth.

In Figure 20 the question of the origin of cosmic ray matter involves the contribution of several different sources – individual stars, the ISM, supernovae, young open clusters, collapsed objects, such as neutron stars and black holes, both isolated and in binary systems – to the matter that is to be accelerated. Resolving this question requires working backward from the composition and energy measurements made in the vicinity of the Earth, through the propagation and acceleration processes, to study the composition of matter in the sources.

### 3.2.2. *Historical Background*

A beautiful book on *Interstellar Matters: An Essay on Curiosity and Astronomical Discovery* written by Gerrit L. Verschuur in 1937 was later updated (Verschuur, 1989). In the introduction and at the beginning of Part I he writes an homage to the pioneer of the research on interstellar matter, Edward Barnard, a self-made astronomer who rose from a life of abject poverty during the (American) Civil War to become the world's first great photographer of the heavens. His greatest challenge was to understand the nature of the dark markings revealed in his beautiful photographs of the Milky Way; dark markings, which had long been believed to be vacancies between the stars, '*holes in the heavens*' through which one could see empty space beyond the stars. Once he began to photograph these markings Barnard was torn between deciding whether they were, in fact, vacancies or, as some astronomers suggested, indicative of obscuring matter between the stars.

The challenge for understanding the interstellar space then started, and in the following we will briefly summarize the first steps.

The interstellar space was suggested to be filled with exceedingly tenuous gas by Eddington (1926). He showed that the stationary $Ca^+$ and Na lines observed in distant stars are owed to $Ca^+$ ions and Na atoms present in the space between the stars.

With the advent of high dispersion spectrographs for stellar spectroscopy, such as the Coudé spectrograph at Mount Wilson Observatory, it was possible to discover other atoms and the first molecules from the interstellar space. From that observatory Dunham & Adams (1937) and Duhnan (1937) observed absorption lines of interstellar $Ti^+$, K, Fe and three very sharp lines at 4300.31 Å, 4232.58 Å, and 3957.74 Å, which were not identified. Later Swing & Rosenfeld (1937) pointed out that the first of these lines coincides with the $R_2(1/2)$ line of the known CH band at 4315 Å. This is the only line in this band that comes from the lowest rotational level ($J = 1/2$ of the ${}^2\pi_{1/2}$ component). This was the first firm identification of an interstellar molecule. This identification was confirmed when three lines of CH in the 3870 Å band were observed in the spectrum of $\zeta$ Ophiuchi by Adams (1941) after the prediction that these lines could arise from the lowest rotational level by McKellar (1940). Moreover, three lines of CN in the band 3883 Å, predicted by McKellar, and observed by Adams showed that CN in the $N = 0$ and $N = 1$ levels is present in the ISM. But the most important result coming out from these pioneering researches is that of McKellar (1940, 1941): he estimated from the observed intensity ratio of those lines a temperature of $\approx$2.3 K! **This was in absolute the first hint of the existence of the 2.7 K cosmic background radiation, at that time not understood.**

The 21 cm line of atomic hydrogen HI was discovered in the interstellar space on March 25, 1951 by H.I. Ewen and E.M. Purcell, veterans of radar research during the second world war at Harvard. This was reported in the Ph.D. thesis of Ewen (1951) – directed by E.M. Purcell – and by (Verschuur, 1989). Seven weeks after Ewen's thesis in Holland, on May 11, such a signal was confirmed by Muller and Oort, by Pawsey (1951) and by Christiansen & Hindman (1951; 1952). This line was for two decades the principal tool in studying the ISM and its large scale distribution in the Milky Way (Muller, Westerhout & Schmidt, 1957; Kerr, Hindman & Gum, 1959; Gum, Kerr & Westerhout, 1960). The advantages of using the 21 cm line are the transparency of the Galaxy to radio wavelengths and the ubiquity of cold atomic gas in the Galaxy. As for the molecular hydrogen,

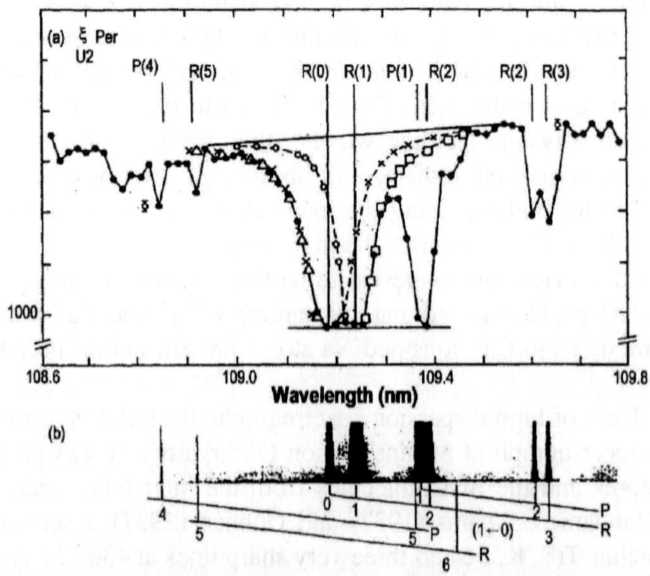


*Figure 21.* The spectrum around 1100 Å, obtained with the Copernicus satellite. One of the strong $H_2$ absorption bands is visible. A laboratory spectrum reduced to the same scale is also reported (Herzberg, 1981, after Spitzer *et al.*, 1973).

the atomic hydrogen remained unexplored until 1970, when Carruthers (1970) observed it directly in UV absorption lines.

Progress in the characterization of the ISM by means of optical and UV spectral data were discussed by Cowie & Songaila (1986). The gas was studied by focusing on background stars whose spectra can be accurately modelled to provide the light source for the absorption line scans. They described the distributions of diffuse gas densities and characteristics of the cold, warm and hot gas in the Galaxy in terms of the elemental abundances, kinetics and distributions of the gas. Particular note was taken of gas in the solar neighborhood and around supernova remnants, and of absorption line data of cosmological significance.

3.2.3. *Morphology and Composition of the Interstellar Medium*

22 years after the first detection of $CH^+$ the next molecule detected was that of OH by a radioastronomical method (Barret, Meeks & Weinreb, 1964). Later, a large number of molecules were identified with radio methods. However, optical instruments firstly detected the presence of molecules in the ISM. In general the molecules without an electric dipole are not amenable to be detected by radio methods. Two such molecules have been detected after the advent of space experiments, namely:

– $H_2$, which is the most abundant molecule in the interstellar medium. The presence of this molecule is shown in Figure 21, where a spectrum around

1100 Å, obtained with the Copernicus satellite, is reported (Spitzer *et al.*, 1973). It is very clear one of the strong $H_2$ absorption bands. For comparison, a spectrum obtained in the laboratory and reduced to the same scale is shown too (Herzberg, 1981).

So the words of Dalgarno (1985) reported in the preface of the NATO Advanced Research Workshop on *Molecular Astrophysics: State of the Art and Future Directions* (Diercksen, Huebner & Langhoff, 1985) are present as never before: *...molecules are unique diagnostic probes of the physical conditions in which they are found.... They influence in a critical way the evolutions of the astronomical systems of which they are a part.*

Many organic molecules have been found in the ISM. Isotopic variants have also been detected, including several deuterated molecules.

The hydrocarbon molecules are all linear with the exception of $C_3H$, which is present in both linear and cyclic forms, and $C_3H_2$. Table 3 shows a list of interstellar molecules that have been definitively identified (Dalgarno, 1992). Molecular ions are contained in the list, attesting to the importance of cosmic rays to the chemistry of dense cloud interiors.

The molecular abundances relative to hydrogen range from $10^{-4}$ for the ubiquitous carbon monoxide to between $10^{-7}$ to $10^{-10}$ for complex molecules. There are substantial variations in the molecular abundances in different clouds and in different regions in a particular cloud.

Of the hydrocarbons, cyclopropenylidene ($C_3H_2$) is found throughout the galaxy both in dense clouds and in apparently diffuse regions. Its fractional abundance relative to hydrogen in dense clouds varies between $3 \times 10^{-9}$ and $3 \times 10^{-10}$ (e.g., Ehrenfreund *et al.*, 2001).

– The CO 2.6 mm line was first detected with the Kitt Peak 11 m telescope of NRAO (Wilson, Jefferts & Penzias, 1970). Since then, the CO(1–0) line has played the same role for the molecular component as the HI 21 cm line did in the earlier exploration of the atomic component of the ISM in galaxies. Since the symmetric $H_2$ molecule does not radiate in the radio energy range, the asymmetric CO molecule, which is the most abundant molecule after $H_2$, provides a superior tracer of $H_2$ in galaxies. The theoretical expectation for $H_2$ density, corresponding to balanced rates between collisional excitation and spontaneous emission, is $\sim 3 \times 10^3$ cm$^{-3}$. Since such densities are also typical of those that actually occur in the molecular gas in the Galaxy, the CO molecule provides a good tracer of the molecular component. Therefore the 2.6 mm line can be considered the analog of the 21 cm line. The Galaxy is transparent at these wavelengths and the CO emission is ubiquitous (Combes, 1991). But the problems arise when one wants to determine the $H_2$/CO conversion ratio. Is it possible to measure this ratio? Is this ratio a universal constant? Combes (1991) in her excellent review defines a factor $X = N(H_2)/I(CO)$, expressed in molecules cm$^{-2}$/(K km s$^{-1}$), being $I$(CO) the intensity of $^{12}$CO(1-0) and $N(H_2)$ the $H_2$ column density. The calibration

TABLE 3
Interstellar Molecules (Dalgarno, 1992)

| Molecule | Name | Molecule | Name |
|---|---|---|---|
| $H_2$ | Hydrogen | $HC_9N$ | Cyano-octatetra-yne |
| $CH^+$ | Methylidyne Ion | CH | Methylidine |
| $C_2$ | Carbon | OH | Hydroxyl |
| CO | Carbon Monoxide | CN | Cyanogen |
| CS | Carbon Monosulfide | NO | Nitric Oxide |
| SO | Sulphur Monoxide | SiO | Silicon Monoxide |
| SiS | Silicon Sulfide | NS | Nitrogen Sulfide |
| HCl | Hydrogen Cloride | PN | Phosphorus Nitride |
| $H_2O$ | Water | $C_2H$ | Ethynyl |
| HCN | Hydrogen Cyanide | HNC | Hydrogen Isocyanide |
| HCO | Formyl | $HCO^+$ | Formyl Ion |
| $N_2H^+$ | Protonated Nitrogen | $H_2S$ | Hydrogen Sulfide |
| HNO | Nitroxyl | OCS | Carbonyl Sulfide |
| $SO_2$ | Sulfur Dioxide | $HCS^+$ | Thioformyl Ion |
| $H_2CO$ | Formaldehyde | $C_2S$ | Dicarbon Sulfide |
| $NH_3$ | Ammonia | $H_2CS$ | Thioformaldehyde |
| HNCO | Isocyanic Acid | HCNS | Isothiocyanic Acid |
| $C_3N$ | Cyanoethynil | HOCO | Protonated Carbon Dioxide |
| $C_3O$ | Tricarbon Monoxide | $C_3S$ | Tricarbon Sulfide |
| $C_2H_2$ | Acetylene | $C_3H_2$ | Cyclopropenylidene |
| $CH_2CO$ | Ketene | $HC_3N$ | Cyanoacetylene |
| HCOOH | Formic Acid | $NH_2CH$ | Cyanamide |
| $C_4H$ | Butadinyl | $CH_3CH$ | Methyl Cyanide |
| $CH_2NH$ | Methanimine | $CH_3SH$ | Methyl Mercaptan |
| $CH_2CN$ | Cyanomethyl Radical | $CH_3C_2H$ | Methyl Acetylene |
| $C_5H$ | Pentynylidyne | $CH_2CHCN$ | Vinyl Cyanide |
| $CH_3OH$ | Methyl Alcohol | $HCOOCH_3$ | Methyl Formate |
| $HCC_2HO$ | Propynal | $HC_5N$ | Cyanodiacetylene |
| $C_6H$ | Hexatrinyl | $CH_3C_4H$ | Methyl Diacetylene |
| $NH_2CHO$ | Formamide | $(CH_3)_2O$ | Dimethyl Ether |
| $CH_3NH_2$ | Methylamine | $CH_3CH_2CN$ | Ethyl Cyanide |
| $CH_3CHO$ | Acetaldehyde | $CH_3C_4H$ | Methyl Diacetylene |
| $CH_3C_3N$ | Methyl Cyanoacetylene | $HC_7N$ | Cyanohexatriyne |
| $CH_3CH_2OH$ | Ethyl Alcohol | $HC_{11}N$ | Cyano-decapenta-yne |

of this ratio has been performed by using $H_2$ in the UV absorption lines in the solar neighborhood, with measurements obtained by Copernicus satellite. Savage *et al.* (1977) derived the $H_2$ column densities in front of 109 stars. Bohlin, Savage & Drake (1978) derived HI column densities across the same lines of sight; therefore they calibrated the total amount of hydrogen relative to the optical $E(B - V)$. The relationship they derived is: $N(\mathrm{HI}) + 2N(\mathrm{H_2}) = 5.8 \times 10^{21} \times E(B - V)$ atoms cm$^{-2}$ mag$^{-1}$.
However, this $H_2$ survey concerns only low extinction lines of sight because of the sensitivity of the Copernicus satellite ($A_V < 1.5$). The direct determination of the ratio $X$ by using CO UV absorption lines is then only relevant for diffuse clouds, where a small fraction of C is in CO (van Dishoeck & Black, 1988; Viala *et al.*, 1988). Thus the only possible solution is that to adopt the gas to extinction ratio from the UV observations and assume that it remains the same for denser molecular clouds, where the content of HI is negligible (Combes, 1991).
An empirical value of $X$ has been obtained by Dickman (1978) from the observations of CO lines ($^{12}$CO and $^{13}$CO) towards line of sight with known extinction, thanks to star counts or IR reddening measurements in the range $1.5 \le A_V \le 10$.
The main argument for the use of a constant value of $X$ is that the telescope beam averages over a large quantity of molecular clumps, which are statistically equal and characterized by a certain density and temperature. Assuming that there is no overlap between clouds at each given velocity, then the integrated emission comes from an optically thin medium of optically thick clumps. In other words, the CO intensity is proportional to the number of clouds in the beam. However, the physical parameters of the 'typical' cloud can vary from place to place in the Galaxy and from galaxy to galaxy. Dickman, Snell & Schloerb (1986) assumed that molecular clouds are all virialized. This yields a relation between the line width, the size and the density. The ratio $X$ is varying as $n^{1/2}/T_b$ where $T_b$ is the CO radiation temperature of a typical cloud. $X$ can then be considered as universal constant since denser regions are also more excited: $T_{ex}$ and then $T_b$ increases with $n$ (Combes, 1991 and references therein).
With this method (Young & Scoville, 1982) found for cloud core regions $X \le 2.7 \times 10^{21}$, for dark nebulae $X \sim 2.5 \times 10^{20}$, and for giant molecular clouds $X \sim 4 \times 10^{20}$ cm$^{-2}$/(K km s$^{-1}$). The empirical value of $4 \times 10^{20}$ cm$^{-2}$/(K km s$^{-1}$) has been adopted for the large scale distribution of molecular clouds and it is valid also for extragalactic distributions.
In the inner Galaxy, Scoville & Good (1987) have found a value of $3.6 \times 10^{20}$ cm$^{-2}$/(K km s$^{-1}$) from an extensive survey of molecular clouds. Figure 22 reports for maps showing the $^{13}$CO in the $l$–$b$ plane, integrated over four intervals of velocities (Bally *et al.*, 1988). Emissions at velocities around zero are not shown because of the confusion with the galactic disk. The main mo-

lecular complexes are indicated. These maps clearly demonstrate the strong confinement of such complexes to the galactic plane.

Figure 23 shows an $l$–$V$ diagram towards the Galactic Center obtained from $^{13}CO$ lines by Bally *et al.* (1988). Most of the mass in the center ($\sim 2 \times 10^8$ $M_\odot$) is in molecular form. Then 10% of the Galaxy's $M(H_2)$ lies within 500 pc of the nucleus, essentially distributed in a thin nuclear disk of radius 200 pc (for a distance of $R = 8.5$ kpc to the Galactic Center). The total mass of $H_2$ is $\sim$(2–3) $\times 10^9$ $M_\odot$. The HI atomic component represents only 2–3% of the total gas mass. In this nuclear disk major HII regions and molecular complexes are found: in Sgr C at negative longitudes and velocities, in Sgr A at the very center, and in Sgr B2 as the dominant molecular condensation in the inner Galaxy (as shown in Figure 22).

Therefore the ISM is a very interesting laboratory for physics and chemistry–physics. Indeed, in it a large number of phenomena occur outside the thermodynamical equilibrium: ionization and recombination, excitation of atomic and molecular levels, formation of spectral lines, thermal exchanges, hydrodynamical phenomena including also shocks, formation and destruction of grains and molecules. ISM plays a fundamental role in the evolution of galaxies, which contain up to $\sim$40% of its mass as interstellar gas and dust. Stars born in ISM and send back into it material enriched in heavy elements through nucleosynthesis processes.

Indeed, it is believed that all the deuterium ($^2$H), most of the helium ($^3$He and $^4$He) and some of the lithium ($^7$Li) is coming directly from the Big Bang. The Standard Big Bang Nucleosynthesis (SBBN) predicts the abundances of these elements (Boesgaard & Steigman, 1985) therefore the comparison with the observed abundances is one of the crucial problems of modern cosmology. Heavier elements, usually called 'metals' are mainly produced in the stars, which form in the ISM, and in turn they give back heavier elements continuously to ISM at relatively low rates via stellar winds, and suddenly at frightful rates via supernova explosions (Wilson & Rood, 1994). For all these reasons the study of ISM is fundamental in understanding a large part of the physics governing the Universe, and these studies can be only performed via space experiments in different wavelength regions.

The ISM in our galaxy is not at all homogeneous. The temperature and density vary by about five and ten orders of magnitude, respectively. The medium is exposed to cosmic rays and starlight, and is pervaded by magnetic fields. For instance, Davies (1981) discussed evidence that magnetic fields pervade the entire ISM including interstellar low and high density gas clouds. Since magnetic fields pervade the Galaxy, they play a crucial role in many astrophysical phenomena. Scattering on aligned dust grains produces the linear polarization of starlight. The pattern of alignment in the polarization of the assembly of stars as projected on the sky indicates a systematic arrangement of magnetic fields in the nearby spiral arms accessible to optical study.

At the large scale, relativistic electrons spiraling in magnetic fields generate radio-synchrotron emission, which is detected throughout the disk of the Galaxy

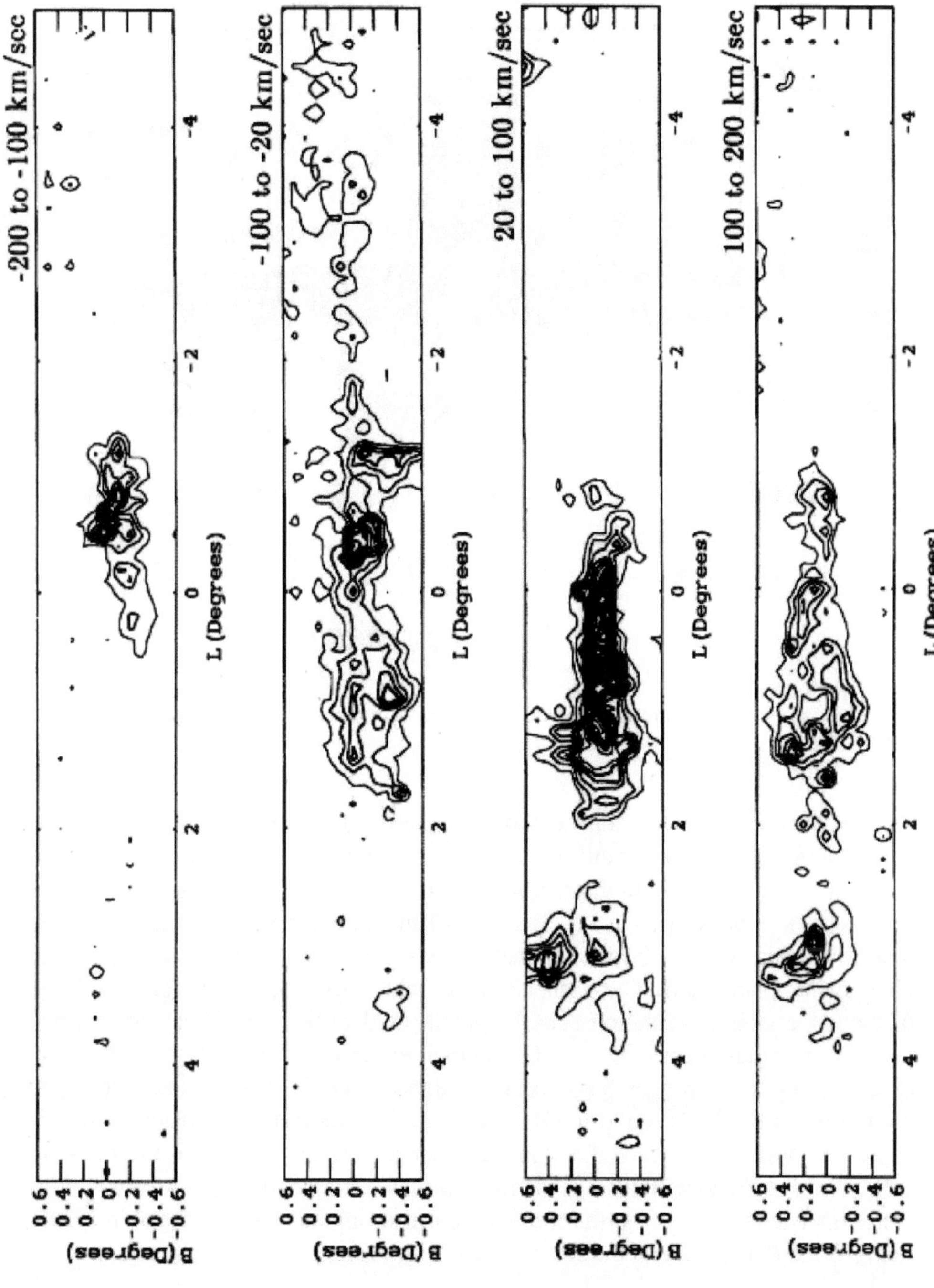


*Figure 22.* The maps showing the $^{13}CO$ in the *l–b* plane, integrated over four intervals of velocities. A 40 km $s^{-1}$ interval centered on null velocity has been excluded to avoid contamination by Galactic disk (Bally *et al.*, 1988).

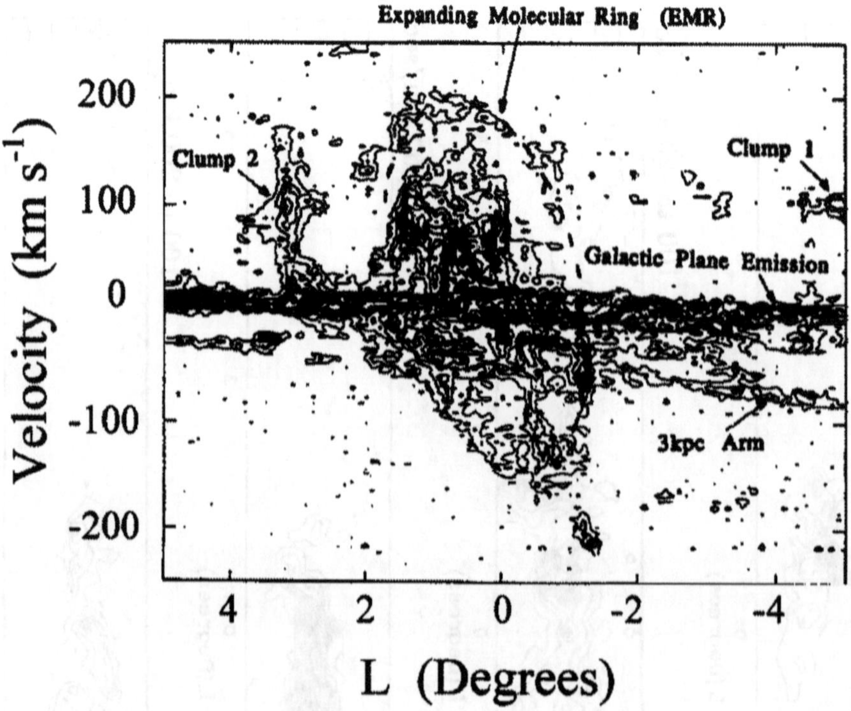


*Figure 23.* An $l$–$V$ diagram towards the Galactic Center obtained from $^{13}CO$ lines by Bally *et al.* (1988).

and possibly in a weak extended halo. Similarly to Faraday rotation of the linearly polarized emission of pulsars and extragalactic radio sources, in the galactic electron and magnetic field distribution it is demonstrated the pervading nature of the magnetic field in the Galaxy. Because of the strong coupling of the magnetic field to the interstellar plasma, magnetic fields play a major role in astrophysical phenomena in the ISM. These phenomena include shocks, gravitational collapse of clouds, and star formation. The magnetic field in the 'seed' gas from which the denser clouds form is (2–3) $\mu$G. Zeeman effect measurements of neutral hydrogen show that stronger fields occur in denser clouds. These data, taken with the micro-Gauss fields found in OH maser sources, indicate that magnetic flux is conserved during gravitational collapse of interstellar cloud from densities of $\sim$5 to $\sim 10^7$ cm$^{-3}$. Magnetic fields appear then to play a major role in the formation of dense interstellar clouds. Furthermore there is a strong indication that the magnetic field's direction is preserved during a cloud's collapse.

The average density of ISM is $\sim 10^{-24}$ g cm$^{-2}$, or about one hydrogen atom per cubic centimeter, corresponding to $\sim$0.025 M$_\odot$ pc$^{-3}$. Therefore, on average, one solar mass of medium is contained in $\sim$40 pc$^3$. The ISM is formed by $\sim$75% of hydrogen, ~23% of helium, and the remaining part of ~2% of heavier elements. Most of these heavier elements are condensed as dust grains having an average density of ~0.001 $M_\odot$ pc$^{-3}$ (Friberg & Hjalmarson, 1990).

The local kinetic gas temperature range from 10 K to $10^6$ K. These temperatures correspond to energies which are much lower than those typical of cosmic rays. The lower temperatures correspond to energies, which are less than those typical for starlight. Both cosmic rays and starlight drive processes including ionization. Therefore the ionization structure sometimes is very far from that in a gas in thermal equilibrium at the gas kinetic temperature. The energy density of cosmic rays is ~0.5 eV cm$^{-3}$, of Galactic magnetic field is ~0.2 eV cm$^{-3}$ ($B_{\mathrm{Gal}}$ ~$10^{-6}$ G) and of diffuse starlight is ~0.5 eV cm$^{-3}$. Accordingly, there is a rough equipartition of energy amongst these components. The description of the interactions of matter with radiation is complicated by the strong variation of opacities at different wavelengths. For example, in the formation of HII regions the photons in the Lyman continuum are heavily absorbed close to the star, causing just a HII region to form, whilst photons at longer wavelengths, which are absorbed by dust, can escape to great distances.

Dynamical processes also affect the appearance of the ISM. The Galactic rotation provides angular momentum which can hinder or delay local contractions. Also the magnetic field can retard the collapse. Rotation produces spiral arms and density waves that compress the medium and may cause cloud–cloud collisions. Density waves may also create shocks waves, which can heat the gas and even disperse clouds. The arms contain stars, which become supernovae generating strong and often very disruptive shock waves. However, as we will discuss later in Section 7.16, we do not need such violent events to form shock waves. Indeed, hypersonic stellar winds from massive evolved stars are powerful enough to sweep up matter and create shock waves. Bipolar outflows, mass loss during early stellar evolution, also can generate shock waves or flows that stir up and possibly can disperse the cloud or cloud core that formed the star (Friberg & Hjalmarson, 1990).

Cosmic ray bombardment, in general, heats the interstellar matter. Whilst, in less obscured regions, the galactic diffuse starlight is responsible for heating the interstellar matter.

Also the formation of the molecular hydrogen ($H + H \longrightarrow H_2 + 4.48$ eV) and the gravitational compression of the gas contribute to the heating. The ISM can also be heated by the intense radiation from massive stars and star forming regions in the vicinity. Of course, cooling of the gas occurs through radiation losses. The heating rate $\Gamma(T)$ is a function of the temperature and composition of the medium as well as of the radiation field and cosmic ray flux. The cooling rate $\Lambda(T)$ is a function of temperature and composition as well as of the probability that the generated photons escape the region. Then for equilibrium the heating must be equal to the cooling: $n^2\Lambda(T) = n\Gamma(T)$, where $n$ is the hydrogen density. Indeed, the heating rate per unit volume is proportional to the ionization cross section, then proportional to the density, whilst the collision rate per unit volume

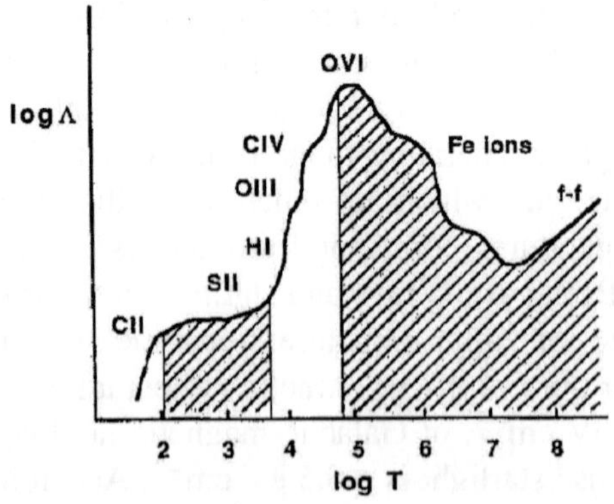


*Figure 24.* The shape of the cooling function $\Lambda(T)$ (Friberg & Hjalmarson, 1990).

is dependent upon the square of the number of colliding particles. The solution of that equilibrium equation then depends on the shape of the cooling function $\Lambda(T)$. This shape is shown in Figure 24. The slope of the cooling function is determined by the temperature dependence of the excitation of different atomic lines. Below 100 K $\Lambda(T)$ increases rapidly as a result of CII cooling. At 100 K, CII cooling saturates and no major coolant turns on until $10^4$ K, when Lyman–$\alpha$ and some other transition start to get excited. For a temperature above $\sim 10^5$ K the cooling rate decreases, since most atoms except iron are completely ionized. Above $10^7$ K free–free emission takes over and the cooling rate coefficient increases again. Radiative cooling becomes important in supernova remnants (SNRs) after they have expanded during an adiabatic phase. The immediate post-shock temperature ranges between $5 \times 10^5$ and $10^6$ K when a SNR has radiated away about half of its energy.

Three phase models of the ISM have been proposed. The temperature which characterizes different components arises naturally from the properties of radiative cooling. Cold ($\sim$80 K), dense, neutral gas then coexists with warm ($\sim 10^4$ K), dilute, neutral or ionized gas and with hot ($\sim 10^6$ K), very dilute gas. These phases are summarized in Table 4 (Friberg & Hjalmarson, 1990: adapted from Jura, 1987).

So the importance of space experiments which can provide crucial measurements for understanding the ISM appears evident. Indeed, methodical investigations of ISM started in 1972 in the soft X-ray range (0.1–10 keV) (McCammon *et al.*, 1983) in order to find the diffuse X-ray background distribution in all the sky. The maps in the ranges 130–180 eV and 160–280 eV showed similar fluxes independent of galactic longitudes, and a general enhancement at galactic poles, especially in the northern hemisphere. Figure 25 shows the diffuse soft X-ray distribution in galactic coordinates, and for comparison the distribution of column density of HI by using the 21 cm line (Clearly, Heiles & Haslam, 1979). The hydrogen column density is anti-correlated with the X-ray intensities; the two X-ray maps are rather similar with the exception of a characteristic in the northern

TABLE 4

Phases of the Insterstellar Medium (Friberg & Hjalmarson, 1990: adapted from Jura, 1987).

| Name | n ($cm^{-3}$) | T (°K) | Filling factor | Diagnostic |
|---|---|---|---|---|
| Hot intercloud Gas | $3 \times 10^{-3}$ | $10^6$ | 0.4–0.7 | Soft X-ray emission, Ultraviolet absorption lines |
| Warm gas | 0.1 | $10^3$–$10^4$ | | If ionized – optical emission lines, pulsar dispersion measures |
| | | | 0.4 | If neutral – 21 cm observations |
| Diffuse clouds | >10 | 50–100 | | Optical & ultraviolet absorption lines, 21 cm observations |
| Cirrus clouds | 10–$10^3$ | 10–100 | | Infrared emission, 21-cm observations, molecules |
| Dark clouds | >$10^3$ | 10 | 0.01? | Dark patches, infrared emission, molecules |
| GMCs | >$10^3$ | 15–40 | | Infrared emission, molecules |
| HII regions | >10 | $10^4$ | | Optical, ultraviolet, infrared, and radio emission lines |
| SNR | >1 | $10^4$–$10^7$ | | Radio to X-ray |

hemisphere, close to $l = 30°$. In the galactic plane the finite X-ray flux is almost constant independent of the galactic longitude.

The Einstein satellite survey, in the band 150–280 eV, showed that stars emit not more than 3% of the total flux (Rosner, Golub & Vaiana, 1985). Non-thermal diffuse mechanisms cannot be responsible of the observed soft X-ray flux (Williamson *et al.*, 1974). Then the most probable source of such a soft X-ray background can be a diffuse thermal emission from a gas (ISM) at a temperature of $\sim 10^6$ K. The existence of such a gas, already suggested by Spitzer (1956) was demonstrated by the presence of interstellar OVI firstly measured by Copernicus satellite (Jenkins & Meloy, 1974). An independent proof of the thermal origin of the soft X-ray emission came from observations with Hakucho satellite of emission lines of high ionized elements, such as OVII (at 0.57 keV) and the measured spectrum was best fitted with a thermal emission of a plasma at $1.4 \times 10^6$ K (Inoue *et al.*, 1979). A further confirmation of the thermal origin of the diffuse radiation was done by Rocchia *et al.* (1984) with the observations of CV, CVI and OVII.

Important results were obtained with the IRAS satellite, especially the detection of large scale extensive stringy emissions between 60 and 100 $\mu$m (Low *et al.*, 1984), which were named IR cirrus. Such well structured extended sources, superimposed on a continuum IR background, are coming from emissions of molecular

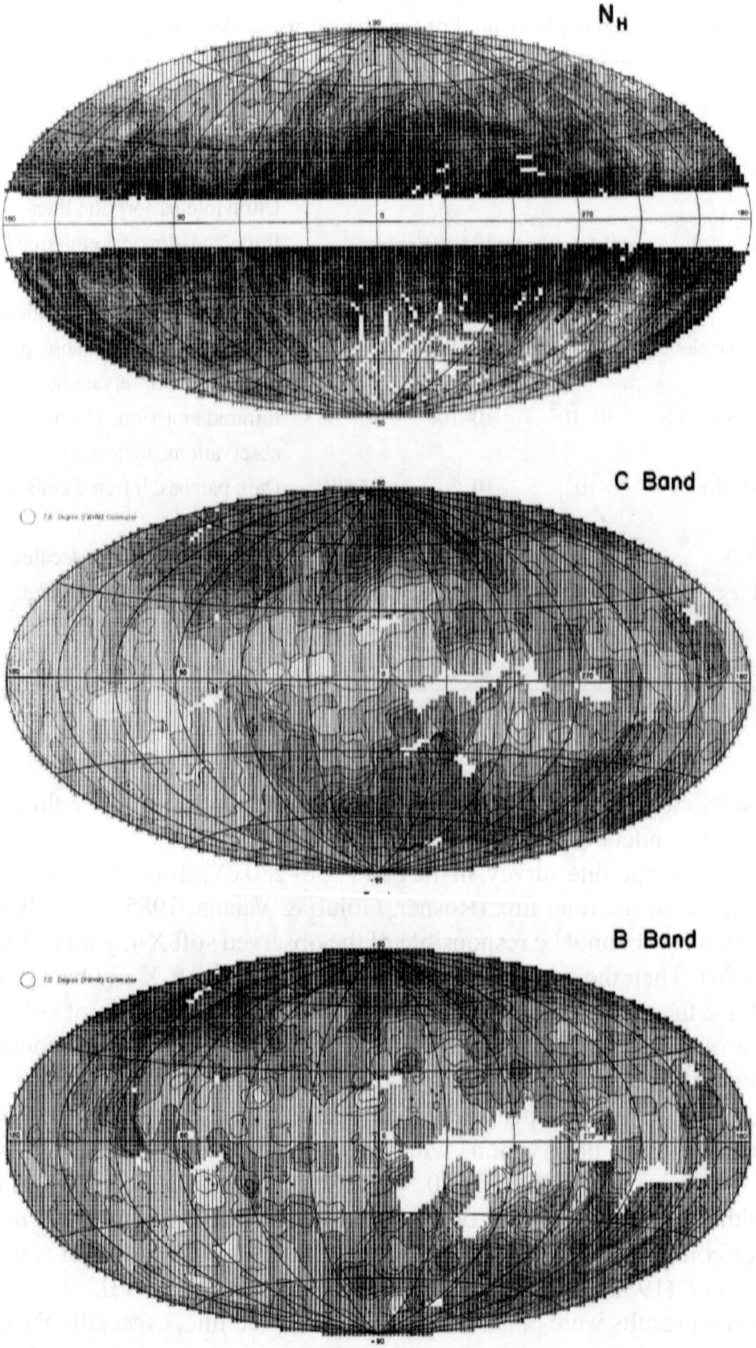


*Figure 25.* The diffuse soft X-ray distribution in galactic coordinates: B-Band (130–180 eV), C-Band (160–280 eV), and for comparison the distribution of column density of HI by using the 21 cm line from Clearly, Heiles & Haslam (1979) (McCammon *et al.*, 1983).

TABLE 5
The Local Interstellar Medium (Knapp, 1990).

| Component | $f_V$ | $\bar{n}_H$ cm$^{-3}$ | $T_e$ (°K) | $f_M$ | Probes |
|---|---|---|---|---|---|
| Hot ICM | 0.5 | 0.005 | $5 \times 10^5$ | 0.001 | [OVI]; X-ray |
| Warm ICM | 0.5 | 0.3 | 8,000 | 0.1? | HI, $H_\alpha$ |
| Warm HI | | | | | |
| HI Clouds | 0.05 | 5–20 | 10–100 | 0.4 | HI, etc. |
| $H_2$ | 0.005 | 300 | 5–30 | 0.5 | CO |
| HII | 0.001 | $10$–$10^4$ | $10^4$ | 0.02 | $H_\alpha$, etc. |
| SNR | | | | | Radio, X-ray |
| Dust | 1.0 | | 5–50 | 0.01 | IR, Extinction |

clouds at high latitudes. Blitz, Magnani & Mundy, (1984) argue that the formation of these IR cirrus depends on the angular size of CO molecular clouds which are producing the stringy radiation, and in the case of the absence of detection of CO, the emission could be owed to HI atomic clouds. The existence of atomic and molecular clouds and how it is possible to pass from one to other it is still an open problem.

### 3.2.4. *The Local Interstellar Medium*

The Local ISM (LISM), in the neighborhood of the solar system, is usually taken to consist of material within the nearest $10^{19}$ atom cm$^{-2}$. The distance to that column density is surprisingly large, ranging from ~30 pc in the galactic plane to as much as 200 pc within 20° of the North Galactic pole. Within that column density contour there is an irregularly shaped volume, wasp wasted in the plane and very extended in the north for which the most prevalent conditions are a temperature of $\sim 10^6$ K, a density of $\sim 5 \times 10^{-3}$ cm$^{-3}$, and therefore a thermal pressure of $\sim 10^4$ cm$^{-3}$ K. The volume of this region is essentially equal to that of a sphere of 100 pc radius. The hot gas within it generates the soft X-ray background via its normal thermal excitation processes (Cox & Reynolds, 1987).

The physical properties of the LISM, observationally defined, are summarized in Table 5 (Knapp, 1990). The LISM within a few hundred parsecs of the Sun is the only part of the Universe where it is possible to use all available techniques to observe all of these components in coexistence. The distribution of those components containing a large fraction of the mass (the HI and $H_2$) or of the energy (the HII regions and SNRs) can be traced across the Galaxy and are therefore those which can readily be observed in other galaxies.

The LISM has been studied by using different types of stars utilized as background sources; in particular white dwarf spectra have been used since their continuum fluxes are strong, especially in the UV range, and with a photospherical structure very small in order to render easier the analysis. Column densities have been measured through the lines of NI, OI, SiII, CII, FeII, and MgII (Bruhweiler & Kondo, 1981, 1982a; Dupree & Raymond, 1982). With the study of MgII from stars within ~25 pc from the Sun, it appeared that the solar system lies at the external border of the local cloud, which is placed between it and the galactic center (e.g., Vladilo, 1986). The densities in the lines of sight to the galactic center and to the galactic anti-center differ of about one order of magnitude in favor of the former. The intensity ratio MgII/MgI, obtained with the Copernicus and IUE satellites in neighboring stars, indicates a temperature of 7,500–10,000 K (e.g., Bruhweiler *et al.*, 1984a,b) which is consistent with the position of the Sun at the external border of the local cloud.

UV explorations of other lines of sight were performed too, by using quasars with proper redshifts (York *et al.*, 1984), extragalactic supernovae (Jenkins, Wallestein & Silk, 1984) and bright stars at high galactic latitude (York, 1983; Savage & Massa, 1987). Most of the material placed at hundreds of parsecs from the galactic plane appears either in neutral form or in low ionization absorption. The gas shows metallic abundances which are typical of the interstellar gas in the neighborhood of the Sun. Therefore it is possible that this material originated in the galactic disk and was ejected outside the galactic plane like a fountain gush (Galactic Fountain) (Shapiro & Field, 1976) or formed through the action of cosmic rays (Chevalier & Fransson, 1984); van Woerden, Schwarz & Hulsbosch (1985) suggested that this material could have been similar to high velocity clouds detected in the 21 cm emission. The higher degree ionized gas shows a different velocity distribution with respect to the colder one. CIV detected within a few kpc could come either from the cooling of hot coronal material (~$10^6$ K) or from direct ionization via cosmic rays and radiation field in the near UV (Savage & Massa, 1987). As reported in Section 2.4.4, results of Spitzer & Fitzpatrick (1992) seem to favor the Galactic Fountain scenario.

A long, complete review on the LISM is that by Cox & Reynolds (1987). Later Egger, Freyberg & Morfill (1996) reviewed the LISM, but essentially the Local Hot Interstellar Medium. The most recent review on LISM is that by Ferlet (1999). We want to comment about the interstellar medium very close to the Solar System and inside it.

The interstellar matter close to the solar system can penetrate within the gravitational field of the Sun and therefore interacts directly with the solar wind and the solar UV radiation. Outside the border of the solar system a non-collisional shock is produced as a consequence of the interactions between charged particles of the ISM, penetrating the gravitational field of the Sun and the particles of the solar wind. Around the shock front a turbulent shell of plasma forms. The interior part of this plasma is the heliosphere. The neutral ISM can penetrate the heliosphere and reach the Solar System before its ionization. Therefore the Solar System is swept by this wind, which can be detected through the diffuse resonance emission of HI ($Ly_\alpha$: 1216 Å) and HeI (584 Å) illuminated by the Sun. The distribution of the intensity of the diffuse radiation beyond the Solar System can be used to determine the densities of H and He in the LISM and the velocity vector, which describes the motion of the Sun with respect to the LISM. The module of this vector and the temperature of the LISM can be deduced by the shape of the $Ly_\alpha$, which represents in some sense the image of the distribution of the velocities of the hydrogen atoms.

Bertaux (1984) summarizes the results of observations made in the solar system by PROGNOZ satellite as follows:

- The modulus of the velocity vector is $v = 20 \pm 1$ km s$^{-1}$ in the direction $\alpha = 254^\circ \pm 3^\circ$, $\delta = -17^\circ \pm 3^\circ$, which is rather different from the direction of the Apex. This means that the LISM also moves with respect to the local reference system;
- The temperature of the LISM is $T_{\rm LISM} = 8000 \pm 1000$ K. The hydrogen density is $N_{\rm H} = 0.04$–$0.06$ cm$^{-3}$ and the helium density is $N_{\rm He} = 0.015$–$0.020$ cm$^{-3}$;
- The high He/H ratio with respect to the cosmological ratio indicates that a significant part of hydrogen is ionized;
- The temperature, density, and degree of ionization of the LISM indicate that the solar ISM is in an intermediate phase of the evolution of the ISM; in other words, the solar ISM is in a phase between a hot light gas and a cold dense gas cloud.

Figure 26 shows the distribution of the interstellar gas in the Solar System, following the results obtained with the PROGNOZ satellite (Bertaux, 1984). The main characteristics of this distribution are:

- a focusing cone of helium in the down wind direction;
- a cavity of ionization, void of atoms, elongated toward the down wind direction.

The space missions ROSAT and Ulysses have been particularly important for the study of the structure and dynamics of the LISM. Indeed, they contribute to the knowledge of the hot LISM in different ways:

- Analyzing low energy X-ray data from ROSAT in the range 0.1–0.4 keV. In this energy range the radiation is absorbed by the interstellar gas at column densities in excess of about $10^{20}$ cm$^{-2}$ – and can therefore be regarded as *local*, by determining the absorption of stellar emission spectra from nearby stars across their lines of sight by intervening gas;
- By direct *in situ* measurements, performed with the Ulysses mission, of those components which penetrate the heliosphere sufficiently far, provided that they can be distinguished from interplanetary material.

The results of LISM, obtained with these two different missions, have been deeply discussed by Egger, Freyberg & Morfill (1996). They provided the following picture, which we summarize:

- From the X-ray spectra and intensities it is possible to say that the solar system is surrounded by a *bubble* of hot gas – whose density, temperature, and pressure are $\approx 0.005$ cm$^{-3}$, $\approx 10^6$ K, and $\approx 1.4 \times 10^{-12}$ erg cm$^{-3}$, respectively

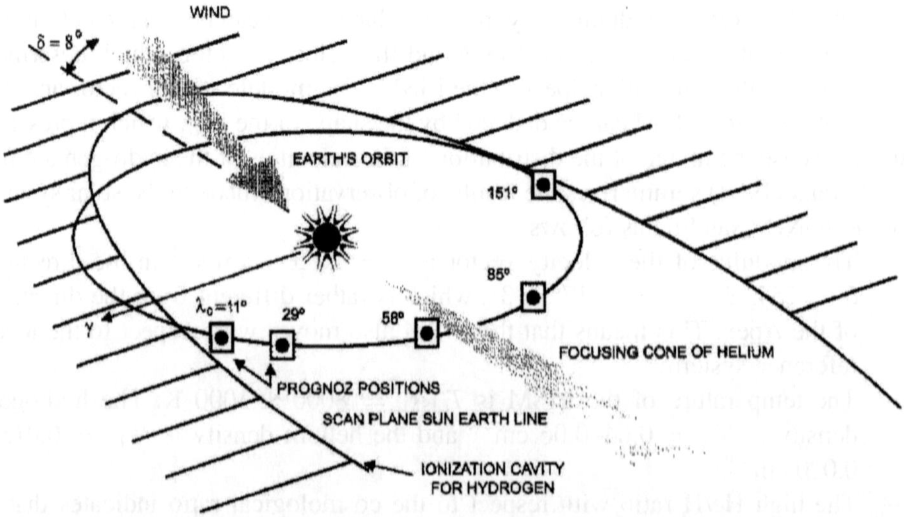


*Figure 26.* The distribution of the interstellar gas in the Solar System, following the results obtained with the PROGNOZ satellite (Bertaux, 1984).

– out to several tens of parsecs. It was assumed that electrons and ions have the same temperature and the gas fully ionized;

– From the *in situ* measurements it is possible to say that more locally, the solar system is embedded in a small warm cloud. Its density is $\cong 0.07$ cm$^{-3}$, temperature $\cong 7,000$ K, pressure $\cong 7 \times 10^{-14}$ erg cm$^{-3}$, column density $\approx 5 \times 10^{17}$ cm$^{-2}$. This gives a mass of $\sim 0.1\ M_{\odot}$; the cloud radius is $\approx 2$ pc and its column density is very small, only $\sim 10^{-6}$ g cm$^{-2}$;

– The transition to the heliosphere is governed by solar UV ionization, snow ploughing of the interstellar gas by the outwardly expanding solar wind and the bow shock. The heliosphere is the region inside the solar wind terminal shock.

The existence of the ISM makes itself felt even far inside the heliosphere by the penetration of neutral gas, dust, plasma waves, shock accelerated particles, and cosmic rays. These are the local probes of the ISM.

Figure 27 shows schematically the solar wind–interstellar medium interface.

The Sun is moving through the Local Interstellar Cloud (LIC). The supersonic solar wind is blowing a cavity called the heliosphere. The fraction of the interstellar neutral hydrogen which penetrates within the heliosphere is decelerated through charge-exchange collisions, invades the interplanetary medium and backscatters the solar Ly$_{\alpha}$-radiation producing an observed emission glow. The interstellar neutral helium is extremely weakly interacting with the plasma and keeps its initial heliocentric velocity, measured *in situ* by Ulysses and found to be equal to the LIC velocity. The transition from supersonic to subsonic flow is supposed to occur at

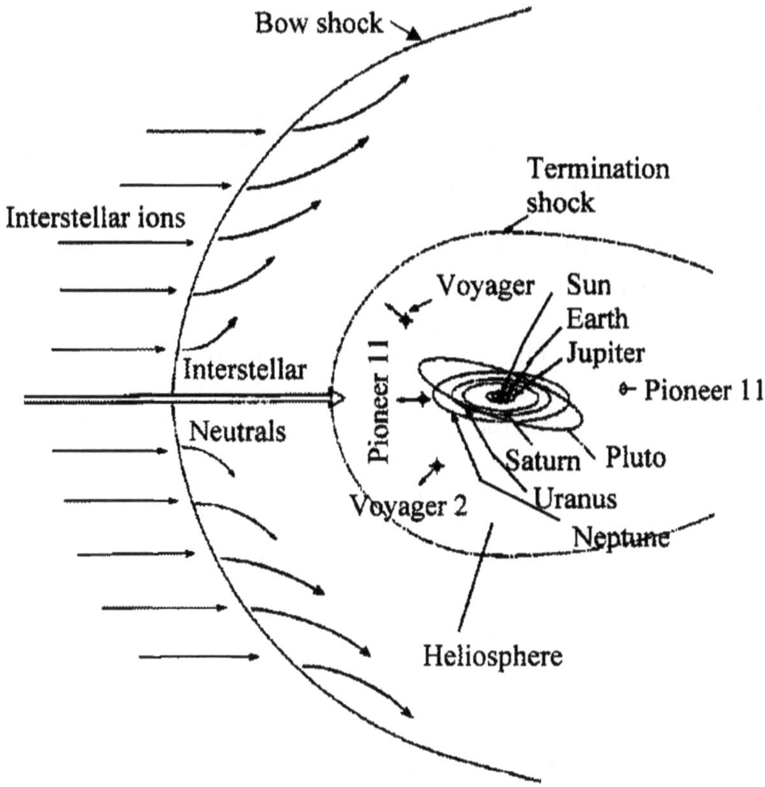


*Figure 27.* The solar wind/interstellar medium interface (Ferlet, 1999).

the so-called termination shock, the first discontinuity the outer probes Voyager and Pioneer 11 should encounter (Ferlet, 1999).

3.2.5. *Interstellar Dust*

From optical extinction measurements, polarization measurements, X-ray scattering and IR emission measurements it is well known that interstellar clouds contain small micron and submicron sized dust particles. The estimate of the mass fraction of this dust gives a value of $\sim 1\%$ based on heavy element depletion measurements and on cosmic abundance arguments.

This dust is extremely important for the chemical, electromagnetic, and dynamical evolution of the ISM. It is the prime catalyst in producing $H_2$ molecules via low-temperature surface reactions. In dense clouds electron–ion recombination occurs mainly on dust particles, which makes the dust dominant in the determination of the charge equilibrium.

In cold clouds dust is the major source of opacity and then responsible for the dynamics of these objects, including, for instance, the possible occurrence of convective instabilities in protoplanetary disks. The dust shields the cloud interior from the UV and soft X-ray background, thus aiding astrochemical evolution.

For reviews on this topic, see the papers by Havnes, Morfill & Melandso (1992), Hartquist, Havnes & Morfill (1993), Morfill, Spruit & Levy 1993) and the book of Morfill & Sholer (1987) on *Physical Processes in Interstellar Clouds*, as well as the successive review of Egger, Freyberg & Morfill (1996).

Fundamental *in situ* measurements of dust have been performed by the Ulysses mission. Indeed, at about 5 AU the Ulysses dust detectors identified unambiguously for the first time interstellar dust particles.

The observed flux is $1.5 \times 10^{-4}$ particles $s^{-1}$ $m^{-2}$, the mean particle mass is $3 \times 10^{-13}$ g, giving a mass flux of $5 \times 10^{-17}$ g $s^{-1}$ $m^{-2}$ (Grün *et al.*, 1994).

The relative velocity between the solar system and the LISM is $v_{ss}$ ~25 km $s^{-1}$. The observed dust mass flux $\rho_{dust} v_{ss}$ – in very good agreement with various earlier estimates – allows to evaluate the interstellar mass density, which is ~1.5% of the mass density of the local interstellar gas.

On the contrary, the dust size distribution is not in agreement with earlier predictions. The measured dust particles have an average diameter of ~10 $\mu$m, whereas other estimates, of course taken from different regions – not the LISM, have yielded values ranging from 0.1 to 1 $\mu$m.

This result is extremely important, since if the small grains are actually absent this would have a number of significant consequences, namely (Egger, Freyberg & Morfill, 1996):

- the available solid surface area would be reduced by a factor 10–100;
- the optical extinction measurements would have to be completely reinterpreted;
- the charge state of interstellar clouds and most of ion–molecule astrochemistry would need to be re-examined;
- grain driven ambipolar diffusion would be significantly changed;
- grain modified shocks would need to be re-investigated, etc..

So this result must be taken with care, in the sense that probably it appear like it is because the measurements were performed during the so-called positive cycles: northerly magnetic field pointing outward, during which small dust grain are unable to penetrate into the inner heliosphere mainly because of the interplanetary magnetic field, dragged out by the solar wind.

This interpretation leaves open the possibility that small grains might be abundant in the LISM.

We want to note this controversial point: indeed, in spite of the wonderful *in situ* measurements, performed with the Ulysses mission, we are not able to solve one of the most important problems connected with the presence of certain grains in the LISM and then with the opacity, the charge state, etc.; we would need the same kind of measurements in the opposite conditions – negative cycles – when northerly magnetic field pointing inward; and this is simply not possible with the Ulysses mission, since it has fixed its trajectory.

This problem is, in a certain sense, similar to that which we very often encounter in photonic astrophysical measurements: cosmic sources are variable; then only with either continuous monitoring during long times and/or simultaneous multifrequency observations is it possible to better approximate their *true behaviour*.

Another possibility, although weaker than the first, for the eventual absence of small grains in the LISM is that the local cloud, being somewhat warmer, more transparent, and possibly relatively short lived by astronomical standards, might have lost the small grains by sputtering, photo-sputtering, surface chemical processes, cosmic ray bombardment, etc..

As pointed out by Egger, Freyberg & Morfill, (1996), in spite of strong improvements in our knowledge of the LISM provided by the ROSAT and Ulysses missions, a number of problems still remain open, namely:

- **the pressure problem**: the hot gas in the bubble in which the local cloud surrounding the solar system is embedded has a pressure which is a factor of 20 higher. This is an unstable situation;
- **the evaporation problem**: the local cloud is embedded in a hot gas. The gravitational escape velocity, a function of the cloud radius, is $\approx 10^3$ cm $s^{-1}$. If the cloud edges are heated by conduction the thermal velocity at the edge could exceed this threshold value and the cloud will evaporate. Of course, it is possible that the cooling may dominate over conductive heating; in this case evaporation can be suppressed. The condition for this is that the cloud radius exceed a critical value that for the LISM is $R_{\rm crit} \approx 23$ pc, which is much larger that the estimated dimensions of the local cloud. Therefore it is unlikely that cooling may suppress evaporation induced by conduction.
  Another possibility is that the thermal conductivity could be reduced by tangled magnetic fields, which could inhibit the mobility of the electrons. Such an effect could be studied in the solar wind plasma;
- **the ablation problem**: unless the local cloud is reasonably at rest with respect to the hot bubble, ablation by the streaming flow could be a major erosion factor. At present it is not possible to quantify this because of lack of information. The problem is very difficult since it is necessary to investigate the interface between clouds and hot flowing inter-cloud plasma. And for this, the ionization and recombination, magnetic field, macroscopic mixing effects, and ambipolar diffusion play fundamental roles difficult to be quantified;
- **the dust size problem**: the size distribution of the dust particles derived from interstellar extinction and polarization measurements is in conflict with the predictions of some astrochemical models. In our opinion this means that models are not good enough to fit data. Alternatively and complementarily to theoretical investigations, it is mandatory to perform new *in situ* measurements;
- **the dust association problem**: the dust–gas frictional coupling length is determined from the collisional momentum transfer of a gas atom colliding with a moving grain. The momentum transfer is significant when the dust

particle has traversed a column density of gas with similar mass as that of the particle. The frictional coupling scale for the local cloud is (Egger, Freyberg & Morfill, 1996) $l \cong 300 \times a_\mu$ pc, $a_\mu$ being the particles radius expressed in microns. Therefore $l$ is much larger than the local cloud dimension and raises the question whether the dust is associated with this cloud or not. The observed mass density supports this association. However, this problem needs further investigations;

- **the dust survival problem**: the local cloud is embedded in an environment, quite well known, but hostile. Then the problem of the survival of such grains against photo-sputtering, cosmic ray sputtering, solar wind sputtering, etc., should be investigated in detail;
- **the hot bubble origin problem**: there are several suggestions for the origin of the hot bubble:
  - the local hot bubble was produced by a supernova explosion occurred $\sim 10^5$ yr ago at 30–50 pc away from the Sun (Cox & Anderson, 1982). The soft X-ray emission measured in the ranges 0.13–0.188 keV and 0.16–0.35 keV (B and C bands, respectively) is fully justified if the supernova energy was $5 \times 10^{50}$ erg and if the remnant material was expanding into a medium with a density of $\sim 0.004$ cm$^{-3}$.;
  - a local superbubble was produced by tens of correlated supernovae exploded some $10^6$ yr ago, with a total energy of $1.5 \times 10^{52}$ erg (Innes & Hartquist, 1984). Such a superbubble has a size $\sim 200$ pc size and the model can roughly fit only the emission in the M band (0.45–0.85 keV) and the observed $O^{5+}$ number density;
  - a large low density ($\sim 0.005$ cm$^{-3}$) hot ($10^6$ K) superbubble, with a total energy stored of $\sim 10^{51}$ erg, can be maintained if very sporadic supernovae explode re-heating the ambient space, the radiative cooling time for such a gas being $\sim 10^7$ yr (Snowden *et al.*, 1990);
  - by using fossil records from arctic ice cores $^{10}$Be concentrations at times about 35,000 and 70, 000 yr ago have been observed (Sonett, Morfill & Jokipii, 1987). This is a strong proof that a recent ($\lesssim 10^5$ yr) supernova explosion occurred within a few tens of parsecs from the Sun. This, together with the ROSAT X-ray results and the absorption measurements, suggests a local superbubble with a spurious late supernova whose remnants have expanded into the tenuous cavity gas.

3.2.6. *Abundances in the Interstellar Medium*

The updated knowledge about the light elements abundances assures that all of the deuterium (D $\equiv ^2$H), most of the helium ($^3$He and $^4$He) and some $^7$Li were produced in the Big Bang. The simplest version of this scheme is referred to as Standard Big Bang Nucleosynthesis (SBBN), as well described by Boesgaard & Steigman (1985). As we have already stated, the comparison between theoretical SBBN and observed abundances is one of the cornerstones of present cosmology.

TABLE 6

Selected isotopic ratios for light elements by number (Wilson & Rood, 1994 and the references therein).

| Relative Abundance | Present day solar system | Protosolar | Nearby ISM |
|---|---|---|---|
| D/H | $3.4 \times 10^{-5}$ | $(2.6 \pm 1.0) \times 10^{-5}$ | $1.6 \times 10^{-5}$ |
| $^4$He/H | $9.8 \times 10^{-2}$ | $0.10 \pm 0.01$ | $8.9 \times 10^{-2}$ |
| $^3$He/$^4$He | $1.4 \times 10^{-4}$ | $(1.5 \pm 0.3) \times 10^{-4}$ | $\leq 10^{-4}$ to $\geq 10^{-2}$ |
| $^7$Li/H | $1.9 \times 10^{-9}$ | | |
| $^7$Li/$^6$Li | 12.3 | | 6.2 to 12.5 |

In the Big Bang theory the Universe consisted of a hot and dense gas, which started to cool because of expansion. Then, in the so-called age of nucleosynthesis, D, He, and Li were produced. Ignoring any dependence on lepton abundance or neutrino families, the relative abundance of more fragile nuclei (e.g., deuterium and $^3$He) rapidly declines with increasing baryon/photon ratio being processed to the more stable species $^4$He. This results in a slowly changing, monotonic dependence of the $^4$He abundance on baryon density $\Omega_b$. The $^7$Li abundance is relatively independent of density, because there are two production paths, as shown in Figure 5.

Because of the relatively low densities present in the nucleosynthesis era of the universe, there is no way to produce carbon and heavier elements (the so-called 'metals'). This step requires the stellar nucleosynthesis, which has been reviewed by Wilson & Matteucci (1992), and by Wilson & Rood (1994).

It is well known that the star formation rate depends, apart from other parameters, on the amount of dense, cold material available (Lada & Lada, 1991). Therefore the ISM in the inner part of our galaxy – whose structure and content is described in the review of Combes (1991) – has undergone more stellar processing and thus there should be a higher abundance of metals. Many determinations of the abundances have been published. It has been demonstrated that high velocity clouds (HVCs) possess O/H and Mg/H ratios roughly 0.1 of the solar value indicating that there has been some stellar enrichment and thus the HVCs are not composed of primordial material (Savage & Sembach, 1996 and the references therein). Table 6 shows the relative abundances for light elements in the present day solar system, protosolar and nearby ISM (Wilson & Rood, 1994 and the references therein).

The measured isotope ratios for selected species are given in Table 7 for a number of locations and sources (Wilson & Rood, 1994 and the references therein).

Savage & Sembach (1996) reviewed the interstellar abundances derived from observations of absorption lines with the Goddard High Resolution Spectrograph

TABLE 7

Ratios for Galactic Center, 4 Kpc molecular ring, carbon stars, Solar System, local ISM, and galaxies (Wilson and Rood, 1994 and the references therein).

| Isotopes Ratio | Galactic Center | Molecular Ring (4 Kpc) | Local ISM | Solar System | Carbon Stars | Nuclei of Galaxies |
|---|---|---|---|---|---|---|
| $^{12}C/^{13}C$ | ~20 | $53 \pm 4$ | $77 \pm 7$ | 89 | >30 | ~40 |
| $^{14}N/^{15}N$ | >600 | $375 \pm 38$ | $450 \pm 22$ | 270 | >515 | |
| $^{16}O/^{18}O$ | 250 | $327 \pm 32$ | $560 \pm 25$ | 490 | 320 to 1260 | ~200 |
| | | | | | >2700 | |
| $^{18}O/^{17}O$ | $3.2 \pm 0.2$ | $3.2 \pm 0.2$ | $3.2 \pm 0.2$ | 5.5 | 0.6 to 0.9 | 8 |
| | | | | | <1 | |
| $^{32}S/^{34}S$ | ~22 | ~22 | ~22 | 22 | | |
| $^{29}Si/^{30}Si$ | 1.5 | 1.5 | 1.5 | 1.5 | | |

TABLE 8

Summary of diffuse cloud gas phase abundances for cosmically abundant elements (Savage & Sembach, 1996, and the references therein).

$[X/H] = \log X/H - \log (X/H)_\odot$

| Element | Halo Abundance | Disk+Halo Abundance | Warm Disk Abundance | Cool Disk Abundance |
|---|---|---|---|---|
| Mg | <−0.28, −0.56 | −0.59, −0.62 | −0.73, −0.90 | −1.24, −1.56 |
| Si | −0.09, −0.47 | −0.23, −0.28 | −0.35, −0.51 | −1.31 |
| S | −0.23, +0.16 | +0.03 | −0.03, +0.14 | ~0.00 |
| Mn | −0.47, −0.72 | −0.66 | −0.85, −0.99 | −1.32, −1.45 |
| Cr | −0.38, −0.63 | −0.72, −0.88 | −1.04, −1.15 | −2.08, −2.28 |
| Fe | −0.58, −0.69 | −0.80, −1.04 | −1.19, −1.24 | −2.09, −2.27 |
| Ni | −0.77, −0.91 | −1.15 | −1.44, −1.48 | −2.46, −2.74 |

(GHRS) aboard the Hubble Space Telescope (HST). They reviewed the precision abundance results for a range of interstellar environments, including gas in the local medium, in the warm local medium, in cold diffuse clouds, and in distant halo clouds, as reported in Table 8.

Through GHRS studies it has been possible to determine the abundances of elements such as C, N, O, Mg, Si, S, and Fe in individual interstellar clouds. These studies have provided new information about the composition of interstellar dust grains, the origin of Galactic HVCs, and the processes transporting gas between the disk and the halo. Precision measurements of the interstellar D/H ratio and of the abundances of r and s process elements have also provided fiducial reference values for cosmological and stellar evolutionary observations and theoretical models, as discussed in details in the books *Primordial Nuclei and their Galactic Evolution* (Prantzos, Tosi, & von Steiger, 1998) and *The Evolution of the Milky Way: Stars versus Clusters* (Matteucci & Giovannelli, 2000).

### 3.2.7. *Distributions of Molecules and Ions in the Milky Way*

After atomic hydrogen, helium and their ions, the Universe probably contains more mass in molecular hydrogen ($H_2$) than in any other species. For instance, in the Milky Way the molecules reside primarily in the large complexes of molecular clouds that contain ~10–20% of the mass in the inner disk. In a continuous exchange of matter the stars supply matter to the ISM through winds and explosions, and the gas coalesces into clouds, while it collapses to form new stars. The state of the clouds and the rate at which they evolve depend on heating and cooling of matter composed almost entirely of $H_2$.

Many years ago, Shull & Beckwith (1982) discussed the interstellar molecular hydrogen. There they discussed, after a swift historical glance, the properties of the hydrogen molecules, the chemical and energy balance of $H_2$, the ultraviolet observations made essentially with the first UV satellite (Copernicus), IR and radio observations and the future perspectives. They also reviewed the theories of excitation: thermal and non-thermal.

Direct observations of the dominant gaseous component in the $H_2$ clouds are generally precluded because the symmetry of the hydrogen molecules implies that there is no permanent dipole moment and there are therefore no strong radio frequency transitions. The permitted electronic transitions in the UV, seen in absorption across the line of sight to nearby bright stars, cannot be studied in the denser clouds owing to the very high opacity of the dust in the UV. As an alternative one can use radio frequency observations of trace gas constituents such as carbon monoxide (CO) – the most abundant molecule after $H_2$ – to probe the interior of these clouds. In the cold ISM the CO molecule is excited to the first rotational level by collisions with $H_2$ molecules. The dipole moment of the $J = 1 - 0$ first rotational transition of CO is very low (0.1 Debye), so that the rate of spontaneous de-excitation is small. The theoretical expectation for the $H_2$ density corresponding to balanced rates between collisional excitation and spontaneous emission is $\approx 3 \times 10^3$ cm$^{-3}$. Because such densities are also typical of those which actually occur in the molecular gas in the Galaxy, the CO molecule provides a good tracer of the molecular component. Therefore the 2.6 mm line can be considered the analog of the 21 cm line. The Galaxy is transparent at these wavelengths and the CO emission is ubiquitous.

However, the determination of the $H_2$/CO conversion ratio is difficult to determine. Indeed, can this ratio be measured? And, once measured, can it be regarded as a universal constant? A deep discussion of this topic has been reported by Combes (1991).

TABLE 9

Total disk mass of HI and $H_2$ in the Galaxy (Scoville, 1992).

| Mass | 0–0.15 | 0.15–1.0 | 1.0–1.6 | Total |
|---|---|---|---|---|
| $M(H_2)$ ($M_\odot$) | $0.4 \times 10^9$ | $1.9 \times 10^9$ | $0.2 \times 10^9$ | $2.5 \times 10^9$ |
| $M(H_2)/M_T(H_2)$ | 14% | 77% | 9% | 100% |
| $M(HI)$ ($M_\odot$) | $1 \times 10^9$ | $0.7 \times 10^9$ | $1.7 \times 10^9$ | $3.4 \times 10^9$ |
| $M(HI)/M_T(HI)$ | 0.4% | ~30% | ~70% | 100% |
| $M(H_2)/M(HI)$ | 0.4 | 2.7 | 0.12 | 0.73 |

Galactic Annulus*

* Annuli in the Galaxy. Radii normalized to the Sun's orbital radius = 1.0. $M_T$ is the total disk mass, most of which is contained in stars.

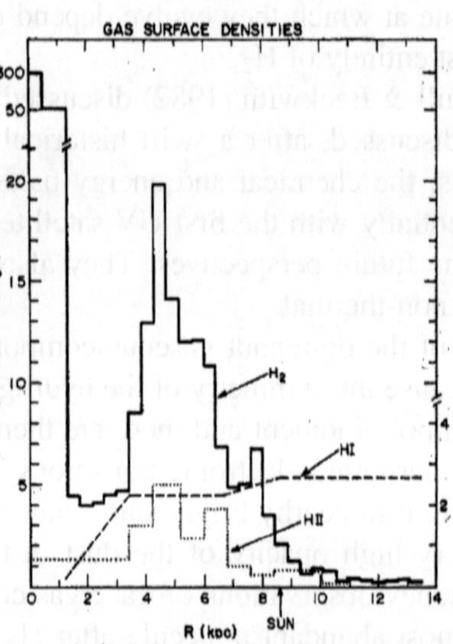


*Figure 28.* Comparison of gas surface densities in the Milky Way disk for molecular ($H_2$), atomic (HI), and ionized (HII) hydrogen (Scoville, 1992).

The galactic distribution of $H_2$ is shown in Figure 28. The Sun is located at a galactic radius of 8.5 kpc. The molecular gas shows a strong concentration in the galactic center and in the molecular cloud ring at 3–7 kpc as does the distribution of giant HII regions. In contrast, the atomic hydrogen shows a deficiency in the center of the Galaxy and a relatively constant density through the galactic disk out to approximately three times the distance of the Sun from the galactic center (Scoville, 1992).

Table 9 compares the $H_2$ and HI masses in the galactic center region, and in the inner and outer (relative to the Sun's location) disk. Overall in the Galaxy the HI and $H_2$ masses are approximately equal. On the other hand, the distribution of the two gas components in the Galaxy are entirely different, approximately 90% of the total $H_2$ mass is found inside the Sun's orbit, whereas nearly 70% of the total HI mass is outside.

The CO surveys of the Milky Way, started in 1973, are reviewed by Combes (1991). In order to construct the composite CO survey of the entire Milky Way, Dame *et al.* (1987) combined data from 16 secondary surveys to produce a 30′ wide effective beam (see Figure 29).

The very high coverage of the sky obtained (20%) is ideal for exploring large scale structures such as spiral arms. However, the corresponding low angular resolution (0.12°) makes these surveys subject to dilution effects, such as intensity not enough to isolate small clouds on the far side of the Galaxy, or nearby small clouds in the confusion of crowded regions. For this reason the higher angular resolution surveys are most suited to creating statistics on the physical characteristics of molecular clouds themselves (for a discussion see the review of Combes, 1991).

Neutral nitrogen and oxygen column densities or limits were derived for 53 path lengths through diffuse interstellar medium and compared with column densities of neutral hydrogen by using data obtained with the Copernicus satellite (York *et al.*, 1983). For neither N nor O is a systematic increase of depletion found as reddening increase. The value found for O/N ratio is 8, and the abundances of N and O are both between 40% and 70% of the solar values. Since also the C shows evidence for systematically greater depletion as reddening increase and all three elements (C, N, and O) show some depletion, the main mass in grains may be an amorphous mixture of C, N, and O. Such grains would have to be well mixed with gas and would have to have similar makeup independent of the radiation or density environment.

With Copernicus satellite observations of weak interstellar lines (Bohlin *et al.*, 1983) and with an accurate evaluation of the MgII (1240 Å) doublet oscillator strength, Murray *et al.* (1984) derived MgII column densities toward 74 stars. Such results show that interstellar magnesium to hydrogen abundance is highly variable, which implies an average of 40% of interstellar magnesium is in the gaseous phase, much larger than the average percentage of iron in the gas phase (2%). No significant correlation is found between magnesium abundance and reddening, directions with similar $E(B - V)$ showing wide variations in magnesium depletion. The magnesium depletions correlate with the mean line of sight density $n$(H), which crudely measures the cloud densities where most absorption occurs. These results imply that solid phase interstellar magnesium may be predominantly grain mantle material. Magnesium depletions do not correspond closely to the strength of the interstellar extinction feature near 2200 Å.

A curve of growth analysis of UV resonance lines of silicon and aluminum suggests that aluminum is more depleted than silicon and that both elements are more depleted in reddened stars than in unreddened stars (Barker *et al.*, 1984).

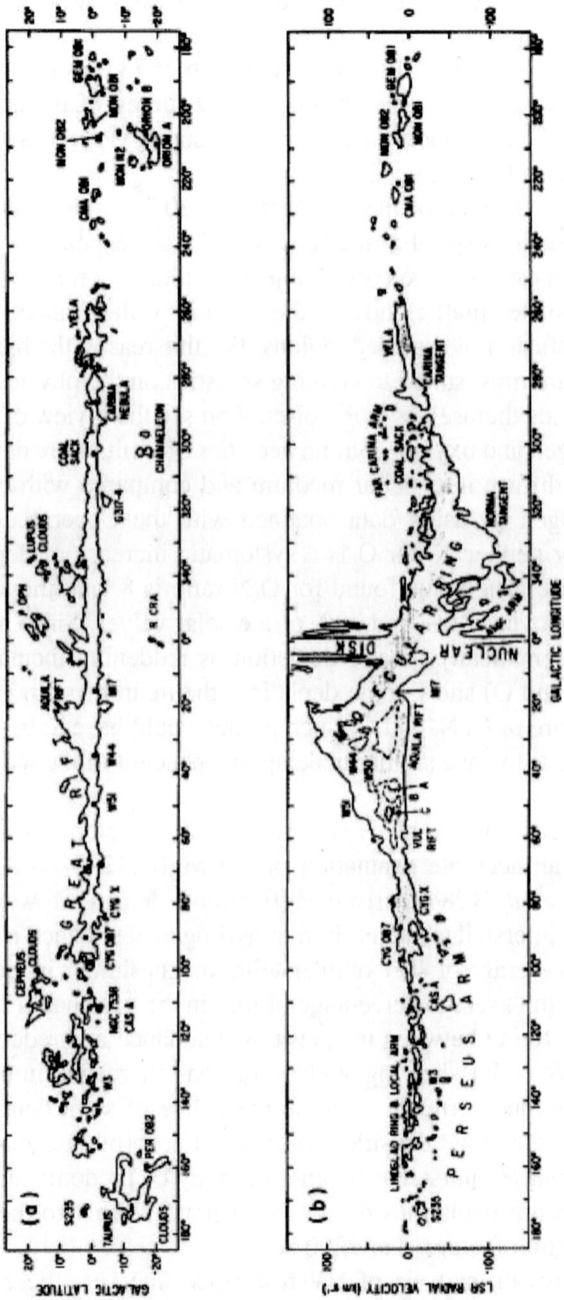


*Figure 29.* An all sky map of the CO emission in the Milky Way: (a) $l$–$b$ diagram, integrated over velocities $|v| < 15$ km s$^{-1}$; (b) $l$–$V$ map integrated over latitudes in the range $-3.25° < b < 3.25°$ (from Dame *et al.*, 1987).

### 3.2.8. *Distribution of $^{26}$Al from Nucleosynthesis*

The nuclear evolution of the ISM assumes that the rate of nucleosynthesis is proportional to the star formation rate, which is believed to be proportional to the mass of interstellar gas. This is equivalent to assuming that the rate of nucleosynthesis per unit mass of interstellar gas is constant (e.g., Clayton, 1984). Another assumption is that the production yields do not depend on the initial abundances and that there is no infall of matter into the galactic disk. Then for an unstable isotope with a lifetime short compared to the time scale of star formation, the mass fraction in the ISM is constant in time. This is, for instance, the case for $^{26}$Al. Similarly, the mass fraction of a stable isotope will increase linearly in time, such as $^{27}$Al. Therefore, the mass fractions of the two aluminum isotopes is $^{26}\mathrm{Al}/^{27}\mathrm{Al} = P_{26}\tau/P_{27}t$, where $P_{26}$ and $P_{27}$ are the production functions for $^{26}$Al and $^{27}$Al, respectively, $\tau$ is the mean life of $^{26}$Al, and $t$ is the time since the beginning of heavy element formation in the Galaxy. The latter equation is strictly valid only if the initial mass fraction of $^{27}$Al was zero; however, the equation is only slightly modified for any reasonable nonzero value for the initial mass fraction (Mahoney *et al.*, 1984). They discovered the presence of $^{26}$Al by using the HEAO 3 $\gamma$-ray spectrometer which detected diffuse emission from the galactic equatorial plane in a narrow line at energy of $1809.49 \pm 0.41$ keV just from the decay of $^{26}$Al. The line flux, from the vicinity of the Galactic Center, is $(4.8 \pm 1.0) \times 10^{-4}$ photons cm$^2$ s$^{-1}$ rad$^{-1}$. The line was interpreted as resulting from the decay of $\sim$3 M$_\odot$ of $^{26}$Al in the present ISM. The measured limit of $\leq$ 3 keV FWHM for the intrinsic line width is consistent with this interpretation.

The observed 3 M$_\odot$ of $^{26}$Al in the ISM correspond to an isotopic ratio of $^{26}\mathrm{Al}/^{27}\mathrm{Al}$ of $\sim 1.0 \times 10^{-5}$, comparable to representative values for the same ratio measured in meteoritic material. Considering the inhomogeneity of both the solar nebula and the ISM, the HEAO 3 observations indicate that the aluminum content of the solar nebula may have been a normal condition of the ISM rather than a transient phenomenon resulting from a nearby supernova that might have triggered the collapse of the nebula.

Later, with the COMPTEL instrument aboard the CGRO, the image of the Galaxy in the 1.809 MeV $\gamma$-ray line confirmed the detection of HEAO 3 satellite, and, moreover, the unexpected irregularity of the apparent emission suggests that only a specific setting of nucleosynthesis conditions produces intermediate mass isotopes such as $^{26}$Al. Most plausible sources are massive stars, through their intense winds and their core-collapse supernovae and novae (e.g., Prantzos & Diehl, 1996; Diehl, 2000). These sources may not be distributed as homogeneously as derived from a stellar disk population alone: $^{26}$Al nucleosynthesis appears enhanced in localized regions, possibly revealing peculiarities of massive star formation and their evolution within dense and otherwise unobservable regions of the Galaxy (Diehl, 1996 and the references therein).

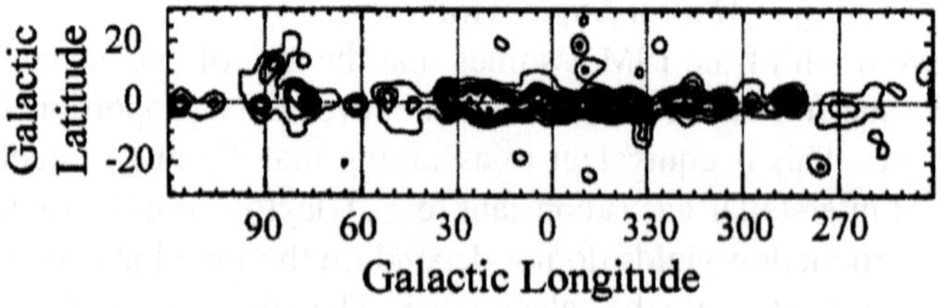


*Figure 30.* Maximum entropy map of the 1.8 MeV line emission of $^{26}$Al (Bloemen *et al.*, 1999a).

Figure 30 shows the COMPTEL Map of 1.809 MeV $\gamma$-rays from observations of the galactic plane, by using the Maximum Entropy deconvolution technique (Bloemen *et al.*, 1999a).

### 3.2.9. *Diffuse Galactic Continuum Radiation*

Cosmic ray protons and electrons interacting with the interstellar matter and low energy photons originate the diffuse $\gamma$-ray continuum emission, which is the dominant feature of the $\gamma$-ray sky. Such an emission has been studied from 50 keV to over 50 GeV by means of observations performed with the OSSE, COMPTEL, and EGRET experiments aboard the CGRO. The OSSE observations show that the diffuse continuum is a composition of three independent components: $e^+e^-$ annihilation line plus continuum, a soft low energy component, and a dominant hard component. The COMPTEL observations provided the first results of the diffuse emission at intermediate energies, and the EGRET provided a clear image of the spatial and spectral distribution of this emission at higher energies. A contribution owed to unresolved distributions of point sources cannot be excluded, particularly from the OSSE and COMPTEL data. However, such data can be best fitted with a sum of bremsstrahlung, $\pi^0$ decay, and inverse Compton components in the energy range from ~1 MeV to ~1 GeV. Above about 1 GeV, the EGRET spectrum is harder than expected from the nucleon nucleon spectrum ($\propto E^{-2.7}$) derived from a cosmic ray proton spectrum as observed in the solar neighborhood (Hunter, Kinzer & Strong, 1997 and the references therein). Figure 31 shows the diffuse $\gamma$-ray spectrum for the inner Galaxy: $350° < l < 30°$ and the curves for the bremsstrahlung, $\pi^0$ decay inverse Compton, and total for the physical model discussed by Strong & Moskalenko (1997, 1998) and by Moskalenko & Strong (1998). The injection spectrum used for the electrons is $\propto E^{-2.1}$, which agrees with detailed modelling of the radio synchrotron data.

### 3.2.10. *Diffuse High Energy Emission at the Galactic Center*

Radio continuum observations of the inner 15′ of the Galactic Center (GC) show two prominent radio continuum structures known as the Sgr A Complex and the filamentary continuum Arc. The Sgr A Complex consists of Sgr A East and its halo and of Sgr A* and its thermal orbiting gas Sgr A West. Sgr A* is unique and

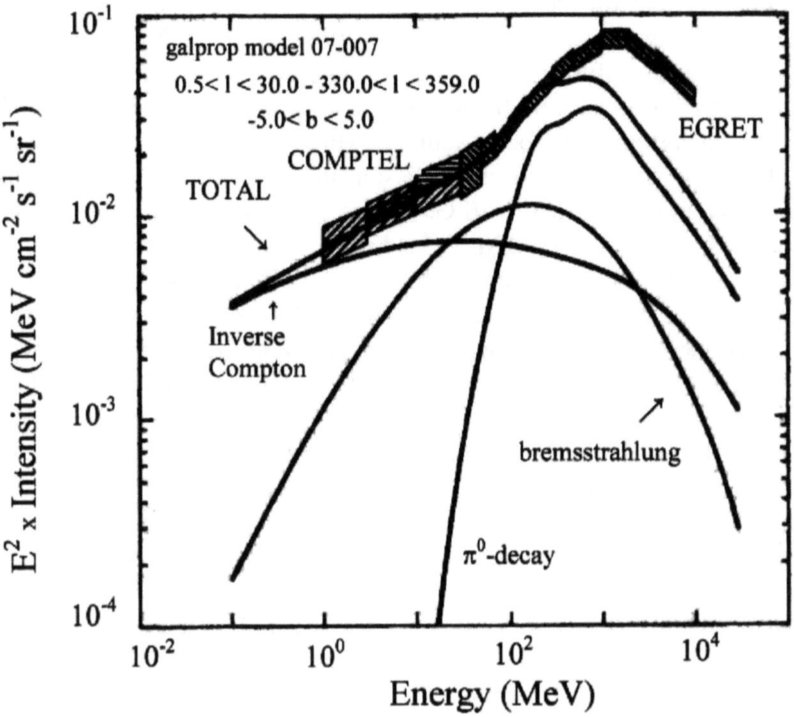


*Figure 31.* The diffuse $\gamma$-ray spectrum for the inner Galaxy: $350° < l < 30°$ and the curves for the bremsstrahlung, $\pi^0$ decay inverse Compton, and total (Hunter, Kinzer & Strong, 1997).

considered by many to be a $10^6$ $M_\odot$ massive black hole at the GC (e.g., Yusef-Zadeh, Purcell & Gotthelf, 1997 and the references therein). The interpretation that Sgr A East could be a standard SNR is questioned by many authors (e.g., Khokhlov & Melia, 1996 and references therein). Indeed, they have considered that Sgr A East is the remnant of a star which is tidally disrupted by a massive black hole, probably by Sgr A*. The explosion energy is estimated to be an order of magnitude greater than a typical supernova. There is also evidence that such an explosion occurred inside the dense molecular cloud M-0.02-0-07, thus depositing $\geq 4 \times 10^{52}$ erg in the ISM (Mezger *et al.*, 1989). The discovery of OH(1720 MHz) masers at the interface of the 50 km $s^{-1}$ molecular cloud and Sgr A East showed definitively that these two are physically interacting with each other (Yusef-Zadeh *et al.*, 1996). The Arc is a non-thermal filamentary source located near $l \approx 0.18°$ and runs in the direction perpendicular to the Galactic plane. The filaments are linearly polarized, showing evidence that they are tracing magnetic field lines and emitting synchrotron radiation. A number of GC molecular clouds appear to outline the linear filaments, supporting the hypothesis that filaments and clouds are physically interacting with each other; in this case the magnetic field strength must be $\geq$ 1 mG in order for the filaments to resist deflection at points of interaction with the clouds (e.g., Yusef-Zadeh & Morris, 1987; Serabyn & Morris, 1994).

The detection of diffuse $\gamma$-ray emission from the Galactic Center (GC) is generally very difficult because of the presence of different objectionable contributors, namely, the background and point sources, which must be removed. The sources contribution dominates the galactic emission in the 0.1–10 keV band; while the ISM emission, in the 30 MeV – 10 GeV band, overwhelms that of point sources. The situation in the 10 keV – 10 MeV band is less clear (e.g., Lebrun *et al.*, 1999a).

X-ray and $\gamma$-ray observations of the Galactic Center (GC) region by ASCA show evidence of 2–10 keV as well as 6.7 and 6.4 keV line emission from the inner 0.2° of the GC. The continuum radiation is coming from a thermal plasma at temperature of 10 keV. The electron density and the thermal energy of thermal gas within the shell of Sgr A East are estimated to be 6 $cm^{-3}$ and $3 \times 10^{50}$ erg, respectively. The electron density and the thermal energy of the hot gas outside the Sgr A Complex are estimated to be ~0.3–0.4 $cm^{-3}$ and ~ $(0.5–1) \times 10^{53}$ erg, respectively (Koyama *et al.*, 1996). The 6.4 keV line emission is peaked on two molecular clouds in the region between the Arc and the Sgr A Complex and in Sgr B2.

CGRO/EGRET instruments show evidence of continuum emission from GC in the 30 MeV – 30 GeV band (Mattox, 1997). The luminosity at the distance of the GC is $5 \times 10^{36}$ erg $s^{-1}$. The source is located within 0.2° of the GC and could be either compact or diffuse within 100 pc of the GC (Thompson *et al.*, 1996; Pohl, 1997). The energy spectrum of this source is best fitted to a power law spectrum having an index of 1.7, which is harder than typical EGRET sources in the Galactic plane.

SIGMA experiment (Paul *et al.*, 1991) aboard the Russian GRANAT satellite is excellent in estimating the sources contribution, while OSSE experiment (Johnson *et al.*, 1993) aboard CGRO is excellent in determining the background. For this reason a program of coordinated observations of the GC region using SIGMA and OSSE experiments was performed (Purcell, *et al.*, 1996). Paltani *et al.* (1999) obtained spectral images from 35 to 1300 keV by using SIGMA experiment in Spectrum-Image mode. They found an excess in the GC observations and derived the spectrum using the same spatial distribution as in Purcell *et al.* (1996): uniform in $l$ from $-40°$ to $40°$, and Gaussian in $b$ with FWHM of 5.4°. The resulting spectrum averaged over the latitude is shown in Figure 32.

The best fit with a power law in the 45–95 keV has a photon index $2.2 \pm 0.04$ and a flux at 100 keV of $(3.4 \pm 0.10) \times 10^{-4}$ photons $cm^2$ $s^{-1}$ $keV^{-1}$ $rad^{-1}$.

Figure 33 shows the spectrum of the diffuse galactic emission at galactic longitude $l = 0°$ (open diamonds), i.e., the net result obtained from subtracting the point source emission measured by SIGMA from the total emission measured by OSSE (Grove, 1996). It is also shown the average spectrum from OSSE observations of the galactic plane at $l = 25°$ and $l = 339°$ (filled diamonds). The uncertainties in the GC spectrum represent the combined uncertainties of the SIGMA and OSSE

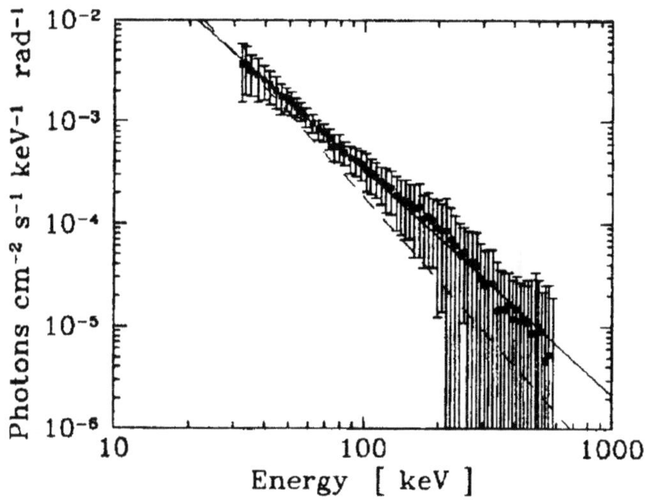


*Figure 32.* The spectrum of the Galactic Center diffuse emission with a power law best fit between 45 and 95 keV. The dashed line is the result of Purcell *et al.* (1996) (Paltani *et al.*, 1999).

measurements and are therefore considerably larger than those for OSSE alone. The spectrum is well described by a model consisting of: i) a single power law with photon index of 2.4, in the 50 keV – 5 MeV band; ii) a narrow line at 511 keV; and iii) a positronium continuum component. Note that, with the exception of the level of the annihilation line and positronium continuum emission, the spectrum at $l = 25°$ from the GC is remarkably similar in shape and intensity to the spectrum at the center (Grove, 1996).

Purcell *et al.* (1997) and Cheng *et al.* (1997) produced the first maps of galactic positron annihilation by combining OSSE (1991–1997), Gamma Ray Spectrometer aboard the SMM (Solar Maximum Mission) (1980–1988) and TGRS (Transient Gamma Ray Spectrometer aboard the WIND mission (1995–1997). Milne *et al.* (2000) augmented the data set with additional data, both recent and archival and re-analyzed to improve the spectral fitting. Such an improvement has enabled the first maps of positronium continuum emission and the most extensive maps of the 511 keV line emission. Bulge and disk combinations have been compared with the 511 keV line data, demonstrating that extended bulges are favored over a GC point source for every disk model tested.

The physical state of stars and interstellar matter in the Galactic Bulge (radius $R$ ~0.3–3 kpc from the dynamical center of the Galaxy), in the Nuclear Bulge ($R < 0.3$ kpc), and in the Sgr A Radio and Giant Molecular Cloud Complex (i.e., the central ~50 pc of the Milky Way) has been discussed in a review by Mezger, Duschl & Zylka (1996).

In a review about *The Galactic Center at High Energies*, Skinner & Kennea (1999) discussed the observations of the Galactic Center (GC) as seen with X-ray

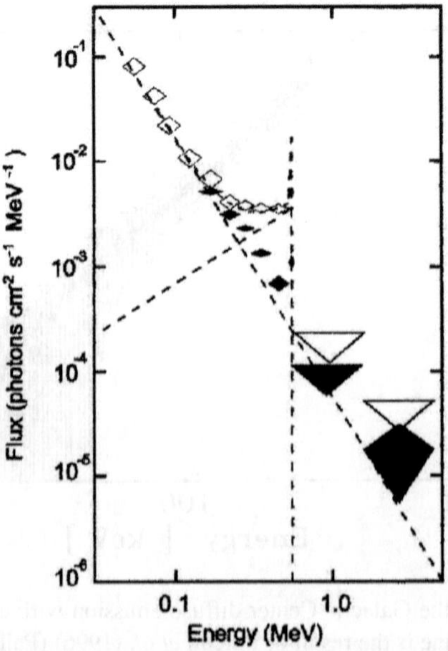


*Figure 33.* Spectra of the diffuse galactic emission at $l = 0°$ (open diamonds) and at $l = 25°$ and $l = 339°$ (filled diamonds). The dashed curves represent a fit to the Galactic Center spectrum, including a power law continuum, an annihilation line, and positronium continuum (Grove, 1996).

and $\gamma$-ray instruments, concentrating on the central ~300 pc diameter, but also discussing some of the exotic objects and phenomena in the surrounding region. Yusef-Zadeh, Purcell & Gotthelf (1997) discussed on the origin of the high energy activity at the GC.

### 3.3. Intra-cluster medium

X-ray astronomy has given a fundamental contribution on knowledge of the physics of clusters of galaxies and in particular of the intra-cluster medium (ICM). After the Einstein X-ray observatory other satellites with experiments in soft and hard X-ray energy ranges have been successfully employed in studying astrophysical plasmas, such as ROSAT, GINGA, BBXRT and ASCA, CGRO, BeppoSAX, CHANDRA, XMM, INTEGRAL, now at the beginning of its observational life.

The BeppoSAX satellite, successfully launched in 1996 (Piro, 1996), recently put out of service, improved the knowledge of clusters of galaxies with simultaneous multifrequency X-ray measurements. Also AXAF-Chandra (Weisskopf, 1990) and XMM-Newton (Jansen *et al.*, 2001) observatories, launched in 1999, are improving a lot the knowledge on the fine structures of those plasmas thanks to their high sensitivity and spectral resolution.

Exhaustive books on these topics are: *X-ray Emission from Clusters of Galaxies* (Sarazin, 1988a), *Physical Processes in Hot Cosmic Plasmas* (Brinkmann, Fabian

& Giovannelli, 1990), *New Horizon of X-Ray Astronomy: First Results from ASCA* (Makino & Ohashi, 1994), *UV and X-ray Spectroscopy of Astrophysical and Laboratory Plasmas* (Yamashita & Watanabe, 1996), *X-Ray Imaging and Spectroscopy of Cosmic Hot Plasmas* (Makino & Mitsuda, 1997), and *Diffuse Thermal and Relativistic Plasma in Galaxy Clusters* (Böhringer, Feretti, & Schuecker, 1999). In the last few years Colafrancesco (1995, 1996, 1997, 1999, 2002a,b,c) and Schindler (2001, 2002a,b) have reviewed the clusters of galaxies as laboratories for cosmology and their evolution, mainly discussing the constraints that the present data on the local and distant distribution of X-ray clusters can set both on the models for galaxy formation and on some aspects of the gravitational instability theory. Blasi (2001a) discussed on the role of cluster mergers starting from multifrequency observations of clusters of galaxies.

### 3.3.1. *Historical Background*

The largest organized structures in the Universe are clusters of galaxies (CGs). They are gravitationally bounded structures containing hundreds of galaxies spread over a region whose size is $\sim 10^{25}$ cm. Their total masses range from a few $10^{13}$ to $\sim 4.6 \times 10^{15}$ $M_{\odot}$ (Schindler, 1996). The first roughly classification of CGs recognizes two classes running from regular to irregular CGs (Abell, 1965, 1975). There is evidence that regular clusters are dynamically more evolved and relaxed than irregular clusters, which contain a large number of spiral galaxies contrary to the regular ones. For a complete classification scheme see, for instance, the book and fundamental paper of Sarazin (1988a,b) and the references therein.

The tendency for galaxies to cluster had been noted long time ago, but only at the beginning of the XXth century Wolf (1906) studied in detail the clusters. An enormous improvement in the knowledge of the properties of clusters was obtained by Abell (1958) when he prepared his statistically complete catalog of rich clusters of galaxies. The richness of a cluster is a measure of the number of galaxies associated with that cluster. Since background galaxies are present it is not possible to definitively state that any given galaxy belongs to a given cluster. Therefore it is not possible to give an exact evaluation of the number of galaxies in a cluster. Richness is a statistical measure of the population of a cluster, based on some operational definition of cluster membership (e.g., Sarazin, 1988a, and references therein).

The Virgo cluster is the nearest reach cluster to Milky Way and the Coma cluster is the nearest very regular cluster. During the first years of the new space era, X-ray emission was detected from the region around M87 placed in the center of the Virgo cluster (Byram, Chubb & Friedman, 1966; Bradt *et al.*, 1967). This was the first detection of X-ray emission outside of our own galaxy. Five years later – 1971 – other X-ray sources were detected in the direction of Coma and Perseus clusters (Fritz *et al.*, 1971; Gursky *et al.*, 1971a,b; Meekins *et al.*, 1971). Since Virgo, Coma and Perseus are three of the nearest clusters, Cavaliere, Gursky & Tucker (1971) suggested that clusters of galaxies could generally be X-ray sources.

The whole sky survey by the Uhuru satellite (Giacconi *et al.*, 1972) confirmed the latter reasonable hypothesis. This survey showed that the clusters are actually bright X-ray sources with luminosities ranging from $10^{43}$ to $10^{45}$ erg $s^{-1}$. These X-ray sources, associated with the clusters, were found to be spatially extended being their sizes comparable to the size of the galaxy distribution within the clusters (Kellog *et al.*, 1972; Forman *et al.*, 1972). Moreover, these X-ray sources were found constant in intensity (Elvis, 1976). Spectral measurements of X-ray emission from several clusters of galaxies (Kellog, Baldwin & Koch, 1975; Davidsen *et al.*, 1975; Scheepmaker *et al.*, 1976) resulted in favor of the thermal origin of the radiation and contrary to the inverse Compton emission by cosmic ray electrons or emission from stellar X-ray population hypothesis. Indeed, their X-ray spectra were most consistent with thermal bremsstrahlung spectra from hot gas. This reasonable interpretation requires that the space between galaxies in the clusters be filled with a very hot ($\sim 10^8$ K) and low density ($\sim 10^{-3}$ atoms $cm^{-3}$) gas, which was called the ICM. It is very important to remark that the total mass of the ICM is roughly equal to the total mass of all the stars in all the galaxies of the cluster. This gas forming the ICM, stored since the formation of the Universe in the large volumes of space between the clusters, had simply fallen into the clusters, like suggested by Gunn & Gott (1972).

The Ariel 5 discovery by Mitchell *et al.* (1976) of an emission feature near 7 keV in the spectrum of the Perseus cluster introduced, in addition, implications concerning the history of the emitting gas. Indeed, this feature was owed to thermally excited iron lines which were found from the Coma, Virgo and Perseus clusters as well by Serlemitsos *et al.* (1977) with the Goddard X-ray Spectroscopy Experiment, in the energy range 2–60 keV, aboard OSO-8. From these measurements, although the values inferred for plasma temperature and iron line equivalent width varied significantly among the three clusters ($\sim$2.85, $\sim$6.95 and $\sim$9.3 keV for Virgo, Perseus and Coma clusters, respectively), and in spite of the different cluster morphologies involved, the deduced iron abundances, relative to cosmic ones, were remarkably similar in magnitude ($\approx$0.4 for the three clusters). This strongly suggested a similar origin for the bulk of the X-ray emission, viz., a thin hot cluster gas of standard abundance, at collisional equilibrium. Moreover, it allowed one to argue that a significant fraction, if not all, of the cluster gas has come from the galaxies, rather than from infall of primordial gas into the cluster (Serlemitsos *et al.*, 1977); the iron abundance in this gas was expected to be well below its solar level (Wagoner, Fowler & Hoyle, 1967).

We would like to remark the importance of:

- the prediction of Cavaliere, Gursky & Tucker (1971) about the possibility that the clusters of galaxies form a class of X-ray emitters; this was experimentally confirmed by the UHURU survey;
- the measurements of Mitchell *et al.* (1976) and Serlemitsos *et al.* (1977): they demonstrated the thermal nature of the ICM and the presence of heavy elements which obviously must come from the galaxies of the clusters in which

thermonuclear processes, occurring in the interior of the stars, produce such elements: these elements are injected in the ICM via galactic winds (Mathews & Baker, 1971), and stripping by galaxy motions through the cluster gas (Gunn & Gott, 1972); Lea & DeYoung, 1976).

### 3.3.2. *X-Ray/Optical Correlations*

Soon after the discovery of X-ray emission from clusters of galaxies many authors started methodical analyses of clusters in other energy ranges and especially in the optical range, in which the clusters were better known, in order to search for correlations between their properties. For this purpose many optical cluster properties have been used, such as the richness, morphology, the galactic content, the core radius, the velocity dispersion, and the central galaxy density; these parameters have been defined in the book by Sarazin (1988a) and in the references therein.

3.3.2.1. *Rich Clusters.* As is well known, the largest X-ray surveys provide only X-ray fluxes – thus luminosities $L_X$ – for a given photon energy range. In addition there are smaller samples of clusters with X-ray surface brightness determination and samples with X-ray spectra.

It was suggested that the X-ray luminosity of a cluster correlates with its velocity dispersion ($\sigma_r$) as $L_X \propto \sigma_r^4$ (Solinger & Tucker, 1972), based on the small sample of known X-ray clusters at that time. They gave a simple model to explain that correlation based on the assumption that the emission comes from the ICM. Even if this correlation has been re-examined several times meanwhile the sample of X-ray clusters was growing, the results are rather similar, at least in the energy range 2–10 keV. Indeed, Quintana & Melnick (1982) found the same relationship for larger samples coming from the HEAO-1 and EINSTEIN observatories. Their relationship for the HEAO-1 data is:

$$L_X(2 - 10\ \text{keV}) \sim 4.2 \times 10^{44}\ \text{erg s}^{-1} \times (\sigma_r/10^3\ \text{km s}^{-1})^4 . \qquad (3.2)$$

This relationship is shown in Figure 34 in a $\log L_X - \log \sigma_r$ plane. For the EINSTEIN data, at lower energies, the power in the latter equation is ~3 (Quintana & Melnick, 1982; Abramopoulos & Ku, 1983).

Generally speaking, the richness of a cluster and its X-ray luminosity measure the mass of stars and of diffuse gas in the cluster, respectively, and it is reasonable to expect these to be correlated. Indeed, this correlation appears in the data, but it is difficult to quantify this correlation since the Abell catalog is incomplete for the richness class 0 clusters, Abell richness are binned in the catalog and the higher richness clusters are generally more distant. However, this correlation was found by Abramopoulos & Ku (1983): the low energy X-ray luminosity increases with richness to the 1.2 power. Soltan & Henry (1983) demonstrated that this correlation is valid also for the very richest Abell clusters. Low luminosity clusters are much less common among the richest clusters.

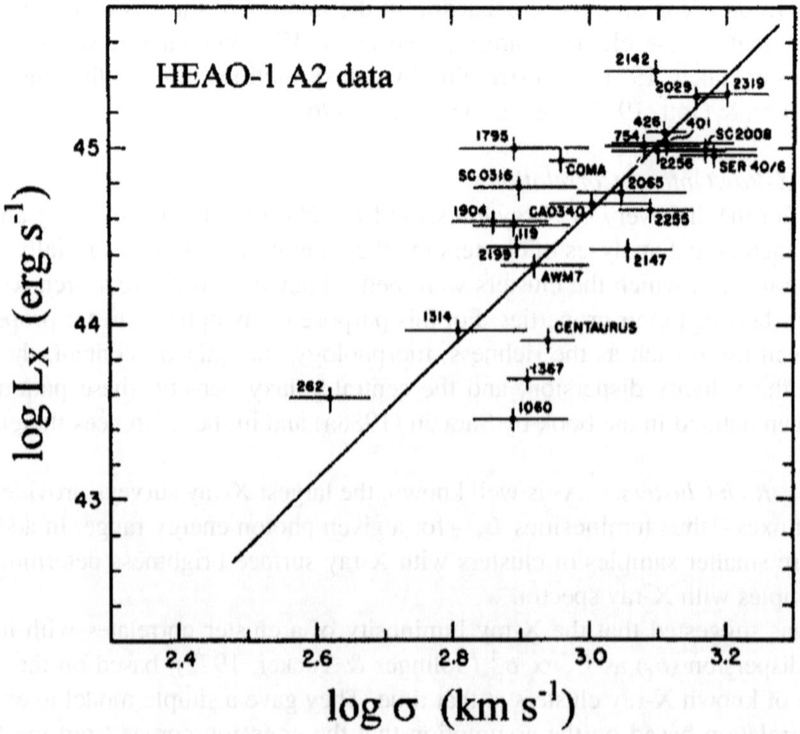


*Figure 34.* The relationship between the X-ray luminosity of clusters observed with HEAO-1 A-2 and its velocity dispersion ($\sigma_r$) (Sarazin, 1988a, after Quintana & Melnick, 1982) (Courtesy of Craig Sarazin).

The optical morphology (Rood & Sastry, 1971; Struble & Rood, 1982) and X-ray luminosity of clusters are also correlated. Indeed, the more regular Rood–Sastry types clusters in general have higher X-ray luminosities than the less regular clusters (Sarazin, 1988a and references therein). Clusters which contain optically dominant galaxies near the cluster center tend to be stronger X-ray sources than those do not (e.g., Jones & Forman, 1984). These authors, since 1978 (Jones & Forman, 1978) argued that all of these optical/X-ray correlations are owed to the correlation of X-ray luminosity and richness plus the tendency of rich clusters to be more regular.

The X-ray luminosity of clusters is well correlated with the projected central galaxy density parameter $\bar{N}_0$ (Bahcall, 1977a). This correlation is tighter than the richness correlation. An explanation could be that thermal X-ray emission depends on the square of the density of the gas; therefore it is most sensitive to the deepest portion of the cluster potential. The correlation in the HEAO-1 A-2 sample is: $L_x \propto \bar{N}_0^{\,3.5}$ – confirmed by many authors (e.g., Mushotzky, 1984a and the references

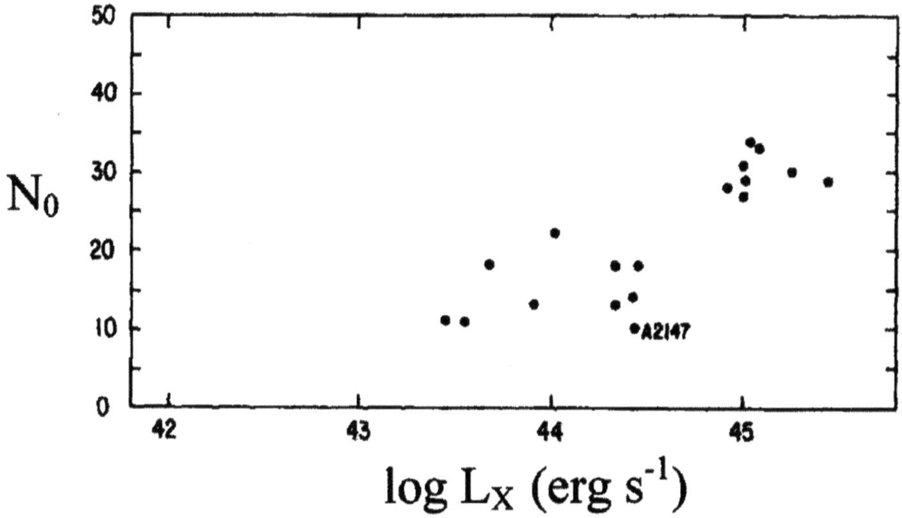


*Figure 35.* The relationship between the X-ray luminosity of clusters observed with HEAO-1 A-2 and the central galaxy density $\bar{N}_0$ (Sarazin, 1988a, after Mushotzky, 1984a) (Courtesy of Craig Sarazin).

therein) – and it is shown in Figure 35. In the case of the EINSTEIN lower energy sample, the power is ~2.5.

The X-ray luminosity of a cluster correlates also with its galactic content: luminous X-ray clusters generally have a small proportion of spiral galaxies (Bahcall, 1977b). The spiral fraction increases with radius in X-ray clusters and the spiral fraction is inversely correlated to the velocity dispersion (Melnick & Sargent, 1977). These correlations are consistent with the theory predicting that spiral galaxies formed in a cluster are stripped of their gas content through interactions with the intra-cluster gas and become S0 galaxies; however these correlations do not prove that such a process has actually occurred. The spiral fraction $f_{Sp}$ – the fraction of cluster galaxies that are spirals – is correlated with the X-ray luminosity as (Bahcall, 1977b)

$$f_{Sp} \sim 0.37 - 0.26\,log(L_x/10^{44}\ \mathrm{erg\ s^{-1}})\,. \tag{3.3}$$

This relationship, shown in Figure 36, was derived assuming a simple model for the stripping of spirals in a cluster. An important example against this relationship is that of the cluster A 194, having a low spiral fraction $f_{Sp}$ ~0.27 (Oemler, 1974; Dressler, 1980). But this fact is not strange: indeed A 194 has a very low X-ray luminosity (Jones & Forman, 1984) and low velocity dispersion; therefore it is unlikely that the spirals were stripped by the ram pressure of intra-cluster gas.

A possible correlation between the sizes of the X-ray sources in the clusters, $r_x$, and the galaxy core radii, $r_c$, has been suggested by Kellog & Murray (1974): $r_x \sim 2 \times r_c$, but, unfortunately, the galaxy core radii had been determined for only a few clusters at that time.

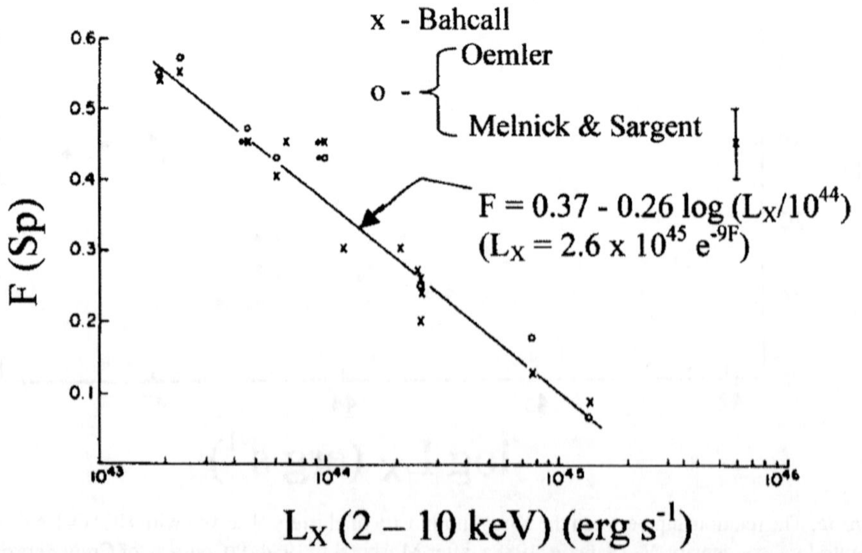


*Figure 36.* The correlation between the spiral fraction in clusters and their X-ray luminosity (Sarazin, 1988a, after Bahcall, 1977b). The line shows the relation from the equation 3.1 (courtesy of Craig Sarazin).

Another correlation between the cluster radius, $R_{\rm Cl}$, and X-ray luminosity has been found in the HEAO-1 A-1 survey: $L_{\rm x} \propto R_{\rm Cl}^2$ (Leir & van den Bergh (1977), and it was confirmed also for the EINSTEIN data (Abramopoulos & Ku, 1983).

Important results have been derived by the spectral surveys of spiral galaxies in clusters performed in X-ray range by the OSO-8, ARIEL 5 and HEAO-1 A-1. The derived spectral parameters and the optical properties of X-ray clusters have been firstly correlated by Mitchell, Ives & Cuhlane (1977), and later by Mushotzky *et al.* (1978), Smith, Mushotzky & Serlemitsos (1979), and Mushotzky (1980, 1984a). The most immediate properties which can be derived from the X-ray continuum emission are the gas temperature, $T_{\rm g}$, and the emission integral, EI, defined as:

$$EI \equiv \int n_{\rm p} n_{\rm e} \, dV \,, \tag{3.4}$$

where $n_{\rm p}$ is the proton density, $n_{\rm e}$ is the electron density and V is the volume of the gas in the cluster. The X-ray luminosity of a cluster is proportional to EI as follows:

$$L_{\rm x} \propto EI \times T_{\rm g}^{0.5} \,. \tag{3.5}$$

The correlation between the gas temperature derived from the X-ray spectra obtained with HEAO-1 A-2 and the cluster X-ray luminosities is shown in Figure 37 (Mushotzky (1984a).

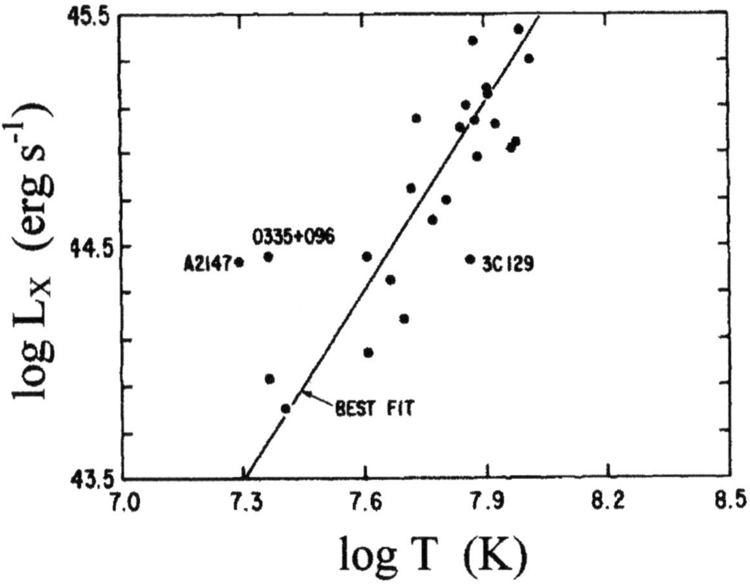


*Figure 37.* The correlation between the gas temperature derived from X-ray spectra with HEAO-1 A-2 and the cluster X-ray luminosities (Sarazin, 1988a, after Mushotzky, 1984a) (Courtesy of Craig Sarazin).

Another correlation is that between the gas temperature, determined from the X-ray spectra, and the velocity dispersion of the galaxies in the cluster. The oldest surveys showed that the temperature varied as:

$$T_{\mathrm{g}} \sim 6 \times 10^7 \mathrm{K}(\sigma_{\mathrm{r}}/10^3 \text{ km s}^{-1})^2 \tag{3.6}$$

if the central dominant galaxy of the clusters (cD clusters) were excluded. Such a correlation would be expected if the gas were gravitationally bound to the cluster, since the the velocity dispersion measures the depth of the cluster potential (e.g., Sarazin, 1988a). Moreover, this correlation supports the view that clusters must contain large masses of unseen material; therefore with the knowledge of these parameters one can derive information on the missing mass problem. The dispersions of the galaxies and gas velocities are similar and proportional to one another; this suggests that both gas and galaxies are bound by the same gravitational potential, which requires a very large mass for the cluster. This correlation is less significant for cD clusters; this means that in these systems a significant amount of gas is bounded within the cD galaxy itself, rather than to the whole cluster. The HEAO-1 A-2 data do not show a very tight correlation, which is shown in Figure 38, and give $T_{\mathrm{g}} \propto \sigma_{\mathrm{r}}$ (Mushotzky, 1984a).

Also the central density $\bar{N}_0$, as defined by Bahcall (1977a), is strongly correlated with the gas temperature: $T_g \propto \bar{N}_0$. It is shown in Figure 39 (Mushotzky, 1984a).

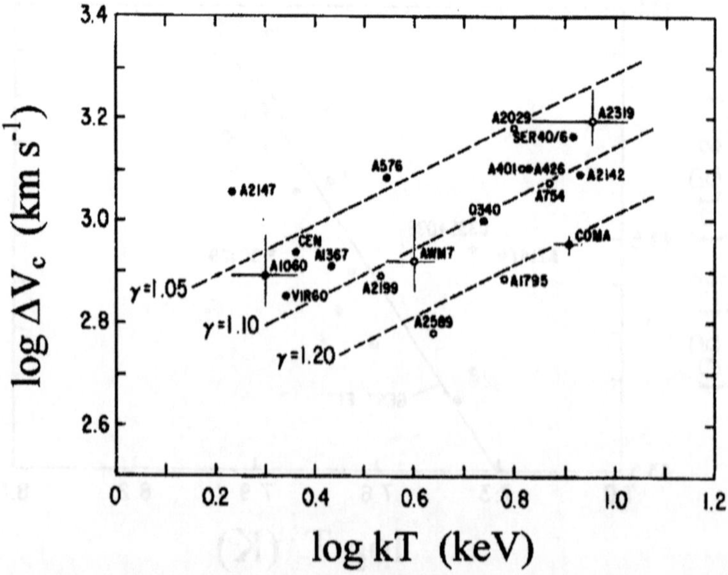


*Figure 38.* The correlation between the gas temperature derived from X-ray spectral observations with HEAO-1 A-2 and the central velocity dispersion ($\Delta V_C = \sigma_r$, at the cluster center. The lines shows the predictions of polytropic models with various indices. The dots are the data for non-cD clusters. The open circles are cD clusters (Sarazin, 1988a, after Mushotzky, 1984a) (Courtesy of Craig Sarazin).

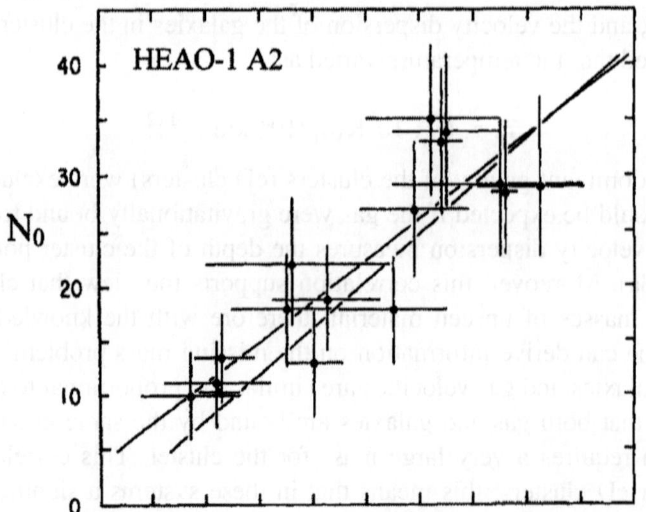


*Figure 39.* The correlation between the gas temperature derived from X-ray spectral measurements with HEAO-1 A-2 and the central galaxy density of Bahcall (1977a), from (Sarazin, 1988a, after Mushotzky, 1984a) (Courtesy of Craig Sarazin).

Felten *et al.* (1966), by using data from the Coma cluster, first suggested that the X-ray emission from clusters was due to diffuse intra-cluster gas at a temperature of $T_g \sim 10^8$ K and an atomic density of $n \sim 10^{-3}$ cm$^{-3}$. At such temperatures and densities, the primary emission process for a gas composed mainly of hydrogen is thermal bremsstrahlung (free–free) emission. For such an emission, the emission integral, X-ray luminosity and gas temperature are related by the formula (3.4).

Therefore the correlation of the X-ray luminosity and gas temperature with optical properties imply their correlation with the emission integral as well.

3.3.2.2. *Poor Clusters.* The X-ray source 2A 0251+413, detected with HEAO-1, was identified with the AWM7 group of galaxies, which is a poor cluster belonging to the list of Albert, White & Morgan (1977) (Schwartz *et al.*, 1980a). The source is extended with FWHM ~9′, assuming an isothermal sphere profile. The spectrum is best fitted by thermal bremsstrahlung of $4.0 \pm 0.5$ keV, with a significant Fe line emission of equivalent width of 0.63 keV centered at 6.6 keV. This temperature is consistent with the measured dispersion velocity $\sigma_r = 600$ km s$^{-1}$ of the central galaxies. This was the first detection in the X-ray range of a poor cluster of galaxies. This detection, together with the detection of other three poor clusters (MKW11, belonging to Morgan, Kayser & White (1975) list, 2A 0335+096, A2666) (Schwartz, Schwartz & Tucker, 1980b) allowed one to consider also the class of poor clusters as X-ray emitters. The derived X-ray temperatures, luminosities, and sizes are similar to those for X-ray emitting rich clusters of galaxies. The authors inferred that these clusters are in a late evolutionary stage (Ostriker, 1978) because of the presence of a dominant galaxy, the absence of spiral galaxies, and the short cooling time inferred for 2A 0335+096: less than 10% of the lifetime of the group.

Confirmation of extended X-ray emission from poor clusters came from the Imaging Proportional Counters (IPC) on the EINSTEIN observatory with the detection of emission from AWM4 and MKW3s, two poor clusters containing dominant galaxies (Kriss *et al.*, 1980, 1981). The 0.25–3 keV luminosities, $10^{44}$ erg s$^{-1}$ for MKW3s and $10^{43}$ erg s$^{-1}$ for AWM4, are typical of those observed for the richer Abell clusters. As in the rich clusters, the mass of X-ray emitting gas in these two clusters is comparable to the visual mass and is ~10–20% of the virial mass. Moreover, these results suggest that poor clusters can collect enough gas to become detectable X-ray sources if they are relatively compact, which the presence of dominant galaxies indicates.

The behaviour of poor clusters in the X-ray range – together with their optical properties which appear simple extensions of the properties of rich clusters to lower richness (e.g., Malumuth & Kriss, 1986), and their radio properties which are also very similar to those of rich clusters (White & Burns, 1980; Burns, White & Hanish, 1980; Burns, White & Haynes, 1981a; Burns, White & Hough, 1981b) – suggest that generally cD galaxies in poor clusters are very similar to cD galaxies in rich clusters and that the main bodies of cD clusters are formed by a process,

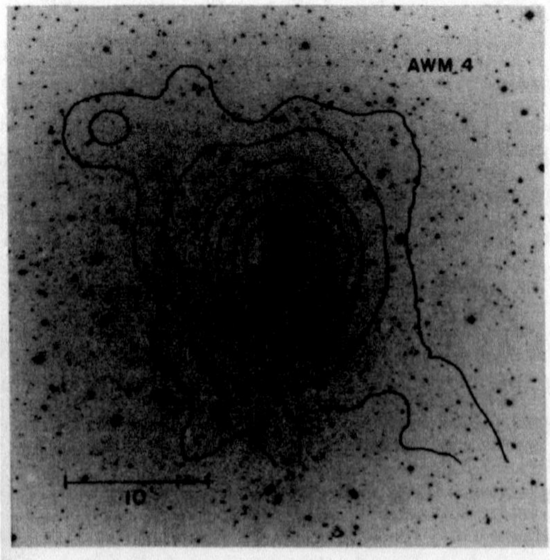


*Figure 40.* The X-ray emission from the AWM4 poor cluster of galaxies taken with the IPC Einstein satellite. Contours of constant X-ray surface brightness are superimposed on the optical image of the cluster. The emission is centered on the cD galaxy (Kriss, Cioffi & Canizares, 1983).

such as mergers, which is not strongly dependent on richness, while the halos are formed by a process, such as tidal interactions, which does depend strongly on richness.

Indeed, Kriss, Cioffi & Canizares (1983) performed X-ray observations of 16 poor clusters containing central dominant galaxies with the IPC of the EINSTEIN observatory. Twelve clusters were detected, and in each case the X-ray emission is centered on the dominant galaxy, as shown in Figure 40 for AWM4.

For the six brightest clusters they found extended X-ray emission that is smooth, centrally peaked, reasonably symmetric, and reaches in one case to nearly 1 Mpc and possibly farther. X-ray temperatures are ranging between 1 and 5 keV. The X-ray surface brightness implies density profiles that are roughly inversely proportional to radius. The low temperatures and high central density observed for the brightest clusters give central cooling times less than a Hubble time, implying that radiative accretion flows may be present. Assuming isothermal conditions and hydrostatic equilibrium, Kriss, Cioffi & Canizares (1983) found total binding masses of $\sim 10^{13}$ $M_\odot$ within 0.5 Mpc. The X-ray gas has ~10–20% of this mass. The mean mass to light ratio for the bright clusters is ~70–100 $M_\odot/L_\odot$. Comparison of the deduced distribution of binding mass with the light distribution of the central galaxies of four clusters shows that $M/L$ rises to over 200 $M_\odot/L_\odot$ in the galaxy halos. Thus these halos, like the clusters themselves, must possess dark matter. Although the dominant galaxies in poor clusters appear to lack the distended, luminous envelopes characteristic of rich cD clusters, they possess many optical and X-ray morphological similarities. In particular, the X-ray data show clearly that

*Figure 41.* The Antennae in the optical range. The HST image on the right hand side (detail) clearly shows colliding dust clouds and numerous nests of young blue stars (image NASA/ESA).

the dominant galaxies sit in the bottoms of the poor cluster gravitational potential wells. This suggests a similar origin for dominant galaxies in poor and rich clusters, possibly through mergers and cannibalism of cluster galaxies.

Among the spectacular discoveries of the IRAS infrared satellite in the mid 1980s was the hitherto unknown class of Ultra Luminous Infrared Galaxies Unlike normal galaxies ULRIRGs emit more than 90% of their energy in the far infrared range between 60 $\mu$m and 100 $\mu$m in the form of dust emission. At the same time, these huge amount of dust block any optical view of processes within. HST high resolution images in the near infrared revealed ULIRGs to be interacting galaxies, sometimes even in an advanced state of merging. Such extremely close interacting systems show an increased rate of star formation.

Figure 41 shows the image of the Antennae galaxies (NGC4038/NGC4039), which are clearly colliding. On the left is shown a ground based telescopic view of the Antennae galaxies, so named because a pair of long tails of luminous matter, formed by the gravitational tidal forces of their encounter, resembles an insect's antennae. The galaxies are located $\approx$30 Mpc away in the southern constellation Corvus. On the right are shown the respective cores of the twin galaxies as the orange blobs, left and right of image center, crisscrossed by filaments of dark dust (photo STScI-PRC97-34a). This natural color image is a composite of four separately filtered images taken with the Wide Field Planetary Camera 2 (WFPC2), on January 20, 1996. The resolution is 15 light years per pixel (picture element). A wide band of chaotic dust, called the overlap region, stretches between the cores of the two galaxies. The sweeping spiral-like patterns, traced by bright blue star clusters, show the result of a firestorm of star birth activity which was triggered by the collision. While stars can be expelled from the disk by elastic collisions during the merger, thereby forming the tidal arms, the interstellar matter experiences inelastic collisions and concentrates mainly in the joint central region. There it is compressed, and as a consequence millions of new stars are forming within in the region of most violent interaction in both galaxies. As many of these star formation regions are still deeply embedded within dust they are not visible in optical light but only at mid-IR wavelengths, where the dust becomes transparent. From the color and luminosity of the dust emission in the FIR and sub mm range, conclusions can be drawn about hidden star forming processes (Staude & Rix, 2001b).

#### 3.3.3. *Updated Status on Clusters of Galaxies*

With the advent of new generation satellites with high sensitivity, spectral resolution and angular resolution in the soft and hard X-ray energy ranges – such as ROSAT, BBXRT, TENMA, GINGA, ASCA, OSSE-CGRO, Beppo-SAX, AXAF-Chandra, and XMM-Newton – the study of clusters of galaxies have had a strong benefit since it is now possible to look in detail at their morphology, temperature, and mass distributions, chemical composition and then to obtain information on their evolutionary status. Moreover, it is also possible to study the interface between clusters and their member galaxies, and in particular:

- production and transport of heavy elements;
- hierarchical distribution of the dark matter, which defines a natural boundary for the galaxy/cluster hierarchy; and
- the role of the intergalactic magnetic fields.

3.3.3.1. *Mass and Temperature Determination.* The mass determination in clusters of galaxies is a fundamental task in understanding the nature of dark matter and the cosmological origin of structures in the Universe. Indeed, the flat orbital velocity distribution in spiral galaxies and the velocity dispersion of galaxies in a cluster provide strong evidence that there is by far more dark than luminous matter in the Universe. The most challenging questions facing astrophysicists and particle physicists are: *how much and what kind of dark matter?*. According to recent estimates some of the dark matter is in the form of ordinary particles, whilst most of it has probably a more exotic character. Excellent reviews on dark matter research are those by Pretzl (1995), Kolb (1996), and Roncadelli (2003).

The cosmological significance of dark matter can be seen through the density parameter $\Omega$, being $\Omega = \bar{\rho}/\rho_c$, with $\bar{\rho}$ the mean mass density and $\rho_c$ the critical density of the Universe [$\rho_c = 3H_0^2/8\pi G = 1.88 \times 10^{-29} h_0^2$ g cm$^{-3}$], with $h_0 = H_0/100$ a dimensionless parameter reflecting the uncertainty in the value of the Hubble constant ($0.4 < h_0 < 1$). Recent evaluations coming from comparative studies of different experiments give for the Hubble constant the value of $59 \pm 6$ km s$^{-1}$ Mpc$^{-1}$ (e.g., Panagia, 1999a) provide a value of $\rho_c \sim 7.3 \times 10^{-30}$ g cm$^{-3}$. As well known, the density parameter determines the geometry and fate of the Universe: for $\Omega > 1$ the Universe is closed, then after the expansion it

TABLE 10
Density parameters in the Universe (Pretzl, 1995).

| | |
|---|---|
| $\Omega_{vis} \sim 0.007^{+0.007}_{-0.003}$ | Visible matter |
| $\Omega_b \sim 0.022 \times h_0^{-2}$ | Baryonic matter (from nucleosynthesis) |
| $\Omega_g \sim 0.03–0.1$ | Spiral galaxies (from orbital velocities) |
| $\Omega_c \sim 0.1–0.3$ | Clusters of galaxies (from velocity dispersion) |
| $\Omega_i \sim 1$ | Inflationary model prediction |

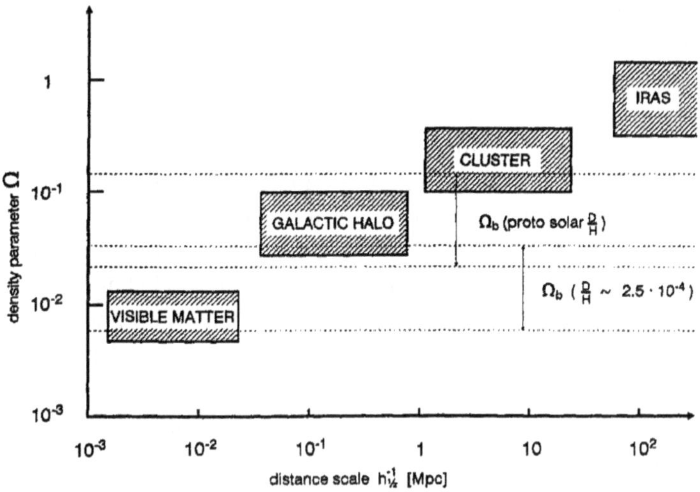


*Figure 42.* The density parameter as a function of the distance scale over which the measurements were made. The baryonic density parameter $\Omega_b(D/H \sim 2.5 \times 10^{-4})$ is also shown. This results from observations made by Songaila *et al.* (1994) on a primordial gas cloud (Pretzl, 1995).

collapses; for $\Omega = 1$ the Universe is flat, and for $\Omega < 1$ the Universe is open, then it expands forever. The density parameters of various forms of matter in the Universe are summarized in Table 10.

Figure 42 shows the density parameters as a function of the distance scale over which the measurements were made (Pretzl, 1995). This figure is the same as Figure 6 (Schramm, 1993), but the graphic presentation. We prefer to repeat for the convenience of the reader.

The amount of dark matter increases with the distance scale and exceeds by far the luminous mass.

To answer to the question 'what kind of dark matter' one has to quantify the baryonic density of the universe. Measurements of the cosmic abundances of light elements such as D, $^3$He, $^4$He and $^7$Li (Geiss, 1993) and the BBN model constrain the fraction of the critical density to $\Omega_b \sim 0.022h_0^{-2}$.

From the density parameters shown in Figure 42 one concludes that:

- there is more dark baryonic than visible matter;
- there is enough baryonic dark matter to explain the dark halo of galaxies;
- non-baryonic dark matter is needed to explain the large density parameters derived from galactic clusters and from IRAS observations.

As noted by White *et al.* (1993) in a fundamental article, baryonic matter constitutes a large fraction of the total mass of rich CGs than is predicted by a combination of primordial nucleosynthesis considerations and standard inflationary cosmology. This cannot be accounted for by gravitational and dissipative effects during cluster formation. They then conclude that either the density of the Universe is less than required for closure, or there is an error in the standard interpretation of element abundances.

Analyzing the absorption of a primordial gas cloud that is illuminated by the light of a distant quasar, Songaila *et al.* (1994) found a value for the deuterium to hydrogen ratio D/H $\sim 2.5 \times 10^{-4}$), which is a factor of 10 larger than the protosolar value from Geiss (1993). Up to now deuterium and $^3$He have only been observed within the chemically active disk of our galaxy. Since deuterium is a very fragile isotope which can easily be destroyed in stars, its primordial abundance can only be estimated by modelling its stellar processing. The deuterium abundance observed in the chemically unevolved environment of primordial cloud by Songaila *et al.* (1994) reflects more accurately the abundance of deuterium produced in the Big Bang. If this is true the baryon density of the universe would have a value consistent with the density of the visible matter (Figure 42) implying that all dark matter, including the one in galactic halos, would be of non-baryonic nature.

Therefore there appears once more very clearly the importance of the knowledge of the light element formation during the Big Bang as well as the mass determination in clusters of galaxies in order to constrain the density parameter. By the knowledge of the baryonic content from the Big Bang nucleosynthesis and by the mass fraction of hot gas within the clusters one can derive the value of Hubble's constant.

In the clusters the gas temperature provides direct information on their gravitational potential and masses. The abundances of heavy elements in clusters of galaxies can be determined from the strengths of the emission lines in the X-ray spectrum. This provides direct evidence on the origin of the ICM and constrains models for star formation and the evolution of galaxies of early type.

ASCA observations of several clusters of galaxies have contributed to determine their masses and temperature profiles. Kikuchi *et al.* (1996), by using a complicated new fitting method, from the energy dependent tail of X-ray spectra of clusters, and $H_0 = 50$ km s$^{-1}$ Mpc$^{-1}$, have derived the radial profiles of the gravitational

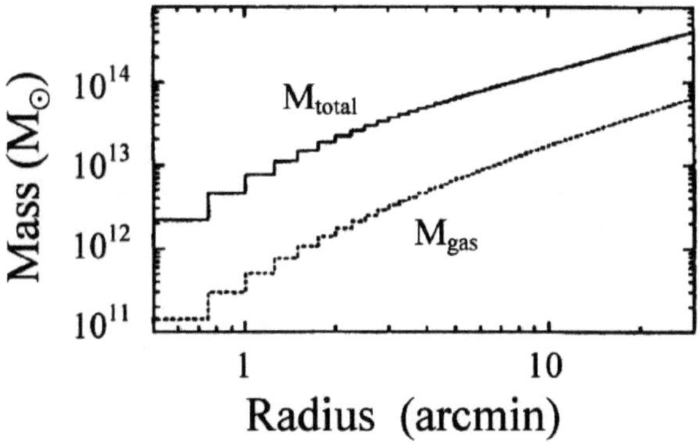


*Figure 43.* Radial profiles of the total gravitating mass and the gas mass based on the best fit parameters obtained from the isothermal fits (Kikuchi *et al.*, 1996).

mass and the baryon mass for each cluster. They found that the amount of baryons, $M_{\text{gas}} + M_{\text{stellar}}$, is about 20% of the total mass in various clusters. The total gravitational mass is calculated assuming the hydrostatic equilibrium:

$$M_{\text{tot}}(< r) = \frac{-kT}{\mu m_{\text{p}} G} \times \frac{d \ln \rho}{d \ln r} + \frac{d \ln T}{d \ln r} \tag{3.7}$$

where $\rho$ and T are the density and temperature of the intra-cluster gas. Figure 43 shows the mass profile for the poor cluster 3A 0336+098.

This result is rather general for the whole sample of the observed clusters. The ratio of $M_{\text{gas}}$ to $M_{\text{tot}}$ becomes larger in the outer regions of the clusters. At $r \sim 2$ Mpc, which is the typical size of the clusters in which significant X-ray emission is observed, $M_{\text{gas}}$ is ~20% of $M_{\text{tot}}$ (strictly valid for 3A 0336+098). Adding the stellar mass [$M_{\text{stellar}} \sim (0.3\text{–}0.2) M_{\text{gas}}$] to $M_{\text{gas}}$ the baryon fraction becomes ~25% of the total mass. Since from ASCA measurements a good isothermality of the hot cluster medium has been confirmed for some clusters, the mass derived from these measurements would be more reliable than the previous estimations. The primordial nucleosynthesis scenario predicts $\Omega_{\text{b}}/\Omega \sim 0.06$. If the present Universe has $\Omega = 1$, ASCA measurements reveal that the baryon fraction is much higher than this in various classes of clusters.

Also the ROSAT satellite has been fruitfully used in studying clusters; indeed it provided by far the most detailed view on the X-ray morphology of clusters, their evolution and mass determination. The first complete X-ray view of the closest cluster Virgo has been obtained by Böhringer *et al.* (1994). The total mass up to a radius of 1.8 Mpc amounts to $(1.5\text{–}6)\times 10^{14}\ M_\odot$. Moreover the gas mass to total mass ratio can be inferred. It was found to be 5–30%. Since this provides a lower limit for the baryon to the total mass ratio in the Universe, a discrepancy with primordial nucleosynthesis appears once more.

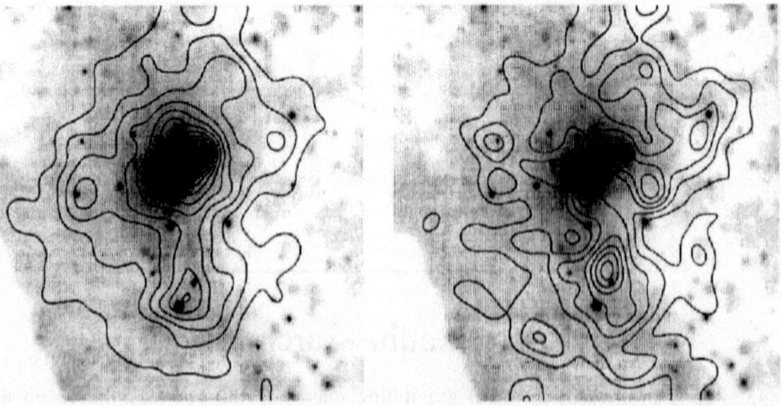

*Figure 44.* X-ray emission of the intra-cluster gas (greyscale) for the Virgo cluster in the ROSAT All Sky Survey in the hard band (0.4-2.4 keV) (Böhringer *et al.*, 1994). Superimposed on X-ray image are contours of galaxy number density (a) of the 1300 member galaxies and (b) of only the galaxies of late type of the Virgo Cluster Catalog (Binggeli, Sandage & Tammann, 1985). The images have a size of 12.8° × 12.8° (from Schindler, 1996)

3.3.3.2. *X-ray Morphology.* The ROSAT satellite has been crucial in studying the physics of CGs because of its high sensitivity and its high spatial resolution with simultaneous energy resolution. As already said, ROSAT results have given the first complete X-ray view of the Virgo cluster. It reveals a complex structure similar to the optical appearance implying that both components – the galaxies and the gas – are in equilibrium in the same potential (Böhringer *et al.*, 1994). Figure 44 shows that the hot luminous gas is extending over most of the optically visible cluster. The X-ray structure of the Virgo cluster is complex with several subclusters around M87, M86, M49 and M60; this indicates that the cluster is still evolving (Schindler, 1996). The contours of constant galaxy number density, taken from the Virgo Cluster Catalog containing 1300 member galaxies (Binggeli, Sandage & Tammann, 1985), are superimposed on the X-ray image (left panel of Figure 44). It is possible to note the good correspondence between the two components: the galaxy distribution shows a similar structure with several subclumps at about the same position. This similarity has a very important meaning because it indicates that both components trace the same potential and are almost in equilibrium in it. Only the number density of the galaxies of late type are reported in Figure 35. They show a completely different distribution. The galaxy density maxima do not correspond with the X-ray maxima. The galaxies of late type are much less concentrated than the early type ones, in agreement with the well known density morphology relationship of galaxies (Dressler, 1980).

Important results from X-ray images of ROSAT are those connected with the verification of the clusters merger hypothesis. For instance, the X-ray temperature map of the cluster A 2256 (Briel & Henry, 1994) confirms the merger hypothesis

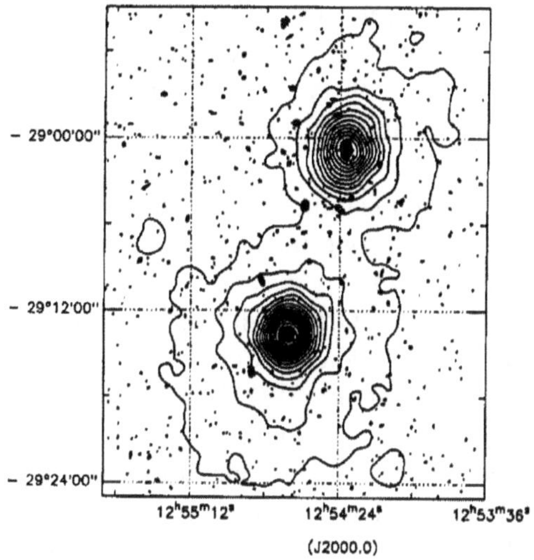


*Figure 45.* ROSAT PSPC image of the cluster A3528 in the hard band (0.5-2 keV). The value of the lowest contour and the contour spacing is $1.36 \times 10^{-3}$ cts $s^{-1}$ $arcmin^{-2}$. The double structure of the X-ray image is clearly visible. Plots of galaxies from the COSMOS/UKSTU Southern Sky Object Catalog (Yentis *et al.*, 1992) are superposed (from Schindler, 1996).

deduced from the ROSAT morphology (Briel *et al.*, 1991) and radio observations (Röttgering *et al.*, 1994); indeed, the temperature map shows regions of different temperature indicating that the cluster is not in equilibrium. Another example of a possible merger of clusters is that of the bimodal cluster A 3528. Indeed, the PSPC-ROSAT image of this cluster in the energy range 0.5–2 keV, shown in Figure 45 (Schindler, 1996) – over which the galaxies from the COSMOS/UKSTU Southern sky objecy catalog (Yentis *et al.*, 1992) are superimposed – clearly indicates its bimodal structure. Although the counts are not enough to deduce a temperature map, Schindler (1996) has found a trend that the temperature in the two inner semicircles is higher than in the outer ones. This temperature trend is an indication that the two subclusters will collide within the next $\sim 10^8$ yr and that the alignment of the two subclusters is not a chance alignment.

Important results are coming from the combined analysis of X-ray data obtained on the cluster A 3921 with GINGA and ROSAT. This is a luminous cluster [$L_x$(2–10 keV) $= 5.1 \times 10^{44}$ erg $s^{-1}$], which appears to be in an early stage of merging with a subcluster (Arnaud *et al.*, 1996).

3.3.3.3. *Chemical Composition.* In order to determine the abundances of heavy elements in the ICM, the X-ray spectra are fundamental. Indeed the knowledge of metal abundances provides basic information about the origin and evolution of the ICM and on the history of star formation and chemical evolution of the CGs. Until the launch of the ASCA satellite information on abundances in CGs referred only to iron; now data referring to other elements are available. Since the unique known source of heavy elements is the thermonuclear fusion in stars, and the unique significant population of stars are in galaxies, one can assume that heavy elements have been ejected from stars into the ISM of galaxies and from galaxies into the ICM.

For thin hot ICM with a temperature higher than several keV the Fe–K emission line is the strongest line among many other elements characteristic lines in collisional ionization equilibrium.

Metal concentration in the central region of several poor clusters has been observed by Fukazawa *et al.* (1994a) in Centaurus, by Matsumoto *et al.* (1996) in Virgo, by Ohashi *et al.* (1994) in AWM7, and by Fukazawa *et al.* (1996) in A262 and MKW9. Typically the iron abundance ranges from 0.3–0.4 solar in the outer region to 0.7–1 solar in the center. None of the hot and rich clusters with the temperature of the ICM larger than 5 keV show significant metal concentration. Indeed, most of the 30 clusters studied in detail with ASCA have an abundance distribution roughly constant over the space. Based on these results, relatively gas poor clusters hosting a cD galaxy have high chance of showing the metal concentration. The compact group of galaxies HGC 62, dominated by two bright early type galaxies, has shown metal concentration (Sakima, Tawara & Yamashita, 1994). By carefully excluding the data in the central (high abundances) regions, Ohashi *et al.* (1996) have derived the mean abundance of iron and silicon for each cluster. These abundances are plotted as function of the ICM temperature in Figure 46. Results from GINGA data reported a negative correlation of the iron abundance with the temperature of the ICM; this correlation was probably owed to the high abundance in the central regions of poor clusters, which have been excluded in the construction of the diagram of Ohashi *et al.* (1996), where this correlation is not present. From the Figure 46 it is possible to note a concentration of iron abundance for ICM temperature KT between 2 and 6 keV. The measured iron abundances all lie between 0.3 and 0.4 solar. Abundances of silicon are larger than those of iron for clusters with $KT > 3$ keV, as reported by Mushotzky *et al.* (1995). In contrast, the abundance of silicon becomes close to that of iron for $KT < 2keV$. This suggests that the supernova types, which generated metals in the ICM, differ between rich and poor clusters: more SNIa in poor clusters.

The constancy of iron abundance is a very important result; indeed, if one may assume that the amount of metal in the ICM is simply proportional to the total mass of galaxies in the cluster, the constant abundance of iron would imply that the mass ratio between galaxies and the primordial intergalactic gas is about the same between rich and poor clusters. This means, in other words, that the efficiency of galaxy formation seems to be constant for different richness of clusters. This result, if definitively confirmed, should be in contradiction with the correlation of iron abundance with the gas temperature. Indeed, before ASCA Forman & Jones (1990) reported that the ratio of the gas to stellar mass, $M_{\mathrm{gas}}/M_{\mathrm{stellar}}$, increases from

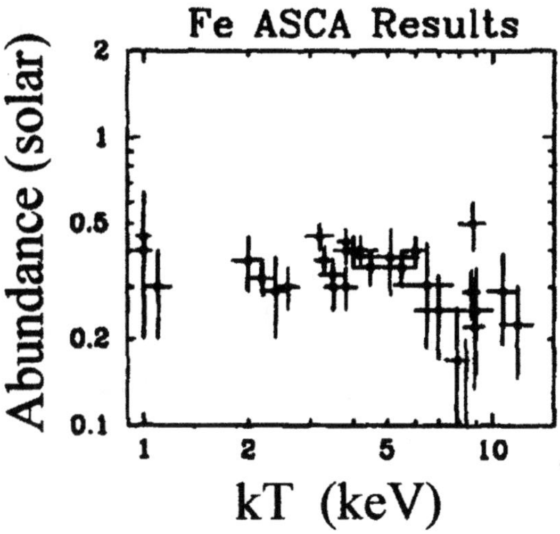


*Figure 46.* Temperature and Iron abundance relation obtained by ASCA observations of clusters of galaxies excluding the central region (Fukazawa *et al.*, 1996).

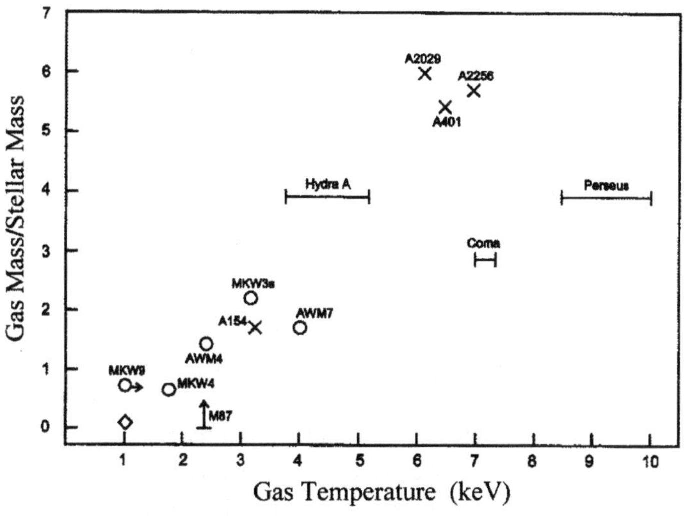


*Figure 47.* Ratio of the gas to stellar mass in clusters of galaxies versus the gas temperature, before ASCA (Forman & Jones, 1990).

unity in groups and poor clusters with low temperature ($T_{\rm gas} \sim 2$ keV) to values of 3–6 in systems with high gas temperatures ($T_{\rm gas} = 6$–$10$ keV).

This correlation, reported in Figure 47, combined with an understanding of the production of heavy elements, predicts a correlation of heavy element abundance with gas temperature. Figure 48 shows quantitative predictions for the correlation of iron abundance with gas temperature. The two solid curves are derived paramet-

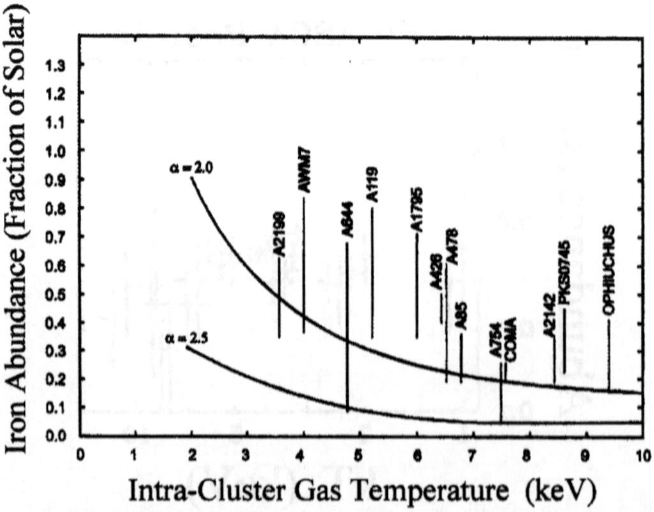


*Figure 48.* Correlation of iron abundance with gas temperature.The two curves assume different initial mass functions (power law exponents $\alpha = 2$ and $\alpha = 2.5$, respectively)(David, Forman & Jones, 1990).

rizing the dependence of $M_{\rm gas}/M_{\rm stellar}$ on $T_{\rm gas}$ and assuming that enriched material is expelled from galaxies only during an early wind phase, during which Type II supernovae can readily drive a galactic wind (David, Forman & Jones, 1990). The two curves assume different initial mass functions (power law exponents $\alpha = 2$ and $\alpha = 2.5$, respectively). The amount of the enriched material equal to that expelled in the wind is produced by stellar evolution and could be liberated by ram pressure stripping. The estimates of supernova yields can explain the observed heavy element abundances in the intra-cluster gas, as shown in Figure 47.

The ejected gas is extremely enriched and is diluted to the observed values by mixing with the predominantly primordial component of the ICM (Forman & Jones, 1990; Forman, Jones & Tucker, 1994). Tsuru *et al.* (1996) have observed distant CGs with the ASCA and have found a significant iron K line from 11 out of the observed 17 clusters. Correlation between abundance and temperature is shown in Figure 48, with the correlation for nearby clusters obtained also with the ASCA. The negative correlation for the distant clusters is clearly seen in the figure. No difference between the correlations for the near clusters and that for the distant clusters can be noted. This result is in agreement with that reported by Forman & Jones (1990), but it has been obtained once more without removing the central high abundant regions. Dividing the whole sample into two sub-samples of $0.14 < z < 0.2$ (average redshift 0.173) and $0.2 < z < 0.55$ (average redshift 0.317), Tsuru *et al.* (1996) have not found differences between the two. Therefore they concluded that no significant evolution is seen in the correlation between abundance and temperature.

3.3.4. *Interface Between Clusters and Galaxy Members*

The interface between clusters and their member galaxies is a useful place for the study of different phenomena occurring because of: i) the behaviour of the galaxy members; ii) the interactions among galaxies; and iii) the influence of the galaxy members on the ICM. In particular, it is possible to study: a) the production and transport of heavy elements; b) the hierarchical distribution of the dark matter, which defines a natural boundary for the galaxy/cluster hierarchy; and c) the role of the intergalactic magnetic fields.

This is just what ASCA is able to do. Thanks to its measurements with the GIS instrument (Kohmura *et al.*, 1993) with a large field of view and a low internal background, the general panorama on GCs is now more clear; indeed, now detailed information on the metallicity of elliptical galaxies, the abundances of elements, the potential and temperature structure in GCs are available. All this new information forces one to invoke the presence of an inter-galactic magnetic field (IGMF) of the order of $\mu$G (Makishima, 1995).

Indeed, the presence of such magnetic fields in CGs has been found by using three different approaches, namely (e.g., Eilek, 1999):

- Radio Halos and Relics. Some clusters show different synchrotron emission which is associated with the ICM rather than with discrete radio galaxies within the clusters. This emission can exist throughout the cluster (haloes), or be peripheral (relics). In both cases this is direct evidence of intra-cluster magnetic fields. Arguments based on minimum energy or minimum pressure, combined with the detection of hard X-rays owed to inverse Compton effect or limits, suggest a value of magnetic field intensity of $B \leq 1$ $\mu$G on large scales (comparable to the size of the cluster core or the cluster itself);
- Faraday Rotation of Background Sources. Comparison of Faraday signals from cluster and non-cluster sources allows a statistical test of cluster magnetic fields (e.g., Clarke, Kronberg & Böhringer, 1999). Recent studies show an excess rotation measure from sources in or behind rich clusters (Kim, Tribble & Kronberg, 1991; Clarke, Kronberg & Böhringer, 1999). Modelling these data in terms of a typical ICM density and scale, and cell size of $\sim$10 kpc, one obtains again a magnetic field of about a micro-Gauss as typical of cluster on moderate to large scales;
- Faraday Rotation of Embedded Sources. Such a method uses high resolution images of radio sources within the cluster. Radio imaging gives evidence on the structure of the magnetic field (e.g., Taylor, Allen & Fabian, 1999). Most of the studies have been performed on cluster-center sources, but recently also on non-central sources (e.g., Feretti *et al.*, 1999). By using this method the results obtained by Eilek, Owen & Zhou (1999) can be summarized as follows: the magnetic field is dynamically significant in the central regions 3C75 and 3C465 of the two clusters A400 and A2634, respectively, as well as in other clusters whose data were collected from the literature. The typical central field strength is $\sim$10–30 $\mu$G. Comparison to pressures derived from

X-ray data shows that $B^2/8\pi \sim nk_BT$ is typical for the cluster center sample. The images can be also used to estimate coherence lengths for the rotation measure. The general trend is that the cluster-core rotation measure has a characteristic length scale of ~10 kpc.

Going back to the impact of ASCA on the knowledge of the physics of CGs, we will try to resume the important role of this satellite in the studies of several GCs such as Virgo, Fornax, Centaurus and A1060, following the paper by Makishima (1996).

3.3.4.1. *Origin of heavy elements.* The iron abundance in the ICM is ~0.3 times solar; this indicates that a significant amount of metal enriched gas was ejected from member galaxies into the intra-cluster space. The hot ISM (HISM) around each elliptical galaxy is then expected to be metal rich, as remnant of these ejection processes. But this is not, as it has been surprisingly shown by Awaki *et al.* (1991), Serlemitsos *et al.* (1993) and Forman, Jones & Tucker (1994); indeed, the HISM nearby elliptical galaxies is metal poor, definitively below solar value (Awaki *et al.*, 1994; Loewenstein & Mushotzky, 1996; Matsushita *et al.*, 1996). Therefore theoretical calculations of stellar mass loss and supernovae must be revisited. One possible explanation of that result could be that the metal enriched gas may have been carried away very efficiently from the vicinity of individual galaxies to the intra-cluster space via ram pressure stripping and thus leaving the HISM relatively metal poor.

About the ICM abundance, ASCA has also given precious results in terms of a spatially rather uniform distribution (Ohashi, 1994, 1996; Mushotzky, 1994). In the cluster A1060 the abundance is uniform at ~0.3 and the temperature roughly constant at ~3 keV from the center to ~400 pc out (Tamura, Fuzakawa & Makishima, 1996). Also the Fornax cluster shows similar properties (Makishima, 1996). These results, together with those on the metallicity near the elliptical galaxies, suggest that the metal rich gas coming from the galaxies was efficiently mixed into the ICM.

The abundance ratios amongst abundant elements have been measured with ASCA by Mushotzky (1994; 1996) and Mushotzky *et al.* (1996). In several clusters the Fe is 1.5–3 times under-abundant compared with O, Ne, and Si (Tamura, Fuzakawa & Makishima, 1996) with respect to solar ratios. This means that a significant fraction of heavy elements present in the ICM was produced by type II supernovae during the early cluster evolution and ejected into the ICM via a strong galaxy wind. After then, most of the metal production has continued in elliptical galaxies via type Ia supernovae and efficiently carried out via ram pressure stripping from the vicinity of individual galaxies.

How these heavy elements are distributed within the ICM is dependent on the cluster; indeed, meanwhile the cluster A1060 shows a uniform distribution of abundances, strong metal concentrations have been detected in the central regions of Centaurus cluster (Fukazawa *et al.*, 1994b; Ohashi, 1994; Ikebe *et al.*,

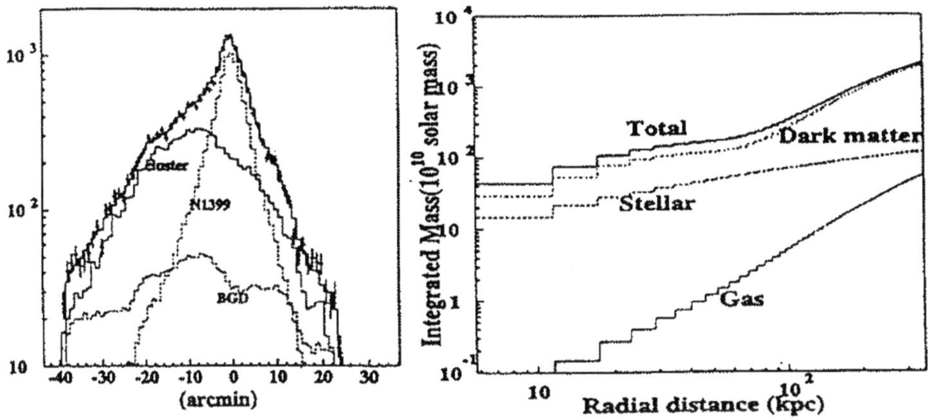


*Figure 49.* (left): The one-dimensional 0.7–3 keV brightness distribution of the Fornax cluster and its cD galaxy NGC 1399, obtained with the Gas Imaging Spectrometer (GIS)-ASCA and fitted with two beta models (Ikebe, 1995; Ikebe *et al.*, 1996b); (right): The derived integrated mass distribution in the Fornax/NGC 1399 system (Fukazawa *et al.*, 1996).

1996a) and Virgo cluster (Koyama, Takano & Tawara, 1991; Matsumoto *et al.*, 1996; Akimoto *et al.*, 1996). Also several poor clusters, such as AWM7, MKW4s and MKW9, show a similar behaviour, although less convincing (Ohashi, 1994; Fukazawa *et al.*, 1996). But cD galaxies are involved in all these cases, and the increase of metallicity is strictly correlated to their vicinity.

The metal concentration is always accompanied by a central cool emission component ($KT \sim 1$ keV) detected in the same region where the abundance is enhanced (Ohashi, 1994, 1996; Fukazawa *et al.*, 1994b, 1996). Therefore in order to study this metal rich cool component it is necessary to have the knowledge of the potential distribution, which is possible to obtain knowing the brightness and temperature profiles.

3.3.4.2. *Potential and temperature structures.* As noted by Makishima (1995, 1996), the most important discovery of ASCA (Tanaka, Inoue & Holt, 1994) on the cluster potential is the detection of a hierarchical X-ray brightness in the Fornax cluster and its cD galaxy NGC 1399. The brightness profile (Figure 49, left) consists of two components, one confined to the galaxy scale and the other extends to cluster extent. The emission is practically isothermal at 1.1 keV for both components, so it has been possible to calculate the integrated mass profiles without any *a priori* assumption on the potential shape. The total mass curve is shown in Figure 49 (right). It is possible to note a shoulder-like structure at ~70 kpc from the center (with $H_0 = 50$ km s$^{-1}$ Mpc$^{-1}$) where the galaxy/cluster interface is inferred to occur. In this interface an excess potential owed to the cD galaxy NGC 1399 attracts an excess amount of ICM. This is the first discovery of galaxy/cluster hierarchy in the distribution of total matter and then of dark matter.

Canizares & Buote (1997) report the studies of the ellipticities of X-ray images, taken with ROSAT and ASCA, of five Abell clusters and two galaxies of early-type in order to discuss the shape of the gravitational potential and the underlying matter distribution. The summary is:

- for the clusters of galaxies:
  - the mass is not distributed like an isothermal sphere;
  - the dark matter distribution is steep ($\sim r^{-4}$) at large radii (for isothermal gas);
  - the dark matter is flattened with ellipticity $\epsilon_{DM} \sim 0.5$;
  - the ellipticity and position angle of the dark matter are consistent with those of the galaxy distribution.
- for the galaxies of early type :
  - the bulk of the mass does not follow the light;
  - dark matter halo is extended and flattened;
  - spectral studies constrain the contribution from discrete sources to the observed X-ray ellipticity;
  - the observed position angle twist may indicate triaxiality.

The cluster findings complement and are consistent with what is being learned from gravitational lensing studies of arcs and arclets (e.g., Allen *et al.*, 1996; Pierre *et al.*, 1996) and with expectations from N-body simulations (e.g., Efstathiou *et al.*, 1988; Navarro, Frenk & White, 1995a,b). Similarly the galaxy results add support to to our knowledge on flattening from polar ring galaxies and agree with expectations of dissipationless collapse (e.g., Dubinski & Carlberg, 1991; Rix *et al.*, 1994; Sackett & Pogge, 1995).

Canizares & Buote (1997) believe that MOND (MOdification of Newtonian Dynamics) cannot obviate the geometrical evidence found for the presence of large quantities of dark matter in clusters of galaxies. MOND (e.g., Milgrom, 1986) attributes the well known flat rotation curves to a breakdown of Newtonian dynamics at very small accelerations. This idea was introduced for an alternative explanation of dark matter problem.

3.3.4.3. *Highlights for Clusters of galaxies.* The knowledge of magnetic field intensity in clusters of galaxies (CGs) is fundamental for understanding the properties of the intra-cluster plasma. CGs with extended radio halo (1 Mpc scale, probably associated with the compenetration of clusters–subclusters) should have a non-thermal emission of hard X-rays, owed to Compton diffusion of relativistic electrons in the CMB.

The coordinated detection of radio and hard X-ray radiation directly provides some of the basic properties of the intra-cluster magnetic field and cosmic ray electrons. These determinations are based on observable quantities, contrary to the only radio measurements, by which it is possible to determine the magnetic field and electron density model dependent.

Before BeppoSAX, only upper limits in the hard X-ray emission from CGs were known. Thanks to its sensitivity, BeppoSAX has measured such a hard emission, removing the previous uncertainties (Fusco Femiano *et al.*, 1999).

The mass determination in CGs is a fundamental task in understanding the nature of the dark matter and cosmological origin of structures in the Universe. X-ray spectra are fundamental in determining the abundance of heavy elements in the ICM. The knowledge of metal abundance is crucial for the knowledge of the origin and evolution of ICM, the history of star formation, and the chemical evolution of CGs.

ASCA measurements (Ohashi *et al.*, 1996) show constancy in the Fe abundance in CGs. If the amount of metals in the ICM is proportional to the total mass in CGs, the constant abundance of Fe would imply that the mass ratio between galaxies and primordial intergalactic gas is the same between rich and poor clusters. The efficiency of the galaxy formation should then be constant for different richness of clusters. *This contradicts the popular correlation of Fe abundance with the gas temperature.*

In a workshop devoted to *Diffuse Thermal and Relativistic Plasma in Galaxy Clusters* (Böhringer, Feretti & Schuecker, 1999) the most important present topics on magnetic fields and on particle diffusion and acceleration have been discussed together with the morphology and physics underlined in the light of multifrequency observations of CGs coming from space and ground based experiments.

In another workshop devoted to *Large Scale Structure in the X-Ray Universe* (Plionis & Georgantopoulos, 2000) the key role of X-ray surveys in cosmology was strongly emphasized.

**Which open problems in CGs still survive?**

In spite of many important results coming from satellites in the last decade, the problems of the production and transport of heavy elements, the hierarchical distribution of the dark matter, and the role of the intergalactic magnetic fields in CGs are still open. Multifrequency simultaneous measurements, with higher sensitivity instruments, in particular those in hard X-ray and radio energy regions could solve such problems.

In this sense the recent results coming from the XMM–Newton satellite are extremely important. For instance, the temperature structure in the central region of the Coma galaxy cluster has been determined (Arnaud *et al.*, 2001). They present a temperature map and a temperature profile of the central part ($r < 20'$ or 1/4 virial radius). The temperature distribution around the two central galaxies is remarkably homogeneous ($r < 10'$) contrary to previous ASCA results, suggesting that the core is actually in a relaxed state. At a larger distance from the cluster center they do not see evidence for recent matter accretion. The temperature radial profile is very weakly decreasing with radius and shows a slight drop beyond $10'$, probably owing to the cold S/W structure. This profile could be very useful in determining the distribution of dark matter in the core.

Neumann *et al.* (2001) show, through the analysis of a complex temperature structure (derived by XMM-Newton EPIC-MOS data), that the NGC 4839 group is falling onto the Coma cluster. Furthermore they measure a displacement between NGC 4389, the dominant galaxy of the group and the extended X-ray emission coming from its ICM. In a simple approximation they explain this displacement by a ram pressure force of Coma's ICM acting on the hot gas of the sub group as it falls onto the Coma center.

Tamura *et al.* (2001) from XMM-Newton/EPIC (European Photon Imaging Cameras) observations of the rich CGs Abell 1795 have found a temperature drop at a radius of $\sim$200 kpc from the cluster center, indicating that the ICM is cooling. Both the EPIC and the RGS (Reflection Grating Spectrometers) spectra extracted from the cluster center can be described by an isothermal model with a temperature of $\sim$4 keV. The volume emission measure of any cool component ($<$1 keV) is less that a few per cent of the hot component at the cluster center. A strong OVIII Lyman $\alpha$ line was detected with the RGS from the cluster core. The O abundance and its ratio to Fe at the cluster center is 0.2–0.5 and 0.5–1.5 times the solar values, respectively. The O abundance is similar to the value found in the center of the Virgo cluster (Matsumoto *et al.*, 1996) based on ASCA spectra, and in the centers of A1835 (Peterson *et al.*, 2001) and Sérsic 159-03 (Kaastra *et al.*, 2001) based on XMM/RGS spectra. The O/Fe ratio at the Abell 1795 center suggests significant contributions from both type Ia and type II supernovae to metal enrichment at the cluster center, since O and Fe are mainly produced in such kinds of supernovae. In the core of Abell 1795 the Fe abundance is a factor of 2 larger that in the outer parts. This is consistent with the general idea that the Fe abundance is higher around the cD galaxy (Fukazawa, 1998; Fukazawa *et al.*, 2000).

Results of UV observations of the central region of Abell 1795 with the optical monitor aboard XMM-Newton satellite are extremely important (Mittaz *et al.*, 2001). They have found evidence of star formation in the central region of such a cluster. However, they have found also evidence for absorption in the cD galaxy on a more extended scale than has been seen with optical imaging. They also report the first UV observation of part of the filamentary structure seen in $H_\alpha$, X-rays and very deep U band imaging. The part of the filament they see is very blue with UV colours; this is consistent with a very early O-B stellar population. This is the first direct evidence of a dominant population of stars of early type at the center of Abell 1795 and implies very recent star formation at the cluster center.

Other important results are coming from the X-ray imaging spectroscopy of Abell 1835 cluster performed with EPIC and RGS on the XMM-Newton observatory. Abell 1835 is a luminous ($L_x \sim 10^{46}$ erg s$^{-1}$), medium redshift ($z = 0.2523$) X-ray emitting cluster. Peterson *et al.* (2001) have found that large amounts of cool gas are present in a multi-phase medium surrounded by a hot ($KT_e = 8.2$ keV) outer envelope. The RGS spectrum shows OVIII, Lyman $\alpha$, and two FeXXIV complexes. The emission measure of the cool gas below $KT_e = 2.7$ keV is much lower than expected from standard cooling flow models, suggesting either a more complicated cooling process than simple isobaric radiative cooling or differential cold absorption of the cooler gas. The problem of cooling flows has been discussed by Fabian (1999) and Allen (1999) and in the references therein.

Re-analysing ASCA and ROSAT data of the Centaurus cluster, Ikebe *et al.* (1999) have found for this cluster the two-phase picture developed by Fukazawa *et al.* (1994b), in which the hot ICM fills nearly the entire cluster volume, wherein a small amount of colder plasma is intermixed and localized near the cluster center. They give a consistent qualitative picture of the two-phase nature observed by invoking the merger process, magnetic reconnection, and a certain magnetic configuration.

## 4. Physics of Cosmic Rays from Ground and Space Based Experiments

### 4.1. Introduction

The subject of High Energy Astrophysics is generally approached through the study of cosmic rays. The reason for this is historical in nature. Since the discovery of this extraterrestrial radiation by Victor Hess (1912) the scientific research involved in trying to discover its nature has been enormous and as a result many separate research fields have been developed. Before particle accelerators came into operation the high energy cosmic rays were the laboratory tools for investigations of elementary particle production, and to date they are still the only source of particles with energies greater than $10^{12}$ eV. The research into the composition of the radiation led to developing the study of the astrophysical environment using the information in the charge, mass, and energy spectra; this field is also known as Particle Astrophysics.

Of great importance was the discovery of high energy photons near the top of the Earth's atmosphere which originated the development of new astronomical fields such as the X-ray or $\gamma$-ray astronomy. But many of these high energy photons have their origin in interactions of the high energy charged particles with cosmic matter, light, or magnetic fields. The particle astrophysics and the astronomical research field have found in this fact a bond to join their efforts in trying to understand the high energy processes which occur in astrophysical systems.

The modern picture of cosmic rays is that of a steady rain of particles moving at speeds close to that of light. The particles are primarily nuclei with atomic weights less than 56, as well as a few nuclei of heavier elements, some electrons and positrons, a few $\gamma$-rays and neutrinos.

The energy spectrum extends over 14 orders of magnitude ($\sim 10^6 - 10^{20}$ eV) and the particle flux rapidly decreases with increasing energy, as shown in Figure 50 (Wynn-Williams, 1992).

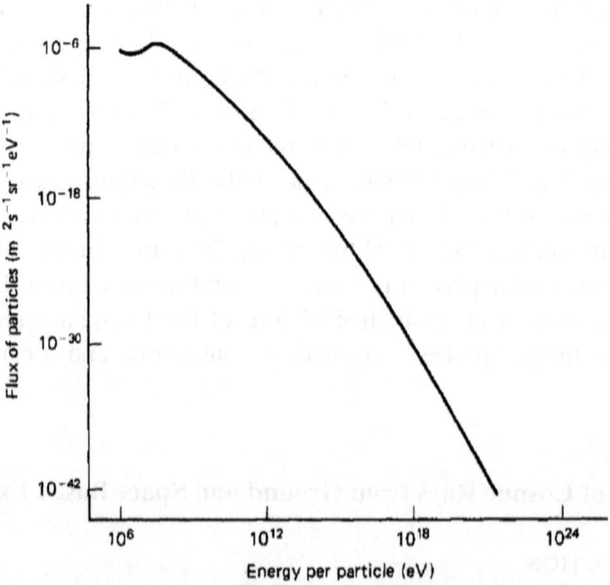


*Figure 50.* The number of cosmic rays versus their energy. At low energy, cosmic rays, which are mainly protons, are detected by space experiments. At energies above roughly $10^{12}$ eV information comes from arrays of ground based detectors (Wynn-Williams, 1992).

Although cosmic rays were discovered a century ago it is puzzling that after so many years the problem of their 'origin' is still an open problem, in spite of the great progress made in any adjacent field of Physics and Astrophysics. As said by Giulio Auriemma (1993), *one may think to the existence of a cosmic conspiracy to hide the source of this phenomenon.* Obviously the main problem is that the bulk of cosmic rays are charged. Therefore we detect at Earth a nearly isotropic flux which does not point back to any particular direction in the Galaxy. Another problem is that the spectrum of cosmic rays is apparently simple and nearly featureless. In fact, if we look at Figure 50, in which a simplified cosmic ray spectrum is reported, we can hardly distinguish their spectrum from a single power law distribution. Such an extended simple power law distribution spanning over about twelve orders of magnitude of energy is one of the more astonishing example of the fractal symmetry in Nature. There are arguments supporting the diffusive shock acceleration (DSA) model introduced by Fermi (1949, 1954), which can give a very natural explanation of the simplicity of the cosmic ray spectrum (e.g., Auriemma, 1993 and the references therein).

The bulk of the cosmic ray particles in the vicinity of Earth have energies around $10^8 - 10^9$ eV, but many particles with much greater energies are also seen. Above $10^9$ eV the number of particles drops steadily with increasing energy all the way to $10^{20}$ eV. We know very little about cosmic rays with energies greater than $10^{19}$ eV because few of them arrive daily on the Earth. There is, nevertheless, great interest in such particles because they have energies vastly greater than can be generated by particle accelerators in Earth's laboratories. They are witness of acceleration processes occurring nearby the most active cosmic sources in the universe. Such energetic cosmic rays have so much momentum that they are practically undeflected as they pass through the Solar System toward the Earth. In contrast, cosmic rays with energies less that $10^8$ eV are strongly influenced by the solar magnetic field and by the ever changing solar wind of ions that stream out from it. Then few particles with such energies reach the Earth.

The chemical composition of the cosmic rays with energies up to $10^{12}$ eV has been well determined using data from space experiments. In general the abundances of the different nuclei follow closely the measured values in the solar system. The over-abundance of the light elements lighter than iron, experimentally found, is the result of spallation collisions during the propagation through the interstellar medium.

For the highest energies the data are derived from studies of Extensive Air Shower (EAS) and owed to the disagreements in interpreting the experimental data by the different experiments it cannot be concluded whether the composition is similar to that at lower energies or whether there is a deficit of heavy nuclei relative to protons (e.g., Müller, 1993). A good general review on the composition of cosmic rays was written by Erlykin (1995).

In their propagation through space the charged cosmic rays are influenced by the magnetic fields in the Galaxy, and for the lowest energy particles also in the solar system. The result is that the distribution of arrival directions as the radiation enters the Earth's atmosphere is nearly isotropic. It is not possible to identify the sources of the cosmic rays by detecting them. However, in the high energy interactions produced at the source, electrically neutral particles such as photons, neutrons, and neutrinos are also produced and their trajectories are not deviated, being directed from their point of origin to the observer. Owing to their short lifetime neutrons cannot survive the path length to the Earth (decay length $\sim$0.009 kpc at 1 PeV) and neutrinos do not interact efficiently in the atmosphere.

It is in this context that the Gamma Ray Astronomy has demonstrated itself to be a powerful tool. The observations made to date have detected $\gamma$-rays from many astronomical objects such as neutron stars, interstellar clouds, the center of our Galaxy and the nuclei of active galaxies (AGNs). One might expect very important implications for high energy astrophysics from the observations at energies greater than $10^{11}\,eV$ of extragalactic sources (e.g., Hillas & Johnson, 1990). The fluxes of $\gamma$-rays at these energies are attenuated because of their interactions with the cosmic radio, microwave, infrared and optical radiation fields. Measurements of the flux attenuation can then provide important information on the distribution of such fields. For instance, the threshold energy for pair production in reactions of photons with the 2.7 K background radiation is reached at $10^{14}$ eV and the absorption length is of the order of $\sim$7 kpc. For the infrared background the maximum absorption is reached at energies greater than $10^{12}$ eV.

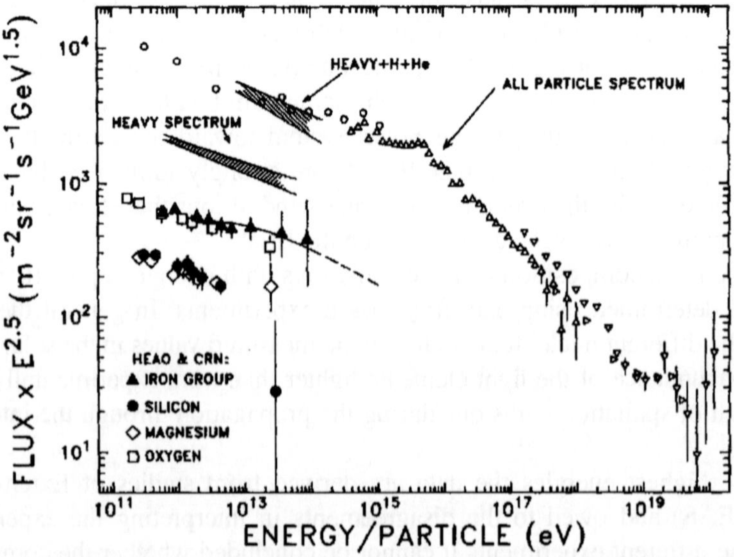


*Figure 51.* The all particle spectrum and spectra of several individual elements.The sum of differential energy spectra of heavy primary nuclei yields the heavy spectrum shown. Adding the data on H and He from JACEE to the heavy spectrum leads to the upper shaded area for the total flux in the $10^{13}$–$10^{14}$ eV region. The solid curve refers to the leaky box fit from iron; dashed is the extrapolation to $10^{15}$ eV (Müller, 1993, after Swordy *et al.*, 1993).

## 4.2. Understanding the Message of Cosmic Rays

In spite of the apparent simplicity of the total cosmic ray spectrum, some deviations from a simple power law spectrum can be distinguished.

For energies greater than $10^9$ eV solar modulation effects can be neglected and the spectrum is well described up to $10^{15}$ eV by a power law with an exponent in the range 2.5–2.7. At this energy the power law breaks, and continues with a steeper slope ($\sim$3) up to $10^{19}$ eV where an apparent flattening occurs. Figure 51 shows the energy spectrum of cosmic rays, roughly starting at $10^{11}$ eV (Müller, 1993 and the references therein). In order to emphasize the variations in the spectrum, poorly seen in Figure 50, the flux reported in figure 51 is multiplied by $E^{2.5}$. This operation, obviously, compresses the spectrum, reducing the decades in the *y*-axis, rendering the features of the whole spectrum much more readable.

A simple inspection of the Figure 51 points out the evidence of a break in the spectrum around $10^{15}$–$10^{16}$ eV. This break is also called the *knee*. The *knee* has been found at the end of the 1950s initially as the steepening in the EAS size spectrum (Kulikov & Khristiansen, 1958). Since then more than 44 years have passed but its origin is still a challenge for the cosmic ray physics. The steepening appears when the intensity of the showers falls down to a value of $\sim$(4–5) $\times 10^{-11}$ cm$^{-2}$ s$^{-1}$ sr$^{-1}$. This intensity delivers one shower above the *knee* per an area of

$\sim$25 $m^2$ in one hour. Owing to this low intensity the *knee* has not yet been studied by direct measurements in space, but by indirect methods which use atmospheric cascades.

Summarizing, the cosmic ray spectrum for protons is at GeV energies close to $E^{-2.75}$, and for He and higher elements close to $E^{-2.65}$ below the *knee* at $\approx 5 \times 10^{15}$ eV, where the spectrum turns down to $\sim E^{-3.1}$, to flatten out again near $10^{18}$ eV, called the *ankle* (e.g., Lawrence, Reid & Watson, 1991; Nagano *et al.*, 1992; Zatsepin, 1995).

The origin of the *knee* can be clarified by the detailed study of the primary spectrum, mass composition and arrival directions of cosmic rays. Reviews of Erlykin (1993, 1995, 1997), Müller (1993), and Hillas (1999) discussed in details such topics. In particular, Erlykin (1995) discussed in a very clear way the mass composition and the origin of cosmic rays. Indeed, the origin of them, independent of all the discussions and models, is focussed around four main possible reasons, namely:

- sources of cosmic rays;
- acceleration mechanisms;
- propagation through the ISM;
- interaction characteristics.

In most models based on one of these four items the mass composition varies with the energy in different ways. Then he discussed the mass composition below $10^5$ GeV, near the *knee* ($10^5$–$10^8$ GeV), and above $10^8$ GeV. He concluded that most of the results indicated that the genesis of cosmic rays is much more complicated than was thought in the past. The most likely tendency for cosmic rays is to grow heavier with the rising energies up to the *knee* region and then become lighter beyond the *knee*. Recently, several reviews have been published about the chemical composition around the *knee* (Alessandro, 2001), about ultra high energy (UHE) cosmic rays (Blasi, 2001b), and about the propagation and clustering of cosmic rays (Stanev, 2001).

A possible interpretation of the *knee* is that it represents the energy at which cosmic rays can escape more freely from the Galaxy or it may indicate a transition between two different acceleration mechanisms. In the first case one might expect an anisotropy effect in the distribution of arrival directions above this energy if the cosmic rays were originated within the Galaxy.

Some years ago Wdowczyk & Wolfendale (1984) examined the Extensive Air Shower (EAS) data available then to search for the presence of an asymmetry in arrival directions favoring the Galactic Plane. They defined the intensity as a function of Galactic latitude by the simple expression:

$$I(b) = I_0[(1 - f_E) + f_E \times a \times e^{-b^2}], \tag{4.1}$$

where $a$ is a normalization factor and $b$ is in radians and determined $f_E$ as a function of assigned energy. Later, Chi, Wdowczyk & Wolfendale (1993) reported the

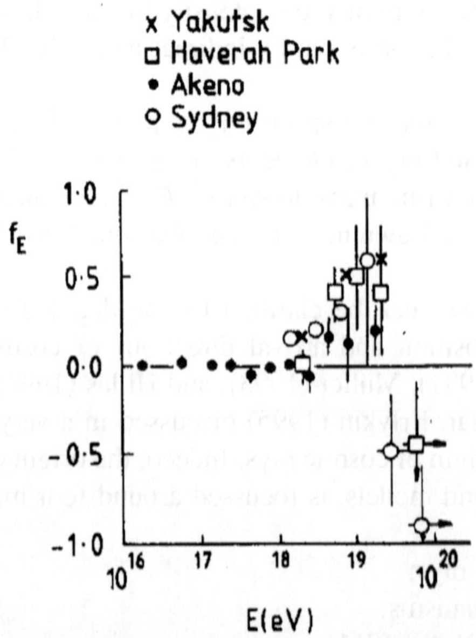


*Figure 52.* Average Galactic Plane Enhancement factor, $f_E$, versus energy for different arrays (Chi, Wdowczyk & Wolfendale, 1993).

compendium of new measurements as shown in Figure 52. It will be noted that the evidence favoring an increase of $f_E$ with energy has strengthened somewhat – at least until above $3 \times 10^{19}$ eV. They regard this result as still comprising the best indication for a Galactic origin of at least some of the particles above $10^{19}$ eV.

Detecting clustering in the arrival directions is an obvious way of searching for 'sources' but it is fraught with statistical dangers. It can be said immediately that if there were strong clustering we would have known about it for many years – and none has been reported. Nevertheless, Chi *et al.* (1992) have examined the world's data in an attempt to at least gain a hint about the presence of sources. The prevalence of clusters across the Galactic Plane suggests to them that some at least are genuine. However, their presence was not universally accepted. But, later with UHE-EAS measurements coming from Fly's Eye, at $3 \times 10^{20}$ eV (Bird *et al.*, 1994) and Akeno, at $2 \times 10^{20}$ eV (Hayashida *et al.*, 1994) some help in recognizing Galactic sources arrived. The situation is shown in Figure 53, in which the clustering of arrival directions of cosmic rays above $3 \times 10^{19}$ eV is shown (Chi *et al.*, 1992) and the Fly's Eye and Akeno measurements are indicated with the letters F and A, respectively (Wolfendale, 1995).

It is obvious that it is possible to invoke dramatic phenomena in distant galaxies on an *ad hoc* basis but there are some limits that can be placed on these phenomena, namely:

– the energetic of the processes – considerable energy outputs are involved;

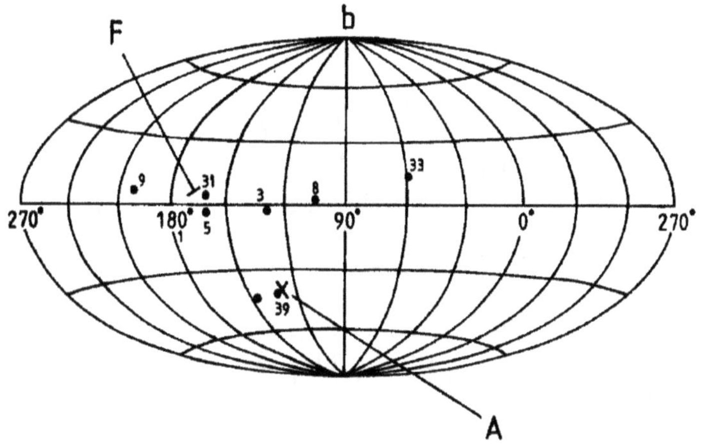


*Figure 53.* Map showing 'best estimate' source directions for the world data above $10^{19}$ eV prior to 1992 (Chi *et al.*, 1992). F and A denote Fly's Eye and Akeno UHE-EAS measurements, respectively (Wolfendale, 1995).

– the energy losses by interaction with the cosmic microwave background. These losses become severe above $\sim 5 \times 10^9$ eV over cosmological dimensions;
– there is the common difficulty of suppressing the attendant emission of $\gamma$-rays, which then cascade through the universe (Wdowczyk, Tkaczyk & Wolfendale, 1972) and can easily cause problems in the range $10^8 - 10^9$ eV where the flux of the so called diffuse extragalactic $\gamma$-rays is well known.

The list of such extragalactic sources, following the opinion of Wolfendale (1995), contains:

– galaxy–galaxy collision (spiral galaxies) where the magnetic fields undergo 'reconnection';
– galaxy–intergalactiic medium bow-shocks (in similar fashion to the Earth's bow shock in the interplanetary medium;
– active galactic nuclei;
– cosmic strings. Such strings intersecting will, in principle, provide cosmic rays of energy up to $10^{24}$ eV (e.g., Chi *et al.*, 1992), but successive some works indicate that the likely fluxes are too low by many orders of magnitude – at least for the more favored string models.

So, the hunt for the origin of UHE cosmic rays goes on.

After the clustering analysis of the world's data on the arrival directions of cosmic rays particles of energies above $10^{19}$ eV by Chi *et al.* (1992), Wolfendale & Zan (1997), adding new data, identified three 'sources' at latitudes beyond $| b |= 20°$: one with a nearby pair of colliding galaxies and the other two sources with objects (also colliding galaxies?) in the Virgo and Ursa Major clusters.

Since the propagation of particles in extragalactic space is subject to interactions with the background radiation, the knowledge of the mean free path for collisions

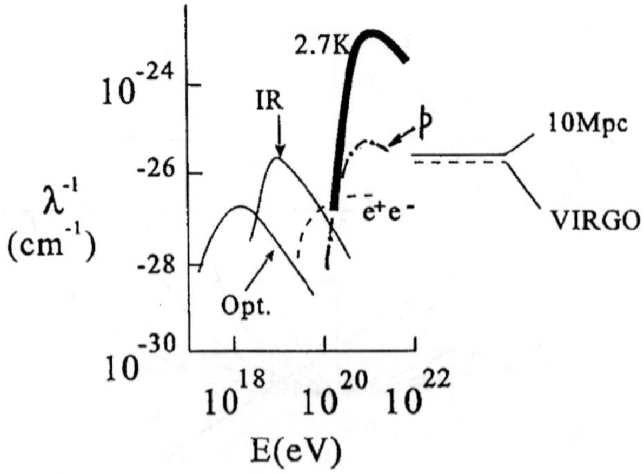


*Figure 54.* Mean free path of protons (Strong *et al.*, 1974) and iron nuclei (Tkaczyk, Wdowczyk & Wolfendale, 1975) passing through extragalactic space. 'p' denotes protons on the 2.7 K radiation field. The rest of the lines are for iron nuclei on the radiation fields indicated. (Wolfendale & Zan, 1997).

of such particles with the radiation fields is important. Figure 54 shows the mean free path for protons and iron nuclei with the optical, IR, and cosmic microwave 2.7 K radiation fields, following the works of Strong, Wdowczyk & Wolfendale (1974) for protons, and Tkaczyk, Wdowczyk & Wolfendale (1975) for iron nuclei.

Wolfendale & Zan (1997) made some comments about the interactions of iron nuclei. The radiation fields adopted were for the intergalactic medium (IGM) in general. However, here we are interested in the situation for the region between the Galaxy and Virgo cluster and other regions of the supercluster of which we are a part. Estimates of the fields have not yet been made, but they will certainly be higher than those adopted in drawing Figure 54 and the attenuation of iron nuclei below $10^{20}$ eV will be correspondingly greater.

Inspection of Figure 54 shows that protons with energies below $\sim 3 \times 10^{20}$ eV, moving rectilinearly, will arrive from Virgo without much loss, but those from further afield will be increasingly attenuated. Wdowczyk & Wolfendale (1975) studied the proton IGM radiation interactions. They used a higher IR radiation field than commonly adopted, specifically the 'high' spectrum of Puget & Stecker (1975). This spectrum had an energy density of $5 \times 10^{-2}$ eV cm$^{-3}$ (to be compared with the more commonly adopted IR energy density of $\sim 1 \times 10^{-2}$ eV cm$^{-3}$). The result was that the attenuation length at $10^{20}$ eV was a factor of 5 shorter than hitherto. In turn, the energy at which the attenuation length is equal to the distance to Virgo cluster, $E_{\mathrm{Vc}}$, is reduced from $\sim 3 \times 10^{20}$ eV to $2 \times 10^{20}$ eV.

Measurements performed with COBE (Puget *et al.*, 1996) have tended to confirm that the general intergalactic IR energy density is higher than commonly adopted. Specifically, they derived IR intensities, in the range 100–400 $\mu$m, only a factor of 2 below the 'high' Puget & Stecker (1975) values. Taken with the inevitably high energy densities towards the center of the Virgo cluster, it appears that a value of $E_{\mathrm{Vc}} \sim 2 \times 10^{20}$ eV is probably appropriate for protons derived some 15 pc away in the Northern hemisphere.

Turning to iron nuclei, these will be severely attenuated for $E > 10^{19}$ eV if coming from Virgo and travelling in straight lines. In point of fact the magnetic deflections would be so great – for an origin of Virgo – that the potential path lengths would be much longer than 15 Mpc and this would make the fragmentation probability very high (Wolfendale & Zan, 1997).

A search for Galactic sources of UHE cosmic rays has been made analyzing catalogs of astronomical objects, which might generate such energetic cosmic rays, specifically pulsars and SNRs. No significant correlation was found (Stanev *et al.*, 1995).

A summary on the status of the search for the origin of the highest energy cosmic rays has been published by Biermann (1999). He mentioned several competing proposals, such as the supersymmetric particles, Gamma Ray Bursts also giving rise to energetic protons, interacting high energy neutrinos and cosmological defects, and then he discussed on the possibility of the propagation of these particles, assuming that they are charged. The distribution of arrival directions of the highest energy particles on the sky ought to reflect the source distribution as well as the propagation history. He remarked that the present status can be summarized as inconclusive. However, he concluded as follows: *If we can identify the origin of the events at the highest energies, beyond* $5 \times 10^{19}$ eV, *the Greisen–Zatsepin–Kuzmin cutoff owed to the microwave background, near to* $10^{21}$ eV $= 1$ EeV, *and if we can establish the nature of their propagation through the universe to us, then we will obtain a tool to do physics at* EeV *energies.*

A good friend of ours, Szczepan Karakuła, during his last 1996 Vulcano Workshop presented a talk on the possibility that EHE photons could be responsible for the highest energy cosmic ray showers. The paper was completed after the sudden death of Szczepan by Włodzimierz Bednarek (his best pupil, and co-worker) and Franco Giovannelli (his co-worker, and friend) and published in the proceedings of that workshop (Karakuła, 1997). The analysis of the path lengths of the EHE cosmic rays in the cosmic background radiation shows that photons are the most penetrating 'particles'. For this reason they are the best candidates for being responsible for the EHE events detected by the EAS arrays. He discussed the interaction of such supposed EHE photons with Earth's magnetosphere and atmosphere. He computed, by the Monte Carlo method, the electromagnetic cascades initiated by EHE photon in the magnetosphere and the cascade initiated by the products of this cascade (secondary photons, $e^{\pm}$ pairs) in the Earth's atmosphere, taking into account the Landau–Pomeranchuk–Migal (LPM) effect. The result was that the effects of cascading in Earth's magnetosphere reduce the influence of the LPM effect on the development of a cascade. As result the final cascade curve initiated by

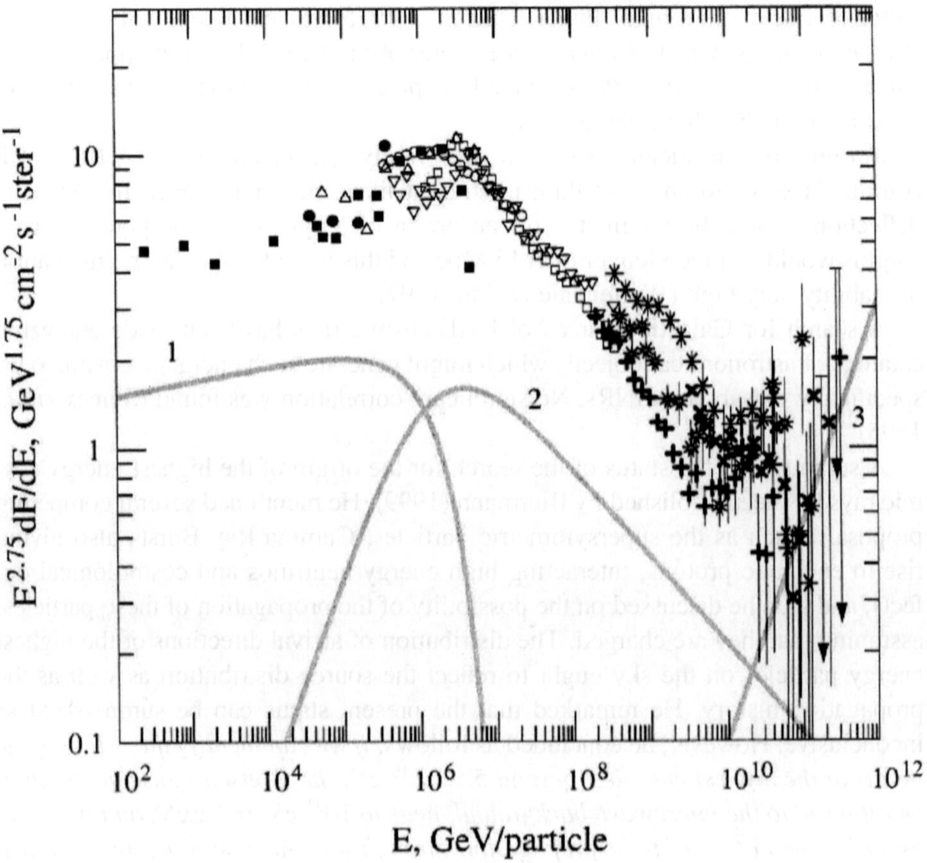


*Figure 55.* The contributions of the spectra of SNRs (curve 1) and unknown sources (curve 2) to the cosmic ray spectrum (courtesy of Todor Stanev).

the EHE photon closely resembles the cascade curve initiated by the EHE proton of comparable energy.

Therefore our opinion is that one must be extremely careful in attacking the problem of the highest energy cosmic rays.

### 4.3. Possible Sources of Cosmic Rays

The current view of the origin of cosmic rays is discussed in many papers (e.g., Biermann, 1995; Erlykin, Wolfendale & Vahia, 1999; Erlykin & Wolfendale, 1999 and the references therein). Figure 55 shows the possible contributors to the production of cosmic rays. Most of them are still unknown.

Current theories argue that most of the lower energy cosmic rays are owed to the explosions of stars into the normal interstellar medium (e.g., Lagage & Cesarsky, 1983). Some of the higher energy cosmic rays are owed to explosions of massive stars into their stellar winds (Völk & Biermann, 1988). For the energy range beyond a few $10^{15}$ eV there is still a general dispute about the origin of the particles (e.g., Biermann, 1994).

In general, the main problem in understanding the origin of cosmic rays is strictly related them being mostly charged particles. The consequence is then that the flux detected at Earth is nearly isotropic because of the galactic magnetic field, which cancels any particular directionality in the Galaxy.

The cosmic particle radiation arriving near Earth covers an enormous range of energies, at least from $10^6$ to $10^{20}$ eV, and apparently includes the nuclei of all known elements, as well as electrons, positrons, and antiprotons. These particles are carriers of rich astrophysical information: their message refers to the sites and mechanisms of particle acceleration, to the physics of compact objects inside the Galaxy and beyond, to the nucleosynthesis history of the material that become cosmic rays, to the processes governing interactions of energetic particles with the interstellar medium, and perhaps to questions of cosmological interest, such as discussed by Müller (1993).

A relative simple explanation of the simplicity of the spectrum of cosmic rays comes from the Diffusive Shock Acceleration (DSA) model (Fermi, 1949; 1954), which gives a power law spectrum with a universal spectral index $\gamma \cong 2$. The observed cosmic ray spectral index at Earth is consistent with $\gamma_{\rm obs} \cong \gamma + 0.6$.

After the aforementioned pioneering works of Fermi many papers have been written on the production and acceleration of cosmic rays, and recently also the sites of their origin have been discussed (e.g., Biermann, 1995 and references therein). He proposes that three sites of origin account for the cosmic rays observed, namely: i) supernova explosions into the interstellar medium, ISM-SN; ii) supernova explosions into the stellar wind of the predecessor star, wind–SN; and iii) radio galaxy hot spots. The predicted spectra – for nuclei and corrected for leakage from the Galaxy – are proportional to $E^{-2.75}$ for ISM–SN (Biermann & Strom, 1993), $E^{-2.67}$ for wind–SN (Biermann & Cassinelli, 1993) below the knee, $E^{-3.07}$ for wind–SN above the knee (Biermann, 1993), and $E^{-2.0}$ at injection for radiogalaxy hot spots (Rachen & Biermann, 1993). The polar cap of the wind–SN contributes an $E^{-2.33}$ component, which, however, contributes significantly only near and below the knee, if at all.

Therefore the main observational properties of cosmic rays can be resumed as follows: i) power law spectrum $N(E) \propto E^{-\gamma}$ from several MeV up to $\sim 10^{20}$ eV, with a slight knee at $\sim 10^{15}$–$10^{16}$ eV; ii) the total energy necessary for the observed cosmic ray flux is $\sim 10^{40}$ erg s$^{-1}$ (e.g., Lipunov, 1995).

In the low energy region, below $\sim 10^{10}$ eV, $\gamma$-ray astronomy comes to the rescue by giving information about the average cosmic ray intensity across various lines of sight. The method is quite simple: the measured $\gamma$-ray intensity is divided by the column density of gas, the argument being that the bulk of the $\gamma$-rays are produced by cosmic rays interacting with the gas in the ISM. Although there are difficulties, associated with discrete sources, the actual gas components (mainly H, $H_2$, $H^+$, and He) and the division between electrons and protons (e.g., Wolfendale, 1995) it is generally agreed that the particles responsible for the $\gamma$-rays detected by satellite-borne detectors are produced in the Galaxy. The energies of the $\gamma$-rays range from $\sim 3 \times 10^7$ eV to $\sim 10^9$ eV and the initiating particles are principally electrons to $\sim 10^9$ eV and protons and heavier nuclei to $\sim 10^{10}$ eV (Wolfendale, 1995). The Galactic origin identification comes from an inferred gradient of intensity, viz., a reduction in intensity with increasing Galactocentric distance (e.g., Strong *et al.*, 1978; Bhat, Mayer & Wolfendale, 1984; Bloemen, 1985).

The nature of the actual sources of cosmic rays is less clear, although SNRs are favored by many, and recently also young open clusters (YOCs) have been suggested as possible cosmic ray emitters (Giovannelli, 1996c; Manchanda *et al.*, 1996) on the basis of measurements in the $\gamma$-ray energy region.

Giovannelli (1997) made an evaluation of the $\gamma$-ray emissivity of SNRs, radio pulsars, isolated black hole, neutron stars and black holes in binary systems, and young open clusters. The contributions of these classes of objects equals the $\gamma$ ray emissivity (for $E \geq 100$ MeV) measured by SAS II satellite ($N_\gamma \sim 1.3 \times 10^{42}$ photons $s^{-1}$) (Strong, Wolfendale & Worral, 1976).

Since $\gamma$-rays are coming from the decay of $\pi^0$s, which are produced by p–p inelastic collisions, the conclusions are that in such classes of galactic objects *in situ* acceleration mechanisms work and then they can be sources of low energy cosmic rays.

At energies to $10^{14-15}$ eV galactic SNR shocks are the most plausible site for nucleonic cosmic ray acceleration (e.g., Harding, 1991). There are reasons for optimism that future VHE $\gamma$-ray observations will produce direct evidence for their expected contributions. Above $10^{14-15}$ eV other sources must be sought, such as winds and SNRs, highly hypersonic winds in young open clusters containing Wolf–Rayet stars, and pulsars. At highest energies, $10^{18-19}$ $eV$ and above, several arguments lead to the conclusion that the dominant sources of such particles are extragalactic (e.g., Wolfendale, 1995).

#### 4.3.1. *EGRET AGNs*

One of the major success of the CGRO was the discovery of high energy $\gamma$-rays from AGNs, whose number was 38 in the first catalog (Fichtel *et al.*, 1994). At longer wavelengths these sources are all bright, flat spectrum radio sources, with, in many cases, optical polarization and rapid optical variability, characteristics of the blazar class of AGNs (Dermer & Schlickeiser, 1992; Antonucci, 1993). In the unified picture of the AGN phenomena blazars are those AGNs which have associated jets with the jet axis oriented near the observer's line of sight. The $\gamma$-ray luminosity of the AGNs measured by EGRET ranges from $10^{44}$ to $10^{49}$ erg $s^{-1}$, assuming the isotropicity of the sources. On the contrary, if $\gamma$-rays are associated with the relativistic jets of the blazars and are thereby beamed, the true $\gamma$-ray luminosity will be several orders of magnitude less (e.g., Bednarek *et al.*, 1990a).

The spectra of the EGRET detected AGNs have a power law shape with slope ranging from 1.5 to 2.6 (von Montigny *et al.*, 1995). There is little or no evidence for a break in the spectra at the highest energies achieved by EGRET, whose actual limit is 10 GeV even for the brightest AGNs. So one of the fundamental observational question is connected with the cut-off in the spectra of AGNs. At the present time it seems that AGNs have intrinsic spectral cut-off below the present threshold of VHE $\gamma$-ray detectors ($\sim$250 GeV). This presents a strong incentive to reduce the threshold of VHE detectors so that this régime can be explored. However, the detection of Markarian 421 (Punch *et al.*, 1992) and its continuing monitoring by the Whipple Observatory (Kerrick, 1994; Schubnell *et al.*, 1996) has made it clear that not all of the $\gamma$-ray blazars are cut off below the threshold of present day VHE $\gamma$-ray detectors (e.g., Zdziarski & Krolik, 1993; Macomb *et al.*, 1995). Therefore, as energy thresholds are lowered and sensitivities of VHE $\gamma$-ray instruments improved, other EGRET detected AGNs will also be detected at those very high energies. Even at the present VHE sensitivity levels there is the possibility that some AGNs could now be detectable in the VHE range, but because of their hard spectra they result in being undetectable at EGRET energies, except by improving the sensitivity in that range, such as, for instance, with the future AGILE or GLAST (e.g., Morselli, 2002).

Although many works on relativistic jets have been developed in the past few years, many questions remain. For example, where in the jet do the $\gamma$-rays originate? In the inverse Compton model proposed by Blandford & Levinson (1995) there is an approximate energy–distance mapping such that the more energetic photons come from greater distances from the central engine. The minimum variability time scale of high energy $\gamma$-rays therefore limits the size of the $\gamma$-ray emitting region. Therefore if time variations are seen in the 100 GeV–multi TeV régime, and if this model is correct, one should expect even more rapid variations at EGRET energies. So once more multifrequency observations of flares in blazars currently hold the greatest promise for exploring the region where the relativistic plasma is generated and the jet focussed and accelerated. Particularly relevant is the correlation between the mm/submm radiation and the GeV/TeV emission. By measuring time delays of variations in brightness as a function of frequency it may be possible to constrain the geometry of the inner jet as well as infer the steepness of gradients in magnetic field, relativistic electron density, and bulk Lorentz factor of the jet flow (e.g., Bednarek *et al.*, 1990a; Marscher, 1993).

The outburst of Mark 421 (Kerrick *et al.*, 1995; Macomb *et al.*, 1995) in which its flux above 250 GeV increased by a factor of $\sim$10 in a matter of days raises the possibility of future correlated broadband, multifrequency observations for these objects and for the other potential targets of the next generation of experiments.

### 4.3.2. *Intergalactic Absorption*

The dominant absorption mechanism for VHE photons as they travel through the intergalactic medium is pair production against low energy photons associated with the intergalactic starlight field. This process, first studied by Gould & Schreder (1966) and recently by Stecker, de Jager & Salamon (1992), produces an attenu-

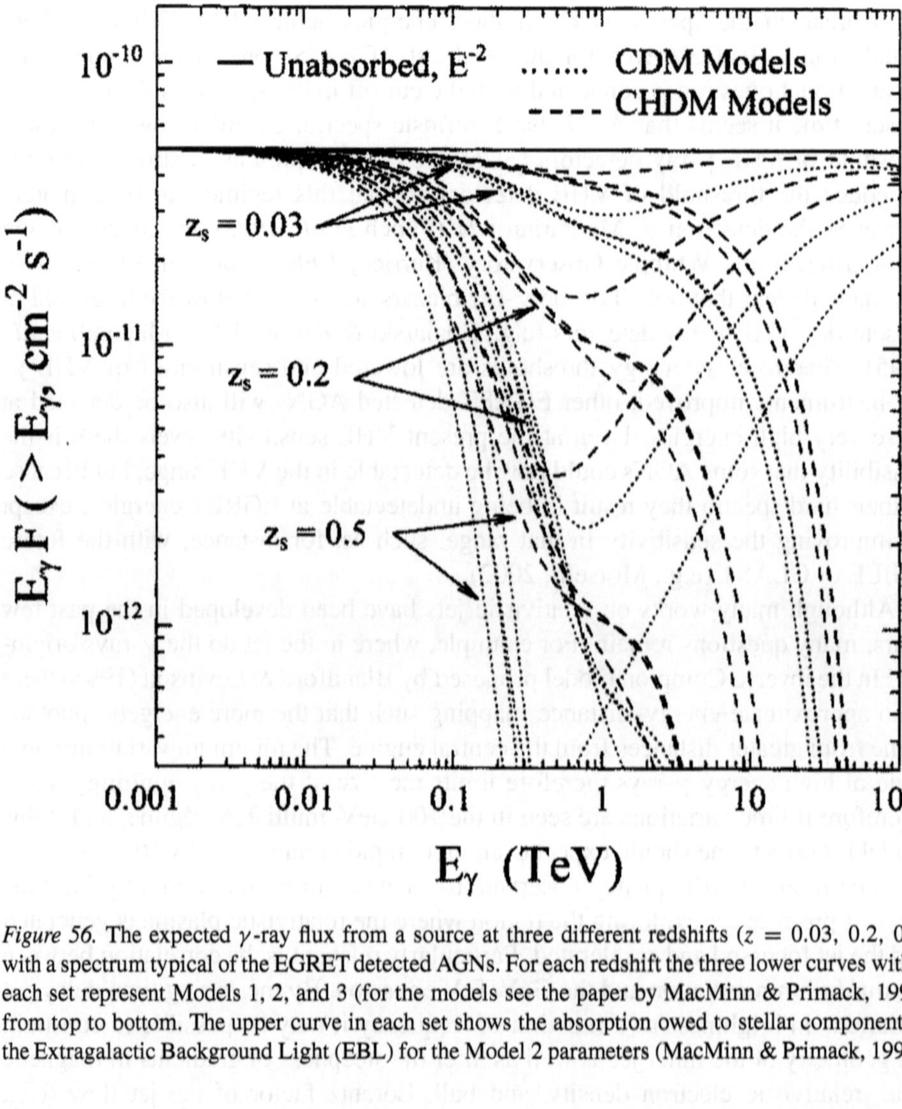


*Figure 56.* The expected $\gamma$-ray flux from a source at three different redshifts ($z$ = 0.03, 0.2, 0.5) with a spectrum typical of the EGRET detected AGNs. For each redshift the three lower curves within each set represent Models 1, 2, and 3 (for the models see the paper by MacMinn & Primack, 1996) from top to bottom. The upper curve in each set shows the absorption owed to stellar component of the Extragalactic Background Light (EBL) for the Model 2 parameters (MacMinn & Primack, 1996).

ation of VHE photons which depends heavily on the density of starlight photons. This density is, in principle, calculable from what is known about the starlight and dust content of galaxies and the time at which galaxies first formed. Such calculations have been performed (MacMinn & Primack, 1996) and show that the spectrum from a source will exhibit a broad absorption dip whose depth and shape is dependent on the epoch of galaxy formation.

Figure 56 shows the expected $\gamma$-ray flux from a source at three different redshifts ($z$ = 0.03, 0.2, 0.5) with a spectrum given by $dN/dE = 5 \times 10^{-8}\, E_{\rm GeV}^{-2}$ cm$^{-2}$ s$^{-1}$ GeV$^{-1}$, which is typical of the EGRET detected AGNs.

Thus if there are several AGNs whose intrinsic spectra reach beyond 10 TeV, then careful measurements of their spectra over at least 3 decades in energy in the VHE range may provide constraints on the epoch of galaxy formation. Of course, if these AGNs exist, a more ambitious use of the intergalactic attenuation of AGN $\gamma$-rays is a proposal for extracting the Hubble parameter from a study of the extinction of AGN spectra at various redshifts (e.g., Salamon, Stecker & de Jager, 1994). The method depends on the presence of the Hubble parameter in the relationship between optical depth and redshift. For the method to be successful many AGNs must be detected and their intrinsic spectra must extend to energies well above 1 TeV.

This is just one of the possibilities of the GAMT experiment proposed by Auriemma *et al.* (1995).

#### 4.3.3. *Galactic Sources*

4.3.3.1. *Possible Origin.* A complete model of the cosmic radiation must explain all characteristics of the different regions of the cosmic ray energy spectrum, it must be consistent with the experimental constraints given by the observations and it must indicate a source of high energy particles with sufficient power to maintain the cosmic ray energy density constant with the time. The models developed tend towards the assumption that both acceleration processes at large scales in the Galaxy and near discrete point sources occur. A detailed explanation of the different acceleration mechanisms proposed for energies below and above the knee in the spectrum can be found in the papers by, e.g., Gaisser (1990a,b,c), Harding (1991), and Biermann (1995).

The mechanism proposed for explaining the existence of cosmic rays with energies up to $10^{14} - 10^{15}$ eV is the shock acceleration. It has been observed to work in the heliosphere. It is based on the first order Fermi acceleration mechanism (Fermi, 1954), which leads in a natural way to a power law in the spectrum of the accelerated particles with spectral index $\sim$2. If the shock is formed by collisionless processes the particles are trapped near the front by scattering from magnetic irregularities and they can travel back and forth across the shock front, gaining some energy on each crossing.

There are several sources of shock waves in the interstellar medium, such as novae, expanding HII regions, and stellar winds (e.g., Axford, 1981a,b; Drury, 1983a), but the supernova explosions are the sources with more energy available (Drury, 1983b). The explosion leads to the formation of a large expanding SNR bounded by strong shocks. The energy input ($\sim 10^{51}$ erg) is enough to maintain the steady cosmic ray density if an acceleration efficiency of $\sim$10% and a galactic supernova rate of one every 30 years is assumed (Drury, 1990). Particle acceleration and consequent cosmic ray enhancement by strong stellar winds has been discussed by Cesarsky & Montmerle (1983). In their model acceleration of particles is achieved by repeated scattering across the shock boundary. For typical stellar parameters, with a wind velocity of $\sim$2500 km $s^{-1}$, the acceleration overcomes the adiabatic losses and the maximum energy reached by a proton is $\sim 10^{16}$ eV. This mechanism of acceleration seems to work in young open clusters, where with the ambient magnetic field of $\sim 10^{-5}$ G and a wind velocity of $\sim 5200$ km s$^{-1}$ a shock front is generated in the dense region around, and the cosmic ray density reaches a value of $\sim 100$ times greater than the average (Polcaro *et al.*, 1991a,b,c) (see Section 7.16).

The maximum energy reached in the stellar wind shock is higher by a factor $\sim 100$ than in the case of supernova shocks and it is much higher than the energy necessary to generate $\gamma$-rays via $\pi^0$ decay ($\sim 1$ GeV). The emergent cosmic ray spectrum is expected to be similar to that of the observed galactic cosmic rays.

4.3.3.2. *Pulsars.* Seven spin powered pulsars have been detected by the CGRO: Crab; Vela; Geminga; PSR B1055-52; PSR B1509-58; PSR B1706-44; and PSR 1951+32 (Srinivasan *et al.*, 1997). The PSR B1509-58 was detected by the OSSE instrument, the other by EGRET. It is important to remark that of two of the firmly established TeV sources are on this list, the Crab Nebula and PSR B1706-44 (Tanimori *et al.*, 1994). For a review see Thompson *et al.* (1997).

As EGRET increased its exposure to the galactic plane, further discoveries of $\gamma$-ray pulsars were likely and they became prime candidates for VHE emission as well. EGRET discovered 271 HE sources, which consist of pulsars, AGNs, and unidentified sources. The number of VHE sources is approaching a dozen: seven CGRO (EGRET and/or OSSE) pulsars; a few AGNs; and a SNR. The VHE $\gamma$-ray astronomy in the CGRO Era has been discussed by Weekes (1997) and Krennrich (1999).

It is also possible that some of the presently unidentified EGRET galactic sources are pulsars, not yet detected as radio pulsars because of sensitivity limits or beaming effects.

After the detection of VHE emission from the Crab Nebula (Vacanti *et al.*, 1991) the ground based air Cherenkov in the visible band became a well established technique for the search for VHE $\gamma$-ray sources in the sky.

Although $\gamma$-rays from these objects most probably originate in the interaction of relativistic electrons with synchrotron photons within the pulsar nebula, the acceleration sites and processes for electrons and hadrons must be closely related. Thus the observations of TeV photons from such pulsar driven SNRs is a confirmation that high energy acceleration processes can be associated with the aftermath of supernova explosions.

4.3.3.3. *Unidentified Galactic Sources and Supernovae of Shell-Type.* Only $\sim 80$ of the 271 EGRET sources ($E > 100$ MeV) reported in the third catalog (3EG) (Hartman *et al.*, 1999) have been identified, because of the large position uncertainties resulting from the poor angular resolution and the structured interstellar background. Irrefutable identifications with a flaring blazar or a pulsar are even fewer. The situation was even worse after the publication of the first (Fichtel *et al.*, 1994) and second EGRET catalogs (Thompson *et al.*, 1995). The first EGRET Catalog (1EG) covered cycle 1 of the CGRO program. The situation was similar to that after the publication of the COS B catalog (Swanenburg *et al.*, 1981). Approximately 90% of its in plane sources were unidentified. In spite of the many theoretical and observational attempts at identifying the unidentified COS B sources, only a few successes are now coming from these efforts, such as Geminga, now identified as a 237 ms pulsar, and 2CG342-02, a 102 ms pulsar (PSR B1706-44).

The 1EG catalog included one solar flare, pulsars, $\gamma$-ray bursts, a normal galaxy, active galaxies, and mostly unidentified sources. The second catalog (2EG) included in addition to cycle 1, cycle 2 and contain new sources detected during cycle 3 or only in the sum of cycles 1,2, and 3 (Thompson *et al.*, 1996).

In our opinion the problem of the identification of the unknown EGRET sources is very difficult to solve, since for their position around the very crowded galactic plane the sources – which in principle can be discrete or diffuse – cannot be resolved with the present generation of instruments, although EGRET greatly improved the angular resolution with respect to COS B. Only attempts can be performed in this sense. Indeed, for instance, Manchanda *et al.* (1996) suggested as a possible class of $\gamma$-ray diffuse emitters the young open clusters containing peculiar objects with hypersonic stellar winds (>2500 km $s^{-1}$). In this way it is possible to account for 7 sources which were, of course, contained in the COS B catalog as well: NGC 6514 (2CG 006-00); Berk 87 (2CG 075+00); NGC 6910 (2CG 078+01); IC 1808 (2CG 135+01); Tr16/Tr18 (2CG 288-00); and NGC 6193 and NGC 6231, associated with two regions of diffuse emission reported by COS B.

Although considering these identifications, most of the EGRET sources still remain unidentified. Are they mostly point sources from compact objects like neutron stars? Are they related to SNRs where one would expect an enhanced cosmic ray density? Do they correspond to matter enhancements like giant molecular clouds being bombarded by cosmic rays of ordinary density? Are they Supernova–OB associations (SNOBs)?

Six or seven of the unidentified EGRET sources are sufficiently near to catalogued shell-like SNRs that at least some of them may be casually linked.

Indeed, Romero, Benaglia & Torres (1999) discussed the possible association of unidentified $\gamma$-ray sources in the 3EG catalog with three classes of cosmic sources, namely SNRs, OB associations, and Wolf–Rayet stars. Their conclusion is that there seem to exist more than a single population of galactic $\gamma$ ray sources. The large number of unidentified EGRET sources, free of any positional coincidence with luminous objects, encourage further studies to find whether there exist a population of exotic objects yet undetected at higher energies.

In fact, VHE $\gamma$-ray astronomy can help in this identification process in several ways.

In the first place they can be detected in the VHE range. Several of the EGRET sources which may be identified with SNRs have relatively hard spectra which extrapolate to flux levels that are easily detectable in VHE range. The expected $\gamma$-ray fluxes from SNR of shell type have been calculated by Drury, Aharonian & Voelk (1994). Their values support those extrapolated from EGRET energies. Similar extrapolation of the EGRET data from Berk 87 (CGRO J 2021+37) to VHE range has been performed by Giovannelli, Bednarek & Karakuĺa (1996). The expected fluxes can be detected by the present generation of VHE experiments and much better by the GAMT experiment (Auriemma *et al.*, 1995), if realized.

One of the long standing prime objectives for VHE $\gamma$-ray astronomy is the desire to establish direct evidence for the sources of cosmic rays. If the source of very high and ultra high energy cosmic rays are extragalactic at distances more than twice the distance to the Virgo cluster (20 Mpc), then the photon window for their exploration effectively closes above $10^{14}$ eV. Thus VHE $\gamma$-ray astronomy has the potential to view the highest photon energy at which distant extragalactic cosmic ray sources may be studied directly.

It is well established that cosmic ray spectrum extends to $10^{20}$ eV, yet the origins of such spectacularly high energy particles remains obscure. Particle energies of this magnitude imply that near their acceleration sites a range of elementary particle physics phenomena is present which is beyond the ability of present day particle accelerators to explore. VHE $\gamma$-ray astronomy may catch a glimpse of such phenomena.

It is becoming increasingly clear that the energy régime covered by VHE $\gamma$-ray astronomy will be able to address a number of significant scientific questions, which include: i) What parameters determine the cut-off energy for pulsed $\gamma$-rays from pulsars? ii) What is the role of shell-type supernovae in the production of cosmic rays? iii) At what energies do AGN blazar spectra cut-off? iv) Are gamma blazar spectral cut-offs intrinsic to the source or due to intergalactic absorption? v) Is the dominant particle species in AGN jets leptonic or hadronic? vi) Can intergalactic absorption of the VHE emission of AGN's be a tool to calibrate the epoch of galaxy formation, the Hubble parameter, and the distance to $\gamma$-ray bursts? vii) Are there sources of $\gamma$-rays which are 'loud' at VHEs, but 'quiet' at other wavelengths?

Grenier (2003) discusses the unidentified EGRET sources in the Galaxy, concluding that nearly two thirds of the sources are hiding in our Galaxy, awaiting identification.

Mattox, Hartman & Reimer (2001) searched for the association of the unidentified EGRET sources with radio sources. They found 46 associations with a high probability of being correct, 37 associations with a 'plausible identification' and 15 more plausible identifications, which were not suggested previously at all.

## 5. Physics of the Discrete Cosmic Sources from Space Experiments

In the previous section we have briefly discussed the physics of high energy cosmic sources, as possible sources of cosmic rays. In this section we want to enter into more detailed discussion of the behaviour of discrete cosmic sources, considered as emitters of photons, following a roughly division into energy ranges, since their emission is typically across almost the entire electromagnetic spectrum. Sometimes the division of sources in high energy cosmic sources or low energy cosmic sources is purely owed to the position of the bulk of their emission in the electromagnetic spectrum: the former emit mostly at high energies, the latter at low energies.

### 5.1. Extragalactic Historical Journey: Gamma and X-Ray Ranges

Space experiments have undoubtedly contributed to knowledge about cosmic sources and in particular have opened a large window to the Universe by means of observations of extragalactic objects giving crucial cosmological information. Because of reasonable limits to the length of the paper we cannot discuss all the physical problems resolved. However, we will discuss briefly some of the most important discoveries using a division into energy ranges.

#### 5.1.1. *Discrete Extragalactic Sources*

Observations of $\gamma$-rays from some extragalactic objects (Dean & Ramsden, 1981) as well as some diffuse emission (Fichtel, Simpson & Thompson, 1978), which appeared isotropic in the scales used, opened a new window onto the Universe. Discrete extragalactic sources were associated with active galaxies and the background was attributed, at least partially, to emission owed to non resolved galaxies at cosmological distances.

There are a large variety of very luminous extragalactic objects, generally defined as Active Galactic Nuclei (AGNs) (e.g., Hazard & Mitton, 1979). This class of objects contains: radio galaxies, Seyfert galaxies, BL Lac objects, and quasars (QSOs). High energy emission was observed from the radio galaxy Cen A (Grindlay *et al.*, 1975), from the Seyfert galaxy NGC 4151 (Perotti *et al.*, 1979) and from the QSO 3C 273 (Swanenburg *et al.*, 1978).

Active galaxies were intensively observed also in radio, IR, optical, UV, and X-ray energy ranges. Figure 57 shows the combined spectrum of the three aforementioned objects, adding also the $\gamma$-ray measurements. The measurements come from the experiments cited in the review paper of Ramaty & Lingenfelter (1982). Such spectra are reported in units of luminosity for an interval of energy. This is convenient since by reporting $dL/d\ln E$ it is possible an easy comparison of the luminosities of the three objects. Their peaks of luminosity are around 0.1 MeV, which is an indication that the observations around this energy can provide direct information on the *central engine* of these objects. A first look at the spectra of the

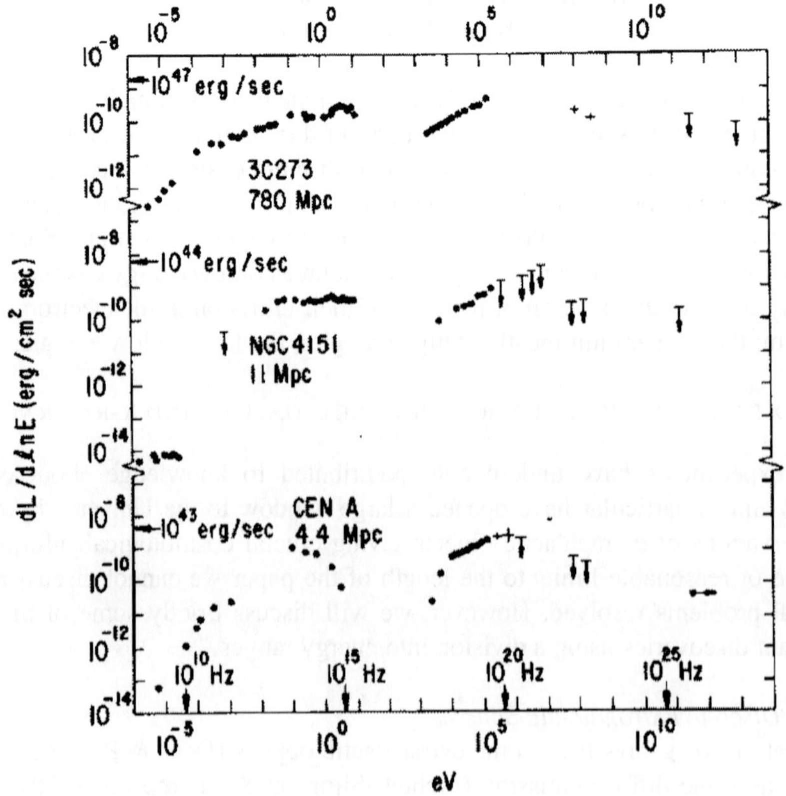


*Figure 57.* Combined spectra of three active galaxies: the radio galaxy Cen A, the Seyfert galaxy NGC 4151, the quasar 3C 273 (Ramaty & Lingenfelter, 1982).

Figure 57 shows that their shapes are very roughly similar, apart from their energy scales. This was a first suggestion in trying to consider such spectra as typical of AGNs, where probably a central black hole is the engine producing such a large amount of energy. The different levels of energy are depending on the mass of the central engine, higher for the quasar 3C 273, lower for the Seyfert galaxy NGC 4151 and even more for the radio galaxy Cen A. The process producing such energy was suggested to be the accretion of matter onto a massive black hole (Lynden-Bell, 1969). Giovannelli, Karakuła & Tkaczyk (1982a,b) calculated the expected spectra from such a process, under the hypothesis of spherical accretion (for simplicity), and successfully attempted the fit of their spectra with the experimental spectra of 3C 273 (Karakuła, Tkaczyk & Giovannelli, 1984).

The luminosity generated by the accreting matter cannot exceed notably the Eddington limit, defined as:

$$L_{\mathrm{E}} = 4\pi G M m_{\mathrm{p}} c/\sigma_{\mathrm{T}} \simeq 1.2 \times 10^{38} (M/\mathrm{M}_{\odot})\ \mathrm{erg\,s^{-1}}\,, \tag{5.1}$$

where $G$ is the gravitational constant, $m_p$ the proton mass, $\sigma_T$ the Thompson cross section, $c$ the light velocity, $M$ the mass of the central collapsed object and $M_\odot$ the solar mass.

If the isotropic luminosity of an accreting object could largely exceed the Eddington limit $L_E$, the pressure of the emerging radiation could prevail over the gravitational force attracting the accreting gas, with the consequence of stopping the accretion. The luminosities of $10^{47}$, $10^{44}$, $10^{43}$ erg s$^{-1}$ imply the masses of $\sim 10^9$, $10^6$, $10^5$ $M_\odot$ for the accreting black holes of 3C 273, NGC 4151 and Cen A, respectively.

Through the observed temporal variations in the X-ray energy range it is possible to obtain the upper limits of the dimensions of the emitting regions (Marshall, Warwick & Pounds, 1981). Indeed, the dimensions of the emitting sources must be $\leq c\Delta t$, where $\Delta t$ is the minimum time scale of the measured variations.

By means of these arguments Bassani & Dean (1981) evaluated the upper limits of dimensions of a number of AGNs, among them were contained also 3C 273 and NGC 4151. They obtained $c\Delta t \leq 10^{15}$ cm. Such values are consistent with the minimum possible dimensions of the emitting region, namely the Schwarzschild's radius, given by:

$$r_S = 2GM/c^2 \simeq 3 \times 10^5 (M/M_\odot) \text{ cm}. \tag{5.2}$$

It is also possible to obtain other information about the nature of the $\gamma$-ray source by the means of the photon–photon production process. Indeed, the optical depth of $\gamma$-rays in the region of production of X-rays can be evaluated through the observed X-ray luminosity and the upper limit of the dimension of source.

Bassani & Dean (1981) argued that for isotropic X-ray emission, the QSO 3C 273 (and other QSOs) would be opaque to all $\gamma$ rays with energies greater than the threshold for the production of $e^\pm$ pairs (0.511 MeV). However, emission around 100 MeV was observed from 3C 273 (Swanenburg *et al.*, 1978). This implies that either the X-ray emission is collimated or that the $\gamma$-ray source is more extended than the X-ray source ($\gg 10^{15}$ cm). The latter case should explain the apparent lack of variability in the 3C 273 luminosity at $\simeq$100 MeV, deduced by the COS B measurements (Swanenburg *et al.*, 1978).

On the other hand, Bassani & Dean (1981) found that X-ray sources in Seyfert galaxies are transparent to all $\gamma$-rays up to energies of order of GeV. Thus, in such objects, X and $\gamma$ ray sources would have the same dimensions.

Some mechanisms can be responsible of the production of the $\gamma$-ray radiation in AGNs. At least a part of $\gamma$-rays could be originated by the same mechanisms producing X-rays (Fabian, 1979), namely:

- bremsstrahlung radiation in a hot gas ($\sim 10^9$ K);
- Comptonization of cold photons;
- synchrotron self-Compton model.

Moreover, there are other mechanisms active only in the $\gamma$ ray region. Indeed, the $e^\pm$ pair production from photon–photon interaction could be important in AGNs, assuming that in the nucleus of our Galaxy the $e^{\pm}$ pair annihilates in a region relatively cold producing a narrow line at 0.511 MeV (Leventhal, MacCallun & Stang, 1978). Such a narrow line (FWHM$\leq$ 3.2 keV) lies on a strong continuum emission, indicating that ~90% of positrons annihilate via positronium. The observed intensity of $2 \times 10^{-3}$ photon cm$^{-2}$ s$^{-1}$ implies an annihilation rate of $\sim 4 \times 10^{43}$ positrons s$^{-1}$ or a luminosity of annihilation radiation of $\sim 6 \times 10^{37}$ erg s$^{-1}$, for an assumed distance to the Galactic Center of 10 kpc.

It is possible that $\gamma$-rays around 100 MeV from 3C 273 are produced by $\pi^0$ decay. Giovannelli *et al.* (1990c) (and the references therein) derived the $\gamma$-ray spectra originated from $\pi^0$ decay during the accretion process around a collapsed object, considering also the photon–photon absorption. They found for 3C 273 a dimension of the X-ray emitting region of $4 \times 10^{17}$ cm, which is in complete agreement also with the more recent big amount of data.

Many observational evidences of highly anisotropic emission and non-spherical structure of several astronomical objects were obtained (Bridle & Perley, 1984; Rees, 1985). In the central region of 12 extragalactic radio sources superluminal motion was discovered (Cohen & Urwin, 1984; Porcas, 1985; Frank, King & Raine, 1985). This means that the particles are moving with Lorentz factor much greater than 1 in the direction of the line of sight (or at small angles). These facts pushed Bednarek *et al.* (1990a) to study the interactions of relativistic particles emitted by the central engine of such sources with the surrounding radiation and matter. In the case of AGNs isotropic background radiation can be produced by ionized gas distributed in clouds (Frank, King & Raine, 1985). Therefore Bednarek *et al.* (1990a) computed the photon energy production spectra for monoenergetic, one-dimensional beam with selected Lorentz factor ($\gamma$) and relative velocity of the beam ($\beta$) through the study of the Inverse Compton Scattering (ICS) of the electron beam with the background photons and the interactions of the proton beam with the matter. They used such spectra for fitting experimental data of the most popular galactic black hole, namely Cyg X-1 (Bednarek *et al.*, 1990b). Bednarek & Calvani (1991) computed also the dependence of the spectrum of the secondary electrons and positrons, which are produced in the interactions of a relativistic proton beam with ambient matter. The best fit to data of 3C 273 shows that the total energy of relativistic protons in the beam is $\simeq 2 \times 10^{61}$ erg. The core of 3C 273 must emit a power of $\approx 1.5 \times 10^{48}$ erg s$^{-1}$, which corresponds to a mass loss of relativistic protons of 1.1 $M_{\odot}$ yr$^{-1}$ in the jet.

### 5.1.2. *X-Rays in Normal Galaxies*

With the launch of the EINSTEIN observatory in 1978 a substantial jump in knowledge of the extragalactic X-ray sky was reached (Giacconi *et al.*, 1979). From that moment, also thanks to the observations performed with other satellites, especially with those having high spectral resolution, it was possible to clarify most of the physical processes in cosmic sources.

The observations of the EINSTEIN observatory allowed a deep study of the X-ray properties of normal galaxies, considered as a homogeneous class. Data from hundreds of galaxies have been taken. Some of the galaxies were detected with sufficient details in order to study their morphology, spectra, and individual sources and then to render possible the comparison with data in optical, IR and radio energy ranges. Such observations showed that normal galaxies independent of their morphology (following the current classification of the galaxies) are X-ray emitters, rather intense, with luminosity ranging from $10^{38}$ to $10^{42}$ erg $s^{-1}$. Although such an emission represents only a small fraction of the total energy emitted by a normal galaxy, X-ray observations are a powerful tool in studying several phenomena, which should be prohibited in other energy ranges. Such phenomena include the final products of stellar evolution (supernova remnants – SNRs – and compact remnants – white dwarfs, neutron stars, black holes) and a warm phase in the ISM, which was discovered in bright galaxies of early type.

In the energy range of the EINSTEIN observatory (~0.2–3.5 keV the spiral galaxies – extended and complex sources – showed a luminosity from $\sim 10^{38}$ to a few $\times 10^{41}$ erg $s^{-1}$ (Fabian, 1981; Fabbiano, 1984).

Observations of the Milky way and the Local Group of galaxies (e.g., Fabian, 1981) indicated that a substantial part of this X-ray emission was owed to individual bright sources, such as accreting X-ray binaries and SNRs, with luminosity in the range $10^{35}$– a few units $\times 10^{38}$ erg $s^{-1}$.

It was also established that stars emit coronal X-rays with luminosity in the range $\sim 10^{28}$–$10^{33}$ erg $s^{-1}$ (see Section 5.6.1) (Vaiana *et al.*, 1981). However, the stars do not significantly contribute to the total X-ray emission of spiral galaxies. Indeed, the X-ray/optical luminosity ratios are greater than expected in a normal star population (Helfand & Caillault, 1982; Fabbiano & Trinchieri, 1985), and the X-ray spectra measured in spiral galaxies are harder than those of stellar emission: $KT > 2$ keV in spirals (Fabbiano & Trinchieri, 1987) and $KT \sim 0.5 - 1$ keV in stars (Helfand & Caillault, 1982). Therefore X-ray emission from something other than stars exist. It can come from the nucleus of the galaxy with non-thermal activity Seyfert-like and from star formation regions, which contribute to the total emission in different proportions.

Amongst the starburst galaxies the bluer galaxies are frequently interacting and show a stronger X-ray emission than redder galaxies and those of intermediate colors. The X-ray emission of these galaxies tends to originate in extended regions, excluding a nuclear origin purely non-thermal. Their spectra, on average, exclude a soft X-ray emission, which indicates that it is not dominated by thermal emission of a gaseous halo (Fabbiano, Feigelson & Zamorani, 1982). Most of the X-ray emission of such galaxies is owed to a number of young SNRs and X-ray binary systems (with X-ray luminosity possibly enhanced because of the low metallicity of the accreting gas), such as observed in the Magellanic Clouds. Although this explication cannot be used in general way, the integrated coronal emission of the young stellar population (see Vaiana *et al.*, 1981 for typical values) can dominate in very young starburst galaxies, for which UV spectra, taken with the IUE satellite, indicate the presence of a large number of OB stars (Fabbiano & Panagia, 1983).

Later Ward (1988) found a relationship between the intensity of the emission Br-$\gamma$ line and X-ray emission in a sample of nuclei of starburst galaxies. This is interpreted by means of a relationship between the number of the ionizing photons produced by the OB stars and the associated population of X-ray binaries, as already indicated by Fabbiano, Feigelson & Zamorani (1982).

The absence of trace of cold ISM in most of the young galaxies (e.g., Faber & Gallagher, 1976) triggered the construction of models in order to explain the gathering of the free gas that stars lose during their evolution (White & Chevalier, 1983). With the actual rate this quantity would be of the orders $10^{9-10}$ $M_\odot$ in the Hubble time in galaxies with optical luminosity of $10^{10-11}$ $L_\odot$ (Faber & Gallagher, 1976).

Later, with the advent of the EINSTEIN observatory, for the first time the presence of a hot gaseous medium was detected in normal elliptical galaxies in the Virgo cluster (Forman *et al.*, 1979). A hot gaseous halo was associated with M 87 in the Virgo cluster (Fabbiano, 1989 and the references therein). Five young galaxies were detected in the Virgo cluster, with X-ray luminosity of (5–70) $\times 10^{39}$ erg $s^{-1}$, roughly a factor of 100 times less luminous than the galaxies in M 87. Forman *et al.* (1979) concluded that the hot gas detected in the galaxies of the Virgo cluster would have been endogenous, since the cooling time of the matter in the interior of the cluster is much too large for allowing accretion onto galaxies.

The discovery of a thermal diffuse X-ray source in the Virgo cluster, centered in M 87, allowed an evaluation of the bound mass of a young galaxy in a radius sensibly greater than the optical nucleus, on which the optical measurements were obligatorily performed. The first evaluation, based on the hypothesis that the gas is in hydrostatic equilibrium in the potential field of the galaxy at a temperature $T \simeq 3 \times 10^7$ K, suggested a very high value for the bound mass: $10^{13} - \geq 10^{14}$ $M_\odot$ (Fabbiano, 1989). The observations of M 87, made with IPC (Image Proportional Counter) of the Einstein observatory – which provided images and high resolution spectral information –, allowed a better measurement of the bound mass and confirmed previous indications on the presence of a massive obscure halo (Fabricant, Lecar & Gorenstein, 1980; Fabricant & Gorenstein, 1983).

Taking into account of X-ray/optical emission ratio of normal galaxies, Trinchieri & Fabbiano (1985) estimated that the integrated emission of such galaxies contributed for about 13% to the extragalactic X-ray background at 2 keV. Assuming that the activity of the starburst galaxies was higher in the past (Bookbinder *et al.*, 1980), such galaxies could be responsible for most of the background. Although galaxies substantially contribute to the 2 keV background, their contribution to the X-ray background at harder energy ranges was not clear. In any case, such a contribution is dependent on spectral characteristics of such emissions (Giacconi & Zamorani, 1987). The spectra of spiral galaxies are consistent with a rather hard X-ray emission ($KT > 2$ keV) (Fabbiano & Trinchieri, 1987), and X-ray binaries can have hard spectra with $KT \simeq 20$ keV. Moreover, it is possible that a hard spectral component could be present in the starburst galaxy M 82 (Schaaf *et al.*, 1989). However, it is not very probable that the galactic contribution would be significantly higher than 10 keV, especially considering that the spectra of far galaxies would be redshifted. The soft X-ray spectrum of the nucleus of M 81 (Fabbiano, 1988) introduces an additional uncertainty in evaluating the contribution of low-activity nuclei to the background in the energy range 2–10 keV (Elvis, Soltan & Keel, 1984), since most of the luminosity of these sources could be emitted in softer energy ranges.

### 5.1.3. *X-Rays in Active Galaxies*

Holt (1980) discussed the impact of the relatively simple spectroscopic techniques on the classification of physical processes occurring in X-ray cosmic sources.

Low resolution wide range spectra obtained with the detectors on board the ARIEL 5, OSO-8,and HEAO-1 satellites, allowed it to be established that clusters of galaxies are strong thermal X-ray sources and provided information on the temperature, composition and quantity of the emitting gas.

The same instruments demonstrated the complexity of continuum spectra emitted by optically thick plasmas in a sample of X-ray binaries. Most of spectral details were provided by the EINSTEIN observatory by means of the Solid State Spectrometer (SSS), with an energy resolution better than a factor 3–10 with respect to proportional detectors, in the energy range 1–10 keV. The spectral resolution of the SSS was sufficient to detect thermal emission lines, which characterize the spectra of several bright stars and SNRs. In spite of the resolution of the SSS, the X-ray spectra of the strongest and enigmatic sources of the whole universe, such as QSOs and Seyfert galaxies showed a continuum spectrum, like a power law, without any particular feature.

In the 1980s, after the launches of the EXOSAT and GINGA satellites, the more sensitive instruments (although of low resolution type) allowed us to extend our knowledge about the spectra of cosmic sources. For instance, the spectra of AGNs started to show small details, which allowed us to sound their inner parts providing the benefit of the knowledge of the nature of their nuclei. This result also opened the way for designing a new generation of instruments for the spectroscopy of the following generation experiments, such as ASCA, and the recent CHANDRA and NEWTON satellites.

First X-ray spectra of an important sample of Seyfert galaxies were obtained with the HEAO-1 satellite. Such results showed that the X-ray spectra of AGNs, with emission lines in a wide optical range, were quite uniform. The best fit of data to a simple power law spectrum gives:

$$F(E) = NE^{-\alpha} \times e^{-\sigma(E)N_{\rm H}}, \tag{5.3}$$

where $\sigma(E)$ is the effective photoelectric cross section for the energy $E$ (expressed in keV) in a neutral gas with solar abundance (Morrison & McCammon, 1983),

$N$ is the flux normalized at 1 keV, and $N_H$ is the equivalent absorption hydrogen column. The average index of energy with the data of HEAO-1 (2–15 keV) was $\alpha \simeq 0.68 \pm 0.15$ (Mushotzky, 1984b), where 0.15 represents the standard deviation of a Gaussian fit to data.

A wider program of survey of AGNs with the EXOSAT was performed between 1983 and 1986, in order to obtain spectra of a larger sample of AGNs with optical emission lines and with hard X-ray emission. These AGNs are collected in the catalogs of ARIEL 5 and HEAO-1 satellites.

Results from 42 objects were obtained. Amongst these, 30 objects were AGNs with emission lines, belonging to the sample of Piccinotti *et al.* (1981) and other sources of the last catalogs of ARIEL5 and HEAO-1 satellites (Turner & Pounds, 1989). Six important results were obtained:

- the emission spectra in the range ~2–10 keV of all the AGNs of the sample can be fitted to a simple power law spectrum. The spectral indices individually measured are not all consistent with a unique value, but they can be approximated by a Gaussian distribution, with average spectral index $\alpha \simeq 0.7$ and standard deviation $\sigma_\alpha = 0.17$. This indicates the intrinsic dispersion of the spectral indices and the error of the measure for each object;
- there is not a clear dependence of the spectral index and the nuclear luminosity over four decades of luminosity;
- roughly 50% of the sample shows an absorption at low energy, which corresponds to an intrinsic column density in the range $10^{21}$–$10^{23}$ cm$^{-2}$. About half of the objects in the sample have intrinsic column densities derived with X-ray measurements higher than those deduced with other measurements, such as the redshift and the Balmer decrement;
- lower luminosity AGNs have, in general, greater column densities;
- Four objects in the sample showed variable absorption in different measurements;
- approximately in 30% of the sample (and in 50% of the non-obscured sources) there is a soft X-ray excess, which dominates below 1 keV. In most cases such an excess demonstrates as additional component of the soft emission, and it shows frequent variations in time scales of several hours or less.

5.1.4. *Iron Lines in Seyfert Galaxies*

As has already been said, observations with the HEAO-1 and EXOSAT satellites showed power law spectra for most of the active galaxies with emission lines, at least for $E \geq 2$ keV.

The case of the Seyfert galaxy NGC 4151 has been an exception for a long time, since when the TENMA satellite detected an emission line around 6.4 keV, which is consistent with the fluorescent $K$–line of iron. This is produced in a dense column which obscures the X-ray source (Matsuoka *et al.*, 1987).

With the larger area of the detectors (Large Area Counter – LAC) of the GINGA satellite (Turner *et al.*, 1989) with respect to those of EXOSAT, it was possible to discover the spectral characteristics of Fe $K$–line during a extraordinary long observation of the Seyfert galaxy MCG-6-30-15 (Nandra *et al.*, 1989).

During the period 1987–89 eight Seyfert galaxies were observed with the GINGA satellite. In these galaxies, Fe $K$–line and/or a cut in the absorption overlapped into a continuum power law spectrum. Important differences among objects were detected too; this demonstrates the importance of such data in the study of the physical nature of the active nucleus and its neighbourhood (Pounds, 1990).

The presence of the Fe $K$-emission line with equivalent width of $\simeq$100–200 eV seems to be a common characteristic of the spectra of individual Seyfert galaxies observed by the GINGA satellite. In some of these galaxies a deep $K$–line in absorption is observed too. Other independent justification of such spectral characteristics was obtained by a re-analysis of data taken by the EXOSAT satellite on AGNs. Nandra, Pounds & Stewart (1991) discussed the presence of Fe lines and/or breaks, which render difficult the fit to a simple power law of the spectra of many sources.

The limited statistics obtainable with the individual spectra measured by the GINGA satellite gives a value of the Fe line energy rather uncertain. It was preferred to identify the emission as the $K$–line of neutral Fe or as a cold line ($\simeq$6.4 keV) instead of a line produced by highly ionized matter (at $\simeq$6.7 keV). The possibility of the existence of a sharp tepid border in some sources, with a very small or null absorption at low energies is a rather complicated problem.

In order to solve such a problem, in a first approximation Pounds *et al.* (1990) considered 12 GINGA observations of 8 different AGNs, Seyfert-like, and made a sum of their spectra to obtain the composed spectrum, called GINGA-12. The selected sample (AKn 120, Mkn 335, MCG-6-30-15, NGC 3227, NGC 4051, NGC 5506, NGC 5548, and NGC 7314) is formed by bright objects in the X-ray range, with the slope of the individual power law spectra equal to the canonical value $\alpha \simeq 0.7$, and showing Fe $K$–line in emission. The best fit of the GINGA-12 spectrum to a power law spectrum gives a value $\alpha \simeq 0.70$, in agreement with the slope of the individual spectra (see Figure 58). This fact confirms the validity of the method used, i.e. the sum of the spectra.

Since a fluorescent $K$–line is visible in each individual spectrum, it is possible to attempt a best fit of the data to a second approximation spectral model, namely a power law spectrum plus a line and a break. The resulting fit gives a value $\alpha \simeq 1.64^{+0.04}_{-0.02}$, with a confidence level of 90%, $N_{\rm H} \simeq 4.1^{+1.2}_{-1.1} \times 10^{21}$ cm$^{-2}$ plus a $K$–line with a break defined as follows:

- the energy of the Fe $K$–line, in the rest reference system, is $6.28 \pm 0.21$ keV. Fabian *et al.* (1989) suggested the idea of the existence of an average energy less than that of the neutral Fe (6.4 keV), owed to the presence of a significant gravitational redshift on the fluorescence produced by the matter close to the central black hole. The equivalent width of the line ($110^{+40}_{-35}$ eV) is also consistent with that expected by the fluorescence of the optically dense

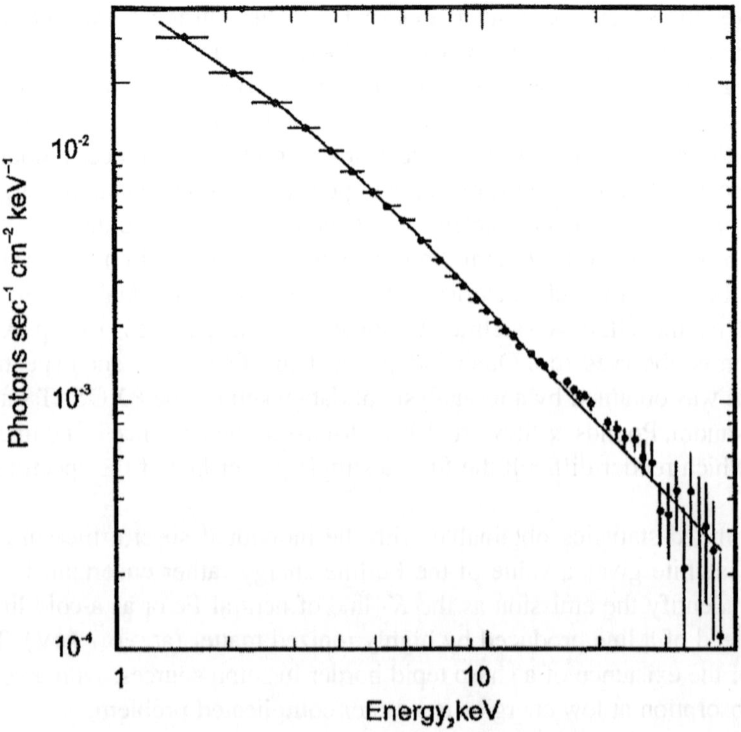


*Figure 58.* Simple power law best fitting the composed spectrum GINGA-12 (Pounds, 1990).

matter, which subtends a large solid angle (e.g., an accretion disk) with respect to the hard X-ray source (George & Fabian, 1991);

– the absorption break of the Fe $K$–line corresponds to a column density very high: $1.2^{+0.7}_{-0.5} \times 10^{23}$ hydrogen atoms cm$^{-2}$ (supposing for the Fe the solar abundance). Such a density is higher than that one can expect from the Albedo component (Lightman & White, 1988; George & Fabian, 1991). This implicates, according to the GINGA-12 spectrum, a small value of the break at low energy. A simple explanation for this could be that the material, across the line of sight, is highly ionized and therefore has a reduced opacity at low energy. The fit for the break gives an energy $E \simeq 7.9^{+0.9}_{-0.4}$ keV, which sustains such an interpretation (Pounds, 1990).

– the hard X-ray tail, which is the third characteristic, represents an excess in the flux for $E \geq 12$ keV (see Figure 58), which can be well modelled by a power law with spectral index $\alpha \simeq 0.45$ in the range 10–35 keV. This was the first evidence of a high energy tail in the continuum spectrum of low luminosity AGNs, and it agrees with the foresight of a reflection bump (Guilbert & Rees, 1988; Lightman & White, 1988). Moreover, the flat spectrum owed to this excess offers a new explanation for the spectral form of the X-ray background radiation (Fabian *et al.*, 1990).

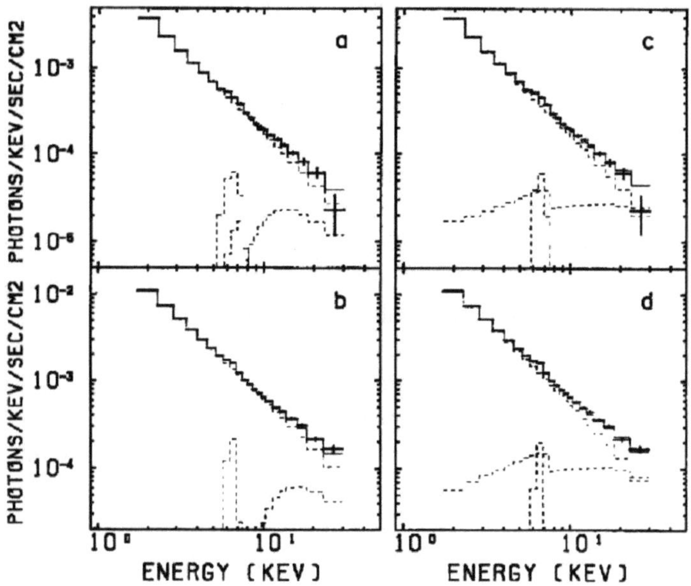


*Figure 59.* Photon spectrum of NGC 7469 and IC 4329A fitted with the partial covering model and broad line [(a) NGC 7469, $\chi^2_\nu = 1.1$; (b) IC 4329A, $\chi^2_\nu = 0.8$] and with the reflection model with broad line [(c) NGC 7469, $\chi^2_\nu = 1.3$; (d) IC 4329A, $\chi^2_\nu = 1.0$]. The components of the model are plotted with dashed lines (Piro, Yamauchi & Matsuoka (1990).

However, a strong controversy on the eventual presence of variability in the slope of the power law spectra of individual AGNs was still open at the GINGA epoch (e.g., Pounds, 1990).

We would like to remark that the composed spectrum GINGA-12 offered an interesting unification in some of the observed properties of AGNs. The variable soft X-ray excess and (possibly) the redshifted Fe $K$–line reveal cold and dense matter at a distance $\leq 100$ gravitational radii, $r_g$ ($r_g = GM_{BH}c^{-2}$) of the central black hole with mass $M_{BH}$. The broadening observed in the Fe $K$–line of several individual sources strengthens such a presence. GINGA-12 data reveal also the existence of reflected components, which are expected by such a cold matter. In some spectra of GINGA there are some signals of the presence of a highly ionized gas across the line of sight and very close to the nucleus, with a minimum Thompson thickness of $\simeq$0.05.

Such kind of behaviour of the Seyfert galaxy spectra were discussed by Piro, Yamauchi & Matsuoka (1990), after GINGA measurements of the bright Seyfert-1 galaxies NGC 7469 and IC 4329A, in which a hard X-ray bump above 8 keV was revealed (Figure 59). This spectral component, detected for the first time, was probably associated to reflection or absorption of the central emission by a very thick ($N_H > 10^{24}$ cm$^{-2}$) cold medium in the environment of the central source, possibly in an accretion disk. This scenario was supported by the detection of iron fluorescent lines at 6.4 keV with equivalent widths of 100–150 eV in both objects. The two possibilities of reflection or absorption by a cold medium with high column density do not exclude each other. Indeed, in the most general case both the absorbed and reflected components should be present, with relative normalization that depends on the distribution of the medium around the central source (Piro, Matsuoka & Yamauchi, 1989). Whereas, Piro, Yamauchi & Matsuoka (1990) did not exclude a scenario based on partial covering and/or reflection by highly dense clouds near the black hole (Guilbert & Rees, 1988) or by an obscuring torus placed outside the BLR (Broad Line Region) (Matsuoka *et al.*, 1990) or by the BRL itself, given the theoretical and observational results suggesting clouds with $N_{\mathrm{H}} > 10^{24}$ cm$^{-2}$ (Ferland & Persson, 1989). Piro, Yamauchi & Matsuoka (1990) then considered the reflection from an accretion disk as the most appealing scenario for the following reasons: first, the reflection model provides a good fit to the spectra with the least number of free parameters. In particular, the column density of the medium is not constrained in a small range around $10^{24}$ cm$^{-2}$, as in the partial covering model, in order to fit the hard X-ray bump. Then an accretion disk can account for several features in the spectrum of AGNs, namely the UV bump (Malkan & Sargent, 1982), the soft X-ray excess (Arnaud *et al.*, 1985), the high energy bump, and the iron fluorescence line. The observed equivalent widths of these lines are in agreement with those expected by fluorescence from a disk illuminated by the X-rays (Basko, 1978). Moreover, the line width observed in NGC 7469 is consistent with that produced by matter orbiting around a black hole in Keplerian motion, as in the case of an accretion disk (Fabian *et al.*, 1989; Bisnovatyi-Kogan & Lovelace, 2001).

These measurements showed also the spectral flattening above 8 keV, which provided the first observational evidence that solved the discrepancy between the spectral shape of the cosmic X-ray background and that of AGNs, under the hypothesis that a large fraction (>70%) of them might have spectra similar to those observed in NGC 7469 and IC 4329A. In this way it was possible to reproduce the spectrum of the X-ray background by the contribution of AGNs distributed in redshifts (Morisawa *et al.*, 1990).

From these data we had a clear indication that future experiments with higher spectral resolutions could have helped in redefining the geometry and kinematics of the emitting regions and their surroundings. Such missions, already programmed in that time, are the ASCA (middle 1990s, CHANDRA and XMM launched in 1999, INTEGRAL launched in 2002, and the future SPECTRUM-X, with the launch planned in 1993, but until now not yet in flight (probably in its original version it will never fly).

#### 5.1.5. *The Class of X-Ray Extragalactic Sources: A Unification*

An inspection of the Figure 57 suggests the possibility of a unique central engine producing energy in AGNs. On the other hand, 'normal' galaxies contribute to the extragalactic X-ray background for about 13%. On the basis of these two facts, Giovannelli & Polcaro (1986) studied a sample of 191 extragalactic objects, with known redshift, which were the X-ray emitters known at the beginning of 1980s, in order to search for general relationships between physical parameters independent of the classes to which the objects belonged. Such a sample was formed by QSOs, Seyfert galaxies, BL Lac objects, elliptical and S0 galaxies, spiral and irregular galaxies. Most of the data came from the Einstein satellite; for this reason the authors considered the equivalent monochromatic X-ray luminosity at 2 keV, as usual for data from this satellite. For the remnant data, coming from other sub-samples, the luminosity was reduced at 2 keV in order to have a global homogeneous sample. A general relationship between the logarithms of the equivalent monochromatic luminosity at 2 keV and redshift was found. In order to avoid selection effects owed to sensitivities of the instruments, they used the most luminous objects, independent of their class, contained in each arbitrary interval of redshift. Such a relationship is a continuous curve, which is practically constant for $z < 0.001$, a flex in the range $0.001 \leq z \leq 0.01$, and for $z > 0.01$, where only AGNs are present, the best fit of the curve is a power law with slope equal to 1.6. This continuous curve suggests a physical continuity of such extragalactic X-ray emitters.

The idea developed by Giovannelli & Polcaro was based on the extragalactic emitters having to be considered only from the point of view of high energy emission, if one wants to study processes at their nuclei. Indeed, processes occurring at other lower energy ranges (i.e. optical, IR, radio) are influenced by the matter around the nucleus, which constitutes a sort of dress of the collapsed object. Therefore if one wants to study the nucleus this must be 'undressed', which is equivalent to studying it at X-ray energies. Moreover, such a radiation is certainly coming from the vicinity of the nucleus, independently of the models one can construct.

The results are reported in Figure 60. The functions $L_{\mathrm{max}}(z)$ of the studied sample (x) and control sample ($*$) at lower luminosity are reported in the upper panel. In the lower panel the same plot is reported, but for $z > 0.01$, with the best fits to data, and data from clusters of galaxies (+), which evidently emit with different processes, and therefore the best fit to data must have a completely different slope. The functions $L_{max}(z)$ for both classes (studied and control) are parallel; this is a strong indication on the presence of a unique kind of central engine, responsible of the X-ray emission.

The fact that such extragalactic X-ray emitters appear different, if observed at different energy ranges, means simply that the central engine is placed in different environments.

Finally, the function $L_{\mathrm{max}}(z)$ could be interpreted as owed to an evolution of the central X-ray source, from a very active phase (that of the AGNs) to a more quiet phase (that of 'normal' galaxies), with an intermediate phase in the range of redshift $0.001 \leq z \leq 0.01$ – to which 'intermediate' objects, such as the *liners*, belong. In this range of redshift the luminosity rapidly, but continuously, decreases by about two orders of magnitude.

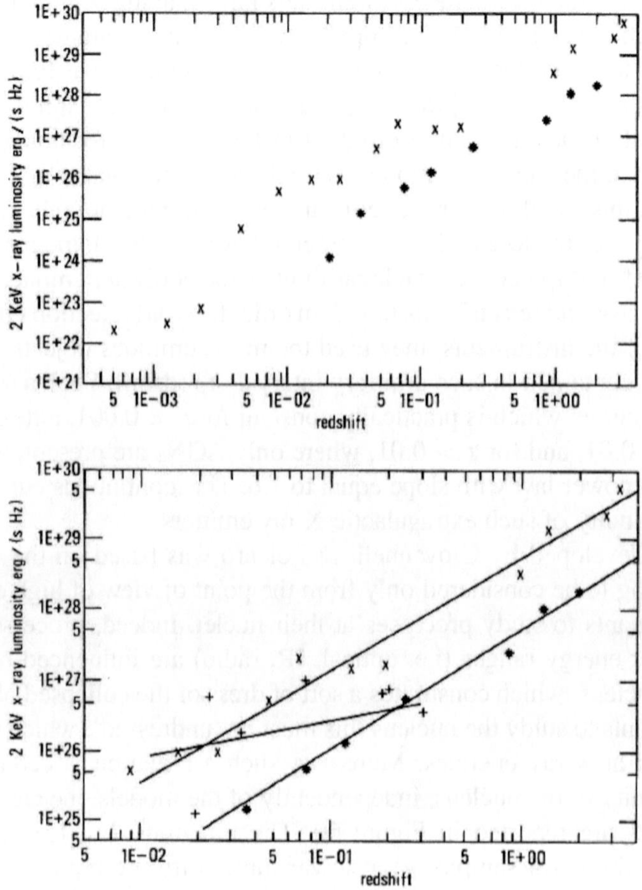


*Figure 60.* Maximum luminosity vs. redshift for two samples of extragalactic objects. Upper panel:(x) the studied sample; (∗) the control sample. Lower panel: (x) the studied sample; (∗) the control sample; (+) the clusters of galaxies sample (Giovannelli & Polcaro, 1986).

Therefore the diagram lg $L_{\max}$ viz., lg $z$ can be interpreted as a sort of *evolutive diagram* of extragalactic X-ray emitters (but the clusters of galaxies, in which the production of X-rays is owed to different processes).

With data from the higher sensitivity experiments, such as those of the ROSAT satellite, it is now possible to check the Giovannelli & Polcaro diagram with more samples of extragalactic objects at lower and lower energies. This will be reported in the Section 7.1.

### 5.1.6. *X-Rays in Clusters of Galaxies*

We have already discussed clusters of galaxies in Section 3.3.2. However, for convenience of readers we summarize here some of the main points related to their X-ray emission.

As is well known, clusters of galaxies (CGs) are gravitationally bounded systems, which usually contain hundreds of galaxies in a region whose dimensions are of the order of $10^{25}$ cm. They can be divided in two classes, namely:

- *irregular*: they have a strong sub-accumulation or any strong central concentration. In both cases there is a high content of spiral galaxies;
- *regular*: they are CGs dynamically more evolved, with a more homogeneous distribution, central condensation and low contents of spiral galaxies.

The CGs constitute a class of X-ray emitters very luminous with $L_x \simeq 10^{43-45}$ erg $s^{-1}$. The emission mechanism is thermal bremsstrahlung. The radiation is produced in a hot gas ($T \simeq 10^8$ K) with low density ($\simeq 10^{-3}$ atoms $cm^{-3}$, which has a distribution similar to that of galaxies and fill up the intergalactic space.

From the X-ray observations one deduces that the mass of the gas is of the same order of that contained in all the galaxies within the cluster. X-ray spectra show the presence of Fe emission lines and other heavy elements. This then indicates that a significant fraction of the gas should have been injected into the cluster from its galaxies.

Observations of the Einstein satellite revealed that in some clusters the gas is in a cooling phase and is accreting from very large central galaxies. The images of clusters indicate that the morphology of the gas reflects the dynamical state of the cluster (e.g., Sarazin, 1988a,b).

When the X-ray emission from CGs was discovered, three possible mechanisms of emission were suggested, namely:

- thermal bremsstrahlung radiation;
- Compton scattering of the background photons with relativistic electrons;
- emission owed to a population of stellar X-ray sources, like those encountered in the Milky Way.

Felten *et al.* (1966) were the first to suggest that X-ray emission in CGs (in particular, in the Coma cluster) was owed to a diffuse gas. Another reason in favor of the thermal bremsstrahlung radiation is that it is not possible to explain the relationships between X-ray and optical emissions with the other two latter mechanisms.

The detection of the lines of highly ionized iron constitutes the strongest proof of the presence of the thermal bremsstrahlung radiation. Moreover, its abundance is similar in all the clusters. Therefore it should be a rare coincidence that the lines and continuum would proceed from two different sources. The observations of the X-ray line showed that such emission, owed to a hot diffuse gas, is of thermal origin. Part of this gas is coming from stars, which probably lie in galaxies and another part is returned to central galaxies, to form again stars, completing the cycle.

## 5.2. Extragalactic Historical Journey: Ultraviolet Range

The sky in the UV range has been carefully described thanks to the IUE (International Ultraviolet Explorer) satellite, although precious contributions were obtained from the first UV experiments on board rockets and the Copernicus satellite, as already mentioned in Section 2.4.3.

### 5.2.1. *Local Group of Galaxies*

The most important information obtained with the observations of the Local Group of galaxies can be summarized as follows:

– i) stellar winds from OB stars in Magellanic Clouds;
– ii) X-ray binaries and other massive binaries in Magellanic Clouds;
– iii) most bright hot stars in M31 and M33.

i) the fluxes and effective temperatures of the OB stars in Magellanic Clouds are approximately equal to those of the same type in the Galaxy.

Usually it is possible to obtain consistent evaluations of the bolometric magnitude $M_{BOL}$ derived from the fluxes measured on the Earth and from the IUE, by means of an appropriate bolometric correction. The determinations made on the Earth are sensible to the effective temperature, whilst those made by the IUE are sensible to the extinction. Therefore the study of the continuum flux of the hot Magellanic Clouds stars has been an important aspect of the work derived from the IUE observations. The extinction, although rather complicated, is very small and the distance is known rather well (Hutchings, Lequeux & Wolf, 1987).

Probably the most important results are coming from the spectra with UV emission lines. The intensities of such lines are lower than those of galactic similar stars; the same occurs in the optical range. In bright stars the resonance lines of SiIV, CIV, and NV are weaker both in emission and absorption and the escape velocities are less than those of the similar stars in the Galaxy. Such a difference is stronger in the Small Magellanic Cloud (SMC) than in the Large Magellanic Cloud (LMC). Such results, as well as those obtained in the optical (Hutchings, 1980), indicate that the stellar winds in stars of the Magellanic Clouds are systematically different with respect to the analogous ones in the Galaxy. Generally, it is assumed that such a difference is owed to the radiative opacity caused by the different abundances of metals in decreasing order in the three galaxies: Milky Way, LMC and SMC. Such a result supports the radiative acceleration model, which assumes that stellar winds are transported in a radiative way.

For the determination of the continuum flux distribution of Wolf–Rayet (WR) stars, Smith & Willis (1983) combined the IUE data with those taken on the ground for 9 WR stars in the LMC and constructed the HR diagram. The WR stars of late N-type (WR-N) seem, in general, colder and more luminous than the correspondent in the Galaxy. In contrast, the young WR-N and WR-C (C-type) seem hotter than those of our Galaxy. The wind velocities are similar to those in the Galaxy, whilst the intensities of the emission lines are greater. Comparing with the Galaxy, in such evolved objects in Magellanic Clouds there is no difference in the abundance of elements; the stellar nucleosynthesis is then the same.

ii) The study of binaries in the Magellanic Clouds has been limited almost exclusively to X-ray binary systems. IUE data allowed the study of the interactions of the stellar winds with the collapsed objects and the dependence with respect to the orbital phase of the binary system. The X-ray binaries in the Magellanic Clouds have an X-ray flux significantly higher than the binaries in the Galaxy (Hutchings, Lequeux & Wolf, 1987 and the references therein). In the case of the system X 0535-66 (Charles *et al.*, 1983) – which is formed by a B star and a fast X-ray pulsar orbiting around in an eccentric orbit with a period of 16.5 days – the UV lines have profiles similar to those of QSOs; this is an indication of high escape velocity of a dense wind. There are important spectral variations around the periastron passage, which is the moment in which the neutron star (X-ray source) probably crosses the shell of the B star (or the equatorial excretion disk of the B star).

All the models, based on the orbital interactions of the two components of X-ray binaries are based on the IUE data.

iii) The observations of supergiant OB stars in M31 and M33 reveal that the lines are weaker and the winds slower than in the corresponding stars in our Galaxy.

It then seems that the strong stellar winds are a characteristic of the stars in our Galaxy. Therefore the idea that the abundance of heavy elements is the main parameter, which determines the existence of such winds could not be totally certain (Massey, Hutchings & Bianchi, 1985).

#### 5.2.2. HII *Regions in Close Galaxies*

The knowledge of the abundance of elements in galaxies provides important information on the history of their stars. The comparison of those abundances with those of our Galaxy allows to improve the knowledge of the evolution of galaxies. The abundances influence many processes occurring in galaxies, since can vary the opacity and the diffusion of radiation.

Some examples of the importance of the knowledge of abundances could be the differences observed in the behavior of stellar envelopes in galaxies, the stellar mass loss, interstellar extinction, and the determination of the fluxes of X-ray sources.

Ultraviolet spectra of HII regions provided the proper information in order to determine the abundances. One of the most important contribution of the IUE satellite has been the observations of the giant HII regions placed outside the Galaxy. Some low resolution spectra of a number of HII regions of close galaxies were taken (Rosa, Joubert & Benvenuti, 1984). The unique emission line clearly detected in most of such spectra has been the CIII] (1909 Å). However, some weaker emission lines were detected in several HII regions of the LMC and SMC (Dufour, Shields & Talbot, 1982). Pagel (1985) summarized the results remarking that the abundance ratio C/N is constant, while the ratio C/H and N/H do not vary as the ratio O/H from one galaxy to another.

The most difficult problem in determining such abundances is the correction of the intensities of UV lines owed to the extinction. However, in the Magellanic Clouds such a problem has been solved. What is happening in these clouds represents a good example of understanding the meaning of such results. In the SMC the carbon is more exhausted than nitrogen and oxygen (with respect to the Galaxy), whilst in the LMC the nitrogen is more exhausted than carbon and oxygen.

From a theoretical point of view it is accepted that N and O are universally generated in the evolution of massive stars ($M > 10$ $M_{\odot}$) and C originates in the evolution of less massive stars ($M = 4$–$10$ $M_{\odot}$). The ratios C/N and C/O then start to have small values, which grow starting from the moment in which the less massive stars begin to evolve. Later, when the stars with masses $M < 4$ $M_{\odot}$ have evolved, the ratio C/N decreases until those stars begin the production of N.

Maran *et al.* (1982) utilized the IUE in order to study planetary nebulae in the Magellanic Clouds and obtained very interesting results. They found that the C abundances are only a bit less than in the Galaxy, whilst the other elements are exhausted following the 'normal' proportions. They also found that in the beams of planetary nebulae (PNe) in the SMC the C abundance is 40 times greater than in its ISM, whilst in the LMC this abundance is 6 times greater than in its ISM. Therefore they found that PNe are the most important sources of C enrichment in Magellanic Clouds, and the processes of its generation in the stars of the Clouds are the same as in the Galaxy. This result is very important for the theory of stellar evolution.

The UV spectrum of HII extragalactic regions is dominated by the star light, which produces a continuum to which sometimes some lines are overlapped. A part of the continuum emission is owed to the contribution of ionized gas. Clearly the hot component of the star clusters embedded into such HII regions is the responsible of the emission in the UV band. The very interesting P Cygni profiles of CIV (1548–51 Å), SiIV (1394–403 Å), and NV (1239–43 Å) (this latter is blended with the interstellar Ly-$\alpha$ line) visible in the HII regions of NGC 604 and IC 132 of the close spiral galaxy M33 reveal the presence of strong stellar winds from hot stars, which would be rather massive (Lequeux *et al.*, 1981). In these two objects there is also a wide emission of HeII (1640 Å), which is one of the characteristics of the Wolf–Rayet stars. The existence of such stars was confirmed in the optical range by Massey & Conti (1983). P Cygni profiles and the characteristics of WR stars are seen also in starburst galaxies. This means that the stellar content responsible of such lines is similar to that in other galaxies, although the angular resolution of the IUE does not allow a spatial discrimination of stars in clusters having ionizing stars (Lequeux *et al.*, 1981).

5.2.3. *Young Peculiar Stars with Emission Lines in the Local Group of Galaxies*

The upper part of the HR diagram is populated by young supergiant stars with emission lines, especially between B0 and B5 spectral type. Such objects play an important role in the development of evolutionary theories of supermassive stars.

Moreover, such stars (with the exception of supernovae) are the brightest stars of the Universe and constitute a good indicator of the distances of extragalactic systems. Most of the radiation emitted by such objects is UV radiation, and so, the IUE has allowed a deep study of their properties.

Such young supergiant stars with emission lines are formed by:

- A sub-class of blue supergiant stars, such as the variable stars S Dor or Hubble–Sandage variables. They are characterized in the optical range by irregular photometric variations of more than one magnitude in time scales from about one year to some decades. Their spectra, highly variable, are characterized by Balmer lines during the maximum luminosity and by the forbidden lines of [FeII] and [NV] during the minimum.
  With the high resolution IUE observations, ground spectroscopic optical observations, and IR photometry it was found, for S Dor variables in LMC during the bright phase, a cold ($T \simeq 8,000$–10,000 K) extended (>10 $R_{\mathrm{star}}$ expanding envelope, with low terminal velocity ($v_\infty \simeq 100 - 200$ km s$^{-1}$). The mass loss rate during such a bright phase is $\dot{M} \simeq 10^{-5}$–$10^{-4}$ $M_\odot$ yr$^{-1}$, which is a factor of 10–100 higher than in the phase of minimum. Therefore, the variations measured in the optical range can be explained as a consequence of the variable mass loss (Hutchings, Lequeux & Wolf, 1987).
  The variables of S Dor type are considered as basic objects for the comprehension of the evolution of very massive stars ($M > 50$ $M_\odot$), and it is assumed that they represent a very short phase preceding that of very massive WR stars (Maeder, 1983);
- A sub-class formed by supergiant Be stars. Contrary to the S Dor variables, such Be supergiant stars in LMC do not show photometric variations, and the mass loss is much more stable. Probably, owing to the high rotational velocity of such stars, an enhancement of the mass loss occurs at the equatorial regions, where a dense and cold wind tends to block the formation of variable stars of S Dor type.

### 5.2.4. *Starburst Galaxies*

The galaxies in which an intense process of star formation occurs are called *starburst galaxies*. Such a process manifests itself with a burst of luminosity. Such galaxies are very interesting since the luminosity owwed to this process dominates their total emission in many energy ranges. Some examples of star formation processes observed in close galaxies provide valuable information on the formation of massive stars and on the new forming galaxies. Such galaxies appear like those formed in the prior phases of the Universe.

Massive stars are responsible of a large part of the radiation of the galaxy. They evolve in a time scale much shorter than the age of the host galaxy. The probability of finding starburst galaxies is then very high. Massive stars emit large quantities of energy in the X-ray, UV, IR and radio ranges (and also in the optical range). In the X-ray range the emission is owed to the radiation generated by the accretion of matter in a binary system with a compact component and to the radiation of the SNRs. In the IR range the emission is owed to dust heated by the absorption of the UV radiation emitted by the massive stars. In the radio range the spectra typically are non-thermal and are produced by the SNRs of exploding massive stars. In the UV range the continuum emission is essentially owed to hot stars. Many of the absorption features can be identified with the lines of SiIV and CIV. Analyzing their structures it is possible to deduce the stellar wind terminal velocities, which are of the order of 2,000–4,000 km $s^{-1}$. Actually, such velocities were measured in NGC 604 (Rosa, 1980) and in other HII regions where WR stars lie (whose existence was already known by means of optical measurements). Sometimes these lines also show P Cygni profiles, which are attributed to mass loss in massive stars of type O. If such a profile is very strong it can be associated either with WR stars or with very massive young supergiant stars.

Probably most of the UV emission is owed to stars with an initial mass of about 20 $M_\odot$ (Kunth & Weedman, 1987). By using UV data Huchra *et al.* (1983) concluded that the observed regions in very blue irregular galaxies of Magellanic type are formed by recent bursts of star formation overlapped to an old stellar population.

Stars with effective temperatures between 30,000 K and 35,000 K and initial masses of about 20 $M_\odot$ dominate the emission in the UV range. The explication could be that the rate of star formation is large enough to keep well populated the upper part of the main sequence, which is that dominating the emission in the UV range.

IUE spectra provided also the possibility of checking the existence of other similar galaxies at high redshifts. Obviously it is necessary a large surface brightness in order to resolve an object like a galaxy at $z \geq 1.5$. Some models support galaxies of primordial type being able to form in a bright phase of starburst, and then both the UV continuum and emission lines must be intense (Kunth & Weedman, 1987).

### 5.2.5. *Globular Clusters*

A globular cluster is formed by $10^5 - 10^6$ stars gravitationally bound in a roughly spherical system. Globular clusters (GCs) possess the characteristics of our galactic halo and, in general, of halos of spiral galaxies. Such clusters are also present in elliptical and irregular galaxies; thus they constitute one of the components more numerous of the Universe. The stars contained in GCs of the Galaxy and of the spiral galaxies are population II stars, which are poor in metals and have an age of $\simeq 1.5 \times 10^{10}$ yr, i.e. they are a class of objects amongst the oldest of the Universe.

Because of the presence of young GCs in the Magellanic Clouds ($\simeq 10^7 - 10^8$ yr), poor in metals (with abundances different with respect to the GCs in the Galaxy), Castellani & Cassatella (1987) extended the definition of population II stars to all metal poor stars, independent of their ages. Such conclusions were reached by means of UV measurements of a large sample of galactic and extragalactic GCs, with a real possibility of resolving individual bright stars in the

Galaxy. Thanks to this the distance of some GCs has been determined in a more precise way. Therefore a better determination of the effective temperature and luminosity of several GCs allowed their more adequate positioning in the HR diagram. The IUE measurements allowed such important conclusions to be reached because the position of the horizontal branch (HB) stars in the theoretical HR diagram is governed by strict theoretical limitations essentially owed to original abundance of helium.

Such measurements allowed also to discover an apparent contradiction: at the time when in some galaxies, recently formed, the GCs interrupted their formation, in other galaxies the GCs followed to be formed (age $\simeq 10^6$–$10^7$ yr). This fact indicates that star formation is strictly bounded to the dynamics of the progenitor galaxies. A possible explication could be that the GCs cannot follow to form in primeval halos where the gas is exhausted, and not even they can form in the disk in late phases owing to large effects of differential rotation (Iben & Renzini, 1983).

### 5.2.6. *Active Galactic Nuclei*

Before the launch of the IUE, only two sources were recognized as UV emitters: the QSO 3C 273, thanks to the measurements of the Dutch satellite ANS (Wu, 1977) and to an experiment of spectroscopy on board a rocket (Davidsen, Hartig & Fastie, 1977) and the Seyfert galaxy NGC 4151 (Hartig & Davidsen, 1978). The IUE, because of its limit in sensitivity, measured only relatively bright AGNs (up to 17th magnitude) and then, the objects measured have low-medium redshifts. However, in spite of this limitation, the IUE expanded the range of energy in which such AGNs can be studied and provided detailed measurements of low redshift AGNs, which together with optical measurements allowed the foresightedness of the structures of the intrinsic UV spectra of high redshift sources.

The spectra of AGNs are characterized by a non-thermal continuum, usually owed to synchrotron emission, and by wide emission lines. Such lines can be of two types:

- permitted resonance lines of abundant ions, which appear mainly in the UV range, with Doppler broadening of $\approx$10000 km $s^{-1}$;
- forbidden lines with Doppler broadening of $\approx$1000 km $s^{-1}$. It is believed that the continuum forms in a very small nucleus, probably supplied by accretion. Such a nucleus is surrounded by a region in which the broad lines form (Broad Line Region: BLR). Such regions have a diameter of $\sim$0.1 pc and contain many dense clouds (electronic density $\simeq 10^{10}$ $cm^{-3}$) and have high velocity. Just in these regions the permitted broad emission lines originate (Malkan, Alloin & Shore, 1987). It is also believed that the BLRs are surrounded by regions in which forbidden narrow emission lines originate (Narrow Line Region: NLR) (Kondo, Boggess & Maran, 1989).

There is a difficulty in understanding the continuum spectra of AGNs, since the comparison of data coming from low redshift objects with those from high redshift objects is needed. Sometimes the wavelength region in which the spectra of the latter highly redshifted objects falls just in the range observable from the ground. Therefore it is difficult to verify if the existing differences are owing to evolutionary effects or to transmission effects in the intergalactic medium or to intrinsic differences.

When it was possible to obtain spectra of different individual sources in a wavelength range of several decades, from the UV to the far IR (FIR), it was possible to see clearly that different phenomena dominate in different energy ranges. In particular many AGNs show a blue bump at wavelengths less than 3000 Å (Oke & Zimmerman, 1979). This prevents the fit of the non-stellar continuum to a simple power law. This blue bump contains most of the emerging energy from many AGNs. The best fit of such an excess is a simple black body with $T \simeq 26000$ K (Edelson & Malkan, 1986). There are other sources in which the data are fitted to models with geometrically thin optically dense accretion disks (Malkan, 1983).

IUE spectra in the range 2000–3000 Å, obtained with a good signal to noise ratio revealed spectral structures poorly contrasted because of blends of many FeII emission lines. Such lines are so close to each other that they can produce the effect of a pseudo-continuum, which contains an integrated flux similar to that of the Ly-$\alpha$ (Wills, Netzer & Wills, 1985). Therefore in constructing models for the BLRs it is necessary to account for such 'unexpected' fluxes of the FeII emission lines. A review on the BLR in AGNs was written by Raine (1988).

The most studied Seyfert galaxy is NGC 4151. In this galaxy the luminosity variations are greater in the UV than in the optical and IR ranges. This fact could be owed to a larger contamination of the star light at longer wavelengths, which could reduce the apparent percentage of the variation of the nuclear luminosity (Ulrich *et al.*, 1985). Typical time scales for doubling the luminosity of the continuum in NGC 4151 are 5–30 days, which puts limits on the dimensions of the emitting source ($\leq$ 0.01 pc). This is in contradiction with the values of the dimensions deduced by the X-ray outbursts, which can develop in less than 12 hours. Therefore it could be possible that the X-ray radiation originates in regions smaller than those in which UV and optical radiation originate, and consequently the latter kind of radiation is produced with different mechanisms from the former type.

It is worthwhile remembering the method of Giovannelli & Polcaro (1986) for exploring the characteristics of the central engine of extragalactic objects emitting X-ray radiation, independently of the models used for a description of such sources. The X-ray radiation must originate from regions very close to the central engine.

The IUE observed in low and high resolution modes the far away bright QSO HS 1700+6416 ($V$ = 16.1 mag, $z$ = 2.72). For the first time it was possible to obtain information on a QSO with a rest wavelength of ~330 Å. Although the intergalactic Ly-$\alpha$ line and continuous absorption enough diminish the continuum at rest wavelengths between 850 and 450 Å, such a continuum grows well defined towards shorter wavelengths of the order of 330 Å. Such a result, applied to other QSOs and AGNs, places the maximum of the continuum at wavelengths $\leq$ 300 Å (Kondo, Boggess & Maran, 1989).

## 5.3. Extragalactic Historical Journey: Infrared Range

The possibility of encountering a new class of 'infrared galaxies' was one of the reasons of the all sky survey in the IR range. The IRAS satellite discovered hundreds of galaxies- which emit $\simeq$95% of their energy in the IR range. The IRAS detected both close and far galaxies with redshifts $z > 0.4$. The IR luminosity of such galaxies is ranging across many decades: from $\leq 10^6$ $L_\odot$ to $10^{13}$ $L_\odot$ (Soifer, Neugebauer & Houck, 1987).

In the IRAS previous evaluations for the PSC (Point Source Catalog) the emission of the quasar 3C 273 was close to the detection threshold. Therefore the perspectives were to attempt the detection of either active galaxies with very high thermal components, such as for instance the Seyfert galaxies, or strong non-thermal sources, such as for instance the OVV (Optically Violent Variables) sources. Indeed, the IRAS measured many AGNs, and discovered without any doubt the phenomenon of star formation bursts, and described in more detail the behaviour of 'normal' galaxies and opened new windows onto the study of 'active' galaxies.

Summarizing, thanks to the IRAS, the number of the extragalactic sources measured passed from some dozens to about 25,000. This allowed a substantial improvement in the study of the extragalactic objects, although without any substantial revolution.

### 5.3.1. *The IRAS Extragalactic Sky*

The all sky survey of the IRAS covered about 96% of the whole sky, with limits of $\simeq$0.5 Jy at 12, 25 and 60 $\mu$m and $\simeq$1.5 Jy at 100 $\mu$m. The main advantage of the IRAS survey has been the uniform calibration better than 10% for all individual sources, in practically all the sky with the galactic extinction virtually neglected at the wavelengths of the survey. For galactic latitudes $| b |> 30°$ the extinction of the galaxies is <1% at 25, 60 and 100 $\mu$m, and several percent at 12 $\mu$m.

The measurements at 60 $\mu$m are the most sensitive for detecting most of the extragalactic objects. About 25,000 galaxies were detected; about 50% of them were already catalogued in optical range. More than 75% of the sources detected at 60 $\mu$m and reported in the IRAS Point Source Catalog (PSC, 1985) for $| b |>$ 30° are extragalactic in origin (IRAS Explanatory Supplement, 1985; Soifer *et al.*, 1986a; Lawrence *et al.*, 1986). Most of extragalactic objects detected by the IRAS are spiral galaxies of late type. Rarely elliptical and S0 galaxies were detected.

The systematic identification of the IRAS sources with optical counterparts (Savage *et al.*, 1987) showed that practically all of them in the FIR range of the PSC – placed at high galactic latitude, and not being manifestation of emission from galactic cirrus (Beichman, 1987) – have a rather bright optical counterpart. Owing to the sensitivity limit of the IRAS the average redshift of the selected objects at 60 $\mu$m in the PSC is $z = 0.03$ and the corresponding luminosity is $\simeq 2 \times 10^{10}$ $L_\odot$. The luminosity of the galaxies range from $\leq 10^6$ $L_\odot$ for the closest dwarf elliptical galaxies to $10^{13}$ $L_\odot$ for the brightest galaxies detected (Kleinmann & Keel, 1987).

The IR bright galaxies with bolometric luminosity $L_{\rm bol} > 2\times10^{11}\,{\rm L}_\odot$ constitute the dominant population of the local Universe, with a space density greater than that of normal galaxies, and optically selected starburst galaxies. The space density of such IR bright galaxies is similar to that of optically selected Seyfert galaxies.

For $L_{\rm bol} > 10^{12}\,{\rm L}_\odot$ the selected IR galaxies have a space density greater than that of the QSOs, which is the unique known class of objects with such a luminosity (Soifer *et al.*, 1986a).

### 5.3.2. *The Predominant Processes at the IRAS Wavelengths*

The three mechanisms probably producing the IR emission in bright extragalactic sources are:

- Photospheric emission of stars: in many young galaxies the stellar photospheric emissions contribute essentially to the 12 $\mu$m emission. At longer wavelengths such a contribution is never relevant;
- Synchrotron emission owed to relativistic electrons in the magnetic fields of the sources. Such radiation dominates the emission observed in luminous objects, such as BL Lac type and QSOs in which optical violent variations are present (OVV: Optically Violent Variables), and can be important also in other type of QSOs and in Seyfert galaxies;
- Emission owed to dust heated by another source: probably the dust heated by various radiation fields is responsible of the IR emission of most of the galaxies detected during the IRAS survey. The IR emission of the dust presents a triple observational characteristic, namely:
  - Extended emission in sources in which it is possible to obtain a sufficient linear resolution;
  - A large IR excess overlapped to the expected photospheric emission or a reasonable extrapolation from the radio spectrum;.
  - The shape of the energetic distribution of the continuum provides a good fit to a Planck function with one temperature, modified because of the emissivity dependent on the wavelength (Soifer, Neugebauer & Houck, 1987).

The small dust grains have a large optical UV opacity with respect to their mass; then they are extremely efficient in the absorption of high energy incident photons, which are transformed into IR photons. The high ratio between the IR luminosity and mass renders the FIR emission an extremely efficient indicator of the cold material in galaxies.

The mass of dust $M_{\rm d}$ necessary to produce a luminosity $L$ by optically thin FIR emission is given by:

$$M_{\rm d} \simeq 10^4 \frac{L}{10^8{\rm L}_\odot}\frac{T_{\rm d}}{40{\rm K}}{\rm M}_\odot\,, \tag{5.4}$$

where $T_{\rm d}$ is the dust temperature (Draine & Lee, 1984).

One of the main problems which are posed is the identification of the mechanism responsible for the production of luminosity in galaxies. It is possible that many galaxies have extremely rapid star formation bursts. Such galaxies are particularly active in the IR range. Although it is accepted that the isolated regions in such galaxies are qualitatively similar to the star formation regions in normal galaxies, the frequent bursts indicate a rate of star formation high enough to prevent the process beyond $10^8$–$10^9$ yr; indeed, if the process were not to be stopped after that time the utilization of a mass greater than that available in the ISM of the galaxy would be necessary, which is absurd.

The most luminous starburst galaxies have an IR luminosity $L_{IR} \gg 10^{11}$ $L_\odot$. For instance, a galaxy with a luminosity of $10^{11}$ $L_\odot$ owed to young stars and with a 'normal' quantity of interstellar matter would consume the whole quantity of the interstellar gas ($\simeq 4 \times 10^9$ $M_\odot$) in a time of $\simeq 5 \times 10^8$ yr (Scoville & Young, 1983).

It is believed that the energy of AGNs is owed to massive compact objects (black holes) placed in their centers or close to them. The luminosity is deduced by starting from the liberation of the gravitational potential energy of the accreting matter onto the compact object, whilst the nature and the emission spectrum depend on the density and optical thickness of the material surrounding the central object. In some sources (e.g., in luminous BL Lac objects) synchrotron emission of relativistic electrons produced by the accretion process is observed. In other cases the emission of a continuum and lines of the dense plasma surrounding the central object is observed. The manifestation of the object in the IR range depends on the degree of reprocessing of the energy in the material surrounding the nucleus. A simple relationship between the observed spectrum and the emission process has not yet been found. One possibility could be simply the absence of such a correlation.

An important review on the accretion problems and on the emission processes is that of Rees (1984).

On many occasions the star formation and AGNs' processes occur at the same time. In particular, in the most IR luminous galaxies both characteristics of star formation bursts and of AGNs often appear. It is a problem in understanding the reasons because two such processes contribute in the same way to the luminosity of many sources.

### 5.3.3. *Spiral and Disk Galaxies*

In the FIR the spiral galaxies constitute the dominant population within the bright galaxies. The two largest spiral galaxies in the sky are M31 and M33, which are, for comparison, inactive galaxies for what concerns their IR emission; indeed the ratio $L_{\rm FIR}/L_{\rm B} \simeq 0.07$ for M31 and 0.2 for M33. Qualitatively the giant spiral galaxy M101 is not different from M31 and M33 for its global properties, although in the FIR it is the most active galaxy of this type, being its $L_{\rm FIR}/L_{\rm B} \simeq 0.4$ (Beichman *et al.*, 1987).

In lower luminosity galaxies ($L < 10^{10}$ $L_\odot$) there is not a direct evidence of a relationship between the FIR emission and the star formation occurring in each galaxy.

In both galaxies M31 and M33, the observations can be interpreted by means of a model in which the IR emission is assumed to be owed to two components:

- one component owed to interstellar dust heated by the diffuse interstellar radiation field;
- other component owed to the activity of star formation in HII regions and/or in complexes of molecular clouds (Soifer *et al.*, 1987).

From the observations of M31, Walterbos & Schwering (1986) separated the two components in order to deduce a luminosity of star formation of $2 \times 10^8$ $L_\odot$ and a diffuse luminosity of $1.4 \times 10^9$ $L_\odot$. Such a result depends on the temperature assigned to the two components.

In M33 the total IR excess was compared with that encountered in individual complexes of HII regions in galaxies. The conclusion of such a comparison was that half of the IR luminosity ($6.4 \times 10^8$ $L_\odot$) is coming from the complexes of star formation and the other half from the dust heated by the diffuse interstellar radiation field (Soifer, Neugebauer & Houck, 1987 and the references therein).

An enhancement of the IR emission is appreciated in barred spiral galaxies of early type. Hawarden *et al.* (1986) discovered that the ratio between the flux density at 12 $\mu$m and that at 25 $\mu m$ in the galaxies is a sensible indicator of the presence of bars. Devereux (1987) noted that 40% of barred spiral galaxies or earlier galaxies, having $L_{\rm FIR} > 4 \times 10^9$ $L_\odot$, posses a considerable nucleus. Puxley *et al.* (1987) suggested that the IR activity originates within the central 2 kpc of the barred galaxies. If the color 12 $\mu$m/25 $\mu$m is a measure of the star formation activity, then the barred galaxies could have a star formation activity stronger than that in non barred galaxies. Such an intense activity would start because of an efficient canalization of the material towards the nuclear regions of the galaxies by means of the bars. Such an effect is not present in barred galaxies of late type. This is a problem to be clarified.

#### 5.3.4. *Irregular, Dwarf, and Blue Compact Galaxies*

Irregular galaxies have roughly the same range of values of the ratio $L_{\rm FIR}/L_{\rm B}$ with respect to normal galaxies (Gallager & Hunter, 1987). In contrast, the observed ratios between the flux density at 60 $\mu$m and that at 100 $\mu$m in irregular galaxies are significantly greater than in 'normal' spiral galaxies for a selected value of the ratio $L_{\rm FIR}/L_{\rm B}$. The observations can be interpreted as the cold component of the disk owed to the interstellar dust, encountered in spiral galaxies, and which is not present in irregular galaxies. This indicates that in such galaxies the fraction of FIR emission – directly associated with star formation – is greater than in spiral galaxies. Such an idea is based on the average value of the color 12 $\mu$m/25 $\mu$m temperature in irregular galaxies being less than in spiral galaxies (Hunter *et al.*,

1987). This is consistent with the model of Helou (1986a), in which the galaxies with colder color 12 $\mu$m/25 $\mu$m temperature are dominated by star formation.

The observations have shown a correlation between the FIR luminosity and $H_\alpha$ in such irregular galaxies (Gallager & Hunter, 1987). This provides one more evidence of the important contribution (probably dominating) of star formation to the FIR luminosity.

Hunter *et al.* (1987) found that the mass of the dust, with respect to the mass of the interstellar gas, is less in irregular galaxies than in spiral galaxies. Then whilst the total emission of the dust is consistent with the observations of a lower metallicity in irregular galaxies, a relationship between metallicity and dust properties was not found. Although in these objects the quantity of dust is lower than in spiral galaxies, the properties of such a dust in the optical range are similar to those in normal spiral galaxies (Soifer, Neugebauer & Houck, 1987). It is probable that the low metallicity of irregular galaxies has no direct correlation with their IR emission (Kunth & Sevre, 1986).

The IRAS observations of compact blue dwarf galaxies support the idea that they are in a period of star formation. The analysis of a sample of such galaxies shows that the ratio between the FIR and blue fluxes ($F_{\mathrm{FIR}}/F_{\mathrm{B}}$) is ranging between 0.5 and $\geq$ 10. Such a range is greater than that encountered in spiral galaxies. However, the dwarf galaxies are relatively weak IR sources, being the average $F_{\mathrm{FIR}}/F_{\mathrm{B}} \simeq 0.4$ (Helou, 1986b).

A large and heterogeneous sample of small (not dwarf) galaxies shows a ratio between the FIR luminosity and CO luminosity ($L_{\mathrm{FIR}}/L_{\mathrm{CO}}$) greater than in large spiral galaxies (Thronson & Bally, 1987). The large value of such a ratio is attributed to there being in small galaxies a very efficient star formation activity, although it should be possible that such a high ratio could be owed to a relative absence of CO.

### 5.3.5. *Elliptical and S0 galaxies*

The IRAS detected few elliptical galaxies. Amongst them the emissions at 100 $\mu$m and at 60 $\mu$m are similar. This indicates that the effective color temperatures are lower than in spiral galaxies (Soifer, Neugebauer & Houck, 1987). The production mechanism of the IR radiation in elliptical galaxies is probably owed to the thermal emissions of the dust.

A strong relationship between the flux detected at 100 $\mu$m and the presence of a strong central radio source has been found.

The emission at 12 $\mu$m in elliptical galaxies represents an excess with respect to what is expected from stellar photospheres (Soifer *et al.*, 1986b). Such an excess could proceed either from the circumstellar shells of the late type stars, which are losing mass, or from the emission produced by grains transitionally heated. Soifer *et al.* (1986b) reached the conclusion that the transitional heating of grains is a phenomenon which can be neglected in elliptical galaxies, since the UV component of the interstellar radiation field is much weaker than in spiral galaxies (Coleman,

Wu & Weedman, 1980). The heating of electrons in a hot gas could provide a notable quantity of energy to the dust of elliptical galaxies.

Since the contribution to the emission at 12 $\mu$m of the transitional heating of grains can be neglected, and the radiation pressure controls the mass loss, the excess at 12 $\mu$m above the photospheric emission provides a good evaluation of the quantity of matter which is injecting into the ISM of the galaxy. In the case of the nuclear protuberance of M31 this evaluation gives a value of 0.015 $M_\odot$ $yr^{-1}$ (Soifer *et al.*, 1986b), which agrees with theoretical foresightedness based on considerations on stellar evolution in old stellar populations (Faber & Gallagher, 1976).

The FIR emission allows an evaluation of the mass of the cold dust component in the ISM of elliptical galaxies, although it is necessary to take into account that such a determination is very sensitive to the temperature adopted for the dust (see equation 5.4). Jura *et al.* (1987) obtained a range of values for 60–100 $\mu$m color temperature between 32 and 42 K, which correspond to the temperatures of grains between 28 and 33 K. With such temperatures the mass deduced for the dust is in the range $10^4$–$10^5$ $M_\odot$. The measurement of the quantity of cold matter in elliptical galaxies, together with the determination of the input rate of matter into the ISM, allows a direct evaluation of the time necessary for accumulating the observed mass. In the spectral bump of M31 it is possible to evaluate $\dot{M}/M \simeq 10^7$ yr (Soifer *et al.*, 1986b). In contrast, the evaluation of the accumulation time based on the total mass, obtained by means of IR measurements, gives values of $10^7$–$10^8$ yr. Such evaluations need an additional answer on the destiny of the interstellar matter, which could 'disappear' to form stars of low mass (Jura, 1977, 1986).

#### 5.3.6. *Radiogalaxies*

In the study of radiogalaxies performed with the IRAS, these has been the fundamental discovery that the broad line radiogalaxy 3C 390.3 emits most of its energy in the IR range ($7 \times 10^{10}$ $M_\odot$) (Miley *et al.*, 1984). The continuum of its spectrum shows an excess around 25 $\mu$m, which can be easily interpreted as owed to a thermal radiation of dust at temperature $T \simeq 180$ K. More than half of the $10^6$ studied galaxies have a 60 $\mu$m luminosity greater than $10^{10}$ $L_\odot$. The most luminous sources in the radio and optical ranges are also very luminous in the IR range. The distribution function of the luminosity apparently does not depend on the optical properties (Soifer, Neugebauer & Houck, 1987).

#### 5.3.7. *BL Lac Objects and OVV Quasars*

Before the launch of the IRAS it was known that objects of BL Lac type and OVV QSOs (Blazars) were bright in the IR range (Rieke & Lebofsky, 1979; Angel & Stockman, 1980).

Since the objects of BL Lac type could have offered a direct view of the central engine of AGNs, a logical consequence was to search for eventual signs of anomalous excess in their emission, probably as the result of a thermal emission in the FIR (Far IR) range. The results of the IRAS measurements of BL Lac objets and Blazars are consistent with a general frame in which a smoothed synchrotron spectrum is generated by a non-thermal central source.

The short life (because of objective technical problems of detectors cooling) of the IRAS (10 months) allowed the repetition for only a few objects, such as for instance 3C 345, 3C 446, OJ 287 and the same BL Lac, which is not a galaxy (Soifer, Neugebauer & Houck, 1987 and the references therein). At the beginning of 1983 a burst which enhanced the optical and NIR (Near IR) fluxes by a factor of 3 was observed in the quasar 3C 345. The amplitude of the burst decayed most at longer wavelengths, and the effects of the burst were still noted at 12 $\mu$m when the IRAS mission started (Bregman *et al.*, 1986). The burst was observed neither at 100 $\mu$m nor at radio wavelengths. The appearance of such a burst with different amplitudes at different wavelengths is consistent with the radiation emitted by a relativistic plasma moving at high velocity with respect to the observer.

### 5.3.8. *Seyfert Galaxies*

These galaxies were discovered before quasars. They are a special class of galaxies with unusually powerful nuclei (Seyfert, 1943). Their most important characteristics are the intense point-like nuclear continuum and strong broad emission lines, which cannot be generated in stellar processes. Such broad lines are produced in the Broad Line Regions (BRL) having $\leq 0.1$ pc diameter, where clouds with high electron density ($10^{10}$ cm$^{-3}$ are rapidly moving. Their are seen only in permitted transitions and have full widths at half-maximum of order of tens of thousands km s$^{-1}$. Forbidden lines with Doppler velocity widths of $\sim$500–1000 km s$^{-1}$ are emitted from a Narrow Line Region (NRL), which is surrounding the BLR and where the electron density is ranging from $10^3$ to $10^7$ cm$^{-3}$. Seyfert galaxies showing broad lines or narrow lines were called Seyfert-1 or Seyfert-2 galaxies, respectively (e.g., Malkan, Alloin & Shore, 1987). Later such a classification was also correlated to radio properties of the galaxies (e.g., Urry & Padovani, 1995; Padovani, 1999). Antonucci (1993, and the references therein) in the unified model for Seyfert galaxies showed that type-2 sources are believed to be intrinsically identical to type-1, but observed at different inclination angles with respect to the symmetry axis of the molecular torus surrounding the compact object (e.g., Ward, 1995). For this reason, at least in optical and UV ranges, the nucleus of Seyfert-2 galaxies is not directly visible.

For this reason the IRAS survey was important for discovering new Seyfert galaxies. The surface density of such galaxies derived from the IRAS data was found to be in agreement with that already known (de Grijp *et al.*, 1985). These galaxies show thermal and non-thermal emissions depending on the wavelength region of observations and on the observed space scale. In some cases an excess or a bump appear in the IR spectrum, which are not easily explained as owing to a reprocessed thermal radiation of the heated dust. In many cases the continuum in the IRAS ranges is smoothed with slopes within the range of the expected values coming from the overlapping of spectra of thermal and non-thermal sources.

In general the IRAS observations can be inserted in a simplified frame in which the Seyfert galaxies have an emission formed by three components, with different influences depending on the objects:

- a non-thermal nuclear component, which originates a power law spectrum up to the X-ray range;
- a cold component, probably owed to star formation bursts;
- a nuclear component in the IR medium, having the peak emission between 20 and 50 $\mu$m (Miley, Neugebauer & Soifer, 1985).

Many IRAS measurements were combined with multifrequency observations of individual objects with the result of accelerating the knowledge of the physics involved in those objects.

The IRAS data from Seyfert-1 galaxies can be fitted to a synchrotron spectrum, such as for instance in the cases of Mrk 79, Mrk 279, MCG 8-11-11 (Spinoglio *et al.*, 1985) and Mrk 507, 5C 3.100, 1Zw 1 (Halpern & Oke, 1987), although in MCG 8-11-11 a bump of thermal emission at 25 $\mu$m was found. In the Seyfert-2 galaxy NCG 1275, which is also identified with a radio source with flat spectrum (3C 84), the IRAS observations show that the IR emission is dominated by a thermal emission overlapped to a synchrotron spectrum (Gear *et al.*, 1985).

### 5.3.9. *Quasars*

The IRAS-PSC contains some dozens of QSOs, although the exact number depends on the adopted positional accordance (Kidger & Beckman, 1986). Generally, the IRAS observations of QSOs provide a continuum which fits with continuity to the continuum determined through measurements in other energy ranges.

At 60 $\mu$m the luminosity can reach $10^{14}$ $L_\odot$, which is comparable with the optical luminosity.

In QSOs with radio emission originating a flat spectrum, the total spectrum is extended without interruptions from the radio, to IR until optical range. This is a circumstantial evidence that the IR emission is probably non-thermal in origin. The luminosity of QSOs with flat radio spectrum are systematically greater by an order of magnitude than in QSOs with sloping radio spectrum and in radio quiet QSOs. Therefore if the QSOs with flat radio spectrum were to have thermal components, probably they should be obscured by a greater non-thermal emission.

All high-redshift QSOs detected by the IRAS have a flat radio spectrum (Soifer, Neugebauer & Houck, 1987). The origin of the IR emission in radio quiet QSOs, with sloping radio spectra, is not clear; probably it is owed to a mixture of thermal and non-thermal components. Probably the thermal emission originates in the heated dust in the NLRs (Carter, 1984).

5.3.10. *Very Luminous IR galaxies*

The IR bright galaxies constitute the most important whole population of galaxies. Since the beginning of the seventies, such objects were identified as IR emitters, in the sense that almost the entire energy was emitted in the IR range (e.g., NGC 253 and MrK 231). The IRAS demonstrated that the IR bright galaxies form a significant fraction of the whole galaxies with bolometric luminosity $L_{\rm bol} \geq 3 \times 10^{10}$ $L_\odot$. The IR bright galaxies, having luminosity in the range $3 \times 10^{10}$–$3 \times 10^{11}$ $L_\odot$, in most cases are crossing a period of intense activity of star formation. In a sample of galaxies analyzed in the FIR range, those having $L_{\rm FIR} > 3 \times 10^{10}$ $L_\odot$ constitute about 50% of the sample and present optical spectra of the same kind of those of starburst galaxies (Soifer, Neugebauer & Houck, 1987, and the references therein). In the range of luminosity $3 \times 10^{10}$–$3 \times 10^{11}$ $L_\odot$ the space densities of the IR and Markarian galaxies are equal (Soifer *et al.*, 1986a).

The luminosities of the Markarian galaxies, in the FIR, are on average sensibly greater than those of normal galaxies (Deutsch & Willner, 1986; Sekiguchi, 1987), whilst the FIR color temperatures of such galaxies are greater than those of 'normal' galactic disks. Such properties indicate that the predominant source of luminosity is the star formation process and the Markarian and IR selected galaxies basically represent the same population for luminosity above the range $3 \times 10^{10}$–$3 \times 10^{11}$ $L_\odot$ (Soifer, Neugebauer & Houck, 1987). The luminosity of these galaxies comes from a non-thermal nucleus and from the peak activity of star formation. Also the conversion of the kinetic energy into IR luminosity can produce, during short periods, the same quantity of luminosity.

The presence of an active nucleus is deduced by the measures of the Br-$\alpha$ line, as occurred in Arp 220, where the measured width was of 1200 km s$^{-1}$ (DePoy, 1987). The presence of a star formation as source of energy is seen by means of the presence of a significant quantity of molecular gas (Sanders & Mirabel, 1985). In many of these IR galaxies the ratio between the FIR luminosity and the mass of molecular hydrogen – $L_{\rm FIR}/M(\rm H_2)$ – is comparable with that encountered in the archetype of starburst galaxies M 82 and NCG 253.

The detection of CO in emission from Mrk 231 (Sanders *et al.*, 1987) shows that the galaxy contains a mass of molecular gas of $1.4 \times 10^{10}$ $M_\odot$, which is a value 4 times greater than in the Milky Way.

In spite of the new inputs from the IRAS, the role of these very luminous IR galaxies in the evolution of the Universe was not yet completely evident. One possibility is the merging of two galaxies containing a large quantity of gas: a big burst of the luminosity is produced as result of the star formation and a black hole pushed by the ISM. The black hole could have been formerly present or could form because of the interaction of the two merging galaxies. An example of a galaxy at this stage of evolution could be Arp 220.

When the luminosity of the system increases the gas either gradually wastes away or is swept away towards the exterior until the material surrounding the nucleus becomes visible. Mrk 231 could be an example of a galaxy in this stage of evolution.

In the case in which the content of the dust becomes neglected the dominant energy should be originated by the process of accretion onto a black hole, and then a 'classic' quasar would appear (Sanders *et al.*, 1988).

#### 5.3.11. *The ISO Extragalactic Sky*

The ISO (Infrared Space Observatory) has been a great success for infrared astronomy and for European space research. One of the most exciting fields studied by the ISO measurements has been that of star formation at galactic scales. The results range from the nearby galaxies, where spatially resolved studies have been possible, to high redshift galaxies where the role of the dust in hiding star formation processes is important.

There has been a discrepancy between the IRAS measurements and ground based mm-continuum and line emission observations. ISOPHOT photometry at 200 $\mu$m show the presence of a cold ($T \sim 15$ K) dust in several nearby galaxies; the mass of this dust is about one order of magnitude greater than seen by the IRAS. This result favors the agreement between the mm and FIR measurements (e.g., Genzel, 1999). From the complete 175 $\mu$m map of the Andromeda galaxy (M31), the bulk of the dust has a temperature of only 16 K, considerably cooler than the 21–22 K previously derived from the IRAS data and also colder than the 19 K found for the Milky Way (Haas *et al.*, 1999). In order to fit the FIR spectral energy distribution the presence of two dust components is necessary: small and large grains, similar to what had been found in the Milky Way. The estimate of the mass corresponding to a 16 K dust component gives a value of $3 \times 10^7$ $M_\odot$, which is about a factor of ten higher than inferred from the IRAS 60/100 $\mu$m data alone. This cold dust mass, if distributed in the plane of the galaxy, would be sufficient to make the disk of M31 moderately opaque in the optical.

A few words are mandatory for commenting on the ISOCAM extragalactic mid-infrared deep surveys. Indeed, such surveys (Elbaz *et al.*, 1999 and the references therein) have detected a population of strong infrared emitters one order of magnitude more numerous that expected if there were no evolution from $z < 0.2$ (IRAS) up to the maximum redshift of these galaxies ($z \simeq 1.5$). The mid-IR cosmic background produced by these galaxies, which is $\simeq 2.35 \pm 0.8$ nW m$^{-2}$ sr$^{-1}$, at 15 $\mu$m above 50 $\mu$Jy, is larger than 30% of the energy radiated in the $I$ band by the optical galaxies detected in the Hubble Deep Field (HDF), which are a factor 100 more numerous. This fraction is much higher than in the local universe ($z < 0.2$) as probed by the IRAS, where all integrated 8–1000 $\mu$m IR emission makes 30% of the optical starlight (Soifer & Neugebauer, 1991). Even assuming a conservative spectral energy distribution (SED), they produce a major contribution to the 140 $\mu$m DIRBE background measured by Hauser *et al.* (1998) ($\simeq 25.1 \pm 7$ nW m$^{-2}$ sr$^{-1}$) and Lagache *et al.* (1999) ($\simeq 15.3 \pm 9.5$ nW m$^{-2}$ sr$^{-1}$). This is both a confirmation of the strong IR cosmic background that was first detected by Puget *et al.* (1996) and the first identification of the galaxies responsible for a large fraction of this background at 140 $\mu$m. Elbaz *et al.* (1999) were able to identify the galaxies responsible for this strong IR emission in the region of the HDF owed to the large number of ground based observations of HDF galaxies. They found that these bright (several times $10^{11}$ $L_\odot$) and massive ($\langle M \rangle \simeq 1.5 \times 10^{11}$ $M_\odot$) infrared galaxies have a typical redshift of $z \simeq 0.7$ and optical colors similar to field galaxies. The fraction of galaxies with morphological signs of interactions is larger at higher redshift. This is consistent with the nature of local luminous infrared galaxies (LIGs, with $L_{\rm bol} \simeq L_{\rm IR} > 10^{11}$ $L_\odot$) found by the IRAS, which often exhibit evidence of galaxy interaction or merging (Sanders & Mirabel, 1996) and indicates that interactions should play a crucial role in the evolution of galaxies as expected from bottom up scenarios.

### 5.4. GALACTIC HISTORICAL JOURNEY: GAMMA RAY RANGE

The pioneer observations of $\gamma$-ray astronomy were performed with detectors embarked on board balloons. But the scientific advances in knowing the $\gamma$-ray sky were obtained from space vehicles, such as: the Orbiting Solar Observatories (OSO-3 and OSO-7), the vehicle APOLLO, the Small Astronomical Satellite (SAS-2), the COS-B, the High Energy Astrophysics Observatories (HEAO-1 and HEAO-3), the Solar Maximum Mission (SMM), a large variety of space probes for measuring $\gamma$-ray bursts, and the more recent Compton Gamma Ray Observatory (CGRO) and BeppoSAX.

For information on the characteristics of these space vehicles, see e.g., *The Astronomy and Astrophysics Encyclopedia* (Maran, 1992).

#### 5.4.1. *Gamma Radiation in the Solar System*

Gamma rays from the Sun and Moon were observed. The solar $\gamma$-ray emission is produced by particles accelerated during the flares, whilst the lunar $\gamma$-rays come from the interactions of cosmic rays with the lunar surface and from the decay of natural long lived radioisotopes. A complete review of this topic is that by Ramaty & Lingenfelter (1982).

5.4.1.1. *Solar flares.* The particles accelerated during solar bursts interact with the ambient solar atmosphere and generate $\gamma$-rays, which manifest both as continuum emission and lines. The emission lines are the product of the interactions of protons and nuclei, whilst the continuum emission is produced by bremsstrahlung of relativistic electrons.

The $\gamma$-ray lines of the solar flares were firstly observed by Chupp *et al.* (1973) with a NaI spectrometer on board OSO-7 satellite. The observed lines were: positron annihilation line at 0.511 MeV; the line of the neutron capture by $^{1}$H at 2.223 MeV; nuclear de-excitation lines of $^{12}$C and $^{16}$O at 4.438 and 6.129 MeV, respectively. Such lines, as well as other de-excitation lines, were observed in a number of flares by the detectors of the satellites HEAO-1 (Hudson *et al.*, 1980), HEAO-3 (Prince *et al.*, 1982) and SMM (Chupp, 1982).

The strongest predicted and observed line in solar flares is that at 2.223 MeV coming from the capture of neutrons aside of the hydrogen: $^{1}H(n,\gamma)^{2}H$. The second intense line in solar flares is that of positrons annihilation at 0.511 MeV.

There are many mechanisms of positrons production, which are important for astrophysics, but in solar flares the 0.511 MeV line is the result of nuclear interactions producing:

- short lived radionuclides (e.g., $^{11}C$, $^{13}N$, $^{15}O$, $^{17}F$) (Ramaty, Kozlovsky & Lingenfelter, 1975);
- $\pi^{+}$, which decay by positron emission;
- $^{16}O$ in the 6.052 MeV level, which decays by electron–positron pair emission.

The initial energy of positrons vary between a few hundreds of keV and a few tens of MeV, but only few positrons annihilate at these high energies (Ramaty & Lingenfelter, 1982).

A variety of $\gamma$-ray lines are produced by de-excitation of nuclear levels. Such levels, in solar flares, are populated by:

- inelastic collisions: e.g., $^{12}C(p,p')^{12}C^{*4.44}$;
- spallation reactions: e.g., $^{20}Ne(p,p\alpha)^{16}O^{*6.13}$;
- non-thermal fusion reactions: e.g., $^{4}He(\alpha,p)^{7}Li^{*0.478}$;
- decay of radionuclides produced by spallation reactions: e.g., $^{16}O(p,p2n)^{14}O(e^{+})^{14}N^{*2.31}$.

In the solar atmosphere lines with two components are produced:

- a narrow component, which results from the de-excitation of the nuclei of the medium excited by interactions with energetic protons and $\alpha$ particles;
- a broad component, which results from the de-excitation of energetic heavy nuclei excited by interactions with hydrogen and helium of the ambient.

The implications of the $\gamma$-ray observations of solar flares concern the timing of acceleration, the confinement of particles at the Sun, the fraction of the total energy of the flare which resides in energetic nucleons, the chemical and isotopic abundances, and the possible beaming of energetic particles. In particular, $\gamma$-ray observations show that some percent of the total energy of the flares is owed to protons and nuclei accelerated in closed magnetic loops – with little probability of escaping into the interplanetary medium – up to tens of MeV in time scale of a few seconds (Ramaty & Lingenfelter, 1982, and the references therein).

A good review on solar flares is that by Bai & Sturrock (1989).

5.4.1.2. *Planetary and Lunar Surfaces.* The most intense $\gamma$-ray line and continuum emission from the Moon are originated by the interactions of galactic cosmic rays with the surface material of Moon. Detailed studies performed by Reedy (1978) on the interactions of secondary particles of the cosmic rays show that the two strongest lines of the lunar surface are at 1.779 and 6.129 MeV, which are coming from the de-excitation of the most abundant nuclei, $^{28}Si$ and $^{16}O$, excited by inelastic scattering of neutrons. The computations predicted also the existence of other lines of less abundant elements: nuclear de-excitation lines at 0.847, 1.369 and 2.210 MeV from inelastic excitation of $^{56}$Fe, $^{24}$Mg and $^{27}$Al, and neutron capture lines on $^{48}$Ti at 6.762 MeV and on $^{56}$Fe at 7.631 and 7.646 MeV. All of these lines were observed with a NaI spectrometer embarked on board the vehicles APOLLO-15 and APOLLO-16 (Metzger *et al.*, 1974; Bielefeld *et al.*, 1976). With such detectors it was possible to obtain data in order to produce the map of a portion of lunar surface. The intensities of the detected lines showed regional variations. Such observations provided important information for the comprehension of the differentiation and thermal evolution of the Moon.

Reedy (1978) suggested that similar observations, made with orbiting spectrometers, could provide maps of the surface composition of Mercury and Mars. Indeed, such planets have atmospheres enough thin to allow the observations of their surface $\gamma$-ray emission.

5.4.2. *Gamma Radiation in the Galaxy*

Many $\gamma$-ray sources have been detected in the Galaxy, as well as diffuse emission. However, in spite of the strong improvement obtained with the CGRO-EGRET experiment with respect to the older SAS-2 and COS B $\gamma$-ray satellites, most of the discrete sources are still not associated with known sources.

Briefly, what we can see from the Galaxy in the $\gamma$-ray range can be listed as follows:

- an intense source of electron–positron annihilation radiation in the Galactic Center (GC);
- the strongest $\gamma$-ray Crab and Vela pulsars;
- some binary systems, among which Cyg X-3 was the first detected;
- a relative large number (271) of discrete sources, mostly not yet identified with known objects;
- a diffuse emission as a result of the interactions of cosmic rays with the ISM.

5.4.2.1. *Galactic Center.* In the first experiments an intense radiation at 0.511 MeV, coming from the electron–proton annihilation was observed in the direction of the GC. The first observations performed with a high resolution Ge detector revealed that the annihilation line is very narrow (FWHM $\leq$ 3.2 keV) and lies on a continuum emission, which implies that ~90% of positrons annihilate via positronium (Leventhal, MacCallun & Stang, 1978). The observed intensity of $2 \times 10^{-3}$ photons cm$^{-2}$ s$^{-1}$ implies a value of the annihilation rate of $\simeq 4 \times 10^{43}$ positrons s$^{-1}$ or a luminosity of the annihilation radiation of $\simeq 6 \times 10^{37}$ erg s$^{-1}$, with a distance to the GC of 10 kpc. Riegler *et al.* (1981) with HEAO-3 confirmed that the emission line at 0.511 MeV is really narrow (FWHM <2.5 keV) and gave more precise information on the localization of the source as well as strong limits on the dimensions of the emitting region. Such measurements demonstrated that the direction of the source is coincident with the GC, within ±4° (observational uncertainty), and that the line is varying with time, diminishing by a factor of 3 in about six months from Autumn 1979 to Spring 1980. Such a variation puts limits on the dimensions of the annihilation region and the positron source, which can be less the light travel distance of $\simeq 10^{18}$ cm. The FWHM$<$ 2.5 keV implies a temperature of the gas in the annihilation region of $<5 \times 10^4$ K and an ionization fraction greater than 10% (Bussard, Ramaty & Drachman, 1979). If the gas were neutral the width of the line would have been broader than observed since it would contain also the Doppler broadening, not owed to thermal motion of the gas, but to the velocity of energetic positrons, which form the positronium in flight by exchange of charge with neutral hydrogen. However, in a partially ionized gas the positrons lose energy in the plasma very rapidly; then they thermalize before the annihilation or the positronium formation. Thus the temperature of the medium must be $<5 \times 10^4$ K, derived from the width of the line. The variations of intensity not only give limits on the dimension of the annihilation region ($\simeq 10^{18}$ cm), but also indicate that the gas density in such a region must be high enough to brake the positrons sufficiently in order to annihilate in less than six months. If the positrons are produced with a kinetic energy of the order of their rest mass, the time necessary for braking via Coulomb collisions is greater than the time necessary to form positronium in such a gas. Both these times are inversely proportional to the gas density. In the case of a braking time of about six months ($\leq 1.6 \times 10^7$ s), the necessary density of the region is $\geq 10^5$ H cm$^{-3}$. Such regions exist in two peculiar warm clouds (Lacy *et al.*, 1980) and the non-thermal compact source must exist in the central parsec of the Galaxy (Kellerman *et al.*, 1977).

The lowering of the line intensity by a factor of 3 in about 6 months excludes the possibility of any multiple extended source, such as cosmic rays or pulsars (Sturrok & Baker, 1979), supernovae (Ramaty & Lingenfelter, 1979) or a primordial black hole (Okeke & Rees, 1980). However, this fact requires the presence of an individual compact source, whose dimensions are $\simeq 10^{18}$ cm, placed at or close to the GC; such a source must have intrinsic variations in time scale $\leq 1.6 \times 10^7$ s. Its minimum luminosity of $\simeq 6 \times 10^{37}$ erg s$^{-1}$ renders this source the most luminous in the Galaxy.

5.4.2.2. *Other Sources with Emission Lines.* Thermonuclear combustion in supernovae and novae (Clayton, 1982) and nuclear interactions of low energy cosmic rays with the interstellar gas (Ramaty, Kozlovsky & Lingenfelter, 1979) can produce in the Galaxy nuclear excitation lines, as well as an additional positron annihilation emission line.

Observations of the galactic emission at 0.511 MeV, performed with a wide field of view ($\geq 50°$), revealed intensities greater than expected from the unique source at the GC. This indicated that probably a diffuse source producing the 0.511 MeV line is present in the Galaxy (Ramaty & Lingenfelter, 1982).

5.4.2.3. *Diffuse Continuum Emission.* The diffuse continuum emission of the Galaxy was firstly observed with the detectors of the OSO-3 ($E > 50$ MeV) (Clark, Garmire & Kraushaar, 1968). Such observations showed a clearly defined band (±15° latitude) with an excess in the intensity across the galactic equator, which was coming from the galactic disk (GD). The observations also showed a strong dependence of the intensity with longitude with a wide peak extending from −30° to +30° longitude around the GC, as a result of an enhancement of the emission from a region within 5 kpc of the GC. The intensity in such a direction was $(1.3 \pm 0.3) \times 10^{-4}$ photons cm$^{-2}$ s$^{-1}$ sr$^{-1}$, which was about 3 times that from other directions in the disk.

After such measurements, surveys with higher sensitivity were performed with SAS-2 (Fichtel *et al.*, 1975) and COS-B (Mayer-Hasselwander *et al.*, 1982) satellites with angular resolution of few degrees. Some sources were detected and resolved, including the Crab and Vela pulsars (Kniffen *et al.*, 1974). The computations of Higdon & Lingenfelter (1976) suggested that roughly half of the observed galactic $\gamma$-ray continuum emission could be generated in young pulsars (age $\leq 10^4$ yr and distances >1 kpc), then not yet discovered. Clearly, measurements with much more sensitivity and angular resolution need in order to determine which fraction of galactic continuum emission proceed from such sources and which one really from the ISM. This was obtained by means of the CGRO.

Some probable sources of $\gamma$-ray continuum emission were suggested as results of the interactions of cosmic rays with the ISM, namely:

- Compton diffusion of star light (Feenberg & Primakoff, 1948);
- cosmic black body photons (Gould, 1965) produced by electrons of the cosmic rays;
- bremsstrahlung radiation produced by the interactions of electrons of the cosmic rays with the interstellar gas (Hutchingson, 1952);
- decay of $\pi^0$ produced by the interactions of nucleons of the cosmic rays with the interstellar gas (Hayakawa, 1952).

Many scientists (e.g., Stecker, 1970) calculated the spectrum of the $\gamma$-rays originated by the decay of $\pi^0$ mesons, which are produced by the interactions of cosmic rays with the interstellar gas. Later the contribution of the $\pi^0$ produced by inelastic proton–proton reactions in a hot plasma was calculated. This is owed to the matter accretion process around collapsed objects as well as the contribution of the e–e and e–p thermal bremsstrahlung (e.g., Giovannelli, Karakuła & Tkaczyk, 1982a,b).

Figure 61 (Ramaty & Lingenfelter, 1982) shows the galactic diffuse $\gamma$-ray spectrum (Hartman *et al.*, 1979; Paul *et al.*, 1978) and the rate of $\gamma$-ray production by cosmic rays nearby the solar system. Also the contribution of the pulsars (Harding & Stecker, 1981) is shown.

5.4.2.4. *Discrete Galactic Sources.* With the first experiments, until the COS-B satellite, a number of discrete $\gamma$-ray sources were detected in the energy range

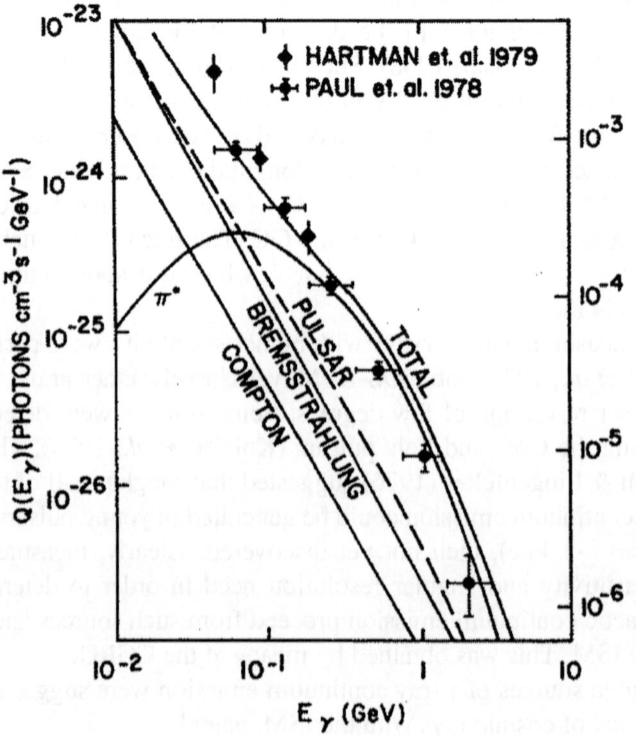


*Figure 61.* Diffuse galactic $\gamma$-ray spectrum (Hartman *et al.*, 1979; Paul *et al.*, 1978) and $\gamma$-ray production rates by galactic cosmic rays in the surrounding of solar system and the contribution of pulsars (from Harding & Stecker, 1981) (Ramaty & Lingenfelter, 1982).

from $\simeq$1 MeV to several GeV, by means of balloons and satellites and for energies >100 GeV with ground based experiments. Such sources include different kind of objects: dense interstellar clouds, pulsars, accreting neutron stars and many unidentified objects.

The proposed mechanisms of emission include the $\pi^0$ decay electron bremsstrahlung, Compton scattering, synchrotron emission, curvature radiation of electrons in an intense magnetic field ($\geq 10^{12}$ G. Such a curvature radiation is produced by high energy electrons which move across the magnetic field lines (Sturrock, 1971).

A $\gamma$-ray emission, with $E > 100$ MeV was first detected from an interstellar cloud, relatively close to $\rho$ Oph, by a balloon borne experiment (Frye *et al.*, 1972a,b). The COS B observations confirmed such an emission, as shown in Figure 62, and moreover detected high energy $\gamma$-rays from the complex of Orion Nebula.

Black & Fazio (1973) suggested that the emission from the $\rho$ Oph cloud could be produced by interactions of cosmic rays with the dense gas of the cloud. Studies, further realized, indicated that the bremsstrahlung of electrons of cosmic rays and the decay of the $\pi^0$, produced by the nucleons of cosmic rays can justify the

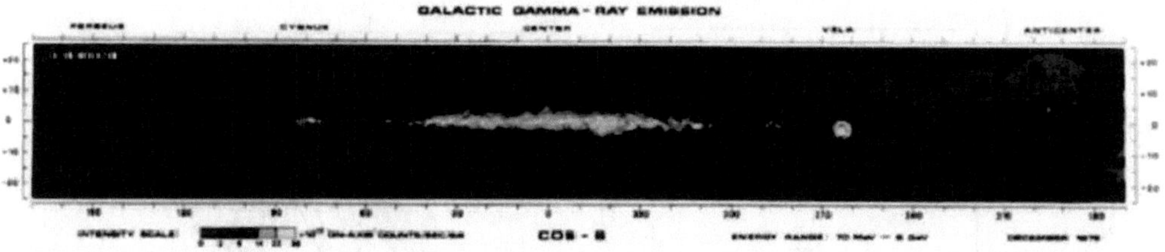


*Figure 62.* The $\gamma$-ray sky as mapped with the European satellite COS-B. The Galactic Plane is bright and the most prominent sources are the Crab Nebula and the Vela Pulsar (courtesy of G. Bignami: from Charles & Seward, 1995).

observed emission in $\rho$ Oph and in the clouds of the Orion Nebula (e.g., Bignami & Hermsen, 1983). Pulsed $\gamma$-ray emission was unambiguously detected only from the Crab and Vela pulsars. However, the spectra and light curves of the two sources are rather different from each other. The processes of pulsed emission were then uncertain. However, the synchrotron radiation process was the most interesting one. The emission is produced by charged particles moving across the lines of a strong magnetic field ($B \approx 10^{12}$ G with curved magnetic lines around the polar caps of the neutron star. The particles are accelerated by the action of the electric fields, which are induced by the rotation of the neutron star.

The source Cyg X-3, which was defined as a peculiar source, was studied in the radio, IR, X-, and $\gamma$-ray ranges (Stepanian, 1981). Its emission shows a modulation of 4.8 hr, which was attributed to the eclipse of the source across the orbital period. The most popular model for such a system is that of a fast rotating neutron star orbiting around a close companion and accreting matter from it. The $\gamma$-ray emission processes include those proposed for the pulsars and the decay of $\pi^0$, which form by the proton-proton inelastic collisions in the accreting matter onto a collapsed object (i.e. neutron star or black hole) (e.g., Giovannelli, Karakuła & Tkaczyk, 1982a,b) or the interactions of electron and proton relativistic beams with the surrounding matter and/or radiation (e.g., Bednarek *et al.*, 1990a). With the theoretical $\gamma$-ray spectra deduced from these interactions and from $\pi^0$ decay and from e–e and e–p bremsstrahlung, Bednarek *et al.* (1990b) successfully fitted to the experimental spectra in the $\gamma$-ray range of the X-ray binary Cyg X-1, which was the most probable galactic black hole candidate, and of the COS B source 2CG 195+04, also nicknamed Geminga. Figure 63 shows such fits. In the left panel the experimental photon differential spectrum of Cyg X-1 (McConnell et al, 1987) is best fitted to the computed spectrum. Dashed line represents one-temperature inverse Compton model (Sunyaev & Titarchuk, 1980). The solid line represents the comptonized black body spectrum ($T = 3 \times 10^4$ K), with a Lorentz factor for the electron beam $\gamma = 600$. The right panel of Figure 63 shows the experimental photon differential spectrum of Geminga, measured by COS B (Masnou *et al.*, 1981), multiplied by $E^2$ and the computed spectra obtained by interactions of a monoenergetic proton beam with a hydrogen cloud for different angles between

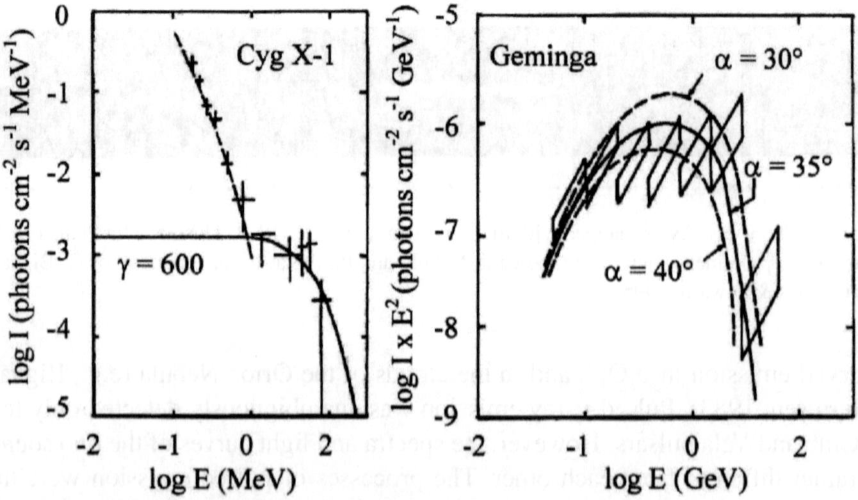


*Figure 63.* Differential photon experimental and theoretical spectra of Cyg X-1 (left panel) and Geminga (right panel) (Bednarek *et al.*, 1990b).

the emitted photons and the beam axis ($\alpha_1 = 30°$, 35°, and 40°) and for a Lorentz factor $\gamma = 30$. The best fit is for $\alpha_1 = 35°$. Such a fit rendered possible to argue the identification of Geminga with a galactic neutron star, placed at a distance of few hundreds parsecs (Tkaczyk, Karakuła & Giovannelli, 1984; Giovannelli, Karakuła & Tkaczyk, 1985), like that reported much later by Halpern & Holt (1992). Even with numerous new data acquired with the experiments of the successive decade, such a fit is still valid as shown by Giovannelli, Karakuła, & Tkaczyk (1996c) and by Giovannelli *et al.* (1997).

Figure 64 shows the second COS B catalog (2CG), whose importance in $\gamma$-ray astronomy has been undoubtedly crucial. Such a catalog contains 25 sources and only four of them were associated with known astrophysical objects, namely the Crab and Vela pulsars, the QSO 3C 273 and the $\rho$ Oph cloud (Bignami & Hermsen, 1983). In such a cloud a probable extended $\gamma$-ray emission was observed (Hermsen & Bloemen, 1983). Later on many attempts in searching for such associations were made by many groups and the situation improved a bit. But only with the advent of the more sensitive and higher angular resolution EGRET experiment on board the CGRO the improvement was significant, but not yet definitive, since the angular resolution of it was not high enough to resolve individual sources in crowded fields, such as for instance the young open clusters, which were suggested to be counterparts of some COS B sources (Manchanda *et al.*, 1996), after the suggestion of Polcaro *et al.* (1990, 1991a,b,c) on the association of the young open cluster Berkeley 87 with the COS B source 2CG 075+00. In order to solve definitively such an association it is necessary to await for the results of the INTEGRAL satellite (launched on October 17, 2002) results.

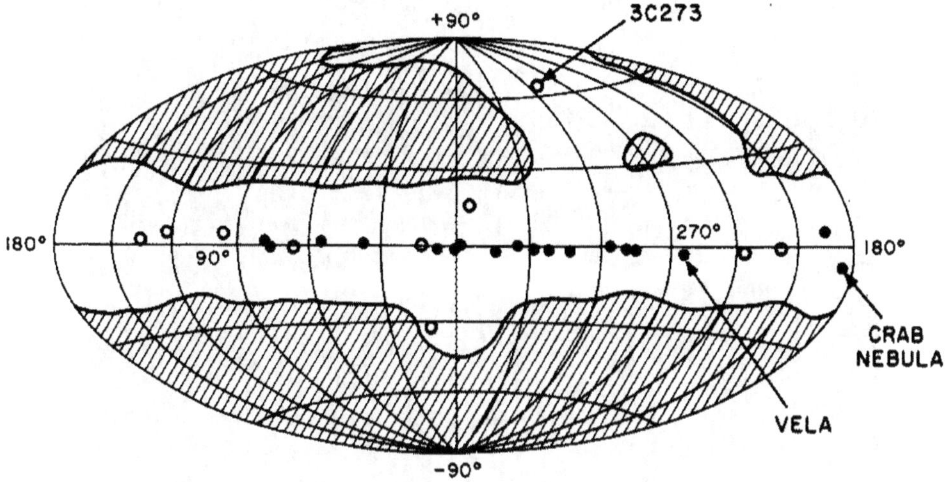


*Figure 64.* The COS B $\gamma$-ray sky. Unshaded regions were explored and the sources detected for $E > 100$ MeV. Black circles indicate sources with intensity $I \geq 1.3 \times 10^{-6}$ photons $cm^{-2}$ $s^{-1}$. Open circles indicate sources with intensity lower than that limit (after Scarsi *et al.*, 1977) (Zombeck, 1990).

All the sources reported in the second COS B catalog (with the unique exception of 3C 273) lie very close to the galactic plane. This fact laid out the question of how many sources have diffuse origin, such as for instance those associated with clouds irradiated by cosmic rays, or with the peaks of the distribution of gas column density (Li Ti Pei & Wolfendale, 1981). With the surveys of HI and CO (Bloemen, 1989), taking into account their diffuse distributions, it was possible to improve the search for discrete sources.

Since the COS B mission ceased operation in 1980 there was no progress at energies greater than 100 MeV until the Compton Gamma Ray Observatory (CGRO) was launched in April 1991. This satellite represents a major step forward because it provided a complete sky map with increased sensitivity and angular resolution. It consisted of four experiments covering five orders of magnitude in photon energy: the Oriented Scintillation Spectrometer Experiment (OSSE) (0.05–10 MeV) (Cameron *et al.*, 1992), the Burst And Transient Source Experiment (BATSE) (0.015–110 MeV) (Fishman et al., 1992), the imaging Compton Telescope (COMPTEL) (0.05–30 MeV) (Schönfelder *et al.*, 1992), and the Energetic Gamma Ray Experiment Telescope (EGRET) (20 MeV – 30 GeV) (Hartman *et al.*, 1992). EGRET is a spark chamber experiment and is of most interest for VHE and UHE astronomy. A general description of the whole CGRO mission is reported by Gehrels, Chipman & Kniffen (1993). A precious book on the first results of the Compton Gamma Ray Observatory is that by Friedlander, Gehrels & Macomb (1992).

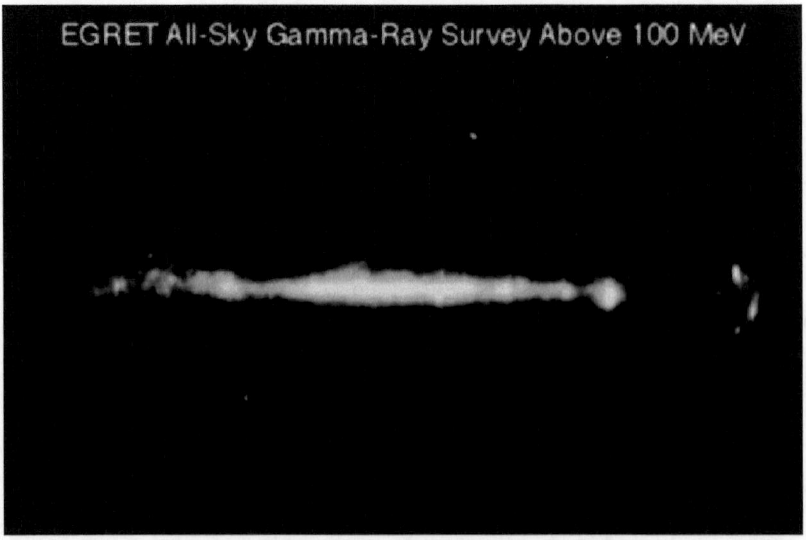


*Figure 65.* The EGRET Sky Map (Kniffen, 1993).

The improvement to the knowledge of $\gamma$-ray astronomy given by the CGRO is enormous, and it is impossible here to review all results. However, we want to summarize those results in Figure 65, where a full sky view by the EGRET detector for photons with energies $\geq$ 100 MeV are reported (Kniffen, 1993).

This map, also reported in the First EGRET catalog (1EG) of 124 sources (Fichtel *et al.*, 1994), shows that 42 sources are within 10° of the galactic plane; approximately half of the out of plane sources are identified with AGNs, and there is a reasonable presumption that many of those remaining $|b| \geq 10°$ sources will be identified ultimately as AGNs. In contrast, only 5 of the EGRET in plane sources are firmly identified with pulsars. Some of the unidentified in plane sources are probably associated with supernova remnants (SNRs), or with young open clusters, as firstly suggested by Polcaro *et al.* (1991b) and confirmed with EGRET measurements by Manchanda *et al.* (1996). Figure 66 shows the Third EGRET Catalog (Hartman *et al.*, 1999).

## 5.5. Galactic Historical Journey: X-Ray Range

### 5.5.1. *Stellar X-Ray Emission*

If the Sun were the prototype of a late-type dwarf star we would have many fewer X-ray observations of the sky than actually observed. Indeed, the Sun is close to the extreme side of the range of luminosity of late-type stars observed by the Einstein satellite, which demonstrate that the coronal luminosity of the Sun is very small compared with those of other stars.

The Einstein satellite observed X-ray emission coming from stars distributed across the entire HR diagram. Figure 67 shows such stars spread on the HR diagram

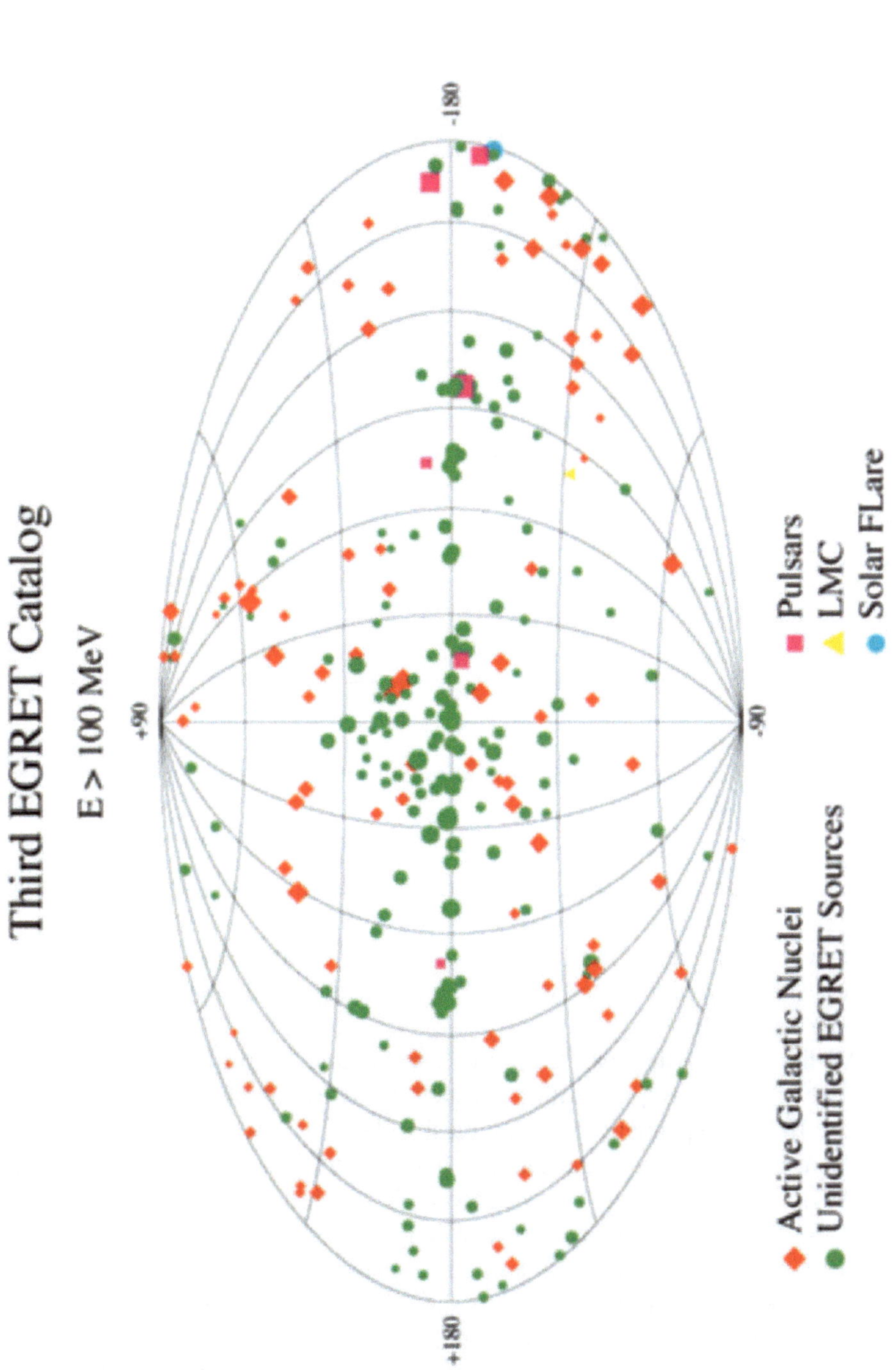


*Figure 66.* The 3rd EGRET-CGRO catalog (Hartman *et al.*, 1999).

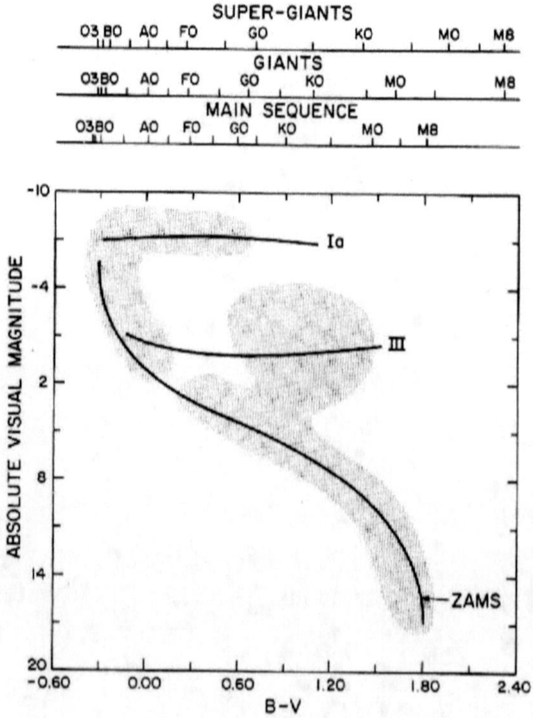


*Figure 67.* HR diagram of X-ray emitting stars, detected by the Einstein satellite (Rosner, Golub & Vaiana, 1985 after Vaiana *et al.*, 1981).

(Vaiana *et al.*, 1981). Analyzing the data there represented, it is possible to remark the following points (Rosner, Golub & Vaiana, 1985):

- Data are consistent for all the X-ray emitting dwarf stars with spectral type between dF (dwarf F) and dM (dwarf M), with rest luminosity ranging from $\sim 10^{26}$ and $\sim 10^{31}$ erg s$^{-1}$.
- All the stars earlier than $\simeq$B5 are X-ray emitters with luminosity between $\sim 10^{29}$ and $10^{34}$ erg s$^{-1}$ with a constant X-ray to bolometric luminosity ratio of $\simeq 10^{-7}$, independent of the luminosity class;
- There is a narrow spectral range – from B8 to A5 – in the Main Sequence (MS) in which there is no evidence of X-ray emission, with the sensitivity of the Einstein, in stars spectroscopically normal and in Am stars. However, there is evidence of X-ray emission in Ap stars;
- Late type giant and supergiant stars show a cut in the levels of X-ray emission, evolving towards later spectral types;
- Pre-Main Sequence stars (PMSSs) are strong X-ray emitters and show a well defined tendency to decrease their average emission levels with the nominal age of stars;

- The X-ray spectra of late-type stars are thermal with only one component and the temperature is ranging from $10^6$ to a few units $\times 10^7$ K.
- As one would expect in the behaviour of solar-type stars, late-type MS stars (Haisch, 1983), and evolved stars in close binary systems show, sometimes, highly variable behaviour in the X-ray energy range. In very young MS stars and in pre MS stars (PMSSs) is very frequent such a variable behaviour (Stern, 1983). However, early-type stars hardly show evidence of such a variability in short timescale, while they show evidence of fluctuations in longer timescales;
- In close binary systems, such as for instance the RS CVn-type stars (Walter *et al.*, 1980) and WUMa-type stars (Cruddace & Dupree, 1984), in which the accretion process does not play a fundamental role, the levels of the X-ray emission are rather higher than in the case of individual stars, whose stellar characteristics (mass, radius, effective temperature) are similar to those of the two components of the binary system (Charles *et al.*, 1983).

5.5.1.1. *Correlations amongst X-ray stellar emission and other stellar parameters*
A large dispersion in the levels of X-ray emission has been observed for stars of any spectral type and luminosity class. Then the question is: Is it possible to determine the physical properties of the stars in order to explain such an experimental fact? Since the total X-ray luminosity is weakly correlated with effective surface temperature of the late-type MS stars, it is necessary to find some stellar parameters – which are not those determining the position of the stars in the HR diagram – for determining the real level of the X-ray emission. Such 'new' parameters were suggested for the observations performed with the HEAO-1 satellite (Walter *et al.*, 1980) and later for the Einstein observations (Walter & Bowyer, 1981) in RS CVn-type stars. In such stars the rotation is a fundamental parameter for the coronal luminosity. Such a relationship between the rotation velocity and chromospheric activity was already known by means of CaII (Kraft, 1967). The discovery of the luminosity–rotation relationship for individual late-type dwarf stars was obtained by analyzing the data coming from the stellar survey of the Einstein satellite (Pallavicini *et al.*, 1981, 1982). However, since the first observations of X-ray emission from early-type stars, it was clear that the bolometric luminosity was the fundamental parameter responsible of the X-ray emission level (Long & White, 1980). Therefore for such stars their position in the HR diagram is a very good indicator of the X-ray emission levels.

Figure 68 shows the relationship between the soft X-ray coronal emission and the rotation of stars of F7-M5 spectral types and III, IV, and V luminosity class. For stars with spectral types from O to $\simeq$ B5, the X-ray emission level is proportional to the bolometric luminosity of the star. Such a constant is expressed by $L_{\mathrm{x}}/L_{\mathrm{bol}} \simeq 10^{-7}$ (Pallavicini *et al.*, 1981).

The transition between the behaviour of early-type and late-type stars (whose emission depends essentially on the rotation) occurs in the intermediate spectral range *A* (Topka *et al.*, 1982). The coincidence between the localization of such

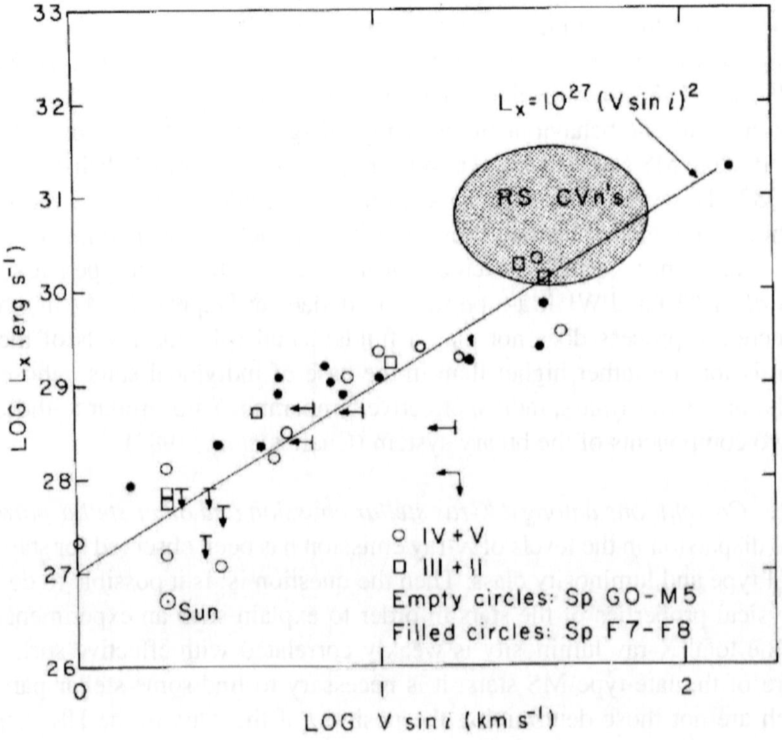


*Figure 68.* Relationship between the X-ray luminosity and rotational velocity of stars with different spectral types and luminosity class (Rosner, Golub & Vaiana, 1985, after Pallavicini *et al.*, 1981).

a transition and the starting of the surface convection authorizes to think that the surface convection phenomenon is a necessary requirement for a solar-like stellar activity.

A particular aspect of the behaviour of the stars is that all of them, independently of the luminosity class, lie in the same regression line of the X-ray luminosity viz., the rotation (see Figure 68). This indicates that rotating stars, which possess external convection zones, emit X-rays independently of the effective gravity (with the exception of cold giant and supergiant stars). However, X-ray data from the first observations showed that the effective gravity plays an important role in young stars. This is because the effective gravity correlates with the bolometric luminosity (and bolometric and X-ray luminosity experimentally correlate). Therefore the effective gravity is correlated with the X-ray luminosity. However, since the theory predicts a relationship between the bolometric luminosity and the X-ray emission levels (at least for models considering instabilities in the radiation driven winds), the X-ray luminosity-effective gravity relationship for such stars is considered as secondary and without physical significance. The theoretical predictions become better if stars of colder spectral types are considered. The results from the Einstein

Observatory revealed that giant and supergiant late-type stars are, probably, very weak X-ray sources (Rosner, Golub & Vaiana, 1985). It is possible to encounter a possible explanation invoking the analogy with the solar coronal holes: in such regions the coronal gas flows to the exterior parts in order to form the stellar wind, and therefore the X-ray emission level is strongly diminished.

5.5.1.2. *Stellar X-ray emission and the diffuse soft X-ray background.* One of the most direct consequences of the stellar X-ray emission is its almost passive role in the contribution to the galactic component of the diffuse soft X-ray background (Gorenstein & Tucker, 1976). Before the Einstein Observatory such a contribution owed to stars was estimated in about 20% at 0.25 keV. This result was one of the points in favor of increasing the sensitivity and space resolution of the future generation detectors of X-ray telescopes. Indeed, with the predicted threshold sensitivity of the Einstein Observatory ($10^{-16}$ erg cm$^{-2}$ s$^{-1}$), the dwarf M stars expected for square degree was $\sim$100.

5.5.1.3. *X-ray emission from very young stars and ISM.* In addition to the passive role, already described, the X-ray emission plays an active role which affects the physics of the interstellar medium in the vicinities of the stars. Such an influence was manifested especially in the case of very young X-ray active objects, such as those encountered in star formation regions. Amongst the possible effects studied is included the X-ray emission as regulator of star formation, owing to the large flux of ionizing photons, coming from objects embedded in regions of molecular clouds (Lepp & McGray, 1983). Such observations stress the importance of the stellar activity in the phase of formation of the solar system and during the formation of the planetary atmospheres (e.g., Feigelson, 1984).

5.5.1.4. *Mass and particles going into the ISM.* It is known that the surface activity of stars can be manifested as an expanding continuous massive wind (Cassinelli, 1979), or, in transitory form, by expulsion of puffs of mass and production of relativistic particles (Coleman & Worden, 1976). As a consequence of such expulsions of matter and relativistic particles the stars are one of the most powerful supply of mass going into the ISM (Abbott, 1982) and can also contribute directly to the formation of the cosmic ray spectrum, by means of the injection of particles accelerated during the stellar flares (Mullan, 1979). Stars can also indirectly contribute to the cosmic ray spectrum by means of acceleration of particles and the shocks within the winds or in their terminal parts (Cesarsky & Montmerle, 1983). This latter possibility is strongly based on the evidence that the elemental composition of galactic cosmic rays heavier than helium is very similar to that of solar energetic particles (Meyer, 1981, 1985).

5.5.1.5. *Conclusions and perspectives of stellar X-ray observations.* The Sun and its vicinities provide a directly observable laboratory in which it is possible to study the magnetohydrodynamics and physics of the plasma, which condition the physics of the X-ray emission. Moreover, they allow the defining of the theoretical frame within which it is possible to interpret the observations. The most important results obtained with these studies were new limits on the models for stellar activity. It was possible to check the theoretical dependence among the levels of activity and stellar parameters (that evidently cannot be deduced by studying only the Sun).

Therefore X-ray observations and the associated models provided additional limits on the theories of coronal heating, and on the generation of the magnetic flux (dynamical theory), which would not have been possible to obtain with solar investigations only. Within a general astrophysical frame stellar X-ray observations allow the extension of solar analogies to other more 'exotic' phenomena (and surely with more astrophysical interest), such as for instance the formation of coronas or accretion disks and the energetics of galactic hot halos. Such an extension is not limited only to the study of late-type stars. The study of the possible instabilities of the radiation driven winds in OB stars, will be useful for attempting the use of these theories to: i.e. instabilities in radiation-driven jets of matter.

This was what the scientific community would have expected from the new generation of satellites, such as ROSAT, whose results will be discussed later.

#### 5.5.2. *X-Ray Emission in Close Binary Systems*

Close binary systems were the first objects to be detected in the X-ray energy region because of their intense emission, high enough to exceed the thresholds of detection even of the first generation X-ray detectors, which were simple proportional chambers, whose sensitivity level was of $\sim 10^{-4}$ photons cm$^{-2}$ s$^{-1}$ keV$^{-1}$ in the range 2–50 keV (e.g., Tananbaum, 1973; Peterson, 1973). Observational techniques in X-ray astronomy were described by Giacconi, Gursky & Van Speybroeck (1968). Two historical conferences well described the situation after UHURU satellite (Giacconi *et al.*, 1972), namely IAU Symposium No. 55 on *X-ray and $\gamma$-ray astronomy* (Bradt & Giacconi, 1973) and *Neutron stars, black holes and binary X-ray sources* (Gursky & Ruffini, 1975), based on a session of the annual meeting of the American association for the Advancement of Science held in San Francisco in February, 1974. Thousands of articles have been written during the 41 years of the life of X-ray astronomy, both experimental and theoretical, as well as hundreds review papers more or less general, and tens of books (e.g., Eggleton, 1976; Sanford, 1979; Giacconi, 1981; Bradt & McClintock, 1983; Lewin & van den Heuvel, 1983; Giovannelli, 1985; Mason, Watson & White, 1986; Giovannelli & Mannocchi, 1987; Gehrels & Share, 1988; Kundt, 1990; Holt & Day, 1993; Johnston, Walker & Bailes, 1996; Swank, Kouveliotou & Hurley (eds.), 1998; Czerny, 2001).

In the 1970s the study of X-ray pulsars stimulated many fields of astrophysics. The measurements of the neutron star masses were especially important as well as their relationships with the nuclear physics and high energy physics, gravitation theory and stellar evolution. It was possible to obtain detailed information about the optical companions of the collapsed objects, and therefore to develop in a coherent framework the evolution of close binary systems (e.g., van den Heuvel, 1976).

The Einstein observatory (HEAO-2 of NASA) (Giacconi *et al.*, 1979) opened a new era in the soft X-ray astronomy (in the keV range) thanks to the X-ray photon focusing telescope, which allowed us to reach high spectral and space resolutions. An Image Proportional Counter (IPC) placed in the focal plane of the X-ray telescope was designed especially for searching X-ray sources in particular selected fields, such as for instance those of the COS-B sources ($\simeq 1°$). The X-ray sources were detected within a precision of $\simeq$ 1 arcmin. With 1–5 IPC observations it was possible to explore one COS-B field, and then with the High Resolution Imager (HRI) it was possible to define the position of the X-ray sources with a precision of 1–2 arcsec. Just to have an idea of the average sensitivity, in 2000 seconds of IPC observations it was possible to detect sources with fluxes of $\simeq 5 \times 10^{-13}$ erg cm$^{-2}$ s$^{-1}$. Comparing with the historical UHURU, a source with a flux of $\simeq 4 \times 10^{-2}$ UEFU (Uhuru Equivalent Flux Units), if converted into the 1–3 keV range of the Einstein, provides a flux of $1.12 \times 10^{-11}$ erg cm$^{-2}$ s$^{-1}$.

Through the study of the X-ray pulsars it was possible also to improve the knowledge of matter transfer processes in binary systems (e.g., de Loore *et al.*, 1984; Giovannelli & Ziółkowski, 1990), to push the theoretical studies on the accretion fluxes, accretion disks, and on accretion with torsional momenta (e.g., Shakura & Sunyaev, 1973; Okuda & Sakashita, 1977; Bahcall, 1978a; Ghosh & Lamb, 1978, 1979; Rappaport & Joss, 1981). Moreover, X-ray pulsations were used as probes of the internal structure of neutron stars and optical companions (e.g., Rappaport, Stothers & Joss , 1980).

By using the data bank of the Ariel 5, OSO-8, SAS-3, HEAO-1, EXOSAT, ASTRON, MIR/KVANT, the series of Japan small satellites (Hakucho, Tenma, Ginga, and Asca), and the experiments of the 1990s on board the German ROSAT, the CGRO-OSSE and BATSE, the Italian BeppoSAX it was possible to improve considerably the knowledge of binary systems both in soft and hard X-ray regions, with strong consequences also on the knowledge of extragalactic objects.

Our opinion is that in the near future the use should be extremely important of dedicated payloads on board small satellites, such as SIXE (Spanish Italian X-ray Experiment) (Giovannelli *et al.*, 1993b, 1999a,b; Gomez-Gomar *et al.*, 1999; Isern *et al.*, 2001) on board the Spanish Minisat Platform (Garcia-Primo, 2001) for a temporal study of X-ray pulsars and other compact objects, in order to definitively understand the physics of mass transfer, the trigger of X-ray bursts, the interior of neutron stars and in general the interactions between the two components in close binary systems.

The trivial definition of X-ray binaries is that they are binary systems emitting X-rays. However, it has been largely demonstrated that X-ray binary systems emit energy in the IR, Optical, UV, X-ray, Gamma-ray and sometimes they show also valuable radio emission.

Observational evidences suggest that X-rays are generated in binary systems because of the accretion onto the collapsed objects of matter coming from the optical companion. The strong gravitational field of the compact star accelerates the accreting material, which provides the necessary energy for the emitted radiation. The accretion mechanism depends on the equipotential surfaces of the binary system. Figure 69 shows such surfaces for a binary system (Ruffini, 1975):

- If the optical star fills its Roche lobe the matter is flowing through the internal Lagrangian point, and on approaching the compact star forms an accretion disk around it because of conservation of angular momentum;
- If the optical star is contained within its Roche lobe the matter accretes onto the compact star via the stellar wind of the optical star.
  Usually the mass accretion rate $\dot{M}$ is greater in the first than in the second case;
- It is possible that the flux of mass pushed onto the compact star via the stellar wind can form an accretion disk close to the compact star beyond the shock front of the stellar wind. Such a disk is unstable and can sporadically change its configuration, with consequent fluctuations of the torsional momentum of the accreting matter (e.g., Giovannelli & Ziółkowski, 1990).

Following the review of Giovannelli (1991), X-ray binary systems can be divided in different sub-classes:

- The High Mass X-ray Binaries (HMXBs) contain a collapsed object: the secondary star (neutron star or black hole) and an optical companion: the primary star with $M \gg 1$ M$_\odot$. They can be divided in two subclasses:
  - Hard X-ray Transient Sources (HXTSs). Their optical counterparts are *Be* stars with luminosity classes ranging between V and III, eccentricity of the orbits $0.2 \leq e \leq 0.5$, and orbital periods $P_{\rm orb} > 10$ d. Their hard X-ray emission has $\overline{KT} > 17$ keV and is extremely variable: $L_{\rm x} \sim 10^{34} - 10^{39}$ erg s$^{-1}$. Their maximum to minimum luminosity ratio is $L_{\rm xmax}/L_{\rm xmin} > 10^2$. With the sensitivities of the first generation detectors, they were not always detectable; this was the reason for their name 'transient sources';
  - Permanent X-ray Sources (PXSs). Their optical counterparts are supergiant OB stars (I luminosity class), eccentricity of the orbits $e \simeq 0$, and orbital periods $P_{\rm orb} < 10$ d. Their hard X-ray emission has $\overline{KT} \geq 9$ keV and is almost constant: $L_{\rm x} \approx 10^{37}$ erg s$^{-1}$. Their maximum to minimum luminosity ratio is $L_{\rm xmax}/L_{\rm xmin} \ll 10^2$. Also with the sensitivities of the first generation detectors, they were always detectable.

  The transfer of mass in both HTXSs and PXSs occurs via stellar wind. In both classes the collapsed objects (neutron stars) are modulating the X-ray emission with their spin periods: $0.069 \leq P_{\rm spin} \leq 1430$ s. For details, see Section 7.6.
- The Low Mass X-ray Binaries (LMXBs) contain a collapsed object: the primary star (neutron star or black hole) and an optical companion: the secondary star with $M \leq 1$ M$_\odot$ is a late-type star. The transfer of mass from the optical

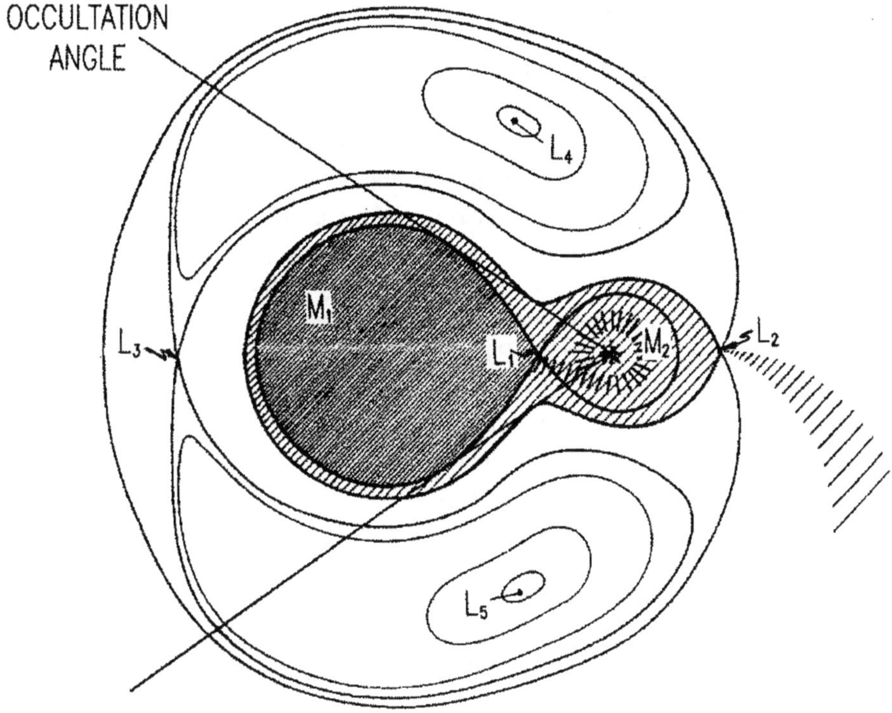


*Figure 69.* Roche lobes and equipotential surfaces of a binary system formed by a normal star $M_1$ and a compact collapsed companion $M_2$. $L_1$, $L_2$,..., $L_5$ are the Lagrange points of the system (Ruffini, 1975).

companion onto the collapsed object occurs via Roche lobe overflow. The matter is escaping from the Roche lobe of the optical star through the internal Lagrangian point and forms an accretion disk around the collapsed object.

LMXBs contain:

- X-ray sources in globular clusters;
- Soft X-ray transient sources (SXTSs). They show soft X-ray emission with $\overline{KT} < 10$ keV. In some systems the emission is pulsed (e.g., Her X-1, 4U 1626-27, GX1+4). SXTSs show Quasi Periodic Oscillations (QPOs) with periods are ranging between $\approx$0.02 s to $\approx$1000 s. At the end of the eighties such QPOs have been observed from about 20 systems (van der Klis, 1989). The orbital periods are ranging from 41 minutes to 11.2 days. The X-ray luminosity is in the range $L_x \sim 10^{36} - 10^{39}$ erg s$^{-1}$. With the advent of the new generation satellites of the last decade hundreds LMXBs have been detected as well as the associated QPOs. They are produced in the vicinity of the neutron stars and therefore they are a good probe for the geometrical and physical properties of collapsed objects. Moreover it has been well established that the bulge sources and bursters, considered in the

past as a distinct subclass of LMXBs, are on the contrary simply LMXBs. For details, see Section 7.7;

- Cataclysmic Variables (CVs) in which the optical companion is a low mass-late-type star and the collapsed object is a white dwarf. The detected CVs are spread roughly around the solar system at distance of $\approx$200–300 pc. Orbital periods range from tens of minutes to about ten hours with the known gap around 3 hours. The mass transfer is occurring either via Roche lobe overflow or via accretion columns, or in an intermediate way depending on the value of the magnetic field. Typical X-ray luminosity ranges from $10^{32}$ to $10^{34}$ erg $s^{-1}$ (e.g., reviews by Giovannelli & Martinez-Pais, 1991; la Dous, 1993, 1994; Ulla, Giovannelli & Martinez-Pais, 1999). For details, see Section 7.9;
- RS Canum Venaticorum (RS CVn) type systems, in which no collapsed objects are present and the two components are a F or G hotter star and a K star. Typical X-ray luminosity ranges from $10^{30}$ to $10^{31}$ erg $s^{-1}$. Usually in the current literature they are excluded from the class of X-ray binaries since historically they were discovered as X-ray emitters only by the second generation of X-ray experiments;
- Anomalous X-ray Pulsars (AXPs). They are X-ray pulsars with properties clearly different from those of the common HMXBs. AXP have spin periods ranging between $\sim$6 and $\sim$12 s, which are monotonically increasing on timescales of $\sim 10^4$–$4 \times 10^5$ yr. The optical counterparts of AXPs are not known. On the basis of the limits in the optical and IR wavelength regions, the presence of a massive early-type companion star, such as OB super giants or Be stars, can be excluded in AXPs. Moreover, no orbital motion signatures are present in their X-ray light curves. A possible exception is 4U 0142+61, for which Hulleman, van Kerkwijk & Kulkarni (2000) reported the discovery of a faint ($R \sim 25$ mag) blue object in its error box. For details, see Section 7.8.

From the study of the $\log N - \log S$ and X-ray luminosity function in the 2–10 keV energy range, and the spatial (3-D) distribution of bright, $L_x \geq 10^{34}$–$10^{35}$ erg $s^{-1}$, X-ray binaries in the Milky Way, Grimm, Gilfanov & Sunyaev (2002), in agreement with theoretical expectations and earlier results, found significant differences betwen the spatial distributions of LMXBs and HMXBs. The volume density of LMXBs peaks strongly at the Galactic Bulge, whereas HMXBs tend to avoid the inner $\sim$3–4 kpc of the Galaxy. In addition, HMXBs are more concentrated towards the Galactic Plane (scale heights of $\approx$150 pc and $\approx$410 pc for HMXBs and LMXBs, respectively) and show the clear signature of the spiral structure in their spatial distribution. The $\log N$–$\log S$ distributions and the X-ray luminosity functions are also noticeably different. LMXBs have flatter $\log N$–$\log S$ distribution and luminosity function. The integrated 2–10 keV luminosities of all X-ray binaries in the Galaxy, averaged over 1996–2000, are $\sim 2.5 \times 10^{39}$ erg $s^{-1}$ (LMXBs) and $\sim 2 \times 10^{38}$ erg $s^{-1}$ (HMXBs). Normalized to the stellar mass and the star formation rate (SFR), respectively, these correspond to $\sim 5 \times 10^{28}$ erg $s^{-1}$ $M_\odot^{-1}$ for LMXBs and $\sim 5 \times 10^{37}$ erg $s^{-1}$/($M_\odot$ $yr^{-1}$) for HMXBs. An important

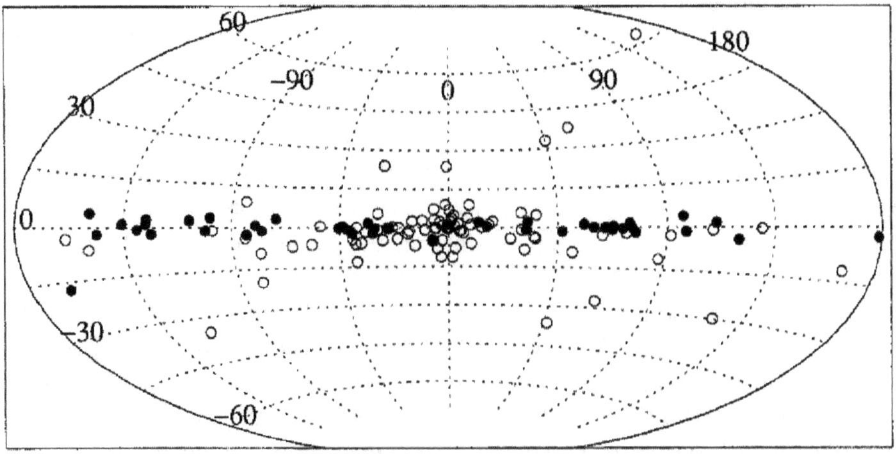


*Figure 70.* Distribution of HMXBs (filled circles) and LMXBs (open circles) in the Galaxy. In total 52 HMXBs and 86 LMXBs are shown. Note the significant concentration of HMXBs towards the Galactic Plane and the clustering of LMXBs in the Galactic Bulge (Grimm, Gilfanov & Sunyaev, 2002).

consequence of this work is that for an outside observer the integrated emission of our Galaxy is dominated by the 5–10 most luminous sources. Figure 70 shows the distribution of the LMXBs and HMXBs in galactic coordinates.

Based on ASCA and Chandra observations of nearby starburst galaxies and RXTE-ASM, ASCA and MIR/KVANT-TTM studies of HMXBs population in the Milky Way and Magellanic Clouds, Grimm, Gilfanov & Sunyaev (2003) propose that the number and/or the collective X-ray luminosity of HMXBs can be used to measure the SFR of a galaxy. They show that, within the accuracy of the present available data, a linear relation between HMXBs number and SFR exists. The relation between SFR and collective luminosity of HMXBs is non-linear in the low-SFR régime ($L_X \propto SFR^{\sim 1.7}$) and becomes linear only for a sufficiently high SFR, ($SFR >\sim 4.5\ M_\odot\ yr^{-1}$, for $M_\star > 8\ M_\odot$).

An excellent review of the formation and evolution of compact X-ray sources have been recently published by Tauris & van den Heuvel (2003).

### 5.5.3. *X-Ray Sources in Globular Clusters*

With the HRI detector of the Einstein observatory it was possible a survey of 30 globular clusters with luminosity $L_X > 10^{34} \times d_{10}^2$ erg s$^{-1}$ ($d_{10}$ is the distance in units of 10 kpc), and with the IPC detector it was possible to reach a luminosity limit $L_X \geq 10^{33} \times d_{10}^2$ erg s$^{-1}$ and to observe 32 clusters. Such X-ray sources had long been identified as extremely interesting objects because of their resemblance to the bright galactic X-ray sources (Canizares, 1975; Grindlay, 1978), and, moreover, the clusters were identified as X-ray burst emitters (Grindlay *et al.*,

1976). Since globular clusters can be studied in the optical region, they constitute a unique possibility for the X-ray astronomy of finding (at least potentially) the physical parameters of the accreting sources, such as the distance and then the luminosity, the mass (or at least the most probable value) and thus the evolution process.

The massive population I binaries can offer the possibility of similar studies, but many difficulties arise owing, for instance, to the heating of the optical companion, which produces a rather large uncertainty in the evaluation of the fundamental parameters, such as the distance (and then the luminosity).

It is evident that the first step in constructing detailed models for permanent or transient X-ray sources is that, obviously, of knowing their physical parameters. The observations performed with the Einstein observatory, having sensitivity and angular resolution 100 times and 10 times, respectively greater than those of the older experiments, provided fundamental knowledge about globular clusters in order to determine their global properties (e.g., their luminosity function), as well as about the individual sources.

The main results obtained can be summarized as follows:

– X-ray sources in globular clusters are binary systems (Clark, 1975; Lewin, 1980) and not massive black holes ($M \approx 1000$ $M_\odot$) (Bahcall & Ostriker, 1975) or intermediate mass black holes ($M \approx 10$–$100$ $M_\odot$) (Grindlay, 1978). Such observations can exclude the presence of isolated supermassive black holes, but not definitively stellar mass black holes ($M \leq 5$ $M_\odot$). The restrictions on the mass and the limits on the free gas in the center of the cluster constrain almost certainly that the X-ray source must be a binary system. The predictions made with the nuclear burst model in explaining the X-ray flares (Joss, 1978) provided strong evidence that such binary systems contain a neutron star;

– X-ray sources have been found in low luminosity ($L_X \simeq 10^{33}$ erg s$^{-1}$) globular clusters, close to the IPC threshold. Such sources are those with the lowest luminosity, which are placed at the end of the luminosity distribution of the compact sources, and are with almost total security binary systems with $10^{32} < L_X < 10^{37}$ erg s$^{-1}$. It is also possible that sources with $L_X \simeq 10^{33}$ erg s$^{-1}$ are diffuse and that the detected X-ray emission is the sum of the individual contributions of the stars at the center of the cluster (Grindlay, 1981). This was one of the problems posed for next generation experiments with higher angular resolution and sensitivity.

Then, summarizing, after the Einstein measurements it was possible to affirm that X-ray sources in globular clusters are very likely compact binary systems (Lewin & Clark, 1980).

With the advent of the more sensitive ROSAT satellite it has been possible to search for low luminosity X-ray sources in globular clusters. Verbunt (2001) analyzed 101 observations from the ROSAT archive to search for X-ray sources in or near 55 globular clusters. New sources have been found in the cores of NGC 362 (a double source), NGC 6121 (marginally significant), NGC 6139, and NGC 6266, and outside the cores of NGC 6205, NGC 6352 and NGC 6388. More accurate positions have been determined for the X-ray sources in some ten clusters. The improved position for the source in NGC 6341 excludes the suggested ultraviolet counterpart. It has been shown that one of the two sources reported near the core of NGC 6626 is spurious, as is the detection of a pulsar period in the PSPC data of this cluster. The central source is resolved in three sources. One source reported previously in NGC 6304 is demoted to an upper limit. For 20 cluster cores better upper limits of the X-ray luminosity have been obtained. From his statistical analysis Verbunt (2001) argued that several sources outside the cluster cores may well belong to the clusters. All spectral energy distributions observed so far were relatively soft, with a bremsstrahlung temperature of $\simeq$0.9 keV. There is evidence, however, that bremsstrahlung spectra do not correctly describe the spectra. The X-ray luminosity per unit mass for the cluster as a whole does not depend on the concentration; the luminosity per unit mass for the core may increase with the cluster concentration.

Eight clusters near which a low luminosity source was discovered with the Einstein satellite have been observed with the ROSAT HRI; three have been observed with the ROSAT PSPC. The more accurate position obtained with ROSAT shows that all sources are within the cluster core, with the exception of the source near NGC 1904, which is well outside the core. The comparison between the Einstein IPC count rates and ROSAT count rates for the three PSPC observations shows that they are similar. This is remarkable, given that at least two of these clusters show variable X-ray luminosity: NGC 104 (Verbunt & Hasinger, 1998) and NGC 6656 (Verbunt, 2001). Amongst the seven ROSAT HRI observations five detect the source at a count rate roughly half that of the Einstein IPC count rate, as expected for a constant source, or slightly below it. Two exceptions are the variable sources in NGC 5272 and NGC 6656, as discussed by Verbunt (2001). The source in NGC 5824 is not detected with the HRI and apparently was at lower luminosity when ROSAT observed it in 1991 and 1994 than during the Einstein observations.

Since the number of faint X-ray sources detectable with ROSAT is very large, it is possible a chance coincidence between a foreground or background X-ray source and a globular cluster. Verbunt (2001) evaluated such a possibility and concluded that it is likely that all sources detected in the HRI observations within $30'$ of the cluster center are indeed cluster members.

### 5.5.4. *X-Ray Emission in SNRs*

Through a violent explosion of a star an enormous quantity of energy is liberated in a very short time (supernova explosion) and a Supernova Remnant (SNR) is formed. Such a remnant contains material of the progenitor mixed with the surrounding ISM.

The results obtained with the observations of SNRs performed with the Einstein observatory provided a strong advance in the comprehension of the physics involved. For the first time images of SNRs were obtained, giving important information about the kinematical and dynamical behaviour of the SNRs, which provided a new impulse for developing more precise theoretical models. Before the Einstein observatory, SNR models were based on optical and radio observations. Thanks to the sensibly better space and spectral resolutions, it appeared necessary to redefine the existing hydrodynamical models. X-ray observations, and especially the X-ray images, substantially contributed to remodelling the SNRs. In this context measurements of young SNRs were remarkably important, since they allowed us to obtain information about their origin because most of their energy is emitted in the X-ray energy range.

What was the state of hydrodynamical models until the moment at which the first Einstein images arrived?

For simplicity we start from an evolutive scheme in which an homogeneous jet of stellar material interacts with a uniform and homogeneous ISM (Woltjer, 1972). In such a scheme we can recognize three phases of the expansion of the SNRs:

– Free expansion (constant velocity);
– Adiabatic expansion (conservation of the energy);
– Radiative expansion (conservation of the momentum).

However, the observations of e.g., HI and CO show that the ISM is neither uniform nor homogeneous. Numerical simulations (Gull, 1973) confirmed also that if the shock wave were uniform it would immediately experience a deceleration and therefore it cannot be adiabatic until the end. Then no one of the three kinds of expansion is real. However, this scheme is still useful. The model frequently used for describing the adiabatic expansion is that developed by Sedov (1959). Such a model was developed in the first decade after the second world war in order to describe the shock wave caused by the atomic bomb explosion. With this model it was possible to calculate the radius of the expanding shock wave as a function of the time, being the shock wave provoked by a point-like explosion occurred in a homogeneous medium. Translating this problem to SNRs, once the radius is known, for instance through the images obtained with the Einstein observatory, it is possible to evaluate its age. However, it appeared evident that by using Sedov's model the calculated ages of historical SNRs were not coincident with those historically known. Therefore it was necessary to readjust such models. Indeed, with the X-ray images it appeared evident that the shock wave of a SNR has frequently an irregular structure and its form is not circular (spherical). It was also found that the luminosity within the same remnant is not constant. On the contrary it shows large variations. It was also difficult to know how much of the detected X-ray radiation is owed to the ISM shocked or to the material coming from the exploded progenitor.

Modelling the measured X-ray spectra of SNRs it was evident the presence of a soft X-ray excess, which implies a structure of the remnant with two temperatures. However, until the end of the 1980s the physical nature of such two-temperature structure was not yet well understood (Claas, 1990).

Three hydrodynamical models born on the basis of Sedov's model and the measurements then available, namely:

- Adiabatic model. This is an extension of the classical Sedov model, in which one takes into account the density gradient of the ISM ($\rho_{ISM} \propto r^{-\delta}$, where $\delta$ is the interstellar density gradient, $r$ is the distance, and $\rho_{ISM}$ is the ISM density). Then this model takes into account, in a certain way, the inhomogeneous character of the ISM;
- Isothermal model (Solinger, Rappaport & Buff, 1975). Contrary to Sedov adiabatic model in which the conduction of the heat is not considered, the isothermal model assumes an infinite conduction ($\partial T/\partial r = 0$). Such a model takes into account also the density gradient of the ISM;
- Inverse shock model (Chevalier, 1982). This model not only considers the shocks in the ISM, but also the jets coming from the progenitor star. In this way it is possible to explain the soft X-ray excess present in the observed spectra.

  The investigations made for modelling the X-ray spectra of SNRs by using data coming from the EXOSAT satellite, taking into account the effects of the non-equilibrium in the ionization, allowed us to conclude that the emission of Si, S, Ar, and Ca is associated with an inverse shock (e.g., in the case of the Cas A SNR) (Smith *et al.*, 1988; Jansen, 1988).

In the epoch of the Einstein observatory, the scientific community was waiting for the launch of the Italian SAX satellite fixed in 1993, and really launched in 1996 (Piro, 1996), and the German ROSAT satellite, launched in 1990 (Trümper, 1990), in order to have spectra with enough resolution to select one of the hydrodynamical models, higher statistics and detection of weaker SNRs. In that time the number of galactic SNRs was 145 (Green, 1984). In Section 7.13 we will discuss more about SNRs considering the data coming from the satellites of the 1990s.

Mathewson *et al.* (1983a,b, 1984) studied the SNRs in the Magellanic Clouds, and suggested four categories:

- SNRs dominated by Balmer lines;
- SNRs oxygen rich;
- SNRs compound by plerions;
- Evolved SNRs.

Weiler & Sramek (1988) reported in two tables the identified characteristics of SNRs and the relationships among supernovae and SNRs.

## 5.6. Galactic Historical Journey: UV Range

The IUE (Internationaal Ultraviolet Explorer) satellite, has been the most long lived and (by a wide margin) the most productive satellite so far, in the history of space astronomy. After going into geosynchronous orbit on 26 January 1978 from the Kennedy Space Center, as a NASA-ESA-UK project, IUE was meant to operate for three years. More than eighteen years later IUE was still working 24 hours a day, harvesting new knowledge for astronomers. IUE was turned off on 30 September 1996. The two last science images taken with IUE on September 27th, 1996 were SWP-58388 low and LWP-32696 high dispersion spectra of Capella, which was also the first IUE target.

IUE payload was equipped with two primaries (P) and two redundant (R) cameras for short (SW) and long (LW) wavelengths. SWP (1150–1980 Å): sensitivity of $2 \times 10^{-15}$ erg $s^{-1}$ $cm^{-2}$ $Å^{-1}$, LWP (1850–3350 Å): sensitivity of $1 \times 10^{-15}$ erg $s^{-1}$ $cm^{-2}$ $Å^{-1}$, LWR (1850–3350 Å): sensitivity of $2 \times 10^{-15}$ erg $s^{-1}$ $cm^{-2}$ $Å^{-1}$, SWR : never operational.

The IUE, with the possibilities of doing high and low resolution spectroscopy, together with its optical monitor, has provided an immense number of excellent results. Its data bank is still used for taking new results. In the books *Exploring the Universe with the IUE Satellite* (Kondo, 1987), and *Ultraviolet Astrophysics Beyond the IUE Final Archive* (Wamstecker & González-Riestra, 1998) the main investigations performed with such a satellite have been collected.

Observations of stars in the IUE range have widened our knowledge of a large number of stars distributed across the whole HR diagram. Figure 71 shows the IUE All sky survey (by object type) of all the images taken between 1978–1992 by IUE with the SWP (1150–1980 Å) and LWP/R (1850–3350 Å) cameras. This is an Aitoff projection of the positions of more than 80,000 images (about 8,000 unique positions are visible) plotted in galactic coordinates. The galactic and ecliptic planes are easily detectable across with other galactic features as well as the Small and Large Magellanic Clouds. Observed objects have been distinguished by IUE object class: solar system: red; galactic: violet; extragalactic: green (Courtesy of the IUE Data Analysis Center (IUEDAC), NASA-GSFC).

The first UV observations of bright early type stars unexpectedly showed that they have fast expanding atmospheres and mass loss. The measurements of the Copernicus satellite revealed that the winds of such stars are super-ionized. It then appeared clear that in the evolutionary studies of massive stars it is necessary to introduce a possible mass loss in a proportion more or less significant depending on their luminosity. We know for certain that stars having similar optical spectra can have, in contrast, UV spectra very different.

Before the UV era, the interest in studying stars of early type was centered on the spectroscopic effects owed to the local thermodynamical non-equilibrium, the study of stellar chromospheres (with the exception of the Sun) was limited essentially to the investigation of the *H* and *K* emission resonance lines of CaII (Wilson & Vainu Bappu, 1957). Whilst with the UV window open to the Universe, the stars caught most of the attention of stellar evolution investigations, dynamics of gases, and observational studies from the X-ray to radio ranges, and it was possible to investigate high excitation lines of many ions. With the first UV experiments on board stratospheric balloons and with the Copernicus satellite it was possible to analyze the MgII doublets (2795–2802 Å). The IUE satellite allowed a more detailed investigation of stellar chromospheres thanks to its high resolution

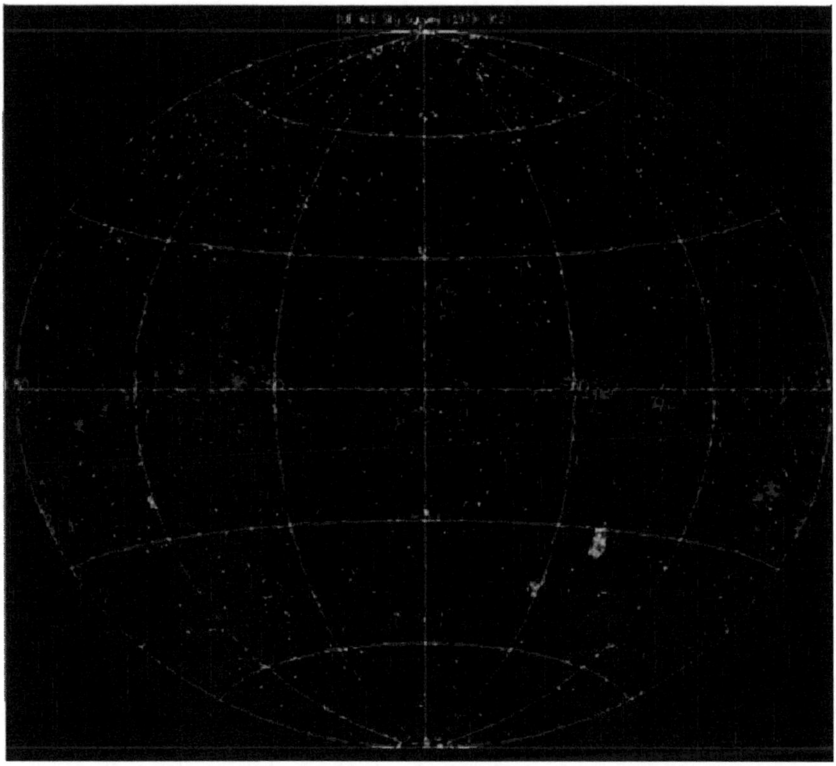

*Figure 71.* The IUE All sky survey (by object type) of all the images taken between 1978–1992 by IUE with the SWP (1150-1980 Å) and LWP/R (1850-3350 Å) cameras. This is an Aitoff projection of the positions of more than 80,000 images (about 8,000 unique positions are visible) plotted in galactic coordinates (Courtesy of the IUE Data Analysis Center (IUEDAC), NASA-GSFC).

(~0.2 Å) and relatively high sensitivity, which allowed also the study of weaker stars.

5.6.1. *Stellar Chromospheres and Transition Regions*

With the IUE observations the study of chromospheric emissions was extended up to the early type stars as primitive as the A spectral type and late type stars such as dwarf M (Brown & Jordan, 1981). Chromospheric emission was detected also in evolved stars (Linsky & Haisch, 1979). It then was possible to find relationships amongst various stellar parameters which characterize zones of sub-photospheric convection with chromospheric emission. The rotation enhances the level of emission, and when the star evolves the stellar wind tends to brake the rotation; therefore the emitted fluxes generally decrease (Hartmann *et al.*, 1984). When the star evolves, the decay of the magnetic activity in the external part can occur faster in the lower than in the upper shells (Simon, Herbig & Boesgaard, 1985).

Detailed studies of the fluxes of the emission lines and their widths suggest that in the stars with spectral types later than intermediate F the chromospheres are heated by means of magnetic fields or by a stellar dynamo (e.g., Ayres & Linsky, 1980), whilst for stars with spectral types earlier than intermediate F, and for weak chromospheres the heating could be dominated by acoustic waves.

Chromospheric properties of individual stars can be considered as those corresponding to solar-like stars, non-solar type stars, or an hybrid type. The solar-type stars have a hot transition region ($T \sim$ few $\times 10^5$ K) and coronal temperature of order of $10^6$ K. The non-solar type stars do not have a hot transition region, but have a cold and massive stellar wind. The stars of hybrid type possess both characteristics: hot transition region and cold massive stellar wind.

The frontier between solar and non-solar type stars in the HR diagram is placed around the spectral type K1 with luminosity class III (Linsky & Haisch, 1979).

The chromospheres of stars in close binary systems are frequently very different from those of the individual stars. A late type star in a close binary system rotates faster than a similar individual star. The loss of angular momentum owed to stellar wind leads to a slower rotation in an individual star, whilst in a binary system the loss of angular momentum is compensated by the large angular orbital momentum. Faster rotations in binary systems enhance the stellar dynamics, with consequent enhancement of chromospheric activities.

After a three year survey in UV and optical ranges of the star Betelgeuse, periodic variations of the luminosity and a difference in phase between photospheric perturbations and chromospheric variations have been detected. This is an indication of a propagating wave (Dupree *et al.*, 1987). The results indicate that the periodic variations can provoke the heating and expansion of the atmosphere and the driving of the stellar wind.

The structure of the external shells of cold star atmospheres was explored in details making observations of binary systems in which such a stars had hot companions (for instance RS CVn systems and $\zeta$ *Aur*-type binaries. The atmosphere of the cold star is observed projected on the hot star during a specific orbital phase.

5.6.2. *Evolutionary Processes in Interacting Binary Systems*

It is estimated that at least 67% of stars lie in binary or multiple systems (e.g., Lipunov, 1989). Moreover, a large part of the objects which are very interesting from an astrophysical point of view belong to binary systems such as novae and X-ray binaries. Therefore the comprehension of the evolution of binary systems is extremely important in understanding the whole process of stellar evolution. The flowing mass in close binary systems play a fundamental role in the evolution of the system. Temperature and density of the flowing gas are important indicators and they manifest spectroscopically in the UV range by means of lines of elements ionized one or more times, such as MgII, FeII, AlIII, CIV, SiIV, and NV.

In practically all the close binaries observed in high resolution mode with the IUE, the flowing gas has been always detected. With the study of the radial velocities of the observed absorption lines it is possible to determine the fraction of the gas lost by the binary system. At the same time a part of the gas flowing from the loosing mass star accretes onto the companion star. This fact is revealed by means of spectral characteristics indicating high temperatures (e.g., SiIV, CIV, and NV), which should be much too high for an individual star with the same spectral type.

The main results obtained with the IUE on close binary systems are:

i) Discovery of high temperature and accretion regions.

The IUE observations revealed the existence of high temperature regions; this was based on the presence of absorption lines of atoms highly ionized, such as SiIV, CIV, and NV in binary systems of Algol type.

In the case of $\beta$ Lyr high temperature lines also appear in emission (Hack *et al.*, 1975). In W UMa type binaries the high temperature regions show emission lines of distinct classes of atoms ionized several times. Rucinski (1985) discovered that the chromospheres of such binary systems are very active. However, their coronal activities are weaker than those in individual stars. He argued also that the gas of the shell in systems younger than W UMa type systems can escape owing to the absence of magnetic lines which could confine it.

The presence of high temperature absorption and emission lines cannot be justified simply in terms of the radiation field of the star considered. Such lines indicate the existence of a non-thermal source. In the case of the Algol system, characterized by absorptions overlapping those of the B star, the high temperature is probably caused by accretion onto such a star (Kondo, Boggess & Maran, 1989);

ii) Mass loss in binary systems.

In most binary systems the Doppler blueshift of the absorption lines indicates that the systems are losing mass. This means that the flux of mass is not simply caused by a gravitational overflowing of the Roche critical equipotential surface. No gravitational forces must be responsible of the expulsion of the gas outside the binary systems (e.g., Modisette & Kondo, 1980);

iii) Discovery of hot companion star.

Parsons (1981, 1983) carried out studies of a number of binary systems in which hot companion stars were encountered. The systems HD 207739, which contains an optically thick variable gaseous envelope, and the 22 Vul of $\zeta$ Aur type, which is a short period binary ($\sim$249 days), with atmospheric eclipses of the G3 supergiant star, became particularly interesting.

### 5.6.3. *Remarkable Results in Individual Binary Systems*

We can summarize some important results obtained from individual binary systems without any pretension of completeness.

i) Gravitational redshift owed to a neutron star.

V Sag (Koch *et al.*, 1986) showed emission lines with a constant redshift of 700 km $s^{-1}$. The persistence of such a redshift is very difficult to explain in terms of matter accreting onto one of the components of the system. In contrast, it can be explained if the lines are produced in a gaseous envelope around a neutron star and the nature of the redshift is gravitational. This was the first case in which a neutron star was spectroscopically detected;

ii) Large scale structures in active K dwarfs.

With the IUE observations of the eclipsing binary V 471 Tau (K2V + DA), the detection of three-dimensional structures of the atmospheric bubbles and the associated star flares was possible (Guinan *et al.*, 1986). Just before and after the total eclipse of the white dwarf component, when it is projected near the K dwarf limb, absorption lines of CII, CIII, CIV, and SiIV appear overlapped on the continuum of the hot white dwarf. Such lines probably form in cold coronal bubbles, which shield the flares in the atmosphere of the K dwarf. Sometimes the bubble can be extended until almost a stellar radius of the K star;

iii) Dynamic mass ejection in U Cep.

The Algol type binary U Cep (B7–6V + G8 III–IV) shows periodically episodes of extremely active mass ejection, during which an optically thick plasma surrounds the less evolved B component. During one of such active episodes in U Cep in 1986, high resolution IUE spectra were taken. Such spectra showed that an optically thick plasma, apparently emanated by the G giant, was shielding most of the B companion, lowering most of the UV flux of the hot star (McCluskey, Kondo & Olson, 1988);

iv) Other works on binary systems.

More detailed information about IUE observations and results from interacting binary systems can be found in the review paper of McCluskey & Sahade (1987). The IUE contributed in a very special way to knowledge of the Cataclysmic Variables (CVs), and this was one of the main reasons for which a IAU colloquium was organized specifically on this topic (Drechsel, Rahe & Kondo, 1987). The IUE also provided important contributions to knowledge of the accretion processes onto collapsed objects in close binary systems, which include the CVs, and the low and high mass X-ray binaries. One of the historical reviews of this topic was that by Cordova & Howarth (1987).

In binary systems containing collapsed objects (white dwarf, neutron star, and sometimes probably a black hole) there has been detected without any doubt the presence of an accretion disk around it, thanks to the study of the UV spectra, in which emission lines of C, Si, and N, highly ionized, and HeII clearly appear.

Figure 72 shows two typical low resolution IUE spectra of the cataclysmic variable (CV) SS Cygni during an outburst (upper panel) and a quiescent phase (lower panel) (Gaudenzi *et al.*, 1990). Many important results on CVs have been obtained with the IUE observations. Amongst them, in our opinion, it is necessary to note the discovery of orbital modulations of the continuum flux and of the fluxes of the emission lines, coming from the accretion disk around the white dwarf of SS Cygni, which opened a discussion on the morphology of such a disk and its origin as well as on the intensity of the magnetic field, which could be sometimes typical of the so-called Polars and sometimes typical of the so-called Non Magnetic CVs (Giovannelli *et al.*, 1984b; Gaudenzi *et al.*, 1986; Lombardi, Giovannelli & Gaudenzi, 1987; Giovannelli *et al.*, 1990a; Martinez-Pais *et al.*, 1994, 1996; Giovannelli & Sabau-Graziati, 1998). Moreover, SS Cygni has been studied in many energy ranges, sometimes with simultaneous observations which gave crucial information not only about SS Cygni itself, but in general about CVs (e.g., reviews of Giovannelli & Martinez-Pais, 1991; Giovannelli, 1996a; Giovannelli & Sabau-Graziati, 1999a, Ulla, Giovannelli & Martinez-Pais, 1999). Such a system has been used as an example of coordinated multifrequency observations (e.g., Giovannelli *et al.*, 1983; Giovannelli & Sabau-Graziati, 1996b).

By means of IUE observations Gaudenzi *et al.* (1990) calculated the mass accretion rate $\dot{M}$ onto the white dwarf of SS Cygni. Such a value practically coincides with that deduced from the equivalent widths of the HeII (1640 Å) emission line by using the model of Patterson & Raymond (1985a,b), which correlates the equivalent widths of the HeII (1640 Å) and HeII (4686 Å) with the mass accretion rate. Gaudenzi *et al.* (1990) found a surprising result: the instantaneous value of $\dot{M}$ during an outburst is less than that during quiescence. With this result they concluded by suggesting that during an outburst of long type (L type) the disk is destroyed, and during an outburst of short type (S type) the disk is partially destroyed; in both cases, the matter partially slowly accretes onto the white dwarf, and partially is expelled via wind (e.g., Giovannelli *et al.*, 1990a). During the following quiescent phase the disk forms again. However, such conclusions are not generally accepted, being the problem of accretion disks, their evolution, and the production of outbursts strongly biased by the fashion in science, as largely commented upon in the Section on the discussion and conclusions in the paper by Giovannelli & Sabau-Graziati (1999b).

Therefore we would like to launch an invitation to: i) the experimental scientific community in reconsidering the philosophy of the observations in favour of deep multifrequency surveys of a few selected cosmic sources rather than a few sporadic measurements of many of them, in order to provide the proper input parameters either to check or to improve the current models. This is just the philosophy of the SIXE (Spanish Italian X-ray Experiment) (e.g., Giovannelli *et al.*, 1999a,b; 2001). Also the suggestion of Nousek *et al.* (1994) is just along this line. However, the observations of a dwarf nova (perhaps SS Cyg itself) through an outburst cycle with the ASCA satellite were never performed. We can add also the suggestion

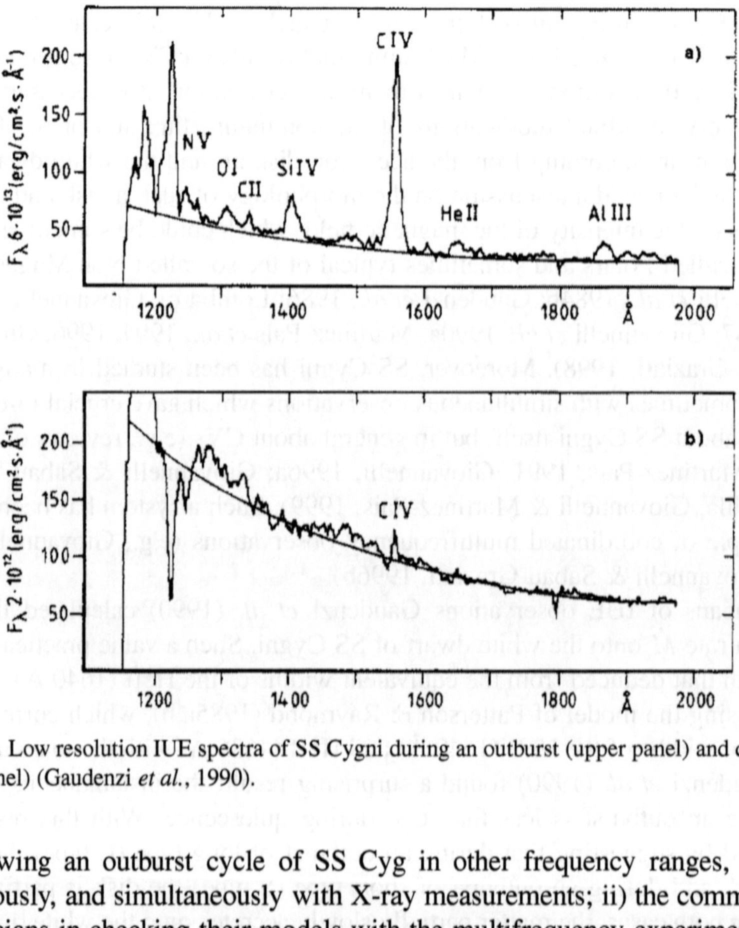


*Figure 72.* Low resolution IUE spectra of SS Cygni during an outburst (upper panel) and quiescence (lower panel) (Gaudenzi *et al.*, 1990).

of following an outburst cycle of SS Cyg in other frequency ranges, possibly continuously, and simultaneously with X-ray measurements; ii) the community of theoreticians in checking their models with the multifrequency experimental data and in not being worried by recognizing the eventual failures; iii) both communities to avoiding following the fashion in science; iv) a part of scientific community in reading papers although they are published in the so-called minor journals. Sometimes they report interesting results a few years in advance with respect to the so-called major journals!

### 5.6.4. *Winds in Young Stars*

The contribution of the IUE observations on knowledge of stellar winds from hot stars has been essential thanks to its better sensitivity with respect to previous UV observatories. It was possible to observe central stars of planetary nebulae and also the most luminous stars in some of the galaxies of the Local Group. The resolution was high enough to identify and measure the parameters of the winds by the analysis of the spectral line profiles. Another positive factor in such investigations has been the very large duration of the IUE mission. Indeed, it has been possible:

- to obtain spectra of hundreds hot stars and then to enhance the knowledge of the properties of this population, and, in particular, the properties of the winds in function of the stellar characteristics;
- to perform repeated observations of selected stars in order to study the variability of the stellar wind;
- to foresee and propose new investigations and to program and develop campaigns of observations with well defined objectives.

An example of the latter point is the *heroic* 1988–89 IUE program of Giovannelli's group on the transient X-ray/Be system A0535+26/HDE245770, which obtained 10 IUE shifts opportunely separated in time and scheduled two of each consecutively, in order to obtain high resolution spectra to be summed for improving the signal to noise ratio owing to the relatively faintness of the Be star. The objective of this multifrequency program was the study of the system around the periastron passage of the neutron star in order to definitively understand the processes triggering the X-ray outbursts. For this reason, besides the IUE observations, also X-ray, optical, and IR observations were programmed. Such a large program was based on the previous multifrequency studies realized by Giovannelli's group (e.g., Giovannelli *et al.*, 1985a) and on the discovery of the narrow absorption lines, which revealed the existence of puffs of material ejected from the Be star towards the neutron star, with measured velocities from 0 to the terminal velocity of the stellar wind ($\sim$630 km s$^{-1}$) (de Loore *et al.*, 1984). During the development of such a program the system experienced the expected X-ray outburst, which was of the type *anomalous* or *casual* (see the review of Giovannelli & Sabau-Graziati, 1992). However, the UV flux was practically constant during all the measurements independent of the X-ray state of the system. The outburst is triggered by the matter accreting onto the neutron star (Giovannelli *et al.*, 1990b). Moreover, probably a temporary accretion disk forms around the neutron star close to the periastron (Giovannelli & Ziółkowski, 1990).

The discussion on the stellar winds from hot stars is still wide open. There are indications that, for instance, the variations on the wind parameters with respect to the stellar type are not consistent with radiative transport. If it were, it would be necessary to consider the radiative acceleration; and for this, there is the agreement of most investigators.

The evidence that the mass loss (and then $\dot{M}$) and terminal velocities of the winds are systematically lower in the OB stars of the Magellanic Clouds than in the Milky Way, indicates a lower metal abundances in the stars of the Clouds than in those of the Galaxy (Hutchings, Lequeux, & Wolf, 1987). The problem of the evolution of the Milky Way has been recently thoroughly discussed in a specialized workshop, whose proceedings have been published by Matteucci & Giovannelli (2000).

The IUE observations revealed the presence of strong winds from the central stars of planetary nebulae (Heap, 1979). Terminal velocities of such winds are between 1000 km s$^{-1}$ and 5000 km s$^{-1}$. P Cygni profiles have been detected in tens of stars. In central stars, whose effective temperature is lower than 63,000 K, the wind is almost always present, and if the surface gravity is $\log g \leq 5.2$ the wind is always produced (Cerruti-Sola & Perinotto, 1985).

Another important result obtained with the IUE observations of stellar winds has been a better knowledge of the dependence of the mass loss rate, $\dot{M}$, on the evolutionary status. In massive young stars $\dot{M}$ enhances of two orders of magnitude from the main sequence to the shell hydrogen burning phase. The $\dot{M}$ of the Wolf–Rayet stars, which probably are in a more advanced evolutionary phase, is even one order of magnitude greater. In such very luminous stars the large enhancement of $\dot{M}$ during the evolution is accompanied by a relatively scarce enhancement of the luminosity (Lamers, 1980). Consecutively Garmany & Conti (1984) found that the terminal velocity/escape velocity ratio in O stars depends systematically on the evolutionary state. Close to the zero age of the main sequence (ZAMS) such a ratio is usually $\leq 3.0$, whilst stars having such a ratio $>3.0$ are normally evolved. Cassinelli & Lamers (1987) combined the IUE and radio observations of a number of hot luminous stars in order to find directly $\dot{M}$. They demonstrated that as the massive stars evolve with a roughly constant luminosity, the value of $\dot{M}$ increases with the effective temperature and decreases with the surface density. The data fit very well with the relationship:

$$\log \dot{M} = -5.5 + 1.6 \log (L/10^6) - 1.0 \times \log (T/T_{\mathrm{ZAMS}}), \qquad (5.5)$$

where $T_{\mathrm{ZAMS}}$ is the effective temperature of a zero age star in the main sequence.

### 5.6.5. *The Interstellar Medium*

We have already discussed the ISM, but we think it useful for the reader to note here some of the crucial points resolved by means of the UV measurements.

i) High resolution survey of OB stars.

During such a survey Shull & Van Steenberg (1985) measured the HI column density of more than 200 stars of spectral types earlier than B2.5 and encountered an average value of the column-density/r ratio of 0.46 $\mathrm{cm}^{-3}$. Such a value decreases for $r > 1$ kpc, probably due to the gas placed between the galactic arms. Van Steenberg & Shull (1988) utilized such a survey in order to measure the abundance of Fe and Si and found that the two elements are present at levels of about 10–1000 times lower than previously expected. Such a factor is correlated with hydrogen column density. This is a result able to be predicted if heavy elements are needed for the formation of grains in dense clouds;

ii) The local Interstellar Medium (LISM)

The LISM was analyzed by means of different kinds of stars, utilized as background sources. Spectra of nearby white dwarfs were used (Bruhweiler & Kondo, 1981, 1982a,b; Dupree & Raymond, 1982), since the nearby convenient objects for the study of the LISM are those having a strong continuum flux, especially in the UV range, with a photospheric structure very small in order to simplify the analysis. The columnar densities can be measured by means of the lines of NI, OI, SiII, CII, FeII, and MgII. Other results were obtained by measuring fast rotating stars of A and B spectral types (e.g., Skuppin *et al.*, 1987) and late-type stars (e.g., Vladilo *et al.*, 1985; Vladilo, 1986). In both types of stars data were obtained essentially from the MgII doublet. Data from MgII doublets of stars placed at ~25 pc to the Sun indicate that the solar system is placed near the external limit of the Local Cloud, which is placed between us and the Galactic Center. The densities in the lines of sight in the hemisphere towards the Galactic Center are one order of magnitude greater than those in the lines of sight towards the Galactic Anticenter.

The values of the $N$(MgII)/$N$(MgI) ratio for nearby stars, obtained with the IUE and Copernicus, indicate a temperature around the Sun of 7,500–10,000 K (Bruhweiler *et al.*, 1984a,b), which is consistent with the position of the Sun near the external limit of the Local Cloud. On the other hand, Bruhweiler & Kondo (1982b) found CI towards $\alpha$ Ps A in the direction of the polarization and concentration of the MgII. This indicates a much lower temperature near the center of the Local Cloud;

iii) Strong and wide CII absorption lines.

One of the first discoveries of the IUE was the existence of strong and wide absorption lines of CII and other elements in the spectra of hot stars of the magellanic Clouds (de Boer, Koorneef & Savage, 1978, 1980). Such observations indicated the presence of an ionized gas at large distances over the Galactic Plane and probably along all the lines of sight towards the Magellanic Clouds. Other lines of sight were explored by using QSOs with proper redshifts (York *et al.*, 1984), extragalactic supernovae (Jenkins, Wallestein & Silk, 1984) and bright stars in the UV at high galactic latitude (Savage & Massa, 1987).

Most of the material, which is at hundreds parsecs over the upper part of the Galactic Plane, appears to be neutral or at a low level of ionization. The gas shows metal abundances typical of the gas around the Sun. Therefore it is possible that such material originates in the Galactic Disk and is ejected in the form of a galactic fountain (Shapiro & Field, 1976) or formed by the action of cosmic rays (Chevalier & Fransson, 1984). It is also probable that such a material is similar to that of the high velocity clouds detected in the 21 cm emission (van Woerden, Schwarz & Hulsbosch, 1985). The gas ionized at a higher level shows a distribution of velocity very different with respect to that of the cold material. CIV was detected across a few kpc (Savage & Massa, 1987). This could come from the cooling of hot coronal material ($10^6$ K) or from direct ionization caused by cosmic rays and for the radiation field in the near UV.

Some stars of the Magellanic clouds were studied in order to investigate the abundances of the gas (e.g., Fitzpatrick & Savage, 1983). The structure encountered was similar to that of the Milky Way, but with lower abundances, which indicates a poorer metal content of the gas in the Magellanic Clouds.

### 5.6.6. *SN 1987 A*

A series of results on the SN 1987A were so important that they would have justified by themselves the existence of the IUE. The SN 1987A, which exploded in the LMC, was observed continuously since its discovery until its decline. On February 24, 1987, Shelton discovered the supernova and the IUE only 3 hours later was already observing the phenomenon both in the UV and in optical region with the FES (Fine Error Sensor) monitor. The SN 1987A, very bright in the FUV, rapidly decayed in the first hours. During the first three days the luminosity decreased by three orders of magnitude. The IUE detected the spectra of two nearby stars. In a photograph of the field just before the explosion, only two stars were present. Then a speculation about the progenitor started. It was possible that such a progenitor could be an unknown object, and since the exploded envelope was thin the SN 1987A would have been a type-I SN, coming from a system with a white dwarf and a companion. However, a more detailed analysis of the photograph showed that the stars before the explosions were three in the direction of the SN 1987A. Gilmozzi *et al.* (1987) and Sonneborn, Altner & Kirshner (1987), independently, concluded that the exploded star was a blue supergiant.

The spectra obtained in high resolution mode also provided important information on the ISM in the Milky Way, in the Magellanic Clouds, and in the galactic halo.

Crucial information was also obtained with the FES. This was the first time that a supernova was followed in the optical continuously with the same instrument and almost simultaneously with the UV measurements. Figure 73 shows the optical light curve of the SN 1987A obtained with the FES data. The maximum luminosity (~2.9 mag) was reached three months after the explosion. A cloud of expanding gas, heated by the radioactivity of the isotopes created in the explosion, was responsible for such a luminosity. There is some strong evidence that the main source of energy for heating the ejected material is the radioactive decay of $^{56}$Co and $^{56}$Fe (Sonneborn, 1988).

Several months after the explosion some high temperature emission lines were observed in the FUV. A detailed analysis of such lines showed that the emitting gas has an anomalous N/C ratio, which indicates that the composition of the gas was changed by nuclear processes in the interior of the progenitor before it were dissipated in the wind. Such lines probably generate in a gaseous envelope formed by the stellar wind present in the red supergiant, which preceded the pre-supernova phase of a blue supergiant. The lines were very narrow, corresponding to a velocity of $\leq 20$ km s$^{-1}$, which is consistent with an origin of the wind in a red supergiant (Fransson *et al.*, 1989). The luminosity of the narrow emission lines had the maximum $\leq 400$ days after the explosion of the supernova. Therefore the radius of the circumstellar shell was $\leq 400$ light days. Reviews on the SN 1987A were that of

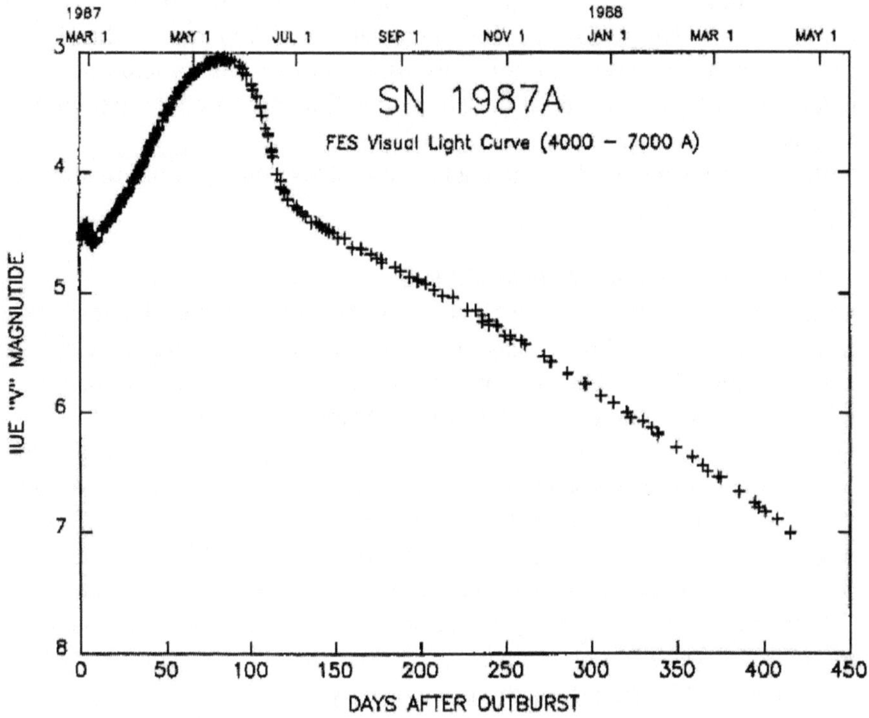


*Figure 73.* Optical light curve of the SN 1987A obtained with the FES (Fine Error Sensor) of the IUE (Sonneborn, 1988).

Arnett, Bahcall & Kirshner (1989) and the addendum to the review of *Supernovae and Supernova Remnants* of Blinnikov, Chugai & Lozinskaya (1988).

## 5.7. GALACTIC HISTORICAL JOURNEY: OPTICAL RANGE

Undoubtedly the Hubble Space Telescope – a long term space based observatory – has provided, and still provides an immense amount of excellent data, which constitute a precious bank of 'unbelievable' photographs of galactic sources up to the farthest objects of the Universe. However, we do not want to discuss this topic here; we prefer to invite the reader to look at the literature (e.g., Benvenuti, Macchetto & Schreier, 1996; Panagia, 1999a,b; 2002a,b and the references therein, where also the New Generation Space Telescope has been presented).

Catalogs and surveys will continue to be important tools for astronomical research, because astronomy deals with large numbers of objects that must be categorized in some orderly fashion to allow future work to be built upon past measurements and discoveries. Then at least a few words are mandatory for remembering the impact of the astrometric satellite Hipparcos on knowledge of the distances of stars. The Hipparcos catalog (~120,000 stars) is probably one of the most important in absolute. The book *Hipparcos Venice '97* (Battrick, 1997) reports the scientific results obtained with Hipparcos satellite. Perryman (1997) reported on the Hipparcos catalogue contents, Mignard (1997) discussed on its astrometrical properties, and Lindegren (1997) discussed on the Hipparcos Catalogue double and multiple systems annex.

In the following we want briefly to note a few highlights coming from the impressive HST and from Hipparcos.

### 5.7.1. *The Bipolar Planetary Nebula M2-9*

The end of a Sun-like star's life was once thought to be simple: the star gracefully casts off a shell of glowing gas and then settles into a long retirement as a burned out white dwarf. On the contrary, it is not so simple. Indeed, now, a dazzling collection of detailed views from the Hubble telescope reveals surprisingly intricate, glowing patterns spun into space by ageing stars: pinwheels; lawn sprinkler style jets; elegant goblet shapes; and even some that look like a rocket engine's exhaust. Figure 74 shows the dying star M2-9, nicknamed the *Twin Jet Nebula* because the two tubes of gas streaming from it behave like exhaust from a jet engine. Because of the nebula's shape and the measured velocity of the gas, in excess of 300 km $s^{-1}$, probably the description as a super-super-sonic jet exhaust is quite apt. (Credit: Bruce Balick (University of Washington), Vincent Icke (Leiden University), Garrelt Mellema (Stockholm University), and NASA).

Deep studies on M2-9 (PK 10 + 18°2) bipolar or 'butterfly' planetary nebula have been performed (e.g., Trammel, Goodrich & Dinerstein, 1995); Schwarz *et al.*, 1997); Doyle *et al.*, 2000, and the references therein).

### 5.7.2. *The Protostar TMR-1 and Its Planet*

Figure 75 shows a NASA HST near infrared image of newborn binary stars (image center): the class I protostar TMR-1 (IRAS 04361+2547) (Terebey *et al.*, 1998). The image was taken on August 4, 1997, with the Near Infrared Camera and Multi-Object Spectrometer (NICMOS) in three wavelengths (1.6, 1.9, and 2.05 $\mu$m). The image reveals a long thin nebula pointing towards a faint companion object (bottom left) which could be the first extra-solar planet to be imaged directly. The brightest objects in the image are the binary protostars, which illuminate an extended cloud of gas and dust (image center) from which the stars formed. So much dust surrounds these protostars that they are virtually invisible at optical wavelengths. However, near infrared light penetrates the overlying dust, revealing the newborn stars within. The faint multicolor cross extending from the neighborhood of the binary is an artifact produced when HST observes bright stars.

At the lower left there is a point of light many times fainter than the binary. Theoretical calculations indicate that this companion is much too dim to be an ordinary star; instead, a hot young protoplanet several times the mass of Jupiter is consistent with the observed brightness. The candidate protoplanet appears at a distance of $\sim$1.4 $\times 10^3$ AU from the binary. The central protostar is itself resolved

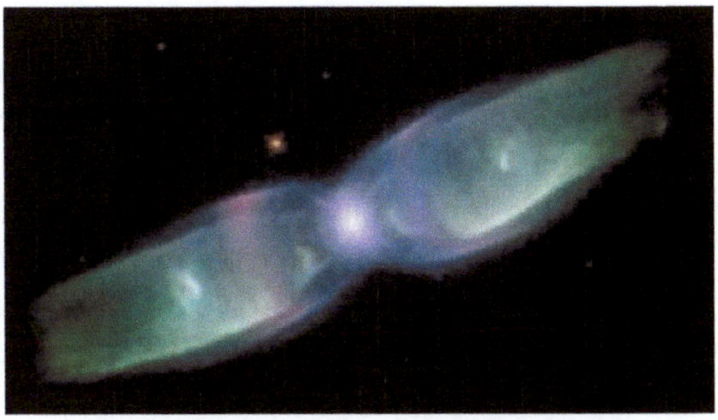

*Figure 74.* The dying star M2-9, nicknamed the *Twin Jet Nebula* (Release Number: STScI-1997-38).

*Figure 75.* NASA HST near infrared image of newborn binary stars: TMR-1 (IRAS 04361+2547) (STScI-PRC1998-19) (Terebey *et al.*, 1998).

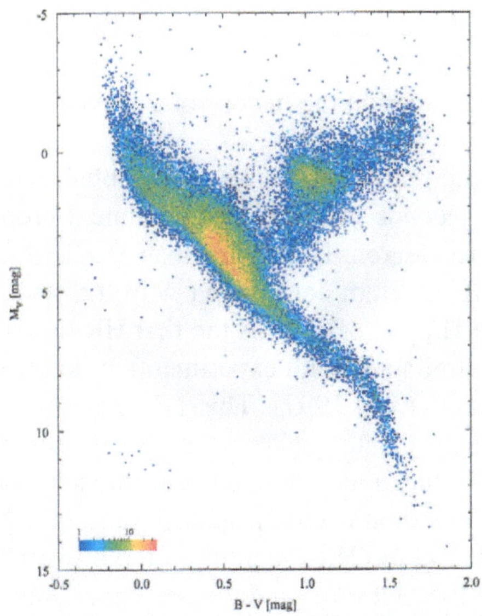


*Figure 76.* HR diagram constructed with the stars of the Hipparcos Catalog, whose absolute magnitude has been determined by Hipparcos parallaxes (Perryman *et al.*, 1997).

as a close binary with 42 AU separation, surrounded by circumstellar reflection nebulosity. A bright streak of nebulosity extends from the binary toward the faint companion, possibly indicating that the protoplanet was ejected from the binary system. Current models predict that very young giant planets are still warm from gravitational contraction and formation processes, with temperatures as high as a few thousand degrees. This makes them relatively bright in infrared light compared with old giant planets such as Jupiter. Even so, young planets are difficult to find in new solar systems because the glare of the central star drowns out their feeble glow. Young planets ejected from binary systems would therefore represent a unique opportunity for studying extra-solar planets with current astronomical technology (Credits: S. Terebey (Extra-solar Research Corp.) and NASA) (Release Number: STScI-1998-19).

### 5.7.3. *The Hipparcos HR Diagram*

Hipparcos was, by present standards, a medium-sized satellite, with a 30 cm telescope sensing simply ordinary light. But it has been described as the most imaginative in the short history of space astronomy. This foresight has been amply repaid. In the long history of stargazing it ranks with the surveys by Hipparchus the Greek in the 2nd Century BC and by Tycho Brahe the Dane in the 16th Century AD, both of whom transformed human perceptions of the Universe. Positions derived from the Hipparcos satellite are better than a millionth of a degree, and recently a thousand times more accurate than star positions routinely determined from the ground. This accuracy made it possible to measure directly the distances to the stars. Whilst it took 250 years between astronomers first setting out on the exacting task of measuring the distance to a star, and a stellar distance being measured for the first time, ESA's Hipparcos mission has revolutionized this long, painstaking, and fundamental task by measuring accurate distances and movements of more than one hundred thousand.

A short but learned paper on the history of star catalogs has been published by Eichhorn (1992).

Hipparcos results are really impressive: on a global scale the internal precision is 0.1 mas ('milli arc seconds') in position and annual proper motions (Lindegren & Kovalevsky, 1995; Lindegren, 1995; Perryman *et al.*, 1995). We want to mention here the wonderful HR diagram constructed with the stars, whose distances have been measured by the Hipparcos. This is the first HR diagram constructed with an homogeneous large set of stars, with experimentally known distances. It is shown in Figure 76 (Perryman *et al.*, 1997). This HR diagram provides powerful confirmation of the astrometric data quality, with a well defined broad main sequence extending to $M = -5$ mag, a well defined giant branch, and a distinct degenerate sequence. Solar neighbourhood 'clump giants' stars in a post-helium flash stage of evolution are located in the HR diagram on the basis of trigonometric parallax distance estimates for the first time, and the first time direct luminosity calibration of M giant is reported.

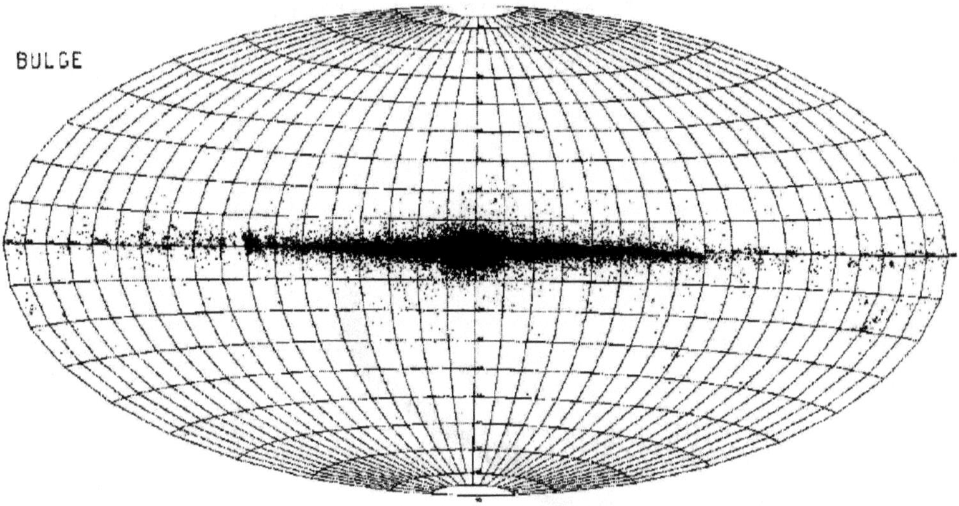


*Figure 77.* Distribution of IR sources at 12 $\mu$m and 25 $\mu$m, detected by the IRAS (Beichman, 1987).

The impact of Hipparcos on the whole of astrophysics has been extremely strong. Indeed, for instance, star evolution can be now studied on the basis of the measured absolute magnitudes of stars, contrary to the pre-Hipparcos era, when different methods for establishing the absolute magnitudes were employed, all these methods being indirect and then subject to large errors.

## 5.8. Galactic Historical Journey: IR Range

The satellite IRAS observed 246,000 sources. About 65% of those are stars, including photospheres and stars with IR excess owed to circumstellar shells of gas or dust (Chester, 1986). The observed sources show a marked concentration in the galactic plane. Figure 77 shows the distribution of the detected sources at 12 $\mu$m and 25 $\mu$m with colour temperature of ~300 K (Beichman, 1987).

Only 44,000 of the stars in the IRAS catalog have been identified with the already known objects. However, a comparison with the Point Source Catalog shows that K and M giant stars, having or not an IR excess owed to mass loss, are the most common stars in the IRAS catalog.

Coté & Waters (1986) and Waters (1987) utilized the colour ($V$–12 $\mu$m) for all the stars of the Bright Star Catalog observed by the IRAS at 12 $\mu$m as a function of the $(B - V)$ colour. Figure 78 shows such a dependence. The lower part of this diagram defines the 'normal' stars, which have not IR excess. The observations of normal stars provide results in agreement with those derived with the photospheric models of Kurucz (1979). The position of stars with IR excess are placed above those of normal stars. In Figure 78 are shown also the spectral types of the Main Sequence. Stars spread along the whole HR diagram are emitting in the IR range.

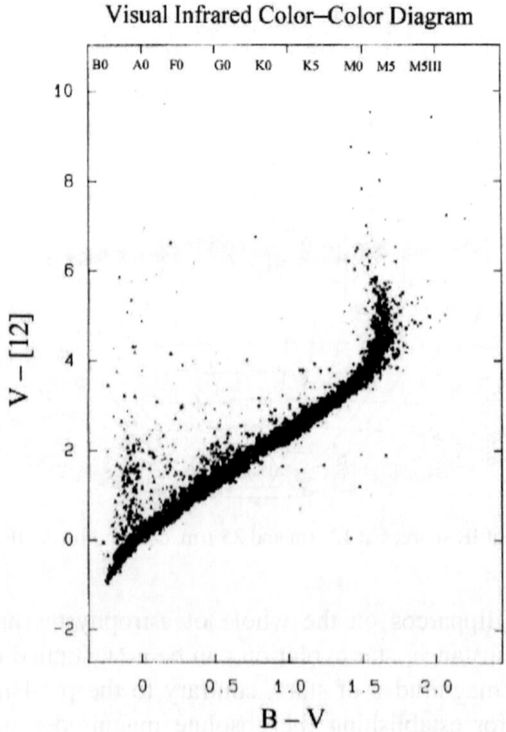


*Figure 78.* The visual infrared color–color diagram for stars in the IRAS catalog (Coté & Waters, 1986). The lower envelope of points defines the locus of bare photospheres. Points above that line have various amounts of excess. Representative main sequence spectral types are given across the top (Beichman, 1987).

Thus, remembering the emissions in the X-ray, UV and optical ranges of normal stars, it appears evident that the stars are emitting energy in all wavelength regions independent of their position in the HR diagram.

### 5.8.1. *Hot Stars with Mass Loss*

a) Supergiant and Wolf–Rayet stars.

Before IRAS it was known that the ionized stellar winds and, on some occasions, dust at $\sim 10^3$ K were responsible of the IR excess around hot stars (Cassinelli & Hartmann, 1977). The emission produced by these mechanisms is stronger than the photospheric one for wavelengths $\geq 25\ \mu$m. Observations of supergiant stars such as for instance, $\zeta$ Puppis (Lamers, Waters & Wesselius, 1984), P Cygni (Waters & Wesselius, 1986) and massive Wolf–Rayet stars (van der Hucht & Olnon, 1985) provided values for the mass loss $\dot{M} \sim 10^{-5}$–$10^{-4}\ M_\odot\ yr^{-1}$;

b) B and A stars.

Most of the dwarf B and giant B stars with optical emission lines (Be stars) have an IR excess owed to expansion envelopes of ionized gas (Waters, 1986a,b,c), whilst many Ae stars have dust shells at $T \sim 50$–400 K (Jaschek, Jaschek & Egret, 1986). In Be stars the mass loss rate through an equatorial disk is of the order of $10^{-8}$ $M_\odot$ $yr^{-1}$, which is approximately 50–100 times greater than that obtained with the UV observations (Waters, 1986b). The existence of an equatorial disk is based on the high rotational velocity of the star. Stars with $v_{rot} \times \sin i \leq 100$–200 km $s^{-1}$ have no IR excess and do not show emission lines (Waters, 1986a).

Starting from the studies realized on the O9.7 IIIe star HDE 245770 (Giangrande *et al.*, 1980), optical companion of the X-ray transient source A0535+26, Giovannelli *et al.* (1982c) obtained a mass loss rate $\dot{M} \simeq 10^{-8}$ $M_\odot$ $yr^{-1}$ and a rotational velocity of $v_{rot} \times \sin i \simeq 230$ km $s^{-1}$ by using the terminal velocity of the wind deduced from CIV and SiIV resonance lines obtained with high resolution IUE spectra and the broadening of the HeII (1640 Å) and HeI (4471 Å), respectively. The system A0535+26/HDE245770 is the most studied among the X-ray/Be systems thanks to the multi-year, multi-frequency observations of the Giovannelli's international group. The knowledge of such a system allowed the checking of the models of the equatorial disk which is responsible of the IR excess, and from this to predict the X-ray emission following the models of Waters (de Martino *et al.*, 1990a,b);

c) Planetary nebulae.

The compact planetary nebulae are bright IRAS sources owed to emission of dust, which is heated by the central star.

Planetary nebulae having different IRAS colours with peaks between 12 $\mu$m and 60 $\mu$m (Pottash *et al.*, 1984), together with oxygen rich planetary nebulae, appear bluer than carbon rich objects in the 25/60 $\mu$m colour (Roche & Aitken, 1986).

It is possible to observe strong evolutive effects in the IR behaviour of planetary nebulae. When the nebula is becoming older, i.e. is becoming larger and more diffuse, the dust emission is less important because:

- the temperature of the dust decreases as consequence of the weaker radiation field;
- the abundance of the dust relatively decreases with respect to that of the gas, probably as result of a process of destruction of grains (Pottash, 1986).

### 5.8.2. *Stars of Medium Spectral Type with Mass Loss*

Many supergiant stars of spectral types F, G, and K have an IR excess owed to massive dust shells with temperatures of 80–90 K and $\dot{M} \simeq 10^{-8}$–$10^{-5}$ $M_\odot$ $yr^{-1}$ (Evans, 1985). Many of these objects are variable stars of RV Tauri type, which were proposed as progenitors of planetary nebulae (Jura, 1986).

### 5.8.3. *Cold Stars with Mass Loss*

A fundamental phase of the evolution of an intermediate mass star ($M \leq 5$ $M_\odot$) is the AGB (Asymptotic Giant Branch) during which a strong stellar wind ($\dot{M} \simeq 10^{-4}$ $M_\odot$ yr$^{-1}$ gives back to the ISM most of the star mass (Zuckerman, 1986).

a) The state of an AGB star is determined by means of the opacity of its shell at a reference wavelength, for instance 9.7 $\mu$m, and the nature of the absorbing dust.

Stars with mass relatively small loss ($\dot{M} \simeq 10^{-7}$ $M_\odot$ yr$^{-1}$) appear as optically bright Mira variables. Stars with mass loss $\dot{M} \simeq 10^{-4}$ $M_\odot$ yr$^{-1}$ gradually become less luminous in the optical and more luminous in the IR range. Theoretical calculations of the energy spectral distribution of oxygen rich stars fit well to the observations for a wide range of opacities (Bedijn, 1986).

Herman, Burger & Penninx (1986) demonstrated that the 12/25 $\mu$m colour of a OH/IR star can be utilized for determining both the bolometric correction and the total optical depth across its shell. Such a result is extremely useful since the Point Source IRAS Catalog contain many sources for which there are not observations from the ground.

OH observations in OH/IR stars allow the determination of the expansion velocity of the shell and the kinematic distance to the object. The combined information coming from OH and IRAS observations allow the determination of the luminosity of the stars, the mass loss rate, and the *gas/dust* ratio of the ejected material (van der Veen, 1987). Moreover, they show that:

- the luminosity function of the OH/IR stars has a peak between 4,000 and 6,000 $L_\odot$;
- the mass loss rate determined in the IR range coincides with the values determined with other techniques (CO, lines in the NIR, OH masers) within a factor 2–3;
- the gas/dust ratio deduced with the fraction of the expelled mass is $160 \pm 40$ for OH/IR stars, and $200 \pm 100$ for Mira type stars (Beichman, 1987).

More massive and younger metal rich stars have a gas/dust ratio lower than older and metal poorer stars (Herman, Burger & Penninx, 1986).

The stellar variability is the key for the comprehension of the mass loss processes. If a giant M star is variable on a large time scale, it shows an IR excess. Also the vice versa is true: if the star shows IR excess, almost always it is variable.

The IRAS data for OH/IR stars show strong evidence of variability at 12 and 25 $\mu$m in time scale of order of 6 months (Beichman, 1987 and the references therein). This fact sustains the theoretical frame in which the mass loss process starts when the value of the amplitude of the atmospheric pulsations becomes enough large in order to allow the condensation of the dust (Jones, Ney & Stein, 1981; Bedijn, 1986). Starting from this moment the radiation pressure onto the dust and the gas grains friction push the gas outside the star. When the mass of the star decreases, the amplitude of the oscillations is greater and the mass loss rate increases with the time. There is also evidence that the mass loss rate and the chemical composition of the losing matter are variable with the time, as well as they are variable from one to another star. For instance, the silicates emission observed in a number of carbon stars implies a variation on the abundance of the expelled material within the last $10^2 - 10^3$ yr (Walker, 1985, 1986). The 25/160 $\mu$m colours of such objects indicate also a richness in oxygen of the expelled material, which implicates a large duration ($10^3 - 10^4$ yr) of the oxygen rich phase (Zuckerman & Dyck, 1986a,b).

b) R Coronae Borealis is the prototype of a hydrogen poor class of supergiant stars, which show irregular outbursts owed to inhomogeneous clouds of the ISM. Many of these hydrogen poor stars show an IR excess (Walker, 1985, 1986). An emission shell at 60 and 100 $\mu$m, with a diameter of $10'$ (8.5 pc!), centered on the same R Cor Bor was found (Gillet *et al.*, 1986). Such an external shell, completely different with respect to the interior one, which is unresolved and responsible of the medium IR excess. The mass loss rate necessary to form the external shell would be 10 times greater than the value deduced now. Such a high value of $\dot{M}$ should have been reached $\simeq 2 \times 10^4$ years ago. The colour temperature at 60/100 $\mu$m (25–39 K) is constant across the whole shell; therefore, R Cor Bor itself cannot be the source of heating. The possible mechanisms of heating would be a transient heating of small grains owed either to an intense radiation field or shocks with the gas in the halo.

#### 5.8.4. *Novae*

The IRAS detection of classical novae maintains the idea that fast novae (like GQ Muscae) produce free–free emission but not dust emission, whilst slow novae (such as V 4077 Sagittarii) produce a dust shell, which expands giving as a result a bright IR source (Dinerstein & Robinson, 1986).

V 4077 Sag showed a dust emission at 900 K. The decrease in its luminosity by one magnitude, when it was observed by the IRAS after 6 and 12 months from the maximum optical luminosity, was attributed to a decrease of the dust shell mass from $4 \times 10^{-6}$ to $4 \times 10^{-7}$ $M_\odot$. (Dinerstein, 1986).

#### 5.8.5. *Supernova Remnants*

The Supernova Remnants (SNRs) show a strong IR excess, over a continuum synchrotron radiation spectrum, between the radio and optical range. The excess is probably owed to the emission of hot dust. With observations of SNRs in the LMC and of selected galactic SNRs, it has been demonstrated that the cooling of the hot gas by means of gas grains shocks can dominate the cooling owed to atomic processes by a factor 10–100 (Beichman, 1987). Therefore the dust plays a fundamental role in the evolution of SNRs, as discussed by Dwek (1981).

Braun (1985) made IRAS maps of the young SNRs Thycho, Kepler and Cas A and of the evolved SNRs Cygnus Loop and IC 443. The IR images are well correlated with those in the optical and X-ray ranges. This fact implies that the

IR emission originates in the interstellar material swept at the border of the SNR and that the emitted dust gets warm, probably because of shocks in the material emitting X-rays. In the case of IC 443 (Braun, 1985; Mufson *et al.*, 1986), the SNR seems to be spread in a series of HII regions already existing.

There are two different mechanisms for heating the dust in SNRs. Marsden (1985) attributes the excess in the Crab and Thycho SNRs to a dust mass of (5–30) $\times 10^{-3}$ $M_\odot$ at temperature of 70–90 K heated by X-ray or FUV radiation. Braun (1985) and Dwek (1986) attribute the IR emission to dust heated by shocks in regions of interstellar medium hit by the wave of the supernova explosion (Draine, 1981).

In Thycho and Kepler supernova explosions, a mass of the expelled material of about 0.3 $M_\odot$ swept a mass of about 13 $M_\odot$ of the ISM.

### 5.8.6. *Protoplanetary Disks*

The discovery of an IR excess around $\alpha$ Lyrae, a main sequence dwarf with unknown mass loss (Aumann *et al.*, 1984) caused a strong interest because such an excess would have been originated by a belt of orbiting solid material, probably of the same kind forming planets.

Gillet (1986) discovered the four stars which better describe the *Vega Phenomenon*: $\alpha$ Lyrae, $\beta$ Pictoris, $\alpha$ Piscis Austrini and $\epsilon$ Eridani. The most important characteristic is an IR excess starting between 12 and 25 $\mu$m with an energy spectral distribution of grains, heated to 60–250 K, which emit as an optically thin black body; this implies emission from grains with dimensions greater than $\lambda/2\pi \simeq 15$ $\mu$m.

A detailed analysis of IRAS data allowed the resolution of the emitting regions for those stars, with the exception of $\epsilon$ Eridani: typical dimensions are about 200 AU.

The measured dimensions and the deduced temperatures are, to the first approximation, in agreement with the idea that the emission is coming from solid material heated by a small fraction of the energy emitted by the central star.

Twenty additional stars have been discovered as Vega-like objects (Aumann, 1985; Sadakane & Nishida, 1986).

This fundamental discovery gave strong scientific arguments for the importance of the new generation orbiting IR telescopes, such as the ISO (Infrared Space Observatory), which could have detected small quantities of interplanetary material.

### 5.8.7. *Emission from the Galactic Disk*

The column density of the dust, within a few degrees from the Galactic Plane, and the energy available for its heating strongly increase with the galactic latitude. Some groups have modelled the energy distribution of the Galaxy as a sum of different components, namely:

a) Hot component.

The energy distribution of the Galaxy shows a wide elbow in emission around 10 $\mu$m. After an extrapolation of the properties of the emission at high galactic latitude towards lower latitudes and the study of the emission detected around HII regions, it was possible to deduce that the small grains play an important role in the energetics of the whole disk (Boulanger, Baud & van Albada, 1985);

b) Warm component.

OB stars in HII regions or embedded in dense molecular clouds heat large dust grains to 30–40 K and are responsible of about 40% of the total IR emission of the galaxy (Cox & Mezger, 1986). The *luminosity/mass* ratio of such a component strongly depends on the proximity of OB stars and is ranging between 7 and 30 $L_\odot/M_\odot$ toward hot spots centered on individual HII regions such as W 51 and M 17 (Scoville & Good, 1987);

c) Cold component.

The coldest material in HI regions (15–25 K) and in molecular clouds ($\simeq$15 K) away from embedded OB stars is essentially heated by stars of B, A, and F spectral types and by old stars in the disk. The dust associated with the cold material is responsible for about 50% of the IR luminosity of the Galaxy. The presence of such a cold material was suspected as a result of the IRAS data at 100 $\mu$m, but it was completely confirmed by submillimetric measurements (Caux & Serra, 1987).

The dust in the HI regions has an emissivity of 1.4–7 $L_\odot/M_\odot$, whilst in the dense regions of molecular clouds, far from the embedded stars, the emissivity is 0.5–2.3 $L_\odot/M_\odot$ (Sodorski *et al.*, 1987; Cox & Mezger, 1986).

### 5.8.8. *The Milky Way as a Galaxy*

a) Energetics of the Milky Way.

The emission of the Galactic Disk at 60 and 100 $\mu$m is concentrated within 1 kpc from the Galactic Center and within a molecular ring of 3–7 kpc (Scoville & Good, 1987).

Cox & Mezger (1986) estimated that the bolometric luminosity of the stars in the Milky Way is $L \simeq 3.6 \times 10^{10}$ $L_\odot$, with one third of this, $L \simeq 1.2 \times 10^{10}$ $L_\odot$, emitted between 4 and 100 $\mu$m, for a distance to the Galactic Center of 8.5 kpc. Scoville & Good (1987) estimated that the emission between 1 and 500 $\mu$m in the molecular ring is $L \simeq 6 \times 10^9$ $L_\odot$. For a total mass of the ISM of $L \simeq 4.6 \times 10^9$ $L_\odot$ (Scoville & Sanders, 1986), the luminosity to mass ratio of the Galaxy is 2.6 $L_\odot/M_\odot$. This ratio can be between 10 and 20 times greater than that corresponding to hot spots in particular molecular clouds (Scoville & Good, 1987). Evaluations of the mass going into stars in the Milky Way give a value of 2.5–7.5 $M_\odot$ yr$^{-1}$, of which 0.3–5 $M_\odot$ yr$^{-1}$ is going to form low mass stars. The exact value is depending on the number of high mass stars which form in agreement with the initial mass function and how many of them form because of induced processes that preferentially form high mass stars (Cox & Mezger, 1986).

b) Star formation and the Galaxy.

The utilization of the IR radiation as direct indicator of formation of high mass stars is very difficult in our Galaxy and in many others, since the ISM is relatively transparent in a scale of 0.1–1 kpc. This means that not all the interstellar dust is heated by OB stars and, moreover, that not all the luminosity emitted by OB stars is converted into IR luminosity. In a scale of about 10 pc within an OB cluster, the relationship between IR emission and star formation is excellent, but this is not valid for scales of 100–200 pc. As an example, 90% of the emitted luminosity, $2.5 \times 10^5$ $L_\odot$, in a region of $9 \times 9$ pc$^2$ ($1° \times 1°$) centered on OMC 1 is owed to the stars in Trapetium (Thronson *et al.*, 1986), whilst in a region of $20° \times 18°$, always centered on OMC 1, only 60% of the total luminosity, $6.5 \times 10^5$ $L_\odot$, is owed to OB stars in Orion A and B (Boulanger, Maddalena & Thaddeus, 1986). Helou (1986b) gives a warning about the cirrus, which can contaminate the evaluations of the star formation rate obtained with the IRAS observations. This is probably not true for very bright galaxies with a dense ISM, which captures all the output energy from the clusters of massive stars. In such galaxies the formation of OB stars is probably an important source of energy. Since the IRAS colours and luminosity to mass ratio of giant molecular clouds are similar to those of bright IRAS galaxies (Sanders *et al.*, 1986), it is possible that a strong star formation is the explanation, certainly not unique, of the High IR luminosity of such galaxies (Scoville & Good, 1987).

We can remark, as written by Beichman (1987) in his important review on IRAS results which we have followed in this section, that: *In the broadest sense, the IRAS legacy is like that of any of the great sky surveys: Herschel's, Palomar's, the Cambridge Radio surveys, or Huhuru's.*

## 6. Multifrequency Astrophysics

High energy cosmic sources are the natural laboratories in which a large fan of physical processes occur. Several physical quantities witness to these processes occurring across the electromagnetic spectrum. Therefore multifrequency observations of cosmic sources are strictly necessary in order to understand their general behaviour. Moreover, since the behaviour of most cosmic sources are variable in time in different time scales both for intensities and spectral shapes, simultaneous observations, with resolutions in time and sensitivity and spectral resolving power as good as possible, are unavoidable. In this section we want to emphasize this kind of experimental approach through examples of multifrequency space based and ground based observations of different sources representative of homogeneous classes of cosmic sources, and theoretical interpretations of the results.

As far as we know, the first workshop devoted to multifrequency observations of galactic cosmic sources was the Frascati Workshop 1984 held in Vulcano island. The proceedings of that pioneering workshop are a clear example on the necessity of this kind of observations in studying cosmic sources (Giovannelli, 1985).

In spite of almost two decades having passed since that time, during which enormous developments in technologies have been realized and then usefully applied to astronomy, most of the 1984 open problems are still open. Although extragalactic astronomy, meanwhile improved a lot, neither cancelled nor reduced the interest in galactic astronomy in contrast to what the powerful *extragalactic scientific lobbies* would like to obtain.

A paper on the importance of multifrequency observations of high energy cosmic sources was published several years ago by us (Giovannelli & Sabau-Graziati, 1994). In that paper we discussed on the improvements of resolution in energy, in position and in sensitivity of detectors, which allowed reaching the excellent results of the present generation of space experiments, as described in the review paper by Giovannelli & Sabau-Graziati (1995). With the last generation of detectors in different energy regions it has been largely demonstrated that all cosmic sources are variable in their emission even if at very different levels in all energy ranges. The sources considered constant (and therefore sometimes used as standard sources) appeared constant because of the limited resolutions in sensitivity (and in time) of the detectors. So we can assume that cosmic sources, and especially the high energy ones, are highly variable both in flux intensity and in the shape of the emitted spectra; this reflects the presence of different processes occurring at different times because of different interactions of the *engines* – producing energy – with the surrounding matter and/or radiation and/or electromagnetic fields (Giovannelli & Sabau-Graziati, 1993). Therefore multifrequency observations, possibly simultaneous, are necessary for the actual comprehension of the intriguing physics governing the behaviour of the high energy cosmic sources, which can be considered the frontier objects between astrophysics and particle physics and then the natural laboratories to test the most advanced theories of the modern physics. Exhaustive books on this topic have been published in the series of Proceedings of the Vulcano Workshops on *Frontier Objects in Astrophysics and Particle Physics* (Giovannelli & Mannocchi, 1989, 1991, 1993, 1995, 1997, 1999, 2001, 2003).

Then it is possible to introduce a *new* branch of science: the so-called **Multifrequency Astrophysics**, which we can tentatively attack from two different points of view, converging toward a better knowledge of the Universe:

- Experimental Multifrequency Astrophysics;
- Theoretical Multifrequency Astrophysics.

**i) Experimental Multifrequency Astrophysics**: it can be developed by means of:

– simultaneous observations with different experiments and/or facilities in different energy ranges (i.e. when it is necessary to obtain the total energy distribution of a source, or to search for energy dependent variations (e.g., Section 6.1.2);
– coordinated observations either when it is not possible for any reasons to perform simultaneous observations or when a certain phenomenon is expected to appear in particular different moments dependent on the energy [i.e. in hard X-ray transient sources, X-ray outbursts appear delayed (transit time of the accreting matter from the optical companion to the neutron star) with respect to strong optical activity which appear either on Balmer $H_\gamma$ line and on FeII lines] (e.g., Section 6.1.3);
– data bases and/or literature (i.e. analysis of data bases in order to find simultaneous observations in different energy regions obtained by chance) (e.g., Section 6.1.4);
– multifrequency observations, not necessarily simultaneous, which can produce results useful to constrain some parameters of the studied systems (see Section 6.1.5);
– multi-site observations in order either:
  - to enhance the probability of success of ground based observations either in the same energy region or not; or
  - to follow a selected source continuously during long time with telescopes at different geographical longitudes; in this case the night–day cycle is absent in all, while it is obviously present in each site (e.g., Section 6.1.6).

**ii) Theoretical Multifrequency Astrophysics**: it can be developed by means of:

– study of wide range physical processes (i.e. manifesting their effects in a wide range of the electromagnetic spectrum): the theoretical spectra obtained can be used to fit multifrequency experimental data. Only if the fits to experimental data, spread over a wide energy range, are good, theoretical models can be accepted (e.g., Section 6.2.1);
– study of narrow range physical processes (i.e. occurring in a particular energy range), which can explain some or all experimental behaviour of cosmic sources (e.g., Section 6.2.2).

In our opinion the race between experiments and theories conducted with these rules is the unique way to proceed for a better and faster comprehension of the physics governing the Universe.

## 6.1. Experimental multifrequency astrophysics

### 6.1.1. *General Remarks*

In our Galaxy the majority of the stars are, probably, in binary systems. However, this is not the only, or even the most important, reason why the binary systems catch the attention of a large community of physicists. Indeed, especially after the beginning of modern Space Astronomy, the close binary systems became fundamental laboratories for studying the high energy processes occurring because of the extreme physical conditions reached in, and at, the collapsed component and the strong interactions with the 'normal' companion, which give rise to a wide fan of phenomena. X-ray binary systems then represent a link between high energy astrophysics and particle physics (e.g., Giovannelli, 1991); therefore they are good examples for correctly addressing this subsection, inviting the readers interested in details to look at the many books and reviews on X-ray binary systems which can be found in the literature. Other targets very useful for the purposes of this subsection are the AGNs in which the physical conditions are similar to those of galactic systems containing collapsed objects, but with a scaling factor of order $10^3 - 10^8$ in mass, dimensions, and energy. It was demonstrated that 'normal stars', spread almost over the entire HR diagram, emit high energy photons of thermal origin in the X-ray energy range (e.g., Pallavicini *et al.*, 1981) from the photospheric, chromospheric, and transition regions. However, in some kinds of PMS stars high energy emission can also be owed to accretion processes of material falling onto the star magnetically channelled onto the poles because of magnetic fields of order of several hundreds Gauss, such as in the T Tauri star RU Lupi (e.g., Giovannelli, 1994). Therefore, these so-called 'normal stars' can also be used in illustrating the necessity of multifrequency studies.

### 6.1.2. *Simultaneous Multifrequency Observations*

In the last two decades a lot of multifrequency observations have been performed by many groups by using space and ground based facilities; therefore it is difficult to choose examples according to their importance in the contribution to the advances of astrophysics. However, for the purposes of this subsection we would like to describe one of them which is familiar to us, namely the detection of Flare-Like Events (FLEs) in the extreme T Tauri star RU Lupi. This star is the best target within this class of objects since it is rather luminous ($V \simeq 10.3$), placed in the empty field of the Lupus-2 dark nebula, visible from the southern hemisphere where the ESO facilities are available. The multifrequency behaviour of RU Lupi have been deeply discussed by Giovannelli (1994). Within a long campaign (1983–88) of multifrequency observations of RU Lupi, simultaneous observations from UV to IR regions have been performed on five occasions as shown in Figure 79 (e.g., Giovannelli *et al.*, (1988; 1995) by using the IUE satellite and 0.6 m, 1.0 m, 1.5 m, 3.6 m ESO telescopes, and sporadically also the X-ray ASTRON So-

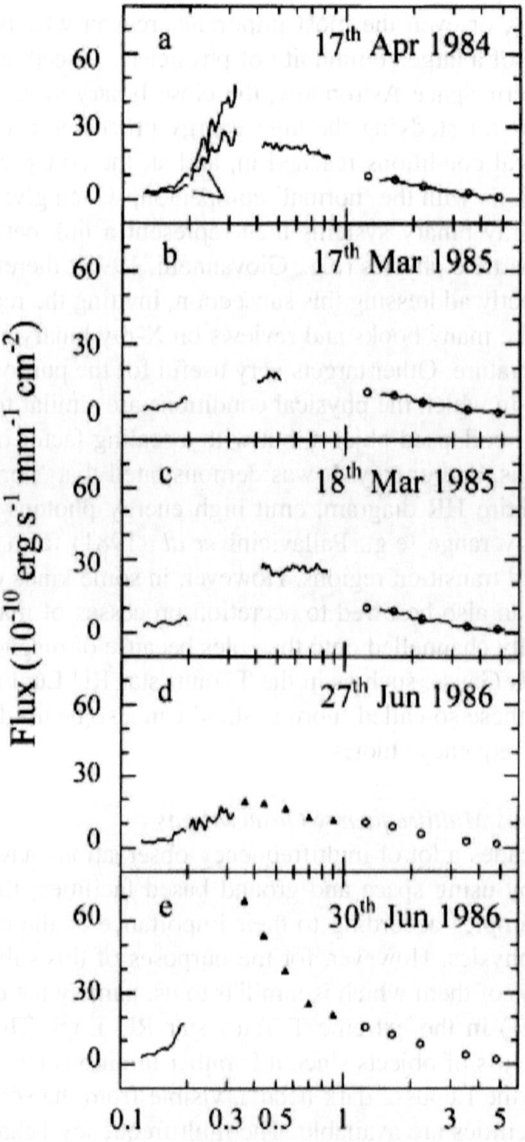


*Figure 79.* Simultaneous multifrequency energy distributions of the T Tauri type star RU Lupi (e.g., Giovannelli *et al.* (1988; 1995).

viet satellite, which provided fundamental measurements that probably definitively clarified the problem of the formation of FLEs, as discussed by Giovannelli (1994).

These results are very exciting since FLEs were detected in two occasions giving the possibility to study, for the first time, the energetic of these phenomenons. Briefly, during the FLE of June 30, 1986, the total flux variation with the respect to the 'quiescent state' obtained on June 27 was $\sim$87% in the total range of measurement (1200–50000 Å). The strongest variation of (314 $\pm$ 47)% was observed in the range 1200–3600 Å. In the UBVRI bands the variations were (263 $\pm$ 22)%, (199 $\pm$ 4)%, (146 $\pm$ 2)%, (105 $\pm$ 3)%, (85 $\pm$ 3)%, respectively. In the band 12500–50000 Å, the relative variation was (42 $\pm$ 4)%. Therefore the longer are the wavelengths the smaller are the variations. Details on the model of RU Lupi, derived by data coming from the long multifrequency campaign are reported in the papers by Lamzin *et al.* (1993; 1996).

6.1.3. *Coordinated Multifrequency Observations*

In X-ray/Be systems a moderate activity of the Be star – such as ejection of 'puffs' of material accreting onto the neutron star – produces an anomalous X-ray outburst, like discussed in the paper by Giovannelli & Sabau-Graziati (1992). The delay between optical activity and X-ray outburst is just the transit time of the material from the Be star to the neutron star. In the case of the transient X-ray source A 0535+26 and its Be companion HDE 245770, amongst the many multi-year multifrequency observations made by our group, Bartolini *et al.* (1981) during the 1981 campaign of spectroscopic observations reported the main event in November–December. Indeed, the star between October 31 and December 17, 1981 showed an unusual activity. $H_\alpha$, $H_\beta$, $H_\gamma$ and HeI (5876 Å) lines were strongly in emission. The emission features were growing from early November to November 22; later they were slowly decreasing except $H_\gamma$ which varied from partially filled-in absorption to clear emission, and HeI (HeI usually is not present in this star). UBV photometry indicated a fading of luminosity, interrupted by a sudden brightening on December 5. They interpreted these observations as indicative of expulsion of blobs of material, overlapping the steady wind, triggered by the stronger gravitational interaction owed to the vicinity of the periastron passage (e.g., Giovannelli & Sabau-Graziati, 1992). Indeed, one member of our group (Adriano Guarnieri) alerted the Hakucho satellite team to the probable incoming X-ray outburst within roughly one week (the transit time of the material from the Be star to the neutron star: $v \simeq 330$ km $s^{-1}$, $d \simeq 1.34$ AU (de Loore *et al.*, 1984). A 0535+26 was observed by the Hakucho satellite and flared-up, in the energy range 1–22 keV, on December 13, 1981 with an intensity of $\simeq$0.2 Crab (Nagase *et al.*, 1982). **We want to remark that this fact was absolutely the first historical witness to predict the future behaviour of an X-ray/Be system!**

6.1.4. *Multifrequency Observations from Data Bases and/or Literature*

In our opinion one of the best and exciting examples of the use of literature for solving important astrophysical problems is that of the supernova which exploded in 1054, originating the very famous Crab pulsar (PSR 0531+21) and Crab Nebula. Until the work of Guidoboni, Marmo & Polcaro (1993) all information about this event was coming from Chinese and Japanese ancient astronomical documents, which had been used to determine the physical characteristics of the explosion, which was reported to occur on 4th July 1054, and this 'guest star' was visible for 23 days. In their work Guidoboni, Marmo & Polcaro (1993) demonstrated, on the basis of a Flemish chronicle, that the explosion of the SN 1054 very probably occurred on April 19, 1054, more than two months before the time reported in the oriental documents. The physical implications are obviously enormous, since in this case the SN 1054 would be a Type II Plateau SN in contrast to the belief that it is a Type II Linear SN; therefore the behaviour of the Crab pulsar and Crab Nebula would be explained in the light of this different nature of the supernova.

One more example of the use of data bases and literature is that of the FLEs' periodicity in the T Tauri star RU Lupi. Indeed, during a deep analysis of the literaure concerning this star, Giovannelli (1994) discovered that eight FLEs were reported in different wavelength regions and epochs largely separated each other. These events were not deeply analyzed in the literature since they were detected during observations planned for other purposes, with the exception of the two FLEs discussed in Section 6.1.2. The FLEs are separated by $\simeq$7 days. Assuming this periodicity as true, the only positive detection in X-ray band (2–6 keV) (Giovannelli *et al.*, 1984c) is just in phase modulo 27 days. In contrast, the several X-ray observations with no detection of any flux are anti-phased. Assuming the positive X-ray detection to be a FLE, and using this FLE, together with the other eight FLEs, it was possible to find the best periodicity for the FLEs as $P_{\mathrm{FLE}} = 27.686 \pm 0.002$ days. This periodicity needs a confirmation with more data. However, since the FLEs are spread over more than 30 years the quoted figure for the periodicity is rather good. This periodicity could be related to the rotational period of the star still unknown. If it were the rotational periodicity, the rotational velocity would be $v \simeq 5.5$ km s$^{-1}$. With such a velocity RU Lupi shows a relationship $\log L_{\mathrm{x}} = 27.2 + 2 \log v$, which agrees to the general relationship found by Bouvier (1990), otherwise violated. Indeed, a wrong rotational period of about 3.7 days has been suggested in the literature following a perverse chain of citations, such as commented on by Giovannelli (1994). With such a rotational period RU Lupi does not obey any of the general relationships valid for the PMSSs.

6.1.5. *Multifrequency Observations without Coordination*

One important example of the use of this kind of measurements is, for instance, that used by Giovannelli *et al.* (1983) in order to find, indirectly, the orbital parameters of the dwarf nova SS Cygni, whose behaviour are extensively discussed in the review of Giovannelli & Martinez-Pais (1991), and by Giovannelli (1996a). By

TABLE 11

Derived parameters of SS Cygni at various orbital inclination angles. S, K, and J refer to Stover *et al.* (1980), Kiplinger (1979) and Joy (1956) data, respectively. The most probable values of the parameters are marked in boldface (adapted from Giovannelli *et al.*, 1983).

| | i (deg) | 50 | 45 | 44 | 43 | 42 | 41 | **40** | 39 | 38 | 35 |
|---|---|---|---|---|---|---|---|---|---|---|---|
| | S | 12.0 | 13.0 | 13.2 | 13.5 | 13.7 | 14.0 | **14.3** | 14.6 | 14.9 | 16.0 |
| $(a/10^{10})$ cm | K | 9.9 | 10.7 | 10.9 | 11.1 | 11.4 | 11.6 | 11.8 | 12.1 | 12.3 | 13.3 |
| | J | 11.8 | 12.7 | 13.0 | 13.2 | 13.5 | 13.7 | 14.0 | 14.3 | 14.6 | 15.7 |
| | S | 0.33 | 0.42 | 0.45 | 0.47 | 0.50 | 0.53 | **0.56** | 0.60 | 0.64 | 0.79 |
| $M_R/M_\odot$ | K | 0.22 | 0.27 | 0.29 | 0.31 | 0.32 | 0.34 | 0.36 | 0.39 | 0.42 | 0.52 |
| | J | 0.44 | 0.57 | 0.60 | 0.63 | 0.67 | 0.71 | 0.75 | 0.80 | 0.86 | 1.06 |
| | S | 0.57 | 0.73 | 0.78 | 0.81 | 0.86 | 0.93 | **0.97** | 1.04 | 1.11 | 1.37 |
| $M_{wd}/M_\odot$ | K | 0.29 | 0.37 | 0.39 | 0.42 | 0.43 | 0.46 | 0.49 | 0.52 | 0.56 | 0.69 |
| | J | 0.40 | 0.51 | 0.54 | 0.57 | 0.60 | 0.64 | 0.68 | 0.72 | 0.77 | 0.95 |
| | S | 0.57 | 0.62 | 0.63 | 0.64 | 0.66 | 0.67 | **0.68** | 0.70 | 0.71 | 0.77 |
| $R_R/R_\odot$ | K | 0.50 | 0.54 | 0.56 | 0.57 | 0.58 | 0.59 | 0.60 | 0.62 | 0.63 | 0.67 |
| | J | 0.66 | 0.71 | 0.72 | 0.74 | 0.75 | 0.77 | 0.78 | 0.80 | 0.82 | 0.88 |
| | S | 2.40 | 2.60 | 2.65 | 2.70 | 2.75 | 2.80 | **2.86** | 2.92 | 2.99 | 3.20 |
| $R_d/R_\odot$ | K | 2.58 | 2.78 | 2.84 | 2.89 | 2.96 | 3.02 | 3.07 | 3.14 | 3.21 | 3.43 |
| | J | 11.20 | 12.10 | 12.30 | 12.50 | 12.80 | 13.00 | 13.30 | 13.60 | 13.90 | 14.90 |

using their optical spectra showing a doubling in the Balmer lines, the authors derived the projected circular velocity of the outer edge of the disk $v_d \times \sin i = 192 \pm 10$ km s$^{-1}$ from the separation of the peaks. Then, assuming the rotation of the disk to be Keplerian, they derived the outer radius of the disk following Robinson's (1976) formula: $R_d/a = (K/v_d \sin i)^2 (1+q)/q^2$ , where $a$ is the separation of the centers of mass of the system's components, $R_d$ is the outer radius of the disk, $q$ is the ratio between the masses of the white dwarf and late type star: $M_{wd}/M_r$, $K$ is the observed semi-amplitude of the orbital radial velocity curve of the emission lines. Then, using the available data of $K$ and $q$ at that time and the derived value of $v_d \times \sin i$, Giovannelli *et al.* (1983) obtained three sets of values of $a \sin i$, and $M \sin^3 i$ for both components of the system, and $R_d/a$. These sets were used to compute the masses of the stars for different values of the orbital inclination angle $i$. In order to evaluate the radius of the secondary, they used Plavec's (1968) mass–radius relationship. With the available values of $q$ they derived the masses, and finally the set of radii of the white dwarf, using the mass–radius relation of Hartle & Thorne (1968). Table 11 shows all these parameters; S, K, and J refer to orbital data reported by Stover *et al.* (1980), Kiplinger (1979) and Joy (1956).

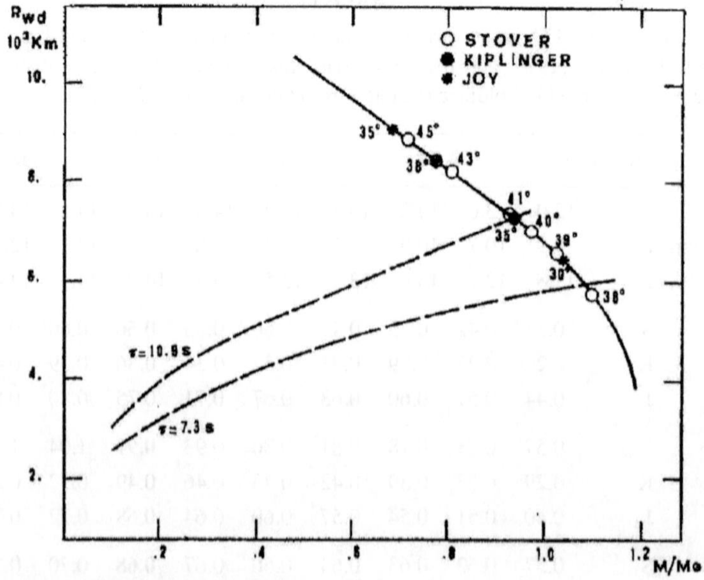


*Figure 80.* The mass–radius relationship for the white dwarf. The continuous line is from Hartle & Thorne (1968) theoretical calculations; dashed lines refer to the maximum and minimum period of optical oscillations detected in SS Cygni. The open circles, black dots, and asterisks refer to the values coming from the data of Stover *et al.* (1980), Kiplinger (1979), and Joy (1956), respectively, for different values of the orbital inclination angle (Giovannelli *et al.*, 1983).

How to choose the actual orbital inclination angle $i$? Giovannelli *et al.* (1983) took into account all the constraints coming from experimental data, namely:

- the detected optical fast oscillations range from ~7.3 s to 10.9 s (Patterson, Robinson & Nather, 1978; Horne & Gomer, 1980; Patterson, 1979, 1981; Hildebrand, Spillar & Stiening, 1981; Giovannelli, 1981). Figure 80 shows the allowed band for the mass–radius values of the white dwarf. The open circles, dots and asterisks represent the values from S, K, and J, listed in Table 11, respectively;
- if the ~9 s soft X-ray pulsation from SS Cygni (Cordova *et al.*, 1980) gives an upper limit for the Keplerian period at the white dwarf surface, the lower limit for the mass of the collapsed object is 0.9 $M_\odot$. With this constraint, entering in Table 11, the only possible value for $M_{wd}$ is that in the column corresponding to $i = 40°$ and in the line with S values.

Then $i = (40^{+1}_{-2})°$ is the most probable orbital inclination angle derived by Giovannelli *et al.* (1983), and the values of the other parameters are: $d = 2.86^{+0.13}_{-0.06} \times 10^{10}$ cm, $M_r = 0.56^{+0.08}_{-0.03}$ $M_\odot$, $R_r = 0.68^{+0.03}_{-0.01}$ $R_\odot$, $M_{wd} = 0.97^{+0.14}_{-0.05}$ $M_\odot$.

From the $H_\beta$ width at the continuum, they derived also a rough evaluation of the inner radius of the disk as ~3.6 $\times 10^9$ cm. These values were later confirmed by Honey *et al.* (1989) and with direct spectrophotometric measurements by Martinez-

Pais *et al.* (1994). In spite of these orbital parameters having been published a long time ago, 1983, and claimed many times by Giovannelli himself in international conferences, workshops even specialized in cataclysmic variables, and lectures, they have not yet been accepted, probably because they were published in a so-called 'minor journal', namely *Acta Astronomica*! The community prefers to use the useless value of $M_{wd} = 1.4\ M_{\odot}$, Chandrasekhar's limit, since the Astrophysical Journal once published that value (Stover *et al.*, 1980)!

### 6.1.6. *Multi-Site Observations*

The most significant example of this kind of observations is that of MUSICOS (MUlti-SIte COntinuous Spectroscopy). It is a project whose goal is to facilitate multi-site, multifrequency observations in stellar spectroscopy (e.g., Catala & Foing, 1988; 1990). One of the goals of this program is to study the rotational modulation of different kind of stars – Be ($P_{rot} \simeq 0.5$–0.8 d), Ap ($P_{rot} \simeq 0.5$–10 d), Ae–Be Herbig ($P_{rot} \simeq 2$–7 d), F, G, K: active ($P_{rot} \simeq 4$–20 d), quiescent ($P_{rot} \simeq 20$–40 d), dKe, dMe ($P_{rot} \simeq 2$–7 d), RS CVn ($P_{rot} \simeq 1$–10 d), T Tauri ($P_{rot} \simeq 1$–30 d) – essentially through the behaviour of Balmer lines, *H* and *K* lines of CaII, and infrared triplet, whose equivalent widths are orbitally modulated (Foing & Char, 1988). Therefore it is possible to obtain direct information on the stellar photosphere and chromosphere, and the non-uniform chemical composition of the surface. The forced rotation in synchronized binaries or geometric distortion of the binary components can also produce some specific modulations during the orbital period. For instance, in the RS CVn system HR 1099 Catala *et al.* (1993) obtained the spatial distribution of a phenomenon linked to the magnetic field. They used the Doppler imaging method (Donati, 1993) consisting in using the effect of localized spots on line profiles, combined with Doppler shifts owed to the star's rotation in order to reconstruct an image of photospheric spots at the stellar surface dependent on the rotational phase. In particular, they obtained:

- the spot configuration and then the comparison with that of Sun and other stars;
- the signature of the chromospheric structure in the CaII H and K lines and the $H_\alpha$ line, for which a correlation with coronal X-ray emission has been found;
- the correlation of photospheric active structures (spots) with the chromospheric structures (plages) in terms of spatial extent (such as on the Sun, plages are not located on top of the spots, but surrounding such spots, or appearing as the precursor of spots?);
- the vertical structure in temperature and density in active regions by which it is possible to derive scaling laws for the energy balance.

## 6.2. Theoretical Multifrequency Astrophysics

In our opinion only the study of 'simple' physical processes can allow the comprehension of the behaviour of cosmic sources. Indeed, sometimes the usual fits to experimental data, although very good, are meaningless from a physical point of view, since they represent only a mathematical approximation of data with more or less complex functions, in which physical parameters play a role of pure mathematical parameters. A beautiful example in this sense is that reported by Lichti *et al.* (1995); they fit a very good experimental multifrequency differential energy spectrum of the quasar 3C 273 from X-ray to $\gamma$-ray energies with a relatively simple function (their formula 6) and represent data and fit in their Fig. 6. This fit simply describes very well the shape of the spectrum. No information on the physics can be deduced from that fit. In contrast, if possible physical scenarios are taken into account, the same spectrum mentioned before can be fitted by using results coming from theoretical studies of the possible physical processes occurring. But this method is still, in our opinion, a bit weak, since it is strongly model dependent. Indeed, it is necessary to imagine a model for the cosmic source, and this model is not always supported, *a priori*, by experimental proofs. In contrast, physical processes – which can occur under certain physical conditions, such as, for instance, density of matter and temperature – which have been thoroughly studied can produce results easily describing cosmic sources without strong *ad hoc* hypotheses about their structure. In this case a good fit is the sign that in the cosmic source those processes definitively occur and then the theoretical results are more likely than in the other model dependent case. In the following we will discuss a few examples of the latter cases.

6.2.1. *Wide Range Physical Processes*

It is widely believed that AGNs are galaxies with a supermassive Kerr black hole in their center. This black hole accretes matter from the surroundings, which forms into an accretion disk. According to the dynamo principle, currents flowing in the disk generate a rotating magnetosphere. The rotating magnetic fields generate jets of plasma which are ejected from the central region perpendicularly to the disk. Knots filled with plasma are accelerated across the jet axis to relativistic energies (e.g., Blandford & Königl, 1979; Marscher, 1980). VLBI measurements have shown the existence of such knots (Krichbaum *et al.*, 1990; Bååth, 1990). This picture seems to apply for 3C 273 (Lichti *et al.*, 1995). If X-rays and $\gamma$-rays are produced by the same process (e.g., inverse Compton scattering), the emission region is of $\sim 1.3 \times 10^{15}$ cm. Since these photons have been observed, they must escape from the production region; then this region must be $>1.5 \times 10^{16}$ cm. A proposed solution to this problem is the beaming of X and $\gamma$ photons (Maraschi, Ghisellini & Celotti, 1992). However, another possible solution of this problem is that suggested by Giovannelli *et al.* (1996c and references therein): X-rays are produced by e–e and e–p thermal bremsstrahlung and $\gamma$-rays are coming from the decay of $\pi^0$s produced by p–p inelastic collisions. If the emission region is the same for X-rays and $\gamma$-rays they must interact. Therefore the dimensions of the emission region are a few units $\times 10^{17}$ cm, in perfect agreement with the upper limit of White & Ricketts (1979).

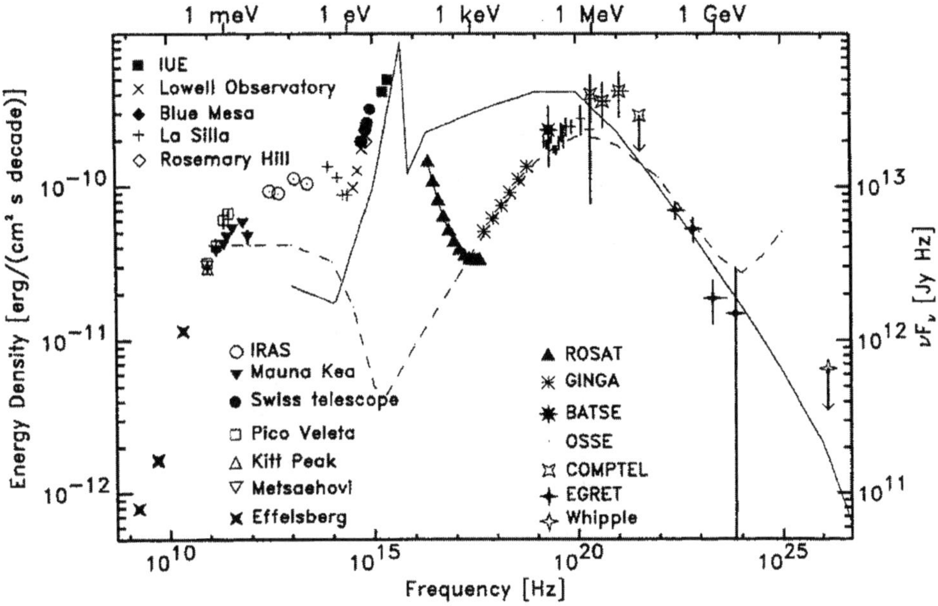


*Figure 81.* Comparison of the multifrequency energy distribution of 3C 273 with different models (see text) (Lichti *et al.*, 1995).

Following the scenario with the jets, the energy density spectrum of 3C 273 from IR to $\gamma$-ray regions (Figure 81) shows at least three, if not four, emission components. Each of them is produced by different emission processes:

- electrons moving in the magnetic fields emit synchrotron radiation which yields the spectrum observed in the radio and in the far IR. Also the emission from dust contributes especially in the IR;
- the prominent emission seen in the optical and UV (the big blue bump) is predominantly emitted from a hot photosphere surrounding the accretion disk;
- X-rays and $\gamma$-rays are produced in optically thin regions by the inverse Compton processes of relativistic electrons with the ambient synchrotron and/or optical and UV photons from the disk corona;
- the optically thin regions filled with relativistic electrons are associated with the super luminal knots observed in the radio band (Lichti *et al.*, 1995).

However, although this scenario provides a rather good explanation, it is not unique, and moreover it seems fairly complicated. Indeed it needs not only a particular morphology of the source, but also many *ad hoc* physical conditions in order to produce the described spread of processes. Based on a leptonic model Dermer & Schlickeiser (1993) calculated an energy density spectrum for 3C 273 (solid line in Figure 81).

It is easy to note that this model fails completely below 1 MeV and it is not very good even at higher energies. Mannheim (1993) developed another model in which

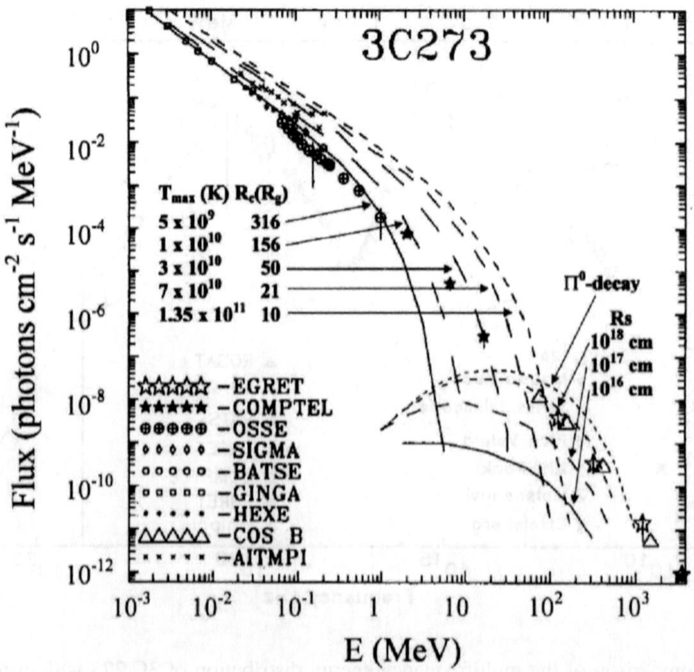


*Figure 82.* Calculated X-ray and $\gamma$-ray spectra overlapped on 'new' experimental data from 3C 273. The best fit of data is still obtained with the 'old' theoretical spectrum having $T_{max} = 5 \times 10^9$ K, corresponding to $R = 316R_g$ (Giovannelli *et al.*, 1996c).

electromagnetic showers induced by hadronic interactions of ultra high energy protons with soft synchrotron photons in the jet are produced (dashed line in Figure 81). This model fits reasonably well the X-ray and $\gamma$-ray regions but it fails completely at lower energies since it is based only on non-thermal processes. Another possible explanation for energies greater than 1 keV is that suggested by Giovannelli *et al.* (1996c): X-rays are produced by e–e and e–p thermal bremsstrahlung, and $\gamma$-rays come from the decay of $\pi^0$s produced by p–p inelastic collisions. Their fit appears to be the best, for the moment, at these energies. Moreover, the theoretical spectra are coming from very simple physical processes (Karakuła, Tkaczyk & Giovannelli, 1984) and 3C 273 does not need particular special morphology and physical conditions, except those characteristics of a massive black hole. Figure 82 shows such a fit.

It is interesting to note how it is difficult to predict any general behaviour of cosmic sources if data are spread only in a relatively narrow energy range. In this case one obviously searches for a process able to explain experimental results in that narrow range. If we take, for instance, the case of the 197 ms $\gamma$-ray pulsar PSR B1055-52 (Fierro *et al.*,1993), it has been measured in the energy range ~1–100 keV and in the EGRET-CGRO range (~30 MeV – 10 GeV) only at energies between 100 MeV and ~4 GeV; then an upper limit at higher energies (Figure 83),

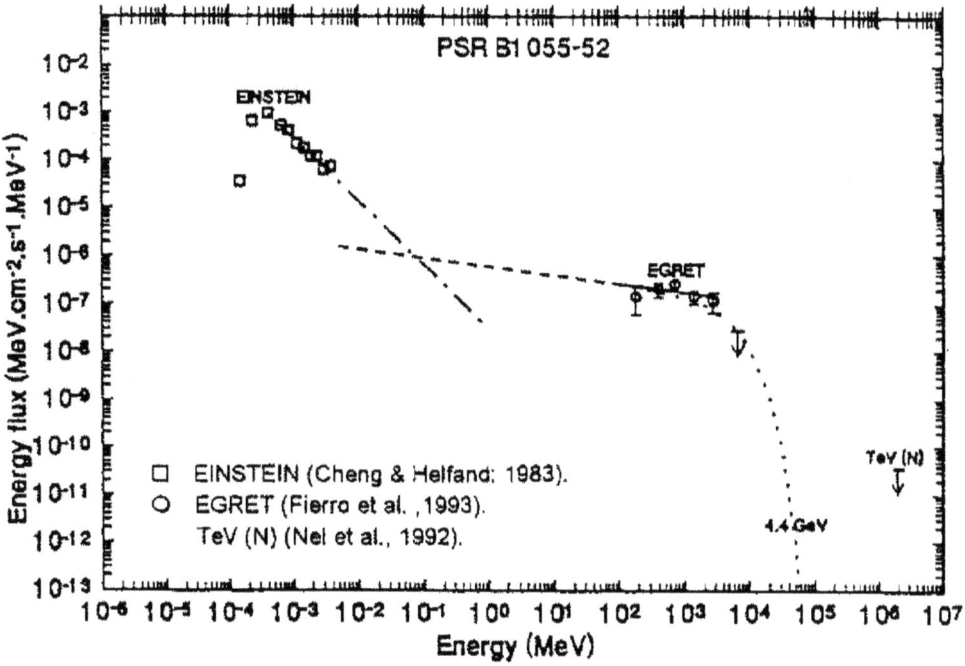


*Figure 83.* Differential energy spectrum of PSR B1052-55 (Nel & de Jager, 1994).

which shows some evidence of a fall off around $E \geq 4$ GeV, and one more upper limit at TeV energies. This TeV upper limit is useful to constrain the maximum characteristic energy $E_0$ by fitting the data to a power law with a rollover of the form: $I(E_\gamma) = KE^{-\Gamma}exp(-E/E_0)$. A high energy cutoff of $E_0 \sim 1$ to 5 GeV is expected from magnetic pair production on the strong magnetic fields near the polar caps of $\gamma$-ray pulsars (Harding, 1981). The limit on the rollover for PSR B1055-52 is shown in Figure 83 with the corresponding value of $E_0 \cong 4.4$ GeV and a power law slope $\Gamma \cong 1.18$.

The EGRET data requires for this pulsar, as well as for Geminga, a sharp rollover of the form $e^{-(E/E_0)^\alpha}$, where $\alpha > 1$ (Nel & de Jager, 1994). This information would have significant implication for the testing of the pulsar emission model. Data at EGRET energies render the aforementioned pulsars similar each other. Indeed, they have a similar age (several hundred thousand years) and almost the same average energy $\overline{E}$, since it is rapidly increasing for pulsars with ages $\tau \leq 10^4$ yr, but saturates for ages $\tau \gg 10^4$ yr (Nel & de Jager, 1994). Whilst if one takes into account only the lower energy ranges, PSR B1055-52 could be considered more similar to the Crab pulsar. Indeed, PSR B1055-52 has a radio interpulse like the Crab, but neither the X-ray nor $\gamma$-ray light curve has a similar structure. Then one possible investigation in this field is that a Compton synchrotron model may be applicable for ages $\tau < 10^4$ yr, resulting in a sensitive dependence of $\overline{E}$ with $\tau$ (the

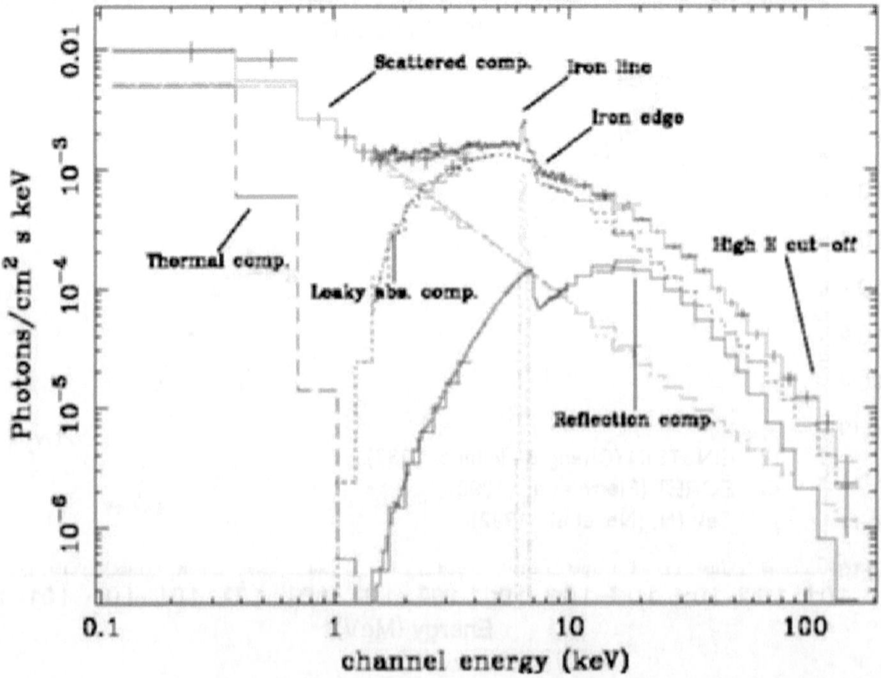


*Figure 84.* The BeppoSAX spectrum of NGC 4151. All the components contributing to the spectrum are shown together with the experimental data (Piro *et al.*, 2000a).

young pulsars show strong optical emission, so that inverse Compton scattering may be important). As $\tau \gg 10^4$ yr, curvature losses dominate. The result is a typical value $\overline{E} \sim 200$ MeV when including the high energy cutoff above $E_0$, and hence the saturation effect as observed (e.g., Figure 4 of Nel & de Jager, 1994).

In the last ten years the increased sensitivity, resolution, and bandpass of X-ray missions have drastically improved our knowledge of the X-ray spectra of cosmic sources, and, in particular, of those of the emission line AGNs. We have moved from an almost featurelesss power law into complex shape, where several broad and narrow features, produced in different sites around the central engine, are imprinted onto the power law. These components span a wide range of energies, sometimes blending or overlapping with each other. An unambiguous determination of each component is then difficult, unless *simultaneous broad band* spectral measurements are secured, as for instance with the BeppopSAX with its 0.1–200 keV energy range. It has been particularly useful to undertake a broad band study of AGNs in the X-ray energy range.

The X-ray spectrum of NGC 4151 is the most complex ever observed in a Seyfert galaxy, as shown in Figure 84 (Piro *et al.*, 2000a).

The spectrum is characterized by features typical of both Seyfert-1 and Seyfert-2 galaxies, making NGC 4151 the best laboratory for the study of these objects. This object has held the attention of our group since many years. Indeed, in 1976

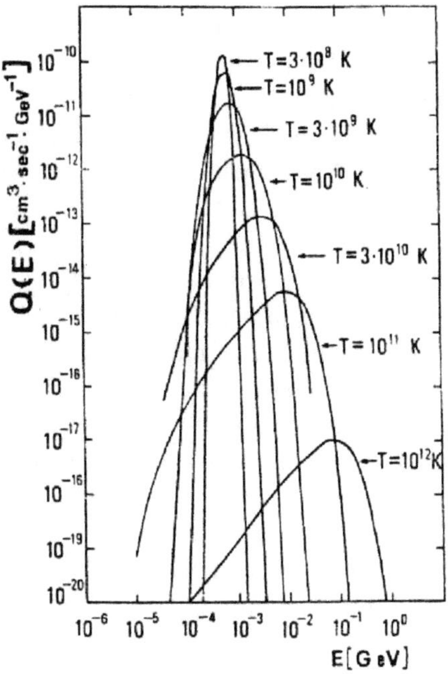


*Figure 85.* The photon energy production spectra from $e^+e^-$ annihilation, for selected temperatures and unit plasma concentration (Karakuła & Tkaczyk, 1985).

we flew an hard X-ray experiment (20–160 keV) with a transoceanic balloon that performed one of the first historical measurements of NGC 4151 in such a range (Auriemma *et al.*, 1978).

### 6.2.2. *Narrow Range Physical Processes*

In our opinion one very interesting example of the study of a physical process manifesting itself in a very narrow energy range is that discussed many years ago by Karakuła & Tkaczyk (1985), but still valid, namely the $e^+e^-$ annihilation. They performed the calculations in the case of Maxwellian momentum distribution of electrons and positrons at same temperature and with isotropic angular distribution. Figure 85 shows the photon energy production spectra for selected temperatures ranging from $3 \times 10^8$ K to $10^{12}$ K and unit concentration of the plasma.

The shape of the calculated annihilation spectra depends on the temperature of the pairs in the plasma. The peak energy of annihilation line indicates a weak dependence on the plasma temperature up to $3 \times 10^9$ K. For greater energies this dependence is almost linear with the temperature. The FWHM of the line is ~200 keV at a temperature of $10^8$ K and increases with it. For very high temperatures $KT \gg m_e c^2$ the annihilation spectra have no relationship with the characteristic 0.511 MeV line. The annihilation spectra have a characteristic thermal cutoff at high energy side. The authors used their spectra to fit experimental results of the annihilation line from the Galactic Center (Leventhal, MacCallun & Stang, 1978; Riegler *et al.*, 1981) and from Gamma Ray Bursts (Mazets *et al.*, 1981). In the first case their analysis confirmed that positrons are produced by a single source (i.e. photon–photon collisions near a collapsed object) and annihilate at great distance from the central source with the ambient gas at a temperature of $\sim 10^5$ K. In the second case the presence of redshifted narrow annihilation line in some GRB spectra gives a strong indication that the bursts occur at the surface of a neutron star containing pair dominant thermalized ($T_{e^+} = T_{e^-}$ plasma at temperature $T \simeq 10^9$ K. In contrast, the observed GRBs with hard tails ($E > 0.511$ MeV) cannot be explained with the annihilation of thermalized positrons because of their thermal cutoff. We have shown the importance of multifrequency astrophysics, discussing several examples both experimental and theoretical. Following our long experience in this kind of astrophysics we hope to have given some tools to the readers in order to facilitate the comprehension of an apparently trivial methodology in programming the multifrequency measurements, and in obtaining and interpreting the experimental data. In particular, we note the extreme importance of studying physical processes independently of particular cosmic sources. Indeed, in this way it is possible to reach good fits to multifrequency experimental data without making strong *ad hoc* hypotheses about the morphology of cosmic sources and their surroundings. We want also to suggest to the community to attempt the comprehension of the physics of cosmic sources in the easiest way, avoiding *a priori* useless complications, since nature is generally less complicated than we can imagine.

## 7. Old and New in Some Crucial Fields of Astrophysics

In this section we want to discuss some particular arguments of astrophysics which have strongly benefitted from the last generation experiments. Of course, the selection of such arguments is biassed by our knowledge and interest. Therefore this section is obviously far from being complete and exhaustive. However, we hope that such a selection could be useful for most of the readers, above all for the abundant literature cited.

### 7.1. Active Galactic Nuclei: Unified Schemes

Active Galactic Nuclei (AGNs) are extragalactic sources, in many cases associated with the nuclei of galaxies. However, in the most distant objects the host galaxy is too faint to be seen. The emission of AGNs is dominated by non-stellar processes in some energy range(s), typically in the optical, although not exclusively. One of the most important characteristics of AGNs is that they emit across practically the whole electromagnetic spectrum, covering almost 20 orders of magnitude in frequency, from the radio to the $\gamma$-ray band.

It is now very well established that AGNs are strong $\gamma$-ray emitters above 100 MeV. At least $\sim$40% of all EGRET sources are AGNs (Thompson *et al.*,

1995; 1996). Moreover, some more AGNs are certainly present amongst the still unidentified EGRET sources, and these make up almost 100% of all extragalactic sources, with the exceptions of the Large Magellanic Cloud and possibly Centaurus A (Padovani, 1999).

The main idea for explaining the emission from extragalactic X-ray emitters, now very popular, was suggested many years ago (Giovannelli & Polcaro, 1986): the *engine* producing high energy radiation is of the same kind for all extragalactic emitters. Mass and mass accretion rates are the unique parameters which differentiate extragalactic emitters containing central black holes from the galactic black holes.

The emission of the extragalactic X-ray sources can be expressed as $L_{\rm TOT} = L_{\rm NUC} + L_{\rm HG}$, where, $L_{\rm NUC}$ is the nuclear luminosity and $L_{\rm HG}$ is the host galaxy luminosity, formed by the integrated emission of its discrete sources. Such components can be derived by using the Giovannelli & Polcaro (1986) diagram.

Hasinger *et al.* (1999) from the combined X-ray surveys from the All Sky Survey (RBS) to the Deepest Surveys (RDS) of AGNs obtained a diagram $\log L_{\rm x}$ vs $\log z$. Taking the brightest objects for an arbitrary binning of redshift ($\Delta z$) one obtains the upper part of the Giovannelli & Polcaro (1986) diagram, $L_{\rm xmax}(z)$, as shown in the Figure 86. If the choice of the brightest object for an arbitrary $\Delta z$ is repeated for each survey with higher sensitivities one obtains a family of curves parallel to that of the aforesaid diagram. This means that the conclusions discussed by the latter authors are still valid, namely, there is a physical continuity between the different classes of compact extragalactic X-ray sources. This strongly indicates the existence of a unique kind of central X-ray source. The numerical continuity of the whole $L_{\rm xmax}(z)$ function should be interpreted as owed to an evolution of the central X-ray source from a very active to a more quiet status. And now this is definitively proved, thanks to the surveys obtained from lower luminosity objects at different redshifts as shown in the Figure 86, in which the points of the upper part of Giovannelli & Polcaro diagram have been superimposed on those of Hasinger, Miyaji & Schmidt, (2000), after Miyaji, Hasinger & Schmidt (2000) (Giovannelli & Sabau-Graziati, 2002c).

The way in which AGNs appear to the observers depends strongly on their orientation: classes of apparent different AGNs might be intrinsically similar (same kind of *engine*), only seen at different angles with respect to the line of sight (e.g., Urry & Padovani, 1995; Padovani, 1998).

More detailed unified schemes have been produced. For instance, in Vagnetti, Cavaliere & Giallongo (1991) and Vagnetti & Spera (1994) and in the references therein the evolutionary unified scheme is based on the changing balance between three optical luminosities, namely:

- nuclear isotropic component;
- relativistic beam component;
- host galaxy component.

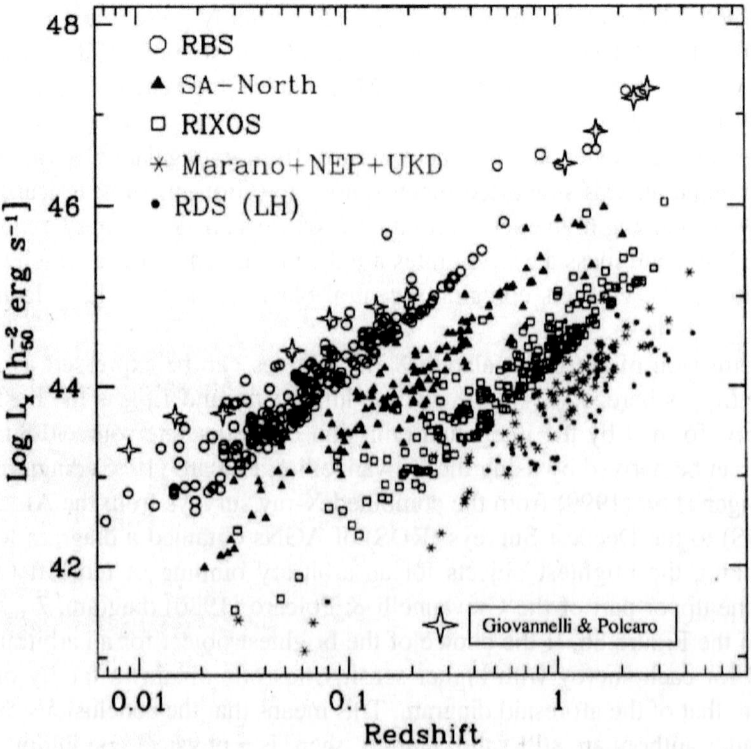


*Figure 86.* X-ray luminosity of different samples of extragalactic emitters versus redshift (Hasinger, Miyaji & Schmidt, 2000). Stars indicate the points of the maximum luminosity diagram of Giovannelli & Polcaro (1986) (Giovannelli & Sabau-Graziati, 2002c).

The intrinsic jet luminosity is assumed to have the same cosmic evolution as the nuclear isotropic component. The bulk Lorentz factor of the beam is able to account for the slower evolution of flat spectrum QSOs. The comparison of the total nuclear luminosity ($L_{\rm NUC} = L_{\rm IS} + L_{\rm BEAM}$) with the non-evolutionary galactic luminosity, ($L_{\rm HG}$), predicts the appearance of a source as a radio galaxy if $L_{\rm HG} > L_{\rm NUC}$.

In order to test the unified scheme for representing AGNs it is necessary to enhance the statistics of the measured objects in order to clearly understand the influence of the beam Lorentz factor, the beam axis orientation versus the line of sight – as already discussed in the case of electron and proton relativistic beams interacting with the matter and/or radiation around (Bednarek *et al.*, 1990a) – and the contributions of the nuclear isotropic component, host galaxy component, as well as that of the beam component. To do this it is necessary to explore experimentally a large sample of AGNs in different wavelength regions.

The open problem in this case is probably not owed to the physics governing such sources, which seems, now, quite well known, but to the methodology of measurements most suitable for obtaining indirectly the physical parameters necessary for testing the theory of unification. These parameters are the beam Lorentz factor, the inclination of the system with respect to the line of sight, fundamental for deriving the actual emission of the source at different energies, which can appear largely altered when observed from the Earth with ground or space based experiments.

For the purpose of sketching the unified model, all active galaxies contain (see Figure 87, centre) (Staude & Appenzeller, 2000): at the center of the galaxy there is a black hole of $\approx 10^8$–$10^9$ $M_\odot$ which attracts gas and dust from the surroundings. Initially, oowing to its angular momentum, this matter gathers together to form a disk around the black hole. Friction or other processes cause the matter to lose kinetic energy, and it spirals slowly into the black hole, following helical paths. As this happens high energy radiation is emitted by the hot gas, and this radiation can escape into space. Supported by strong magnetic fields two gas jets can shoot out into space simultaneously and almost perpendicularly to the disk plane in the direction of both poles, at virtually the speed of light. Strong jets end far beyond the galaxy in so-called 'Hot Spots', which are generally surrounded by extended bubbles of intensive radio radiation. In the interior of the jets electrons are moving at almost the speed of light. As they do so, they move away from the black hole on corkscrew-like paths across the magnetic field lines, emitting synchrotron radiation primarily in their direction or movement. The synchrotron radiation thus forms a cone, similar to the cone created by a car headlamp, with the jet as the central axis.

Now, in the unified model the different appearances of active galaxies are interpreted as a purely geometrical effect. In fact, if one looks at these systems from different angles, different areas supply the major portions of the observed radiation.

If one of the jets is pointing directly at the observer, the synchrotron radiation emitted by the electrons will dominate the spectrum. Objects with these spectral characteristics are known as BL-Lacertae objects and blazars. Blazars – a word constructed from BL-Lacertae and quasars – are primarily typified by the quick variability of their radiation. This is attributed to fluctuations in the ejection of the jet particles.

If the jet axis is viewed from a greater angle, then the radiation of the hot spots and radio bubbles will dominate. This has been observed in the majority of quasars with radio emission and in radio galaxies.

However, a large portion of the quasars and other active galaxies only emit weak radio radiation, or none at all. These are referred to as radio quiet quasars. To explain their appearance within the unified model as well, one would have to postulate the existence of a further component: a thick ring of dust, which surrounds the black hole and the central disk. If one looks directly onto the edge of this torus it covers the black hole and the central region of the disk. In this case, moreover, we do not register the optical and UV radiation, which is typical of quasars.

Figure 87 shows the radio galaxy Cygnus A. Near the center of the radio image a schematic diagram of the unified scheme for active galaxies, surrounded by a dense dust torus, is shown. The black hole at the center of the galaxy produces two jets. The observed spectra, shown outside the radio picture, represent selected AGNs

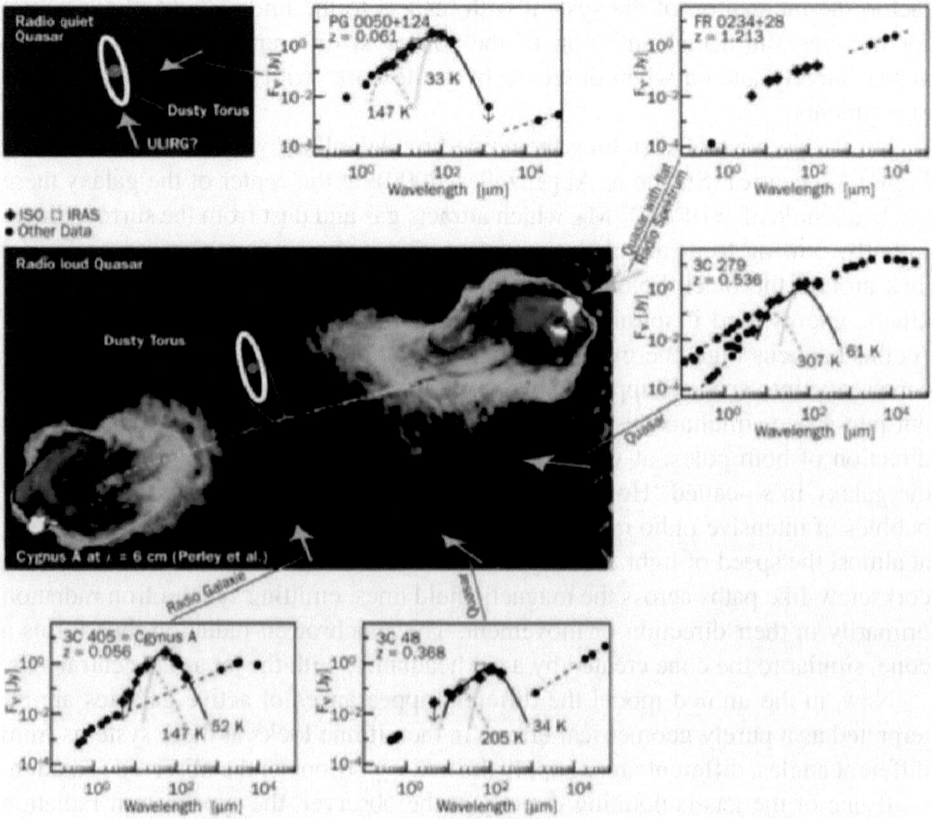


*Figure 87.* The VLA 6 cm radio image of the galaxy Cygnus A. A sketch of the unified scheme for active galaxies, surrounded by a dense disk torus is shown. The observed spectra outside the radio picture correspond to different active galaxies seen at different angles (Staude & Appenzeller, 2000).

placed at different redshifts and their shapes depend on the angle from which the active galaxy is viewed (Staude & Appenzeller, 2000).

A workshop on *AGN in the year 2000* was held in Trieste on May 2000, and the proceedings (Celotti, 2001 and the references therein) reported the highlights of this topic.

## 7.2. Blazars

Blazars are the most extreme variety of AGNs known. Their signal properties include irregular, rapid variability, high optical polarization, core-dominant radio morphology, apparent superluminal motion, flat ($\alpha_r \leq 0.5$) radio spectra, and a broad continuum extending from the radio through the $\gamma$ rays (e.g., Urry & Padovani, 1995). Blazar properties are consistent with relativistic beaming, that is bulk relativistic motion of the emitting plasma at small angles to the line of sight, as originally proposed by Blandford & Rees (1978), which gives rise to strong amplification and collimation in the observer's frame. It then follows that an object's appearance depends strongly on the orientation. Hence the need for 'Unified Schemes', which look at intrinsic, isotropic properties, to unify fundamentally identical, but apparently different, classes of AGNs. The blazar class includes flat spectrum radio quasars (FSRQ) and BL Lac-like objects. These are thought to be the beamed counterparts of high and low luminosity radio galaxies, respectively. The main difference between the two blazar classes lies in their emission lines, which are strong and quasar-like for FSRQ and weak or in some cases outright absent in BL Lac-like objects. The current view is that there is actually a continuity of at least some properties between the two classes, so the distinction between a BL Lac and an FSRQ can be somewhat blurred (Padovani, 2002). Owing to their peculiar orientation with respect to our line of sight, blazars represent a rare class of objects, probably less that 5% of all AGNs (Padovani, 1997). As consequence all existing blazar samples were relatively small, also because it was difficult their identification at high fluxes. Therefore, as stated by Padovani (2002), *our understanding of the blazar phenomenon is mostly based on a relatively small number of intrinsically luminous sources, which means we have only sampled the tip of the iceberg of the blazar population.* For example, the radio luminosity function, LF, of FSRQ derived by Urry & Padovani (1995), although based on 52 sources (the best at that time), included only one source at $L_r < 10^{26.5}$ W Hz$^{-1}$, which is the power that coincides roughly with the predicted flattening of the LF according to unified schemes. Moreover, only in the limited range $10^{26.9} < L_r < 10^{27.7}$ W Hz$^{-1}$ were the statistics good enough to have more than one source per bin! The need for a deeper, larger blazar sample is then obvious.

Padovani (2002) reviewed the ongoing deep surveys, discussing their limits and limitations. He presented some results on the evolutionary properties and luminosity function of faint blazars as derived from the Deep X-ray Radio Blazar Survey (DXRBS), which agree well with the predictions of unified schemes at low powers.

Blazar emission is dominated by strong and highly variable non-thermal radiation across the entire electromagnetic spectrum. Multifrequency ground based observations, combined with data from high energy astronomical satellites, have been used to derive the broad band spectral energy distribution (SED) of blazars, that is the source intensity as a function of energy, usually represented in the $\nu f(\nu)$ vs $\nu$ or $\nu L(\nu)$ vs $\nu$ plane. These measurements are consistent with the widely accepted scenario where blazar emission is owed to synchrotron radiation whose power increases with energy up to a peak value above which it drops sharply. At higher energies the spectrum is dominated by inverse Compton emission (ICE), which also smoothly raises until it reaches a second luminosity peak. According with the scenario that the observational characteristics of the blazars are owed to the emission from a relativistic jet seen at a very small angle with respect to the line of sight, the position and the relative power of the synchrotron and inverse Compton peaks directly depend on important physical parameters such as the intensity of the magnetic field, the maximum energy at which electrons can be accelerated, and the

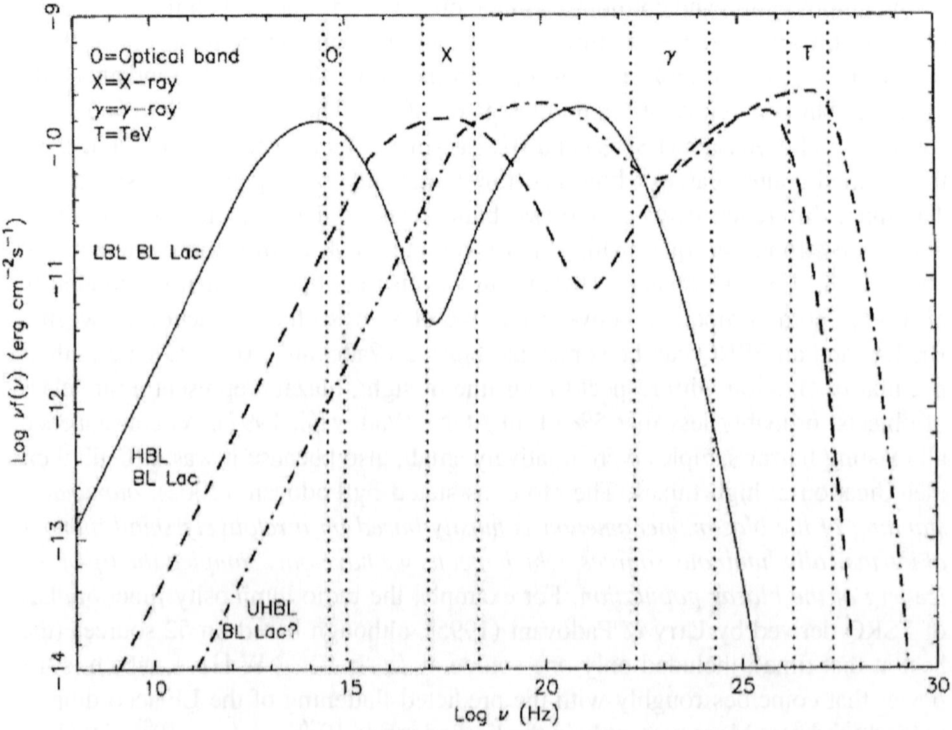


*Figure 88.* The spectral Energy Distribution of BL Lac-like objects. It is shown as a sequence of Synchrotron Self-Compton spectra peaking at different energies. For LBLs, the maximum of their synchrotron power output occurs in the IR-Optical band, whilst for HBLs the peak occurs in the UV or X-ray band. In UHBLs the synchrotron component might even reach the $\gamma$-ray band. Note that for the same peak luminosity the radio power decreases by orders of magnitude in going from LBLs to HBLs and possibly to UHBLs (Giommi *et al.*, 2002).

relativistic motion and orientation of the emitting plasma. The synchrotron peak is located at energies ranging from less than $\approx$0.1 eV (or $\nu \approx 10^{13}$ Hz to well over 10 keV (or $\nu \approx 10^{18}$ Hz) or even 100 keV in flaring states, demonstrating the existence of a wide variety of physical and geometric conditions in blazars. These are the reasons for which the SED of blazars is subject of an intense research activity.

Figure 88 (Giommi *et al.*, 2002) shows the expected emission from Synchrotron Self-Compton models (SSC) tracing a hypothetical sequence of blazar SEDs that ranges from LBL (Low energy BL Lac) sources where the synchrotron peak frequency ($\nu_{\rm peak}$) occurs at low energies to HBL (High energy BL Lac) sources where $\nu_{\rm peak}$ reaches the X-ray band and up to to the extremely large $\nu_{\rm peak}$ energies of the, possibly existing but still unseen, Ultra High Energy (UHE) peaked BL Lacs (UHBLs). As shown in Figure 88, within the broad band energy spectrum of blazars the X-ray region is particularly important, since at these energies a variety of different spectral components exist and have been seen. These include that flat and rising Compton component, the transition between the two régimes, and the high energy end of the synchrotron spectrum, which is produced by very energetic electrons, sometimes even extremely energetic. Then, these crucial observations, together with other multifrequency data, allow the determination of the overall spectral shape and therefore the estimation of important physical parameters. This is just what has been done by Giommi *et al.* (2002). They present a catalog of 157 X-ray spectra and 84 SEDs in blazars observed with the BeppoSAX during its first five years of operations. The SEDs have been constructed by combining BeppoSAX LECS, MECS, and PDS data with multifrequency photometric data, mostly non-simultaneous, obtained from NED and other large databases, including the GSC2 and the 2MASS surveys. All data from BeppoSAX have been taken from its archive and homogeneously reprocessed. They present for each source the $\nu f(\nu)$ vs $\nu$ plot, and for every observation they give the best fit parameters of the spectral model, which better describes the data. The energy where the maximum of the synchrotron power is emitted spans at least six orders of magnitude ranging from $\approx$0.1 eV to 100 keV. A wide variety of X-ray spectral slopes have been found, depending on whether the synchrotron or inverse Compton component, or both, are present in the X-ray band. Thanks to the wide bandpass of the BeppoSAX an important result has been obtained: it was possible to detect and measure, with a good accuracy, continuous spectral curvature in many objects whose synchrotron radiation extends to the X-ray region. This convex curvature, which is described by a logarithmic parabola law better than other models, might be the spectral signature of a particle acceleration process that becomes less and less efficient as the particles energy increases.

However, such SEDs for blazars have already been found for several sources, as for instance the BL Lac object Mkn 421 and the Quasar Hosted Blazar PKS 0528+134 (Takahashi *et al.*, 1997), with the simultaneous ASCA observations together with those of the CGRO and Wipple, as shown in Figure 89.

Left panel of Figure 89 shows the spectrum of Mkn 421 (Macomb *et al.*, 1996), a fairly typical X-ray selected BL Lac object (XBL). Its spectrum is soft (steep). The extrapolation of this spectrum from the ultraviolet range suggests that the synchrotron emission extends into the X-ray range (see, e.g., Makino, Fink & Clavel, 1992; Tashiro, 1992; Edelson *et al.*, 1995). Right panel of Figure 89 shows the broadband spectrum of PKS 0528+134 (Collmar *et al.*, 1997), a strong lined quasar hosted blazar (QHB). In contrast to Mkn 421, the X-ray spectrum of PKS0528+134 is harder than the UV spectrum, implying that the X-rays are already the onset of the second, HE, component. Takahashi *et al.* (1997) discussed also on the photon indices of the spectra obtained with the ASCA observations. They found that the X-ray spectra of XBLs are the softest, with the power law index $\alpha \sim 1 - 1.5$, and they form the highest observable energy tail of the low energy peak. X-ray spectra of the QHBs are the hardest ($\alpha \sim 0.7$) and are consistent with the lowest observable energy end (‘onset’) of the high energy peak. For the radio selected BL Lac objects (RBLs), the X-ray spectra are intermediate. It is worthwhile noting that the photon

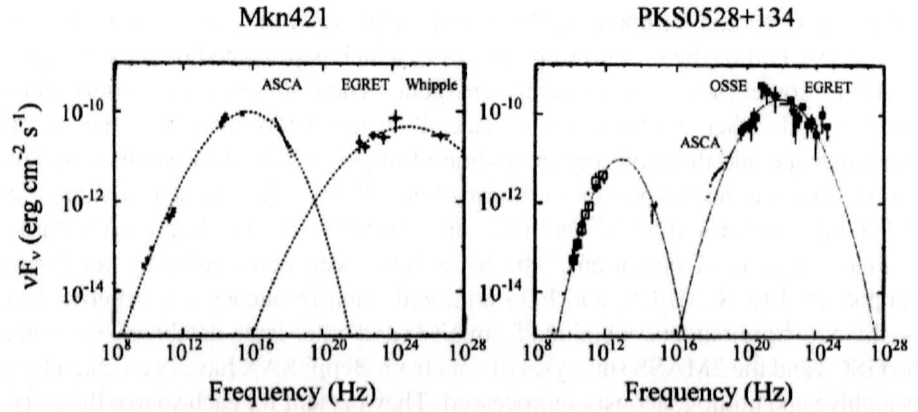


*Figure 89.* Left: The multifrequency spectrum from the simultaneous observations of the X-ray selected BL Lac object Mkn 421. The solid line shows the results of the fit to a third order polynomial function for these two components. Right: as the left, but for the Quasar Hosted Blazar PKS 0528+134 (after Takahashi *et al.*, 1997).

TABLE 12
Black holes in the Milky Way Galaxy (adapted from Blandford & Gehrels, 1999; Filippenko *et al.*, 1999; Casares, 2001)

| Source Name | Identification | Companion | f(M) | $M_{\rm Opt}$ ($M_\odot$) | $M_{\rm BH}$ ($M_\odot$) |
|---|---|---|---|---|---|
| Cygnus X-1 | HD226868 | O9.7 Iab | 0.24 | 24-42 | 11–21 |
| GS2023+338 | V404 Cyg | K0 IV | 6.26 | $\sim 0.6$ | 10–15 |
| GS2000+25 | QZ Vul | K3-K5 V | 4.97 | $\sim$0.7 | 6–14 |
| H1705-250 | V2107 Oph | K3 | 4.86 | 0.3-0.6 | 6.4–6.9 |
| GROJ1655-40 | N Sco 1994 | F6 IV | 3.24 | 2.34 | 7.02 |
| A0620-00 | V616 Mon | K3-K5 V | 3.18 | 0.2-0.7 | 5–10 |
| GS1124-68 | GU Mus | K3-K4 V | 3.10 | 0.5-0.8 | 4.2–6.5 |
| GROJ0422+32 | V518 Per | M2 V $\pm$ 2 | 1.21 | $\sim$0.3 | 6–14 |
| 4U1543-47 | | A2 V | 0.22 | $\sim$2.5 | 2.7–7.5 |
| GRS1009-45 | N Vel 1993 | K8 V $\pm$ 2 | 3.17 | $\sim$0.6 | 4.4 |

index for XBLs is often variable on a short timescale (e.g., Takahashi *et al.*, 1996), whilst for QHBs the index remains almost constant when the flux changes.

An updated book on Blazar Astrophysics is that by Giommi, Massaro & Palumbo (2002), which reports the proceedings of the Conference on 'Blazar Astrophysics with BeppoSAX and Other Obervatories'.

### 7.3. Macro-quasars and micro-quasars

Following the current definitions, a macro-quasar (macro-QSO) is an AGN, such as 3C 273, with a giant black hole of mass of $\sim 10^7 - 10^9$ $M_\odot$ as the main source of energy emission, whilst a micro-quasar (micro-QSO) is a galactic X-ray binary with radio jets. The list of known galactic micro-QSOs contains ten objects. They are shown in Table 12, in which the mass function of the binary system and the masses of the collapsed object and the optical companion are reported too (Blandford & Gehrels, 1999; Filippenko *et al.*, 1999; Casares, 2001). In addition, there are several more objects which are possible candidates for being included into the list (e.g., Ziółkowski, 1997a; Casares, 2001 and the references therein).

The idea of the same *engine* responsible for the emission in extragalactic macro-QSOs and galactic micro-QSOs has received strong experimental support in the last decade. Indeed, comparing the two classes of macro-QSOs and micro-QSOs, one notes that they show:

- similarities in the X-ray spectra (Tanaka & Lewin, 1995):
  - the power law components have similar slopes;
  - the intensity of the power law components is highly variable on a short time scale;
  - the slope of the power law for a given source remains approximately constant in spite of large variations of the source intensity;
  - the soft component, believed to originate in the inner region of the accretion disk, is usually seen in both types of objects. In the micro-QSOs it is observed as an ultra soft X-ray component, whilste in the macro-QSOs as an UV bump;
- common disk luminosity vs radio luminosity relationship (Falcke & Biermann, 1996);
- common disk luminosity vs jet length (empirical) relationship (Sams, Eckart & Sunyaev, 1996);
- common disk luminosity vs brightness temperature of the jet (empirical) relationship (Sams, Eckart & Sunyaev, 1996).

The above similarities suggest that a substantial amount of the underlying physics might be the same for macro-QSOs and micro-QSOs. Moreover, most of the galactic micro-QSOs (perhaps all of them) seem to contain black holes. Two of them (those with superluminal jets), namely GRO J1655-40 (Orosz *et al.*, 1997; Zhang *et al.*, 1998; and GRS 1915+105 (Morgan, Remillard & Greiner, 1996; Greiner, Morgan & Remillard, 1996; Zhang, Cui & Chen, 1997), possibly contain rapidly rotating Kerr black holes.

Therefore a possible easier line of investigation for understanding the physics of quasars is that of investigating micro-QSOs, since they are typically $\sim 100$ times brighter than macro-QSOs, as seen from the Earth, and the corresponding variability time scales are $\sim 10^6 - 10^8$ times shorter than in macro-QSOs.

The question of possible connection between rapid rotation and the highly relativistic nature of the jets remains open. Also the mechanism of the jet acceleration remains still unknown, although efforts have been done by many authors (e.g., Li, Miller & Colgate, 1997; Kirk, 1997; Ostrowski, 1997; 1999; 2000).

Looking far into the future, a most compelling test of General Relativity is the promise of a phenomenon that is believed to happen quite often in the Universe: the merging of two black holes, which occurs most often when two galaxies collide. The final cataclysm is accompanied by a powerful burst of gravitational radiation, which could be an impressive test of Einstein's theory.

A beautiful example of a micro-quasar is that of GRS 1915+105. It was discovered by Castro-Tirado, Brandt & Lund (1992) by using the WATCH all sky monitor (ASM) on the Granat satellite. Since then it has been one of the most luminous X-ray sources in the sky. Superluminal motion from this source, for the first time in the Galaxy, was detected by Mirabel & Rodriguez (1994). A two-sided jets with an intrinsic velocity of $0.92 \pm 0.08\,c$ and an inclination of $70° \pm 2°$ was detected. Similar jet parameters were measured from other ejection events in 1997 October/November (Fender *et al.*, 1999).

The RXTE has been monitoring GRS 1915+105 frequently and a rich pattern of variability has emerged from these data with timescales from years down to 15 ms (e.g., Belloni *et al.*, 2000, and the references therein). Vilhu *et al.* (2001), analyzing RXTE data, found a good fit with a thermal Comptonization model. There is a strong correlation between the inner disk temperature and the disk luminosity. The hard Comptonized luminosity does not depend significantly on the total luminosity. The spectral hardness of the Comptonized radiation, the fraction of seed soft photons scattered by the Comptonizing cloud, its Thompson optical depth, and the fraction of the total power dissipated in the optically thin hot plasma all strongly anticorrelate with the luminosity. They found that the inner disk radius is almost constant and that the hot Comptonizing corona shrinks at high luminosities.

In contrast, analyzing nine OSSE-CGRO observations from 1995 to 2000, eight of them being simultaneous with those of the RXTE, and two RXTE-OSSE spectra with the lowest and highest X-ray fluxes, Zdziarski *et al.* (2001) found that the OSSE data show a power law-like spectrum extending up to $\geq$ 600 keV without any break. They interpret this emission as strong evidence for the presence of non-thermal electrons in the source. The broadband spectra cannot be described by either thermal or bulk motion Comptonization, whereas they are well described by Comptonization in hybrid thermal/non-thermal plasmas.

Ueda *et al.* (2002) performed from 2000 April 16 to April 25 one of the largest multi-wavelength coordinated campaign of GRS 1915+105, mostly in the 'plateau' (or low/hard) state, covering the wide energy band in radio (13,3–0.3 cm), NIR ($J$, $H$, and $K$ bands), X-rays and $\gamma$-rays (from 1 keV to 10 MeV). They found that the IR-emitting region is located far from the black hole by $\geq 10^{13}$ cm, whilst its size is $\leq 10^{12}$ cm constrained from the time variability. They found good correlation between the quasi-steady flux level in the NIR band and in the X-ray band.

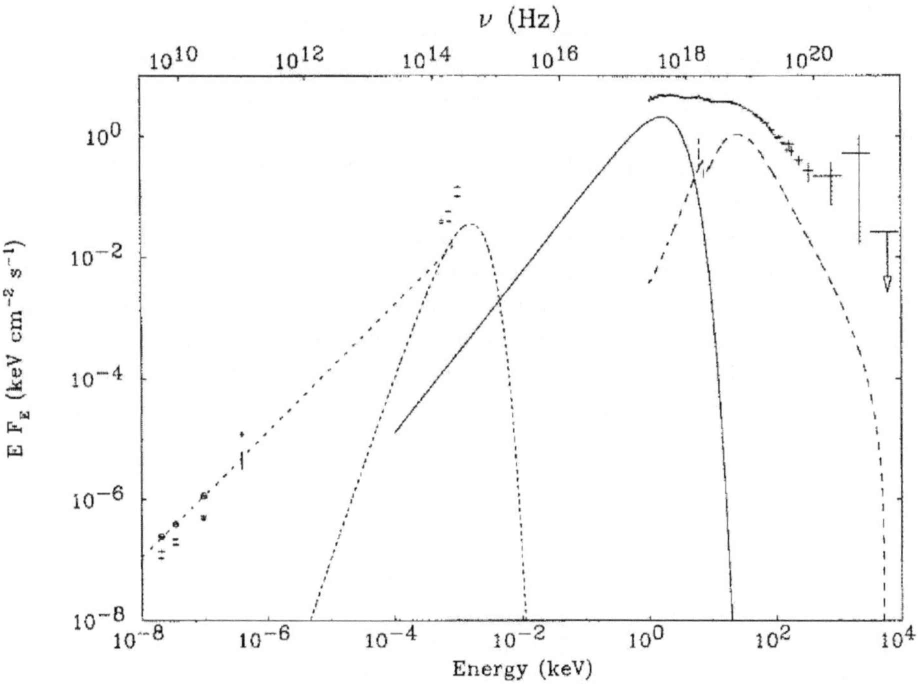


*Figure 90.* Spectral energy distribution from the largest multifrequency campaign of observations of the micro-quasar GRS 1915+105 in the plateau state (except for the 94 GHz data) (Ueda *et al.*, 2002).

From this they estimated that the reprocessing of X-rays, probably occurring in the outer parts of the accretion disk, accounts for ~20–30% of the observed *K* magnitude in the plateau state. The OSSE spectrum in the 0.05–10 MeV band is represented by a single power law with a photon index of 3.1 extending to ~1 MeV with no cutoff. They can model the combined GIS-PCA-HEXTE spectra covering 1–200 keV by a sum of the multi-color disk model, a broken power law modified with a high energy cutoff, and reflection component. The power law slope above ~30 keV is found to be very similar between different states in spite of large flux variations in soft X-rays, implying that the electron energy distribution is not affected by the change of the state in the accretion disk. Figure 90 shows the total energy distribution of the micro-quasar GRS 1915+105 obtained during such a coordinated campaign of observations.

Yadav (2001) presented evidence of a direct accretion disk jet connection in GRS 1915+105 based on the analysis of RXTE/PCA data with a 'spike' in X-ray light curves. He found that the radio emission increases as the hardness ratio increases during the low hard state. He suggested that the 'spike', which separates the dips with hard and soft spectra, marks the beginning of the burst phase when the luminosity of the soft X-rays (5–15 keV) increases by a large factor (~10).

This produces a major ejection episode of the synchrotron-emitting plasma termed 'baby jets', which are associated with IR and radio flares of about half an hour period widely reported in the literature. Subsequent short, but frequent, soft dips produce overlapping faint flares which result in an enhanced level of quasi-steady emission.

Therefore the necessity of having simultaneous multifrequency observations appears once more evident.

Three workshops devoted to micro-quasars have been held in the recent past: Workshops on Micro-quasars in Greenbelt (1997), in Paris (1988), and in Granada (2000). The proceedings of this last workshop (*Micro-quasars* by Castro-Tirado, Greiner & Paredes, 2001) represent an useful updated tool for people interested in such a topic. However, two more workshops have been held in such a topic: Workshops on Galactic Sources with Relativistic Jets in Jodrell Bank (1996) and Milton Keynes (1998).

## 7.4. Radio Pulsars, Millisecond Pulsars

Radio pulsars discovered by Hewish *et al.* (1968) are rotating magnetic neutron stars (Pacini, 1967) whose radio emission is highly directional. Radio beams are generally not aligned with the rotational axis and therefore their rotation produces a 'lighthouse' effect observed on Earth as radio pulses. The rotating magnetic pulsar then emits a magnetic dipole radiation at the rate (Ostriker & Gunn, 1969):

$$L_{\rm mdr} = 2/3c^3 (B^2 R^6 \Omega^4 \sin^2 \alpha) \,, \tag{7.1}$$

where $B$ is the polar field strength at the neutron star's surface, $R$ and $\Omega$ are the radius and the angular rotational velocity of the neutron star, and $\alpha$ is the angle between the angular momentum axis and the spin axis. The magnetic dipole radiation is emitted at the expense of the rotational energy of the pulsar, which is the cause of the apparent spin down, corresponding to the rotational energy loss:

$$\dot{E} = -I\Omega\dot{\Omega} \,, \tag{7.2}$$

where $I$ is the moment of inertia of the neutron star. If the observed spin down is indeed caused by magnetic dipole radiation, the two formulas can be equated. Then $\dot{E} = L_{\rm mdr}$.

The observable quantities are the rotational period $P$ and the spin down rate $\dot{P}$. Then substituting these quantities to $\Omega$ and $\dot{\Omega}$, it is possible to express the pulsar magnetic field in terms of the observable quantities, by using typical values for the moment of inertia and the radius of the neutron star (Ziółkowski, 1997b):

$$B = 3.2 \times 10^{19} (P\dot{P})^{1/2} \text{ G} \,. \tag{7.3}$$

The rate of energy loss is (Thompson, 2000):

$$\dot{E} \simeq 4 \times 10^{46} \dot{P} P^{-3} \text{ erg s}^{-1} . \tag{7.4}$$

The line open field line voltage is (Thompson, 2000):

$$V \simeq 4 \times 10^{20} \dot{P}^{1/2} P^{-3/2} \, V \simeq \dot{E}^{1/2} . \tag{7.5}$$

Another important parameter expressed in terms of the observable quantities is the apparent age of the pulsar $\tau = P/2\dot{P}$. The value of $\tau$ gives the present time scale of the spin down of the pulsar; then it is called the 'spin down' age of the pulsar. If the magnetic field of the pulsar is not substantially changed since its birth, $\tau$ gives the true age of the pulsar. Another estimate of the age is possible through the age of the associated SNR. This is the 'kinematic age' based on the proper motion and the distance to the galactic plane. Similarly, the estimate of the magnetic field strength based on the cyclotron lines detected in X-ray pulsars or the spin up and spin down properties of X-ray pulsars provide results in reasonable agreement with those obtained from the equation (7.3). This equation and the spin down age are then useful tools for estimating magnetic fields and ages of pulsars, but with the hypothesis of the constancy of the magnetic field intensity.

Pulsars born with a fast period (order of few tens of millisecond) and strong magnetic fields ($B \geq 10^{12}$ G). The Crab pulsar is the most popular example: $P =$ 33 ms, $\lg B(\text{G}) = 12.6$, $\lg \tau(\text{yr}) = 3.1$, and its true age is 949 yr. Owing to spin down the pulse period is increasing, so that an 'old' pulsar, with age of $\sim 10^7$ yr, has a rotation period of order of seconds. According to the classical picture the magnetic field of a neutron star decays on a similar timescale (Gunn & Ostriker, 1970). Owing to both processes the efficiency of the radio emission mechanism is expected to decrease, and after reaching the critical value of the ratio $B/P^2$ the mechanism switches off completely (Sturrok, 1971). In the $\lg B - \lg P$ diagram the line $B \propto P^2$, across which the switch off condition is satisfied, is called the 'death line'. After crossing such a line the neutron star is no longer a pulsar and this part of diagram is called the 'graveyard'. It occurs at an age of $\sim 10^7$–$10^8$ yr. The pulsars in the graveyard are not only radio quiet, but are unobservable in other energy regions also. The exception are dead pulsars that are members of binary systems and now receive the matter transferred from their binary companions. Such systems are seen as low mass X-ray binaries (LMXBs).

The behaviour of pulsars were discussed in the book *Timing Neutron Stars* (Ögelman & van den Heuvel, 1989).

The first millisecond pulsar PSR 1937+21 discovered by Backer *et al.* (1982) immediately became a sensation because of its unusually fast rotation ($P = 1.56$ ms). Soon afterwards two more millisecond pulsars were discovered: PSR 1953+29 ($P = 6.13$ ms) and PSR 1855+09 ($P = 5.36$ ms). These new fast pulsars had very weak magnetic fields ($\lg B(\text{G}) \sim 8.5$) and very long spin down ages ($\tau \sim 2 \times 10^8$ $- 5 \times 10^9$ yr). Their properties seemed paradoxical, since weak magnetic fields and long spin down ages would indicate very old objects. At the same time, the very rapid rotations would indicate very young aged objects. The apparent paradox was soon solved (Radhakrishnan & Srinivasan, 1982) under the hypothesis that these new objects were descendants of the old dead pulsars, which were spun up, and so resurrected, to a new life as result of reprocessing through a binary evolution. Therefore the term 'recycled pulsars' was coined.

Recycled radio pulsars are then very old neutron stars, which spent some time in the graveyard, but were resurrected by spin up associated with the accretion of matter from their binary companions. Their progenitors were, therefore, LMXBs. For reviews of this topic, see the papers by Ziółkowski (1997, 1999). The spin periods of recycled pulsars are similar to those observed in LMXBs and are consistent with the expected outcome of the accretion spin up process. A magnetized neutron star which accretes matter from the companion is expected to adjust its rotation to the so-called *equilibrium period.* This period is defined as a period at which the accelerating accretion torque (spin up torque) and the braking propeller torque (spin down torque) balance each other, and the spin period, in the first approximation, remains constant (e.g., Ghosh & Lamb, 1979). In a simplified description this corresponds to the situation when the outer edge of the magnetosphere rotates with Keplerian velocity. The equilibrium period is directly related to the magnetic field strength and to the accretion rate (e.g., Ziółkowski, 1999 and references therein). By using the Eddington limited accretion rate $\approx 10^{18}$ g s$^{-1}$, one obtains the expression for the shortest rotational period to which a neutron star of a given $B$ could be spun-up by accretion:

$$P_{\rm eq} \approx 0.7 B_{12}^{6/7} \text{ s} . \tag{7.6}$$

This equation defines the so-called 'spin up line' in the $\lg B - \lg P$ diagram.

There is a strong observational evidence (Bhattacharya, 1995; Ziółkowski, 1997b) that magnetic fields decay in the neutron stars accreting the matter during the binary evolution.

Figure 91 shows the magnetic field versus the spin period for known radio pulsars (Ziółkowski, 1999).

In spite of about 1000 'second' pulsars being known, after the recent discovery of a few hundreds of new pulsars (D'Amico, 2000) and the class of millisecond pulsars (e.g., Bailes & Lorimer, 1995) is becoming quite populated, the problem of the continuity between the two classes still deserves further studies both experimental and theoretical, as well as the problem of the so called *death line* whose slope is model dependent. Indeed, the idea of the *death line* was coming from the experimental evidence on the complete absence of radio pulsars with periods exceeding few seconds. In this picture, beyond this boundary, pulsars with low spin rate cannot accelerate particles above the stellar surface to high enough energies to initiate pair cascades through curvature radiation, and the pair creation needed for radio emission is strongly suppressed. But this line has been violated at least once; then a re-consideration of such a picture is mandatory.

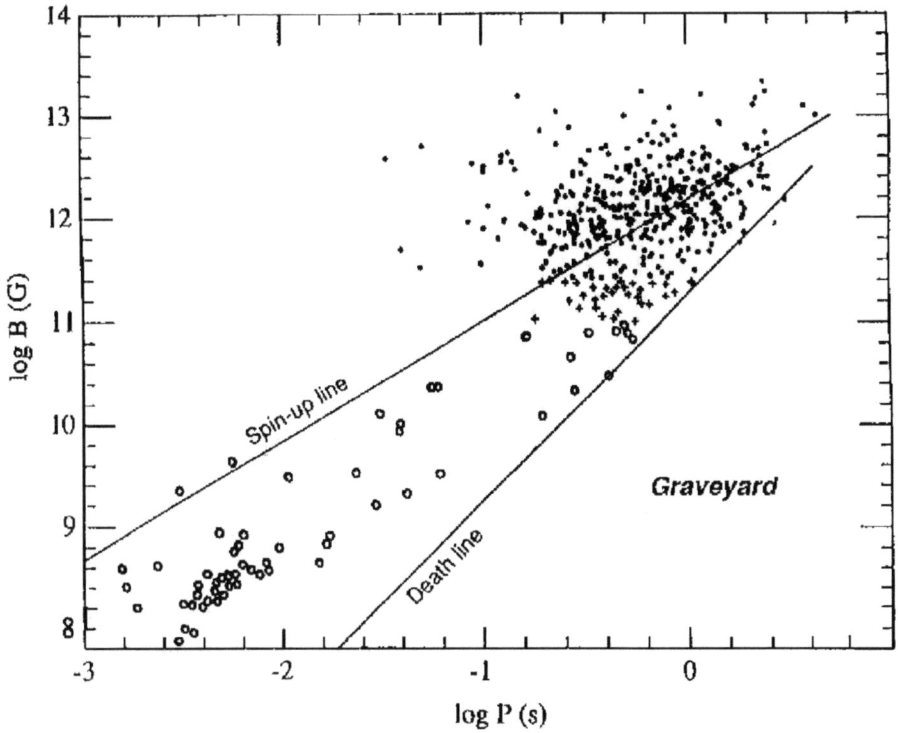


*Figure 91.* The magnetic field $B$ versus the spin period $P$ for known radio pulsars. Young pulsars are shown with dots, recycled pulsars with open circles, possible recycled pulsars with crosses (Ziółkowski, 1999).

Baring & Harding (1997) postulate the existence of another pulsar *death line* corresponding to high magnetic field $B \sim 10^{13}$ G in the upper part of the $\dot{P}$–$P$ diagram, a domain where few radio pulsars are observed. The origin of this high $B$ boundary is owed to the suppression of magnetic pair creation $\gamma \rightarrow e^+e^-$. Above this boundary, pulsars are expected to be radio quiet, but perhaps still X-ray and $\gamma$-ray bright.

The X-ray observatories of the 1990s, such as ROSAT, ASCA, BeppoSAX and RXTE have achieved important progress in neutron star and pulsar astronomy. The identification of Geminga, the discovery of X-ray emission from millisecond pulsars and the identification of cooling neutron stars are only a few of the fascinating results. Becker (2000) briefly reviewed the X-ray emission properties of rotationally powered pulsars and their wind nebulae as observed by the experiments aboard the former satellites.

Millisecond pulsars form a separate group amongst the rotationally powered pulsars. They are distinguished by their small spin period ($P \leq 20$ ms) and high rotational stability ($\dot{P} \approx 10^{-18} - 10^{-21}$) and consequently they are very old objects with spin down ages up to $10^9$–$10^{10}$ yr and magnetic field strengths of order of $10^8$–$10^{10}$ G.

ROSAT, with a significantly higher sensitivity compared with previous X-ray satellites, allowed for the first time to detect X-ray emission from objects as faint as millisecond pulsars. However, although 10 of the 34 detected rotationally powered pulsars belong to the small group of millisecond pulsars, the origin of the detected X-ray emission for most of them is not yet known (Trümper & Becker, 1998; Becker & Trümper, 1999). Five of the ten millisecond pulsars (PSR 1957+20, PSR J10912+5307, PSR J0751+18, PSR J1744-1134, and PSR J1024-0719) are identified by ROSAT only by their positional coincidence with the radio pulsar, and in view of the low number of detected counts do not provide much more than a rough flux estimate. These objects are so faint that the sensitivity of AXAF-Chandra and XMM-Newton is needed to detect enough photons required for a detailed spectral and temporal study in the soft and hard bands beyond 2 keV. More detailed results have been found for the other 5 millisecond pulsars (PSR 1821-24 in M28, PSR 1937+21, PSR J0218+3242, PSR J0437-4715, and PSR J2124-3358), which all provide important empirical information on the pulsar's X-ray emission mechanisms, as discussed by Becker (2000).

However, putting the observed emission properties of the detected millisecond pulsars in a somewhat wider frame, Becker & Trümper (1997) found that the X-ray luminosity of the detected millisecond pulsars show the same linear relationship with the same X-ray efficiency as the Crab-like pulsars, as shown in Figure 92, indicating that the bulk of their emission is mainly owed to non-thermal processes. As remarked by Becker (2000), the occurrence of power law spectra in PSR 1821-24 and PSR J0437-4715 and the similarity between the radio/X-ray pulse profiles seen also for PSR J0218+3242 and PSR J2124-3358 may be considered as providing additional evidence for a non-thermal origin of the millisecond pulsars' X-ray emission.

Becker (2000) also discussed the other groups of rotationally powered pulsars and concluded that the current findings of the pulsars' emission properties show that young pulsars with ages of less than ~2000 yr appear Crab-like (i.e. bright synchrotron nebula, sharp X-ray pulses with high pulsed fraction), whereas ~$10^4$–$10^5$ yr old pulsars resemble more the emission properties observed for the Vela pulsar (i.e. X-ray emission beyond 0.5 keV dominated by the emission from the pulsar driven synchrotron nebula, soft pulses only observable from the Vela pulsar). However, although these emission properties are found to fit very well for the young rotationally powered pulsars (i..e. radio pulsars) there is strong evidence that they are not representative for the whole sample of young neutron stars, which exist in our Galaxy. Topics which are discussed in respect to this are soft $\gamma$ ray repeaters and anomalous X-ray pulsars.

Figure 93 shows the period versus the period derivative for many of the known pulsars, showing three of the derived parameters, namely spin down age, surface

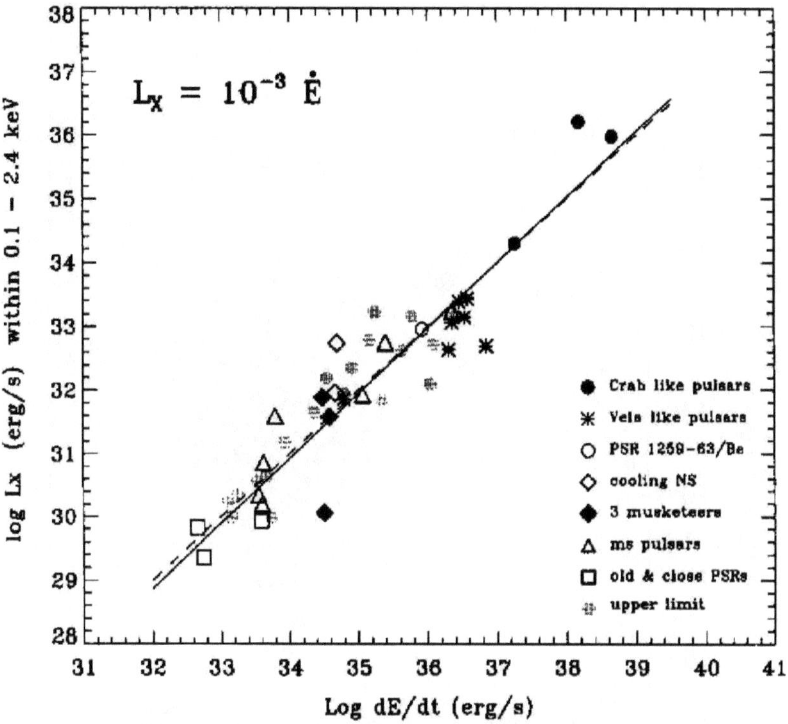


*Figure 92.* X-ray luminosity of the ROSAT detected rotationally powered pulsars versus the pulsars' spin down energy $\dot{E}$. Thermal spectral components have been subtracted. Although field pulsars and millisecond pulsars form well separated populations, they obey the same X-ray efficiency (see Becker & Trümper, 1997 and discussion therein). The close correlation between $L_X$ and $\dot{E}$ strongly suggest that the bulk of the observed X-rays is emitted at the expense of rotational energy, as it is observed for the radio and $\gamma$-ray emission (Becker, 2000).

magnetic field, and open field line voltage, which are reported too (Thompson, 2000).

Thompson (2000) discussed the high energy emission from active pulsars. He found that non-thermal emission is seen from some pulsars across much of the electromagnetic spectrum, implying acceleration of particles to high energies. The power for this process usually comes from the rotation of the magnetized neutron star. The luminosity of these pulsars peaks in the X-ray to $\gamma$-ray bands. The light curves for some of these pulsars suggest particle acceleration and radiation from a hollow surface above one magnetic pole of the star. Multifrequency observations have been used to construct and test models of high energy pulsar radiation. Although no single model for high energy emission from pulsars has emerged, the observations with the present generation of high energy telescopes have channelled all models in certain directions:

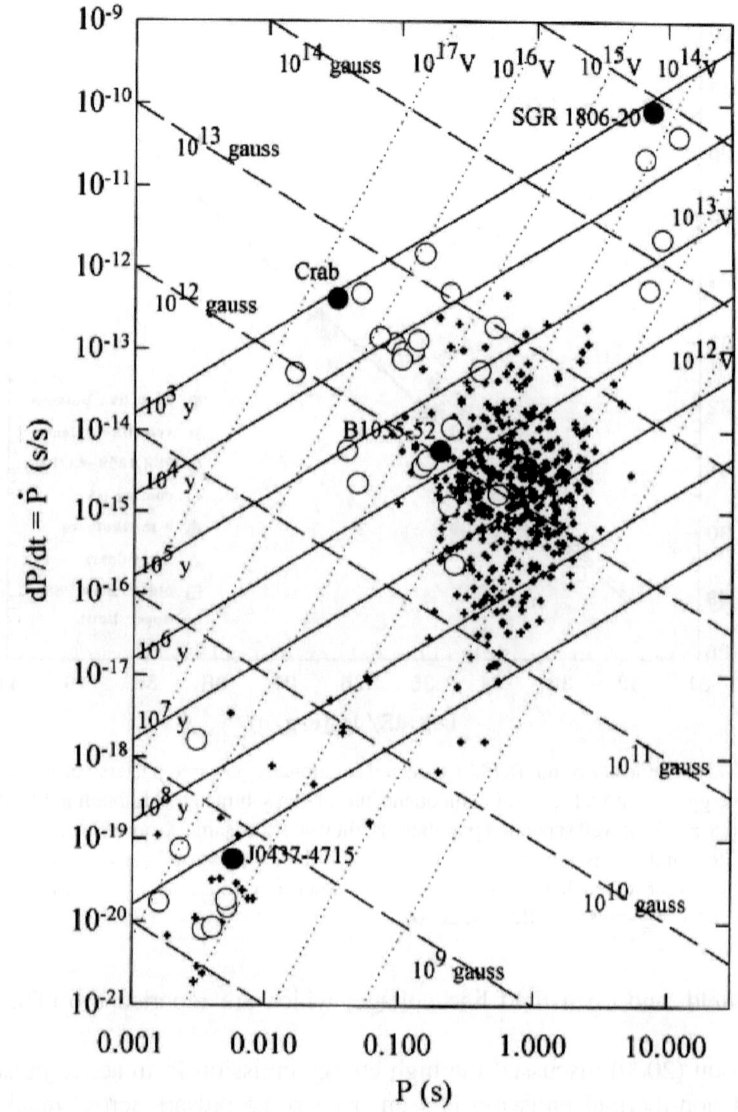


*Figure 93.* Period versus the period derivative for many of the known pulsars, showing three of the derived parameters, namely spin down age, surface magnetic field and open field line voltage are reported too (Thompson, 2000).

- the particle acceleration by the strong electric fields takes place somewhere above the magnetic poles;
- the principal high energy emission processes are synchrotron radiation, curvature radiation, and inverse Compton scattering, with thermal emission a secondary contributor;

- the high energy pulsed emission is probably associated with a single magnetic pole of the neutron star.

Thompson (2000) pointed out several open questions about spin down pulsars:

- Where in the magnetosphere are the particles accelerated?
- What is the shape of the beam?
- Are there many more radio quiet pulsars to be found in X-rays and $\gamma$-rays?
- How do these recent discoveries affect our thinking about supernovae and neutron star formation?

These open questions can be solved by improving the studies of the X-ray archives of the ROSAT, ASCA, BeppoSAX, and RXTE, while AXAF-Chandra, and XMM-Newton will give significant improvements in sensitivity and resolution. In the $\gamma$-ray range the study of the CGRO archives and the new data from the INTEGRAL mission will help in filling some gaps in such a low energy $\gamma$-ray band. The next major step in high energy $\gamma$-rays will come with the GLAST experiment (e.g., Kamae *et al.*, 2000).

## 7.5. X-RAY BINARY SYSTEMS

We have already discussed (Section 5.5.2) X-ray binaries (XRBs) in general. Now we want to remark several facts and the main behaviour of XRBs following the most recent literature, separating the discussion devoted to HMXBs from that devoted to LMXBs, and to the new class of Anomalous X-ray Pulsars (AXPs).

But before this we want to recall several fundamental facts which determined the history of X-ray astronomy and then of Modern Astrophysics.

### 7.5.1. *When and How a Nobel Prize Was Born*

In our opinion it is extremely important to note in this section some of the fundamental steps which originated the modern high energy astrophysics. The progenitor of all X-ray measurements was obtained with the launch of a rocket at the White Sand Missile Range, New Mexico (USA), on June 18, 1962, strongly wanted by the historical group formed by Riccardo Giacconi, Herbert Gursky, Frank R. Paolini, and Bruno Rossi. The original article describing the discovery of the first extra-solar X-ray source was submitted to Physical Review Letters in October 12, 1962 (Giacconi *et al.*, 1962). And in honour of this historical group we dare to reproduce a few lines of the original article, which opened a new fundamental window to the Universe.

*The experiment was intended to study fluorescence X-rays produced on the lunar surface by X-rays from the Sun and to explore the night sky for other possible sources. On the basis of the known flux of solar X-rays we had estimated a flux from the Moon of about 0.1 to 1 photon cm*$^2$ *s*$^{-1}$ *in the region of the sensitivity of the counter. The rocket launching took place at the White Sand Missile Range, New Mexico (USA), at 23:59 MST on June 18, 1962. The Moon was one day past full and was in the sky about* 20° *east of south and* 35° *above the horizon. The*

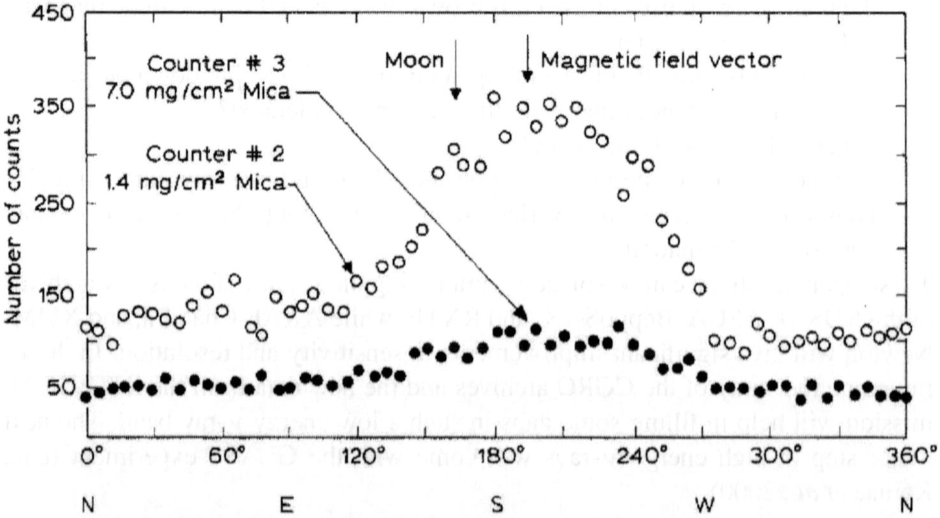


*Figure 94.* Number of counts versus azimuth angle. The numbers represent counts accumulated in 350 s in each 6° angular interval (Giacconi *et al.*, 1962).

*rocket reached a maximum altitude of 225 km and was above 80 km for a total of 350 s. The vehicle travelled almost due north for a distance of 120 km. Two of the Geiger counters functioned properly during the flight; the third counter apparently arced sporadically and was disregarded in the analysis. The optical aspect system functioned correctly. The rocket was spinning at 2.0 rps around the longitude axis. From the optical sensor data it is known that the spin axis of the rocket did not deviate from the vertical by more than 3°; for purposes of analysis, the spin axis is taken as pointing to zenith. The angle of rotation of the rocket corresponds with the azimuth Φ and is measured from the north as zero and increasing to the east. The data were reduced by using the optical aspect information to determine the azimuth as a function of time. Each complete rotation of the rocket was divided into sixty equal intervals, and the number of counts in each of these intervals was recorded separately.*

*The total data accumulated in this manner during the entire flight are shown in Figure 94 (Figure 1 in the original article) for the operating Geiger counters. The observed region of the sky is shown in Figure 95 (Figure 2 in the original article).....* .

After a thorough discussion of the impossibility of the radiation responsible for the asymmetry of the background being unable consist of charged particles, Giacconi *et al.* (1962) wrote: *Thus we conclude that the bulk of the observed radiation is not corpuscular, but electromagnetic in nature.* And after a discussion of the possible location of the source they wrote: *From Figure 2 (here Figure 95) we see that the main apparent source is in the vicinity of the galactic center at GT azimuthal angle of about* 195°. *We also see that the trace of the GT axis lies close to the galactic*

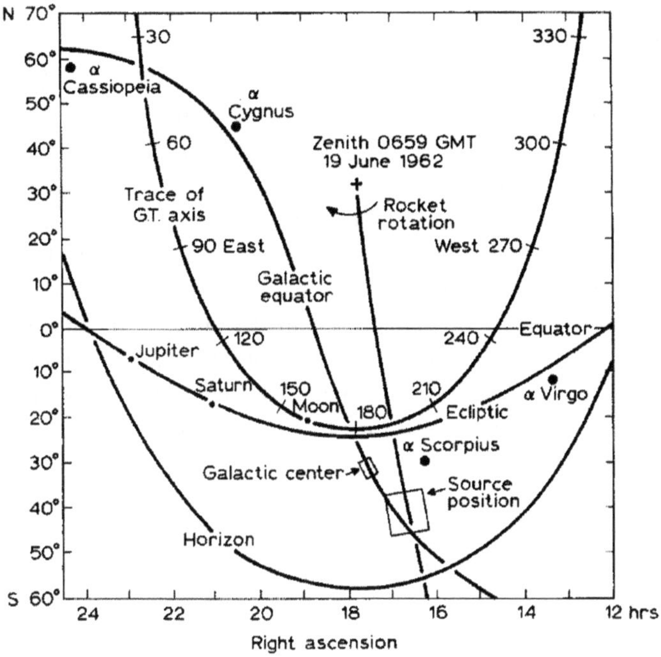


*Figure 95.* Chart showing the portion of the sky explored by the counters (Giacconi *et al.*, 1962).

*equator for a value of the azimuthal angle near* 40°, *which is the region where the background radiation is recorded with greater intensity. This apparent maximum of the background radiation is the general region of the sky where two peculiar objects, Cassiopea A and Cygnus A, are located. It is perhaps significant that both the center of the galaxy where the main apparent source of X-rays lies, and the region of Cassiopea A and Cygnus A in which there appears to be a secondary X-ray source, are also regions of strong radio emission. Clark (1962) has pointed out that the probable mechanism for the production of non-thermal component of the radio noise, namely, synchrotron radiation from cosmic electrons in the galactic magnetic fields, can also give rise to the X-ray we observe......... With this one experiment it is impossible to completely define the nature and origin of the radiation we have observed. Even though the statistical precision of the measurements is high, the numerical values for the derived quantities and angles are subject to large variation depending on the choice of assumptions. However, we believe that the data can best be explained by identifying the bulk of the radiation as soft X-rays from sources outside the solar system. Synchrotron radiation by cosmic electrons is a possible mechanism for the production of these X-rays. Ordinary stellar sources could also contribute a considerable fraction of the observed radiation.*

### 7.5.2. *The Challenge of Exploring the Sea of XRBs*

We would like to start this short section in which we want to provide, especially for the youngest readers, fundamental historical elements, which probably can create a sort of original atmosphere which the oldest of us lived through in those times. Such *old times* appear distant, even though only 40 years passed. This span of time is just the working life of most scientists, but it represents only an infinitesimal fraction of time in the scale of the Universe. It is important for one of us (FG) to remember the figure of Professor Bruno Rossi, who was his teacher during a few monographic courses he gave at the University of Rome (in those times the unique in Rome, and now named 'La Sapienza') in the sixty years during his frequent trips to his original country, and the first memorable Erice School in 1969, where he delivered over the updated knowledge on the new born X-ray astronomy, and together with him, amongst others, Professor William A. Fowler, who held the stage with his lofty nuclear astrophysics course.

Bruno Rossi (1973), in his introductory remarks of the IAU Symposium No. 55 on *X- and Gamma Ray Astronomy*, said, "*I am sure you all remember the climate of skepticism in which X-ray astronomy began. Which explains why the avowed purpose of the rocket program which opened up this new field was an attempt to detect an X-ray fluorescence from the Moon, rather than a search for extra-solar X-ray sources; although the hope of finding such sources was in fact the main motivation of the program. And you remember the surprise when the rocket did, in fact, detect a galactic X-ray source many orders of magnitude stronger than anything one had expected (Figure 94). In retrospect we see that the little confidence people had in the worth of X-ray astronomy was owed primarily to the failure to reckon with the possible existence of hitherto unknown celestial objects whose emission would lie almost entirely in the X-ray band of the spectrum; although it is also true that most astrophysicists had been too conservative in extrapolating toward the high energies the electromagnetic spectra of known peculiar objects, such as supernova remnants.*

*The comparatively large X-ray fluxes reaching the Earth from celestial sources explains why so many observations of crucial significance could be carried out even before the launching of the first X-ray satellite (UHURU, note of the authors), at a time when X-ray astronomers had at their disposal only rockets and balloons, with the well known limitations of these carriers. Just to mention some of the highlights, there was the identification, by the lunar occultation method, of the source in Taurus with the Crab Nebula, the identification of the source in Scorpio with a faint apparently insignificant star; the measurement of several X-ray spectra; the discovery of a variability of some X-ray sources; the discovery of an X-ray pulsar; the discovery of the first extragalactic sources. Nor must one forget the discovery of a diffuse, nearly isotropic X-ray background and the fairly accurate measurement of its spectrum. Rockets and balloons will continue to play an important role in the future. But the performance of UHURU has made it abundantly clear that the availability of satellites marks the opening of a new era for X-ray astronomy. And here again we find that scientists had been too conservative in their expectations. The main purpose of UHURU, as originally conceived, was to carry out a general X-ray survey of the sky, adding weaker sources to the existing catalogues and determining source positions with good accuracy. UHURU is successfully fulfilling this assignment. But, in so doing, it has discovered new and highly significant features of individual sources; and this not by virtue of any sophisticated instrumentation, but because of its ability of scanning the sky at a very low rate and of going back to any desired source over and over again, on command from the Earth.*

*Gamma-ray astronomy has developed at a much slower pace than X-ray astronomy, mainly because of the technical difficulties of detecting very weak $\gamma$-ray fluxes. But a promising beginning has been made in this field too.*"

One of the most impressive features of the binary X-ray sources, which has appeared evident since the first historical measurements performed with experiments aboard sounding rockets, consists in a sharp differentiation in the kind of X-ray spectrum they emit. In one class of sources the X-rays are emitted in pulses of great regularity, which recall many features of pulsars. In another class, although variability down to time scale of a few milliseconds is observed, no regular pulsations are present in the spectrum with the possible exception of train of pulses of radiation. The difference between these two classes of sources is better understood by a direct inspection of the Figure 96, coming from X-ray data of a rochet flight from Goddard Space Center (Rothschild *et al.*, 1974).

The flight lasted 330 s with ~70 s observations of the pulsating X-ray source Her X-1, ~40 s on Cyg X-3, and ~60 s on Cyg X-1. A direct comparison between the data of Her X-1, Cyg X-3 and Cyg X-1 clearly shows the marked differences in the X-ray signals in the three sources: those in Her X-1 are regularly pulsating with a period of ~1.2 s, those in Cyg X-3 are not regularly pulsating, and those in Cyg X-1 are bursting with time scales down to a few milliseconds.

Leach & Ruffini (1973) proposed a classification to identify all the pulsating binary X-ray sources with neutron stars, and bursting sources, with short time variability but no regular pulsations, with black holes. And they were essentially right. However, the absence of X-ray regular pulsation can be ascribed to the collapsed object being either a black hole or a neutron star deprived of an off-axis magnetic field of such an intensity as to modulate the X-ray emission of the accreting material. The identification with a neutron star is clearly impossible if the collapsed object is recognized to have a mass larger than the absolute upper limit of the neutron star critical mass (Rhoades & Ruffini, 1974).

Later such a classification was supported by experimental evidence. However, the crucial point of this classification consists in the clear possibility of determining from direct observations the value of the critical mass of a neutron star against gravitational collapse: the pulsating sources are expected to have masses up to this critical value, and all the bursting sources masses larger than this critical value.

For a deeper discussion see the paper *The Physics of Gravitational Collapsed Objects* by Remo Ruffini (1975) and references therein.

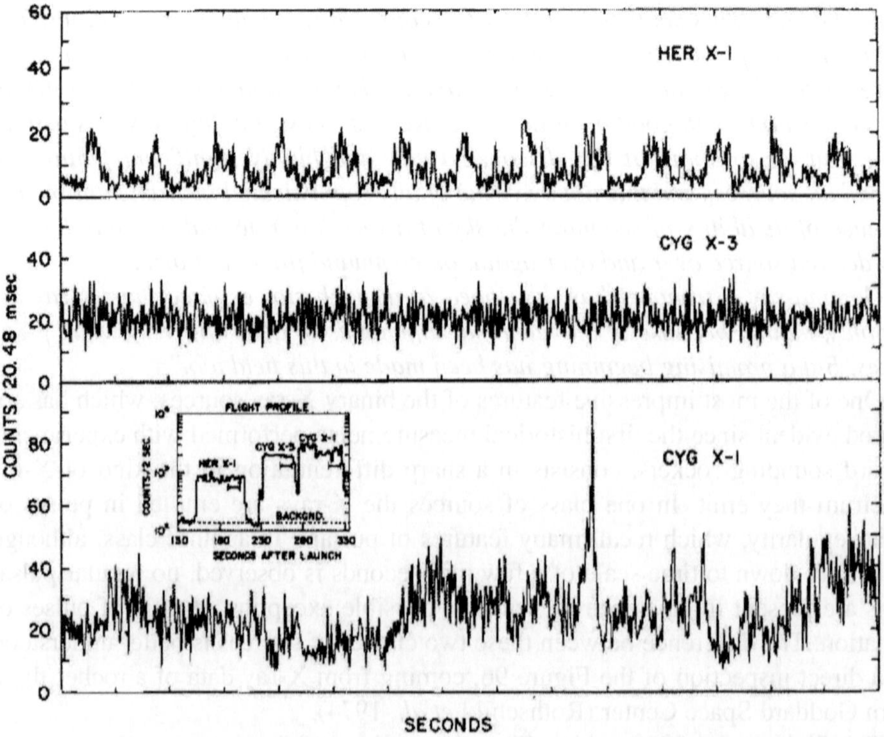


*Figure 96.* Rocket flight X-ray data of Her X-1, Cyg X-3 and Cyg X-1 shows the marked differences in the X-ray signals in the three sources: those in Her X-1 are regularly pulsating with a period of ~1.2 s, those in Cyg X-3 are not regularly pulsating, and those in Cyg X-1 are bursting with time scales down to a few millisecond (Rothschild *et al.*, 1974).

One of the first classifications of galactic X-ray sources recognized two classes: the first including the so-called Sco X-1-like sources and the second including the pulsating sources, showing evidence of a binary nature (e.g., Tananbaum, 1973). The Sco X-1-like sources show intensity changes of order 50% on time scale of minutes to hours. Their spectra are exponential suggesting thermal bremsstrahlung, and their temperatures vary on the same time scale as the intensity, ranging from 50 to 150 million degrees. Two of the objects of this class, Sco X-1 and Cyg X-2, have similar optical counterparts – blue stars with UV excess, variable intensity – flicker and flare, complex emission and absorption lines, and overall electromagnetic emission dominated by the X-rays. Such sources have also radio counterparts. The pulsating X-ray sources are dominated by intensity changes of factors of 2 or more on time scales of seconds or less. The spectra tend to be flat and are often cut off at low energies. There is considerable evidence linking a number of these sources to binary systems.

Probably the most significant of the UHURU results for the galactic X-ray sources has been the discovery of pulsations from Cyg X-1 (Schreier *et al.*, 1971),

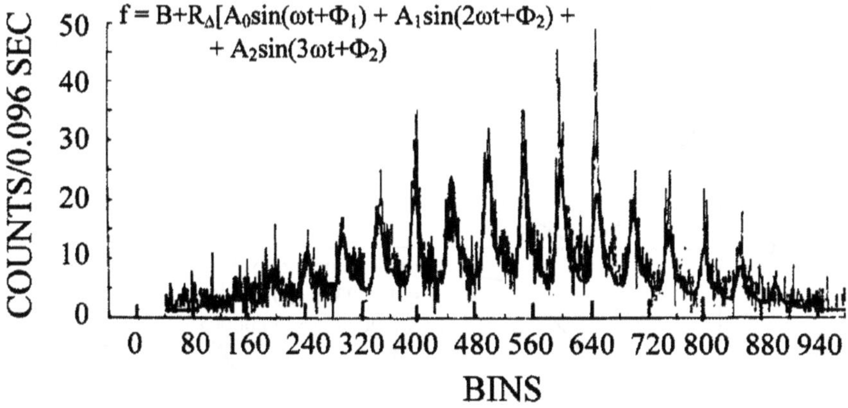


*Figure 97.* Counts accumulated in 0.096 s bins from Cen X-3 during a 100 s pass on May 7, 1971, modulated by the triangular response of the collimator. The functional fit obtained by minimizing $\chi^2$ is shown as the heavier curve (Schreier *et al.*, 1972a).

which since those times has been the best galactic black hole candidate. They detected substantial variations in X-ray intensity on time scales from 100 ms to 10s of seconds. Similar X-ray variability was also reported by scientists at MIT (Rappaport, Doxsey & Zaumen, 1971), Goddard Space Flight Center (Holt *et al.*, 1971), and Naval Research Laboratory (Schulman *et al.*, 1971). These concurrent facts allowed the consideration of compact objects smaller than $10^9$ cm for X-ray sources (Tananbaum *et al.*, 1972). With the relatively good X-ray location determined by an MIT rocket flight (Rappaport, Zaumen & Doxsey, 1971) and by UHURU, a radio source was discovered by Braes & Miley (1971) and by Hjellming & Wade (1971a,b). It was the precise radio location that led to the identification of the optical counterpart of Cyg X-1 by Webster & Murdin (1972) and by Bolton (1972), with a 9th magnitude B0 supergiant star HD226868, which forms, together with Cyg X-1, a 5.6 d spectroscopic binary system. Conservative mass estimates such as 12 $M_\odot$ led to a mass in excess of 3 $M_\odot$ for the unseen companion. If such a companion is the compact X-ray source then it could be a neutron star more massive than previously considered or it could be a black hole.

Two other historical X-ray sources are Cen X-3 and Her X-1, which were identified as binaries solely from their X-ray properties. Figure 97 shows the counts accumulated in 0.096 s bins during a 100 s pass on May 7, 1971 (Schreier *et al.*, 1972a; and 1972b for errata).

Cen X-3 was found to be an occulting binary system with the period of the observed transitions being 2.087112 ± 0.00004 d. By measuring the arrival time of individual 4.8 s pulses, shown in Figure 97, they determined that the pulsation frequency is Doppler shifted in phase with the 2.087 d occultation cycle. Such an occultation cycle is shown in Figure 98.

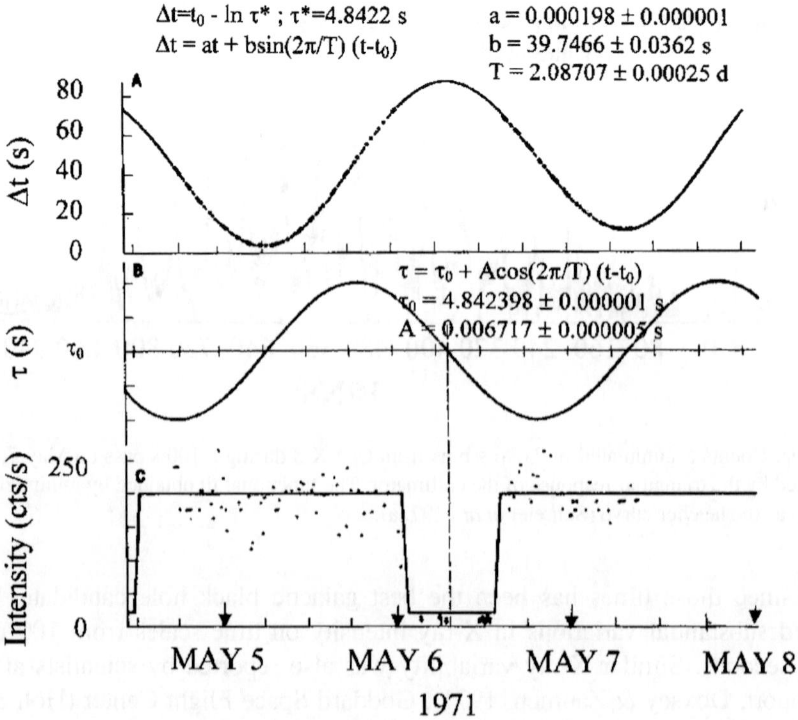


*Figure 98.* (Upper panel): The difference $\Delta t$ between the time of occurrence of a pulse and the time predicted for a constant period is plotted vs. time. A best fit function and the values of the parameters are given. (Central panel): The dependence of the pulsation period $\tau$ on time as derived from the best-fit phase function. Note the coincidence of the null points of the period function with the centers of the high and low intensity states. (Lower panel): The intensity observed from Cen X-3 on May 5–7, 1971 with the UHURU satellite (dots) and the light curve predictions (Schreier *et al.*, 1972a).

With a model of an occulting binary system, they determined the projected orbital velocity 415.1±0.4 km s$^{-1}$, the projected orbital radius $(1.191 \pm 0.001) \times 10^{12}$ cm, and the mass function of the system $(3.074 \pm 0.008) \times 10^{34}$ g. Figure 99 schematically shows this system with a compact object orbiting around a central optical star.

From the Doppler velocity the lower limit of the optical star mass is $M_{\rm opt} \geq 15$ M$_\odot$ and, following calculations of Wilson (1972), the mass of the compact star $M_{\rm cs} \approx 10$ M$_\odot$. Krzeminski (1974a,b) made a plausible estimate of the distance, 8 kpc of the optical counterpart of Cen X-3: a B0II star with $V_{\rm average} = 13.35$, $B - V = +1.07$, and $U_{\rm B} = -0.04$, later nicknamed Krzeminski's star. Detailed analysis of the amplitude of the optical light curve, using the X-ray measured mass function and a range of observed eclipse durations, provided the following results for Cen X-3/Krzeminski's star: $0.6 \leq M_{\rm X} \leq 1.8$ M$_\odot$, $16.5 \leq M_{\rm opt} \leq 20$ M$_\odot$, and $15 \leq V_{\rm opt} \sin i \leq 40$ km s$^{-1}$ (Avni & Bahcall, 1974; 1975).

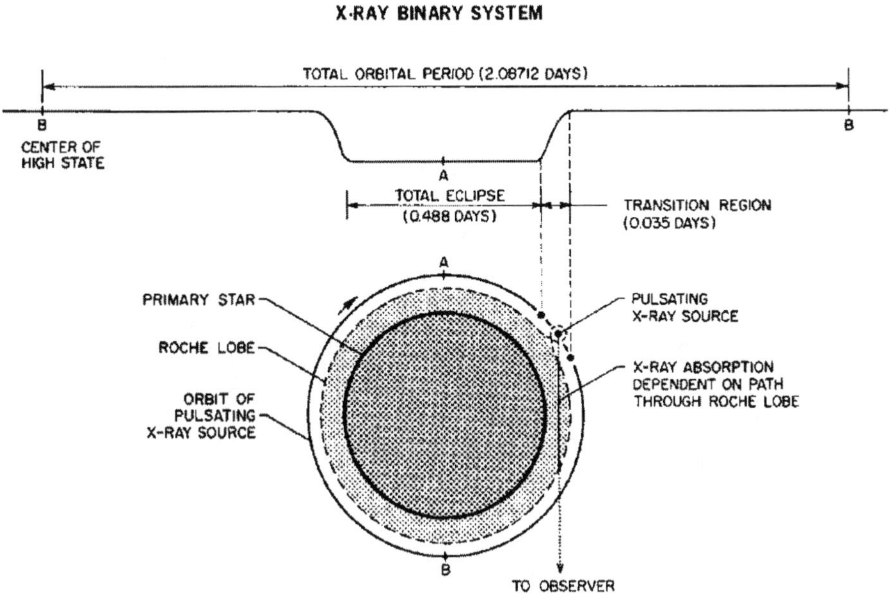


*Figure 99.* Sketch of an occulting X-ray binary system (Tananbaum, 1973).

Very detailed reviews of these topics were published by Giacconi (1978) on the observations of compact sources and by Bahcall (1978b) on the optical properties of binary X-ray sources. And in general the book on *Physics and Astrophysics of Neutron Star and Black Holes* (Giacconi & Ruffini, 1978), which reports the papers discussed during the Course LXV of the International School of Physics 'Enrico Fermi' in Varenna on July 14–26, 1975, is an exhaustive, now historical, book, probably the best compendium ever published in such a topic.

Figure 100 shows the pulsating source Her X-1, which pulses with a 1.24 s period and it is essentially totally pulsed, like Cen X-3. The heavier lines represent all data taken from its discovery in November 1971 through March 1972 (Tananbaum *et al.*, 1972). Also this source shows an occultation cycle with a period of $1.70017 \pm 0.00004$ d, which is the orbital period, as shown in Figure 101. The data show a further ~35.7 d on off intensity cycles.

The results of the phase analysis, reported in Figure 101, show that the arrival time of the pulses differs from the time expected under the hypothesis of a constant period by an amount which can be represented by a sinusoidal function of time. The period of this function was found to coincide with the period of the observed intensity variations (also shown in the same figure) and the phases of maxima and minima were also in coincidence. Therefore this period variation was interpreted as the Doppler shift in frequency owed to the orbital motion of a star around an occulting companion. It is interesting to remark that the projected diameter of the orbit can directly be measured in *light seconds* from the delay in arrival times between maxima and minima: 26.4 s (Tananbaum *et al.*, 1972).

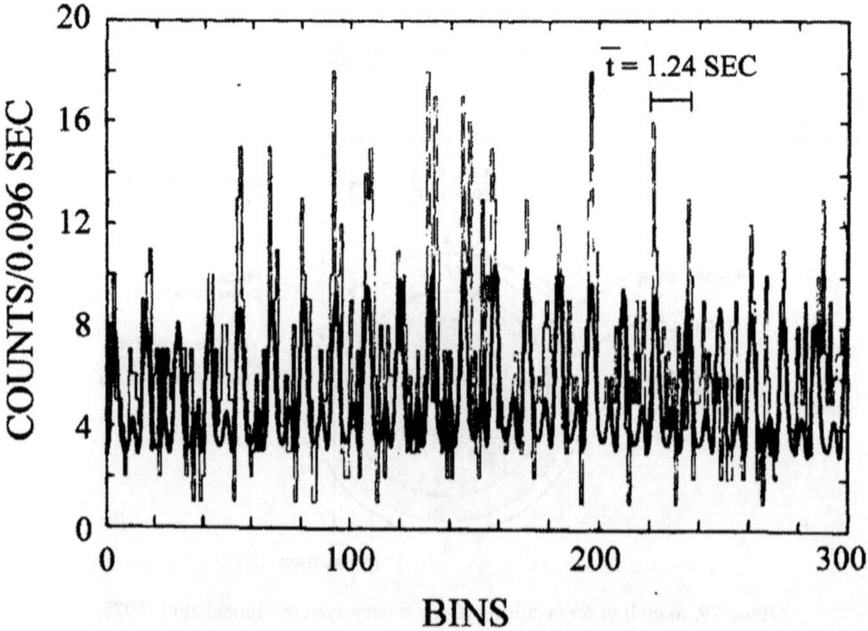


*Figure 100.* Counts accumulated in 0.096 s bins from Her X-1 during the central 30 s of a 100 s pass on Nov 6, 1971. The heavier curve is a minimum $\chi^2$ fit to the pulsation of a sine function, its first and second harmonics plus a constant, modulated by the triangular response of the collimator (Tananbaum *et al.*, 1972).

Middleditch & Nelson (1976) described in detail the 35 day structure of Her X-1/HZ Her system, which is rather complex. From their optical observations of HZ Her, they revealed weak intermittent pulsations at frequencies near to, but in general not identical to, the expected Her X-1 frequency. These pulsations are correlated with both the binary and 35 day phases. They constructed a simple geometrical model of the binary system and calculated the expected amplitude and frequency of pulsation as a function of binary phase, for any assumed equipotential surface of HZ Her, mass ratio and orbital inclination. By comparing these calculations with the data they concluded that the surface of HZ Her is pointed towards Her X-1, HZ Her fills its Roche lobe (and thus must be co-rotating with the orbital motion), Her X-1' spin is pro-grade (the same sense of the orbital motion), the mass ratio is $M_{\rm HZHer}/M_{\rm HerX-1} = 1.68 \pm 0.10$, $M_{\rm HZHer} = (2.18 \pm 0.11)\ \mathrm{M}_{\odot}$, $M_{\rm HerX-1} = (1.30 \pm 0.14)\ \mathrm{M}_{\odot}$, $i = 87^{\circ} \pm 3^{\circ}$.

Their model was unable to predict any of the observed 35 day behaviour. Only later, with the increase of data coming from different experiments (e.g., from ASTRON X-ray Soviet satellite: Giovannelli *et al.*, 1989; Sheffer *et al.*, 1992) was it possible to affirm that pulse profile variations with the 35 day cycle occur because of a gradual occultation of the X-ray generating region by the precessing accretion

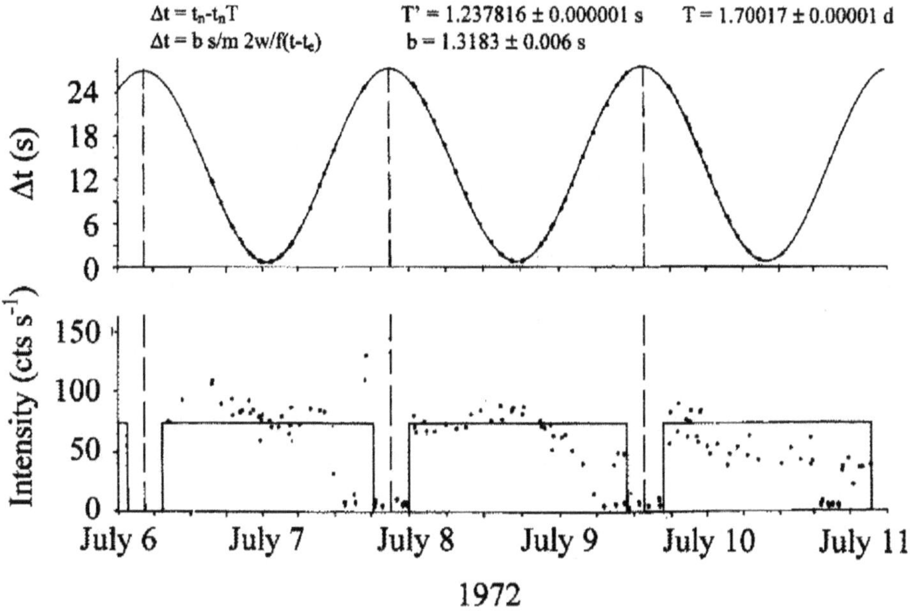


*Figure 101.* Five days of X-ray observations of Her X-1 in July 1972. The 2–6 keV intensity observations show three occultation cycles. The data give individual sighting intensities with only upper limits observed during the eclipses. The eclipse centers are given by the vertical dashed line as determined from the Doppler variation of the 1.2 s pulse period shown in the upper panel, and the rectangular envelope is a schematic representation of the 1.7 d cycle of highs, lows, and transitions (Tananbaum *et al.*, 1972).

disk. The disk precession should be in the direction of the neutron star's rotation and the orbital motion.

In October 1997, on the occasion of the sixty fifth birthday of Riccardo Giacconi, the University of Rome 'La Sapienza' awarded him with the *Honorary Degree in Physics*, and concurrently a close number of his past and present collaborators gathered in Rome where a meeting on *Exploring the Universe: A Festschrift in Honour of Riccardo Giacconi* was held in his honor. The book from such a meeting (Gursky, Ruffini & Stella, 2000) covers the history of X-ray Astronomy, which in its first steps we have briefly described in this Section.

## 7.6. High Mass X-ray Binary Systems

HMXBs are young systems, with age $\leq 10^7$ yr, mainly located in the galactic plane (e.g., van Paradijs, 1998). A compact object – the secondary star –, mostly a magnetized neutron star (X-ray pulsar) is orbiting around an early type star (O, B, Be) – the primary – with $M \geq 10\ M_\odot$. The optical luminosity of the system is dominated by the early type star. About 10% of the X-ray flux emitted in the vicinity of the compact star is intercepted by the primary star and reprocessed into optical radiation (e.g., Bradt & McClintock, 1983). As a function of the nature of the secondary star, HMXBs are usually divided into two subclasses, namely Be/X-ray binaries and O-B/X-ray binaries if the optical stars are Be or O-B supergiant stars, respectively (e.g., van Paradijs, 1983). Most of the HMXBs are X-ray/Be systems (see the catalog of Liu, van Paradijs & van den Heuvel, 2000), which contains 130 systems with orbital periods ranging from 4.8 hr to 187 d.

#### 7.6.1. *HMXBs: A Cauldron of Physical Processes*

In 2000 the known X-ray pulsars were 95 (Giovannelli & Sabau-Graziati, 2000a) and the orbital periods were known only for about three dozens of them. This number is subject to slowly increase, thanks to new discoveries made by the on flight satellites.

HMXBs contains the group of the permanent sources and that of transient sources (X-ray/Be systems), whose components are an X-ray pulsing neutron star – the secondary – and a giant or supergiant OB or a Be star – the primary –, respectively. Moreover, some low mass X-ray Binaries (LMXBs) containing an X-ray pulsar and some pulsars belonging to Magellanic Clouds are contained too in the sample of 95 systems. The X-ray/Be binaries are the most abundant group of massive X-ray binaries in the galaxy, with a total inferred number of between $10^3$ and $10^4$. The ones which do occasionally flare up as transient X-ray/Be systems are only the 'tip' of this large 'iceberg' of systems (van den Heuvel & Rappaport, 1987). The mass loss processes are owed to the rapid rotation of the Be star, the stellar wind and, sporadically, to the expulsion of casual quantity of matter essentially triggered by gravitational effects close to the periastron passage of the neutron star. The long orbital period ($>$10 d) and a large eccentricity of the orbit ($>$0.2) together with transient hard X-ray behavior are the main characteristics of these systems. Amongst the whole sample of systems containing 95 X-ray pulsars, only a few of them have been extensively studied. Among these, the system A 0535+26/HDE 245770 is the best known thanks to concomitant favorable causes, which rendered possible twenty six years – since its discovery as X-ray source by the Ariel-V satellite (Rosenberg *et al.*, 1975) – of coordinated multifrequency observations (e.g., Giovannelli & Sabau-Graziati, 1992; Burger *et al.*, 1996; Piccioni *et al.*, 2000). Accretion powered X-ray pulsars usually capture material from the optical companion via stellar wind, since this primary star generally does not fill its Roche lobe. However, in some specific conditions (e.g., the passage at the periastron of the neutron star) and in particular systems (e.g., A 0535+26/HDE 245770), it is possible the formation of a temporary accretion disk around the neutron star behind the shock front of the stellar wind. This enhances the efficiency of the process of mass transfer from the primary star onto the secondary collapsed star, as discussed by Giovannelli & Ziółkowski (1990) in the case of A 0535+26. An interesting sketch of the possible accretion processes is reported in Figure 102 where also the time scales of the different processes occurring in the binary system are indicated (Giovannelli & Sabau-Graziati, 2001).

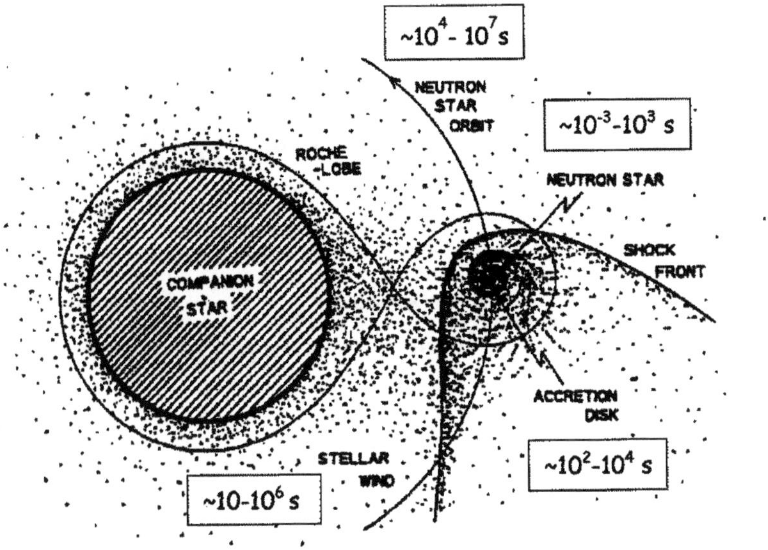


*Figure 102.* Sketch of mass accretion onto a neutron star in wind-fed pulsar. A temporary accretion disk can be formed near the neutron star behind the shock front of the stellar wind in some specific conditions (e.g., the passage at the periastron of the neutron star) and in particular systems (e.g., A0535+26/HDE245770). The timescales of the physical processes occurring in different parts of the HMXBs are indicated (Giovannelli & Sabau-Graziati, 2001, after Nagase, 1989).

The range $\sim 10^{-3}$–$10^3$ s for the rotation of neutron star; $\sim 10^2$–$10^4$ s for the accretion disk; $\sim 10$–$10^6$ s for the stellar wind; $\sim 10^4$–$10^7$ s for the orbital period. In order to study processes occurring in this range of time scales ($\sim 10^{-3}$–$\sim 10^7$ s) one requires experiments with high time resolution ($\ll 10^{-3}$ s) and with the possibilities of long observing time $\geq 10^7$ s), such as the possibilities of SIXE (e.g., Giovannelli *et al.*, 1993b; Giovannelli *et al.*, 1999b, 2001; Isern *et al.*, 2001).

Optical emission of HMXBs is dominated by that of the optical primary component, which is not, in general, strongly influenced by the presence of the X-ray source. The behavior of the primary stars can be understood in the classical (or almost) framework of the astrophysics of these objects, i.e. by the study of their spectra which will provide indications on mass, radius, and luminosity. Both groups of HMXBs differ because of the different origin of the mass loss process: in the first, the mass loss process occurs via a strong stellar wind and/or because of an incipient Roche lobe overflow; in the second group, the mass transfer is probably partially owed to the rapid rotation of the primary star and partially to stellar wind and sporadically to expulsions of a casual quantity of matter, essentially triggered by gravitational effects because of periastron passage where the effect of the secondary collapsed star is more marked.

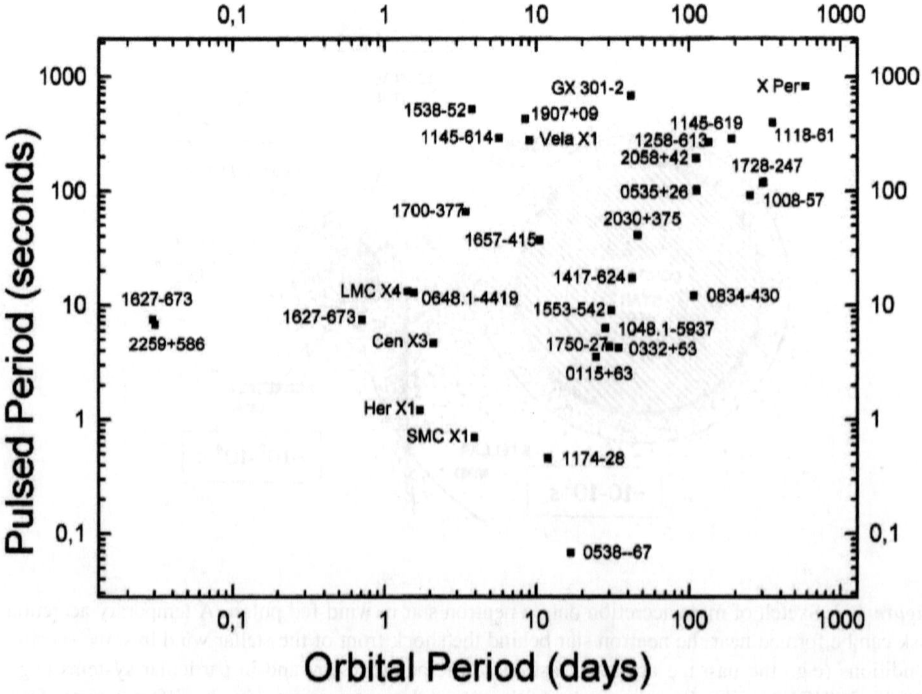


*Figure 103.* The dependence of the spin period versus the orbital period of X-ray pulsars (Giovannelli & Sabau-Graziati, 2001, after Corbet, 1984, 1986).

A relationship between orbital period of HMXBs and the spin period of the X-ray pulsars is shown in Figure 103, (Giovannelli & Sabau-Graziati, 2001, updated from Corbet, 1984, 1986). It allows to recognize three kinds of systems, namely disk-fed, wind-fed ($P_{\rm pulse} \propto P_{\rm orb}^{4/7}$), and X-ray/Be systems ($P_{\rm pulse} \propto P_{\rm orb}^{2}$).

Most of the systems having a Be primary star are hard X-ray ($KT > 10$ keV) transient sources (HXTS). They are concentrated on the galactic plane within a band of ~3.9°. The orbits are quite elliptic and the orbital periods large (i.e. A 0538-66: $e = 0.7$, $P_{\rm orb} = 16.6$ d (Skinner *et al.*, 1982); A 0535+26: $e = 0.5$ (Finger *et al.*, 1994), $P = 111.0$ d (Priedhorsky & Terrell, 1983). The X-ray flux during outburst phases is of order 10–1000 times greater than during quiescent phases. The stars belonging to the first class which do not present such strong variations in X-ray emission, in contrast, can be named 'standard' high mass X-ray binaries. In X-ray/Be systems, the primary Be star is relatively not evolved and is contained within its Roche lobe. The strong outbursts occur almost periodically in time scales of the order of weeks–months. Their duration is shorter than the quiescent phases. During X-ray outbursts, spin up phenomena in several systems have been observed (i.e. A 0535+26 and 4U 1118-61 (Rappaport & Joss, 1981)). The observed spin up rates during the outbursts are consistent with torsional accretion owed to an accretion disk (e.g., Ghosh, 1994). So the formation of a temporary accretion disk around the collapsed object should be possible during outburst phases (e.g., Giovannelli & Ziółkowski, 1990).

The most exciting historical results from one of the HXTS are those coming from A 0535+26 (Giovannelli *et al.*, 1985a; Guarnieri *et al.*, 1985). They found UV and optical features of HDE 245770 – the optical companion, nicknamed *Flavia's star* (Giovannelli & Sabau-Graziati, 1992) – as indicators of X-ray activity of A 0535+26 – the collapsed companion. They found experimentally that the X-ray flaring activity of A 0535+26 is preceded by modifications in the optical spectrum of the companion HDE 245770, of about one week ($\approx 0.6 \times 10^6$ s). This roughly corresponds to the transit time of puffs of material expelled by the Be star at $\sim$300 km s$^{-1}$ in order to reach the neutron star, being the dimensions of the orbit of $\sim$1.34 AU (de Loore *et al.*, 1984). The $H_\gamma$ line seems to be the best indicator of the activity going on. Moreover, enhancements of optical luminosity were observed in four occasions before X-ray outbursts (Bartolini *et al.*, 1983; Maslennikov, 1986). Narrow absorption components present in the blue wings of SiIV and CIV resonance lines indicate a variable mass loss superimposed on a steady wind of $\sim 10^{-8}$ $M_\odot$ yr$^{-1}$ (Giovannelli *et al.*, 1984a; de Loore *et al.*, 1984). They are strong probes in studying the physics and dynamics of the mass transfer at periastron and the subsequent X-ray flaring.

X-ray outbursts are triggered by the eccentric orbital motion ($P_{\rm orb}$ = 111 d: $\approx 10^7$ s), such as is demonstrated for A 0535+26. A multifrequency monitoring of a number of other X-ray/Be systems around the passage at the periastron could then give the conclusive answer to many of the open problems on the Be/X-ray transients. At optical wavelengths, a monitor of these systems in the optical and X-ray energy ranges, during an orbital period, could give the definitive proof of the existence of the optical indicators of the following X-ray flaring activity. Within the framework of such a program in the optical range, the system A 1118-615 has shown interesting behavior in relation with the X-ray flaring activity. $H_\alpha$ equivalent width is strictly related to the X-ray activity (e.g., Coe *et al.*, 1994); short and long term variability in $H_\alpha$ equivalent width has been also found by Villada *et al.* (1992; 1999), and by Polcaro, Villada & Giovannelli (1993) in correlation with the hard X-ray emission of the system. The significant variation of $H_\alpha/H_\beta$ flux ratio shows a general decrease of the ionization level after the large 1992 X-ray outburst. The derived scenario is that of a large extended envelope surrounding Hen 3-640 (the optical Be companion), expanding because of processes intrinsic to the optical star. When the density around the compact object reaches a value large enough to generate a temporary steady accretion disk around the A1118-615 pulsar, the mass transfer efficiency and thus the X-ray flux sharply increase. In this scenario any exact periodicity in X-ray outburst is predicted, as is experimentally proved. It is possible to consider this system as the prototype of a new sub-class of 'atypical' X-ray/Be systems in which X-ray outbursts are not triggered by periastron passage mechanism like for instance in the so-called 'typical' X-ray/Be systems

A 0535+26-like. By using the Corbet (1986) $P_{\mathrm{spin}}$–$P_{\mathrm{orb}}$ diagram and that between the orbital period and the $H_\alpha$ equivalent width (Reig, Fabregat & Coe, 1997), but using its average value (~70 Å) and not the maximum as suggested by them (~107 Å), one deduces a long orbital period of ~350 d for A1118-61. The experimental discovery of atypical and typical X-ray/Be systems is very important from the point of view of their origin. Indeed, such systems can be formed after an explosion of a supernova in a medium mass (10–20 $M_\odot$) binary-system progenitor as demonstrated by Giovannelli *et al.*, 1993a,c; 1994) and confirmed with the analysis of a larger sample of HMXBs by Giovannelli & Sabau-Graziati (2000a,b). The atypical and typical X-ray/Be systems would be the remnants of symmetric supernova explosions (low eccentricity: $e \leq 0.3$) and asymmetric supernova explosions (high eccentricity: $0.3 \leq e \leq 0.8$), respectively.

Other optical indicators of X-ray emission from a binary compact companion, accreting matter from the Be star gas envelope, are the HeI lines at 5876 Å and 10830 Å (Apparao & Tarafdar, 1994). They found these lines in emission from several Be stars. The observed flux of the HeI (5876 Å) is compatible with that of an HII region formed in the envelope of a Be star (earlier than B1) owed to the EUV or X-ray emission from a compact binary companion. However, in their small sample of Be stars, the expected X-ray emission associated is rather low. These stars cannot then be observed in surveys, but for longer times (e.g., in pointed mode) in order to check the validity of their reasoning. Alternatively, to check the presence of such HeI emission lines, measurements in stronger X-ray emitters, where their binary nature is definitively certain, and then the presence of the neutron star, can be performed. The correlation between HeI lines and Balmer lines, especially $H_\alpha$, simultaneously obtained could be another powerful low energy indicator of high energy processes. X-ray/Be systems usually show an IR excess owed to thermal free–free emission from an ionized gaseous envelope around the Be star (e.g., Persi *et al.*, 1979; Giovannelli *et al.*, 1984a; de Loore *et al.*, 1984; Gnedin *et al.* (1988). Part of this excess could result from interaction between the stream of accreting matter and the neutron star plasma, with the plasma waves being excited and subsequently transformed into radiation at ~1% efficiency (Gnedin *et al.*, 1981, Gnedin, Khozov & Larionov, 1981). In the particular case of A 0535+26, Gnedin *et al.* (1983) found IR variations with periodicity close to the X-ray period (104 s), which supports the hypothesis that a part of the IR radiation could be generated near the accreting neutron star via maser nature. The IR emission is then also a powerful low energy indicator of high energy processes.

What we can say about the inverse problem?

We know that neutron stars can behave either as the so-called group of radio pulsars or as the so-called group of accreting neutron stars. All neutron stars rotate, but the main mechanisms of rotation are different: the rotation is determined by rotational energy loss in the case of young and old radio pulsars, whilst the rotation is determined by accretion in the recycled radio pulsars and in the accreting neutron stars. The equilibrium period is determined by the magnetic field intensity

TABLE 13
Classification of the rotating neutron stars (Ziółkowski (2002).

| Class | $P_{spin}$ |
|---|---|
| RADIO PULSARS | |
| Newborn | ~10–100 ms |
| Recycled | 1.6–536 ms |
| Old ($10^{6-7}$ yr) | ~1,000–8,400 ms |
| WEAKLY MAGNETIZED | |
| ACCRETING NEUTRON STARS | 2.5–3.8 ms |
| STRONGLY MAGNETIZED | |
| ACCRETING NEUTRON STARS | 0.061–1,413 s |
| MAGNETARS | |
| Soft $\gamma$-ray repeaters | 5.16–8 s |
| Anomalous X-ray pulsars | 5.45–11.77 s |

and by the mass accretion rate, as: $P_{\rm eq} \approx 5 \cdot B_{12}^{6/7} \cdot \dot{M}^{-3/7}$ s. If the accretion rate is at the Eddington limit, the equilibrium period is a minimum and expressed by $P_{\rm eq} \approx 0.7 \cdot B_{12}^{6/7}$ s. Therefore, roughly speaking for equilibrium periods of $\sim$1 s and $\sim$1 ms, the magnetic fields are $B \sim 10^{12}$ G, and $B \sim 10^{9}$ G, respectively. The class of radio pulsars contains the youngest pulsars, whose spin periods (33.4–267.4 ms) and derivative spin periods have been measured. Their associated log $B$ range from 11.96 G to 12.88 G. The radio pulsars also contain the so-called recycled pulsars. These can be sub-divided in three classes depending on their spin periods and log $B$. The ranges of values of spin periods and log $B$ characterizing the three classes are: i) 1.6–10 ms and log $B$ = 7.88–9.64 G for 52 objects, ii) 10–100 ms and log $B$ = 8.66–10.37 G for 16 objects, iii) 100–536 ms and log $B$ = 10.07–10.83 G for 12 objects (e.g., Ziółkowski, 1997b, 1999). In the class of the 95 accreting pulsars, whose spin periods range from 0.061 to 1413 s, the log $B$ range from 11 to 12 G. However, a new class of super-strongly magnetized neutron stars ($B \sim 10^{14}$–$10^{15}$ G), namely magnetars, can be considered. The list of candidates contains five soft gamma repeaters (presumably younger magnetars) and eight anomalous X-ray pulsars (older magnetars). Most magnetars are associated with young supernova remnants and all of them rapidly spin down (on time scale of $10^3$–$10^5$ yr) (e.g., Ziółkowski, 2002).

The possibility of classifying the rotating neutron stars in four classes, as shown in Table 13 is then evident.

The values of magnetic field strengths of $10^{11}$–$10^{13}$ G in X-ray pulsars are inferred from the pulsations. They require anisotropic in-falling radiation, the cyclotron lines observed in a number of pulsar spectra (Nagase, 1989), and the observed changes in pulse period (Ghosh & Lamb, 1979; Joss & Rappaport, 1984; Prince *et al.*, 1994; Finger & Prince, 1997). The strengths of the magnetic field are good probes for the physics of the accretion flow and for the kinetic energy of the in-falling matter converted into radiation. A good estimate of the temperature observed in X-ray pulsars ($T_{\rm eff} \sim 10$ keV) comes from the formula $T_{\rm eff} \sim (L_{\rm x}/\sigma A_{\rm cap})^{1/4}$ keV, where $L_{\rm x}$ is the X-ray luminosity, $\sigma$ is the magnetic Thompson scattering cross section, and $A_{\rm cap}$ is the heated polar cap area. However, the pulsar spectra are not of blackbody type; then, in order to understand the emission, an accurate description of the radiating plasma is necessary. Experimentally, the magnetic fields of X-ray pulsars can be determined through the cyclotron lines which can appear in the X-ray spectra. Indeed, in X-ray pulsars, for $B_{\rm n} \ll B_{\rm crit} \equiv m^2c^3/e\hbar = 4.413 \cdot 10^{13}$ G, the energy spacing between Landau states reduces to the cyclotron energy $\Delta E \simeq \hbar\omega_{\rm B} = 12$ keV $\cdot(B/10^{12})$ G, where $\omega_{\rm B} = eB/m_{\rm e}$ is the electron cyclotron frequency and assuming the gravitational redshift equal to $\sim$1. Since $KT \leq \hbar\omega_{\rm B}$ in the atmosphere of X-ray pulsars, electrons will mainly occupy the ground Landau state ($n = 0$) in a one-dimensional distribution. Moreover, the rate of collisional excitation to higher Landau states is much less than the cyclotron radiation rate from excited states; the population of the levels may then be far from thermal equilibrium, and will be dominated by the radiation field (e.g., Harding, 1994). Cyclotron line emission regulates the cooling of the atmosphere: i) for $B > 10^{12}$ G, the electron temperature can be maintained at few times $\hbar\omega_{\rm B}$; ii) for $B < 10^{12}$ G, the ion temperature is high enough to excite the electron Landau levels via collisions; the electron temperature becomes enough high ($KT_{\rm e} \approx 10^9$ K) to produce $\gamma$-rays. The spectrum from the atmosphere is a Doppler broadened cyclotron line, with practically no continuum contribution from bremsstrahlung.

Thus the detection of absorption line features due to cyclotron resonance scattering (CRSF) in the X-ray spectrum is a definitive method for estimating the strength of magnetic field at the neutron star surface. The first measurements of cyclotron lines from binary systems were obtained by Trümper *et al.* (1978) from Her X-1 and by Wheaton *et al.* (1979) from 4U 0115+63. From GINGA observations, the CRSFs were discovered in the spectra of several X-ray binaries: Her X-1, 4U 0115+63, 4U 1538-52, X0331+53, Cep X-4, 1E 2259+586, Vela X-1, 4U 1907+09, GX 301-2 (Nagase, 1994 and the references therein). The cyclotron resonance energies measured from these nine X-ray pulsars are in the range 7-40 keV, corresponding to a magnetic field strength of $(0.6\text{–}3.5) \cdot 10^{12}$ G at the neutron star surface, taking the gravitational redshift to be $\sim$1. The luminosities of these nine pulsars were less than $4 \times 10^{37}$ erg s$^{-1}$ during the observations of the CRSFs. In more luminous X-ray pulsars, such as Cen X-3 and SMC X-1, such a feature was not observed. A correlation between the fundamental resonance energy ($E_{\rm B}$) and the cutoff energy ($E_{\rm c}$) of the power spectrum has been found: $E_{\rm B} \simeq 2E_{\rm c}$. The absorption line features are generally broader than the energy resolution of the detectors (proportional counters). In the case of Vela X-1 the feature is pulse-phase

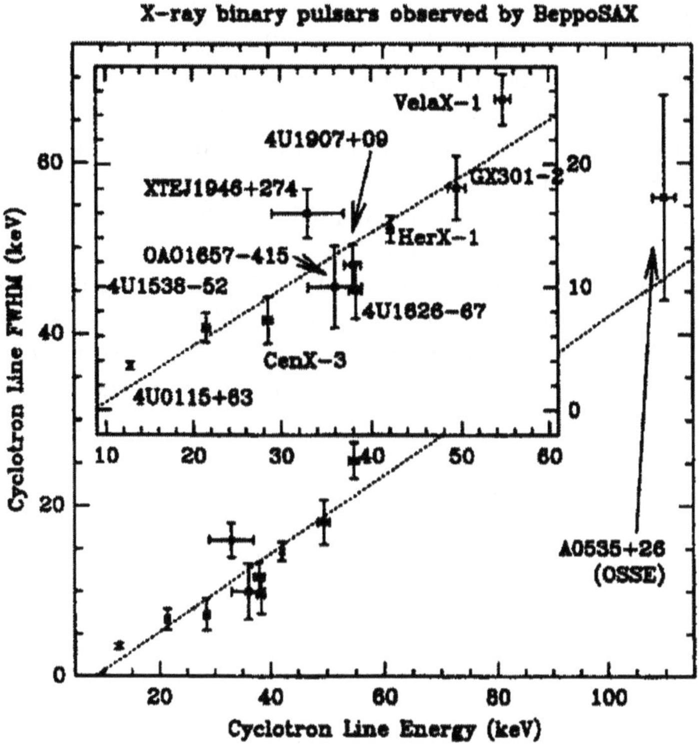


*Figure 104.* Cyclotron features FWHM versus centroid energy for several X-ray pulsars (Orlandini & dal Fiume, 2001).

dependent. The correlation $E_{\rm B} \simeq 2E_{\rm c}$ suggests that the cutoff of the spectrum at high energies relates to the cyclotron resonance scattering. Therefore, the cutoff energy could be a good indicator of the cyclotron energy and then of the magnetic field intensity for those pulsars from which the CRSF is not detectable. The scatter in the magnetic field intensity for the known X-ray pulsars is rather small; indeed it is $B = (1–4) \times 10^{12}$ G. This is an important track for the investigation on magnetic field evolution of neutron stars. Cyclotron lines have been measured by BeppoSax satellite from several X-ray pulsars, namely Cen X-3, 4U1626-67, Her X-1, Vela X-1, and A0535+26 (dal Fiume *et al.*, 1998). Later cyclotron lines have been found also in other X-ray pulsars (e.g., Orlandini & dal Fiume, 2001). A relationship between the cyclotron line energy and the FWHM of the line is present, as shown in Figure 104.

One of the most important results obtained with the BeppoSAX came from the observations of the pulsating transient X 0115+63, whose X-ray spectrum showed for the first time four cyclotron harmonic features at 12.74, 24.16, 35.74, and 49.5 keV (Santangelo *et al.*, 1999).

Recently, from the measurements obtained with the RXTE satellite, Coburn *et al.* (2002), looking at the data base, found all accreting X-ray pulsars showing cyclotron lines in their spectra. Then, they searched for correlations among the spectral parameters, concentrating on how the cyclotron line energy relates to the continuum and therefore how the neutron star magnetic field influences the X-ray emission. As expected, they found a correlation between the CRSF energy and the spectral cutoff energy. They found also not only that the width of the cyclotron line correlates with the energy, as suggested by theory, but also that the width scaled by the energy correlates with the depth of the feature, as already found by dal Fiume *et al.* (1998) and Orlandini & dal Fiume (2001) (see Figure 104). Coburn *et al.* (2002) discussed the implications of these results, including the possibility that accretion directly affects the relative alignment of the neutron star spin and dipole axes. Table 14 shows the pulsar magnetic fields, as they derived, assuming the radius and mass of the neutron star of 10 km and 1.4 $M_{\odot}$, respectively and scattering near the surface of the star.

### 7.6.2. *Quasi-Periodic Oscillations in Transient Pulsars*

The discovery of Quasi-periodic Oscillations (QPOs) in the power spectra of the flux from the 42 s X-ray binary pulsar EXO 2030+375 (Angelini, Stella & Parmar, 1989) offers a new tool for investigating the physics of the accretion in these systems. This discovery is similar to that previously found in LMXBs (see Section 7.7). Observations of the transient X-ray pulsars A 0535+26 and EXO 2030+375 have revealed QPOs at either the Keplerian orbital frequency at the magnetosphere radius, or its beat with the neutrons star spin frequency. Therefore QPOs can provide direct information about the poorly understood accretion disk–magnetosphere boundary region. Unlike most unpulsed sources, for many X-ray pulsars the basis system variables of spin period, orbital parameters, and surface magnetic fields have been determined. In addition, pulse timing allows measurements of accretion induced torques which can provide a measure of mass accretion rate unaffected by distance or spectral uncertainties. Therefore the study of QPOs in X-ray pulsars is a useful tool for testing either QPO models either providing new insights into accretion physics. Finger (1998) discussed such QPOs as probe for gaining new insight into X-ray pulsar systems.

The existence of nine systems (GRO J1744-28, SMC X-1, X 0115+63, V 0332+53, Cen X-3, X 1626-67, EXO 2030+375, A 0535+26, X Per) in which QPOs have been detected allowed to Ghosh (1998) to discuss their diagnostic potential in X-ray pulsars for probing:

- accretion flows;
- conditions in the inner accretion disks;
- properties of accreting torques exerted on the neutron stars;
- the nature of the coupling between stellar magnetic fields and accretion disks.

These QPOs have frequencies in the general range 10 mHz $\lesssim \nu_{QPO} \lesssim$ 400 mHz, considerably lower than those of both the low (∼6–50 Hz) frequency QPOs and the

TABLE 14
Pulsar Magnetic Fields (adapted from Coburn *et al.*, 2002).

| X-ray Pulsar | $E_c$ (KeV) | B ($10^{12}$ G) |
|---|---|---|
| Her X-1 | $40.4^{+0.8}_{-0.3}$ | 4.5 |
| 4U 0115+63 | $11.6^{+0.2}_{-0.4}$ | 1.3 |
| Cen X-3 | $30.4^{+0.3}_{-0.4}$ | 3.4 |
| 4U 1626-67 | $39.3^{+0.6}_{-1.1}$ | 4.4 |
| XTE J1946+274 | $34.9^{+1.9}_{-0.8}$ | 3.9 |
| Vela X-1 | $24.4^{+0.5}_{-1.1}$ | 2.7 |
| 4U 1907+09 | $18.3^{+0.4}_{-0.4}$ | 2.1 |
| 4U 1538-52 | $20.66^{+0.05}_{-0.06}$ | 2.3 |
| GX 301-2 | $42.4^{+3.8}_{-2.5}$ | 4.8 |
| 4U 0352+309 | $28.6^{+1.5}_{-1.7}$ | 3.2 |

discovered high (~300–1200 Hz) frequency QPOs observed in LMXBs (e.g., van der Klis, 1995; 1997, respectively for reviews), although QPOs at both ~400 mHz and ~40 Hz (Zhang *et al.*, 1996) have been reported from the bursting X-ray pulsar GRO J1744-28.

The basic diagnostic value of the studies of the QPOs in X-ray pulsars resides in that if their timescales are comparable to the dynamical timescales in the gravitational fields of NSs, the corresponding radii are:

$$r_d \sim (GM_{NS}/4\pi^2\nu_{QPO}^2)^{1/3} \sim 10^8 - 10^9 \text{ cm}. \tag{7.7}$$

QPOs are then expected to provide information on the processes occurring in the inner accretion disks around the strongly magnetized NSs in these X-ray pulsars, since these disks are expected to be terminated at radii of the above order by the strong stellar field. Indeed, all models proposed for both the QPOs shown by X-ray pulsars and the low frequency QPOs shown by LMXBs on their so-called horizontal branch (Horizontal Branch Oscillations: HBOs, see Section 7.7) have invoked the dynamical timescale at a characteristic radius in the inner accretion disk as one basic timescale. It is obvious that this characteristic radius is much smaller for LMXBs than for X-ray pulsars. This is consistent with our knowledge that the NSs in LMXBs have magnetic fields much weaker than in X-ray pulsars. In addition, magnetospheric beat frequency models (MBFMs) assume the spin period of the neutron star as a second basic timescale.

Extension of these arguments to dynamical timescales of about milliseconds or even smaller in the close vicinity of NSs led to expectations of observing KHz-QPOs around weakly magnetic NSs in LMXBs. These were fulfilled with the discovery just of the KHz-QPOs by RXTE measurements (e.g., van der Klis, 1997).

In the MBFM the QPO frequency is interpreted as the beat frequency between that of the stellar rotation, $\nu_s$, and that of Keplerian rotation, $\nu_K$, at the inner edge of the accretion disk. It is given by (Alpar & Shanam, 1985; Shibazaki & Lamb, 1987) $\nu_{QPO} = \nu_K - \nu_s$ and it was largely used in HBOs in LMXBs and also, by analogy, to QPOs in X-ray pulsars.

If all QPOs in X-ray pulsars are interpreted in this way, then $\nu_K$ is in the range ~50–2500 mHz (Ghosh, 1998), and the corresponding inner radius of the disk, $r_0$, which is obtained from the relation $\nu_K = 1/2\pi \times (GM_{NS}/r_0^3)^{1/2}$, lies in the range $\sim 10^8$–$10^9$ cm, typically smaller than the length scale $r_d$ introduced above by a factor of 3. With this interpretation the speed parameter $\omega_s \equiv \nu_s/\nu_K$ for the rotation of the NS, which is of fundamental importance in accretion torque theory (Ghosh & Lamb, 1979; Ghosh, 1994, 1995) can be obtained immediately.

However, a warning is necessary. In fact, the observations are available completely for this interpretation only for EXO 2030+375 and A 0535+26. For the other there is either lack of information or contrary evidence. Indeed, the radius $r_0$ scales with the accretion rate $\dot{M}$, the magnetic moment $\mu$ of the star, and the star mass $M$ as $r_0 \propto \dot{M}^a \mu^b M_{NS}^c$.

Since the exponent $a$ is negative for all known accretion disk models, $r_0$ always decreases as the X-ray luminosity increases, so that $\nu_{QPO}$ should increase with increasing X-ray luminosity. This has been observed only in the two aforementioned sources EXO 2030+375 (Angelini, Stella & Parmar, 1989) and A 0535+26 (Finger, Wilson & Harmon, 1996).

### 7.6.3. *Gamma Rays from Massive Binaries*

Observations of $\gamma$-rays from massive binaries (Cyg X-3, LSI 303°, Cen X-3, and Vela X-1) by the COMPTEL and EGRET detectors of the CGRO have provoked a discussion on their propagation inside such binary systems (Bednarek, 1997; Bednarek & Giovannelli, 1999). If VHE $\gamma$-rays have been injected somewhere inside HMXBs, they may initiate the inverse Compton pair cascade in the anisotropic radiation of the massive companion, provided that the inverse Compton scattering (ICS) losses of secondary cascade $e^{\pm}$ dominates over other energy losses. The spectra of the escaping secondary cascade $\gamma$-rays depend on the angle of observation, but in a quite broad range of the angles this dependence is not strong. This may create problems with the detection of orbital modulations of $\gamma$-ray signals from such massive binaries. Indeed, in 10 hours of observations of Cen X-3 in TeV region, there is no evidence for periodicity at the X-ray spin period either from a site in the region of the neutron star, or from any other potential site in the orbit (Chadwick *et al.*, 1998). A significant fraction of energy escaping from a HMXB ($\geq 80\%$) – the other part is absorbed by the HMXB – produces a continuum $\gamma$-ray spectrum, and annihilation and nuclear $\gamma$-ray lines generated on the surface of the primary star by the infalling secondary $\gamma$-rays. These spectra and the features can be measured by the instrumentation (SPI and IBIS) on board the INTEGRAL.

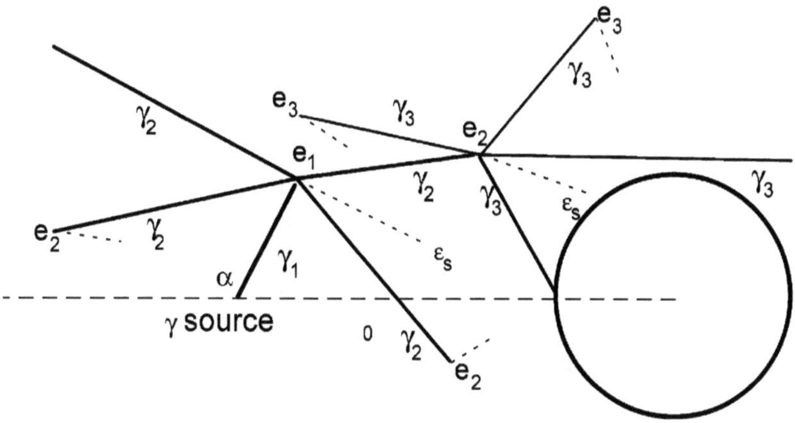


*Figure 105.* $\gamma$-ray cascade in a massive close binary system. The center of the primary star, with radius $r_s$, is connected with the dashed line to the $\gamma$-ray source, not necessarily coincident with the center of the secondary star. From this source, $\gamma_1$ is emitted and interacts with the $\varepsilon_s$ photons, coming from the primary star, producing $e^{\pm}$ pair ($e_1$). This pair cools and emits secondary $\gamma$-rays ($\gamma_2$), etc. (Bednarek, 1997; Bednarek, Giovannelli & Sabau-Graziati, 1999).

The cascade initiated by VHE $\gamma$-rays in massive binary systems is shown in Figure 105 (Bednarek, 1997; Bednarek, Giovannelli & Sabau-Graziati, 1999).

In Figure 106 the spectra of the secondary $\gamma$-rays escaping from the system in a direction forming an angle $\theta$ with the dashed line of Figure 105 are shown, for different angles $\alpha$ (0° and 180°) and different energies of the primary $\gamma$-rays ($E_\gamma = 10^5, 10^6$ MeV); the parameters of the binary system are those of Cyg X-3 (Moffat & Marchenko, 1996), namely $T_s = 9 \times 10^4$ K and $r_s = 2 \times 10^{11}$ cm. The emerging $\gamma$-ray spectra have been computed as a function of the $\alpha$ and $\theta$ angles, as well as the amount of $\gamma$-rays colliding with the primary star surface and their spectra (Bednarek, 1997; Bednarek & Giovannelli, 1999).

Giovannelli, Bednarek & Sabau-Graziati (2001 and the references therein) assumed that VHE $\gamma$-ray photons are injected isotropically from a point source, which elliptically orbits around a massive star. They computed the $\gamma$-ray power seen by the observer at twenty different locations of the VHE $\gamma$ ray source on the elliptic orbit. These powers are obtained by averaging over one hundred Monte Carlo simulations of cascades initiated by VHE $\gamma$-rays isotropically injected at specific location defined by the phase $\omega$. The dependence of the $\gamma$-ray light curves on the eccentricity of the orbit for different locations of the observer, defined by the $\cos\alpha$ and for $\Phi = 0$, is shown in Figures 107. For small angles $\alpha$ the power emitted at different phases of the source of the primary $\gamma$-rays can change by a factor of $\sim$4 (Figures 107a,b). For large angles $\alpha$ the $\gamma$-ray power do not depend significantly on the phase of the $\gamma$-ray source, even for eccentric orbits (Figure 107c). The light curves are also almost independent of the eccentricity of the orbit between $e = 0$ and $e = 0.3$ for different angles of observation $\alpha$. Figures 108 show the dependence of the $\gamma$-ray light curves on the phase $\Phi$ of the observer for

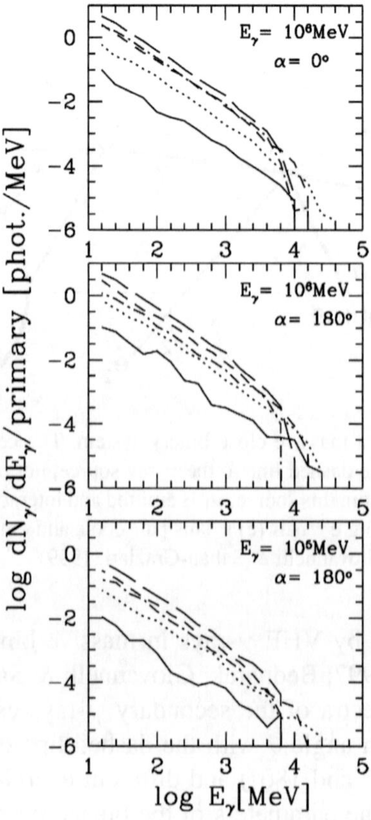


*Figure 106.* The average spectra of secondary cascade $\gamma$-rays which escape from the vicinity of the primary star at different angle $\theta$. Specific curves show the spectra of $\gamma$-rays escaping at the angles $\cos\theta = 0.8 - 1$ (full line), $\cos\theta = 0.4 - 0.6$ (dotted line), $\cos\theta = -0.2 - 0$ (dashed line), $\cos\theta = -0.6 - -0.8$ (long-dashed line), and $\cos\theta = -1 - -0.8$ (dot-dashed line). The cascade is initiated by monoenergetic $\gamma$-ray photons with energies $E_\gamma = 10^5$ and $10^6$ MeV, injected at the distance $x_0 = 2r_s$ and at angles $\alpha = 0°$ and $180°$ (Bednarek, 1997; Giovannelli, Bednarek & Sabau-Graziati, 2001).

the eccentricity of the orbit of the $\gamma$-ray source $e = 0.3$. The light curve for the observer located at $\Phi = 90°$ is shifted with respect to the light curve for $\Phi = 0°$ and the escaping $\gamma$-ray power is comparable. This is consistent with the fact of a weak dependence of the $\gamma$-ray light curve on the eccentricity of the orbit of the $\gamma$-ray source (Figures 107). Again, for large angles $\alpha$, the light curve is flat and shows no significant variability of the $\gamma$-ray power escaping from the system for different phases $\omega$.

Figures 109 show the dependence of the $\gamma$-ray light curves on the distance of the $\gamma$-ray source from the massive companion. The results are shown for the semi-major axis of the orbit $a = 3r_s$, and $6r_s$. The $\gamma$-ray light curves are independent of $a$ in the investigated range of values.

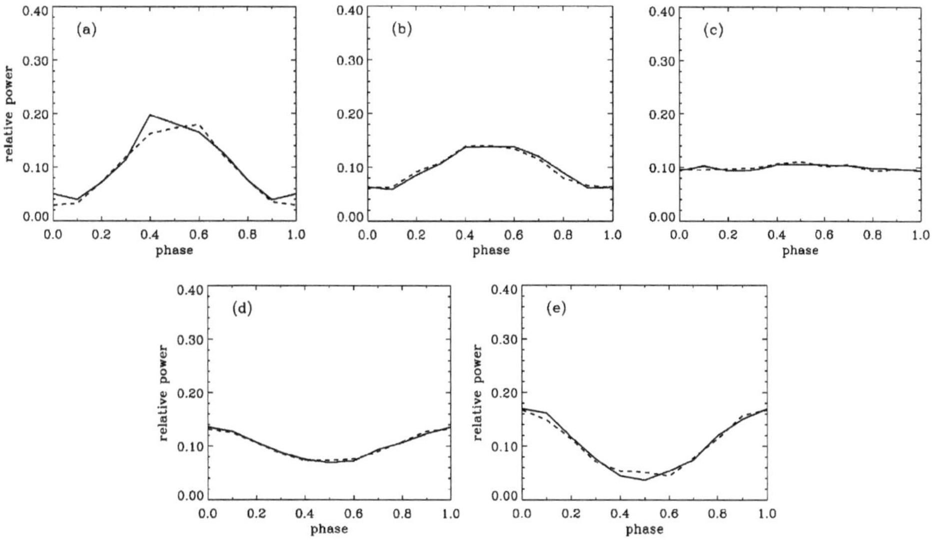


*Figure 107.* The relative $\gamma$-ray power (above 10 MeV) escaping to the observer (located at phase $\phi = 0^0$ and angle $\alpha$) as a function of the phase $\omega$ of a point source emitting VHE $\gamma$-rays with energy $E_\gamma = 10^{12}$ eV. Specific figures show the power emitted at the cosine angles: $\cos\alpha = 0.8 - 1.0$ (a), $0.4 - 0.6$ (b), $0.0 - 0.2$ (c), $-0.4 - -0.2$ (d), $-0.8 - -0.6$ (e). Full curves give the results for the eccentricity of the orbit: $e = 0$ and dashed curves for $e = 0.3$. The semimajor axis of the orbit is $a = 3r_s$ and the parameters of the massive companion are: $T_s = 9 \times 10^4$ K, $r_s = 2 \times 10^{11}$ cm (Bednarek & Giovannelli, 1999).

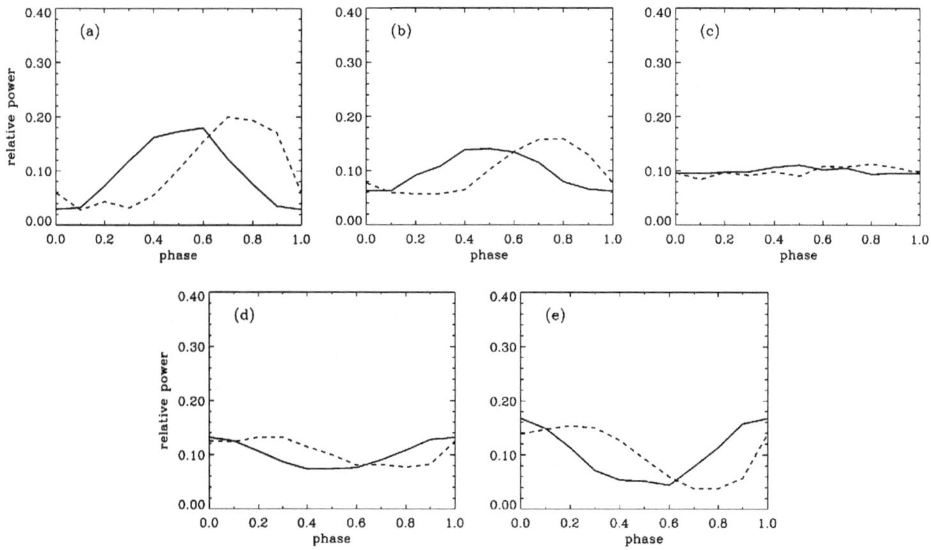


*Figure 108.* The comparison of the $\gamma$-ray light curves seen by the observer located at $\phi = 0$ (full curves) and $\phi = 90°$ (dashed curves). The specific figures correspond to the cosine angles as in Figure 105. The eccentricity of the orbit is $e = 0.3$ and the parameters of the massive star and compact object are as in Figure 105 (Bednarek & Giovannelli, 1999).

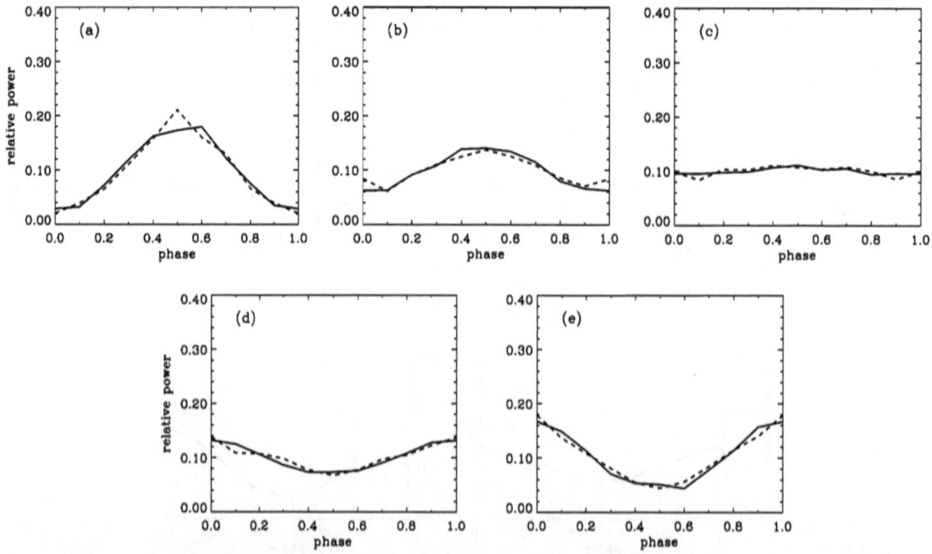


*Figure 109.* The dependence of the $\gamma$-ray light curve on the semimajor axis is investigated. The other parameters are as in Figures 105 and 106. The full curves show the light curves for $a = 3r_s$ and the dashed curves for $a = 6r_s$ (Bednarek & Giovannelli, 1999).

### 7.6.4. *HMXBs: Multifrequency Laboratories*

In this discussion about HMXBs we have shown how good targets they are for a number of measurements in different energy ranges, from the IR to VHE $\gamma$-ray, in order to understand many crucial problems on the physics of the collapsed objects and their interactions with the optical companions. For these purposes two of the most significant energy ranges are the optical and X-ray ranges. Simultaneous measurements in these ranges can solve, in a first approximation, most of the problems still open in understanding the physics of the accretion of matter onto collapsed objects, the interactions with the stellar winds from the optical companions, the mechanisms triggering the X-ray emission and outbursts, and others.

In spite of the many space and ground based multifrequency experiments and the multitude of data of excellent quality obtained and partially still not analyzed, the question of the nature of the accretion flow to the neutron star and the resultant accretion torque in X-ray/Be binaries is still open. It is amazing that we have so little understanding of the class that forms the majority of known accretion powered pulsars. A better understanding of the basic properties of the torque owed to accretion from stellar winds before we can quantitatively account for the spin behavior of wind-fed sources is essential; whilst, for the disk-fed sources it should be important to identify the essential physical processes that determine the magnetic pitch and the spin down torque.

The X-ray source in HMXBs provides a unique active probe of the winds of massive stars. In fact, the binary system orbits around its center of mass, then our line of sight continuously changes, allowing the map of the distribution of

X-ray absorbing gas in the system. Meanwhile the strong X-ray flux alters the local dynamics of the wind, then it is possible to study the physics of radiatively driven winds, and thanks to X-ray variability in different time scales, it is also possible to study the dynamics of radiatively driven winds. Thus the study of winds provides precious information also about the accretion of matter onto the collapsed objects for systems clearly wind-fed. In the opposite case, in systems where the accretion is completely disk-fed we do not know completely the dynamics of the Roche lobe overflow process, and in particular, the fraction of matter and angular momentum lost from the binary system is still poorly known. And, moreover, wind accretion and Roche lobe overflow are the extreme cases of the same accretion process. An increase of the number of systems deeply studied simultaneously in different energy regions is then necessary in order to improve the statistics of the different cases of accretion, from 'pure' wind-fed to 'pure' Roche lobe overflow.

For this purpose we are rather lucky, since our Galaxy contains an adequate number of HMXBs whose emissions are the highest measurable; this renders such systems the most powerful laboratories for testing theories of collapsed objects and plasma physics. In order to go on in such a sense, a change in the philosophy of the experiments is mandatory. A continuous long term monitoring of a 'few' systems, representative of homogeneous class of systems, is then necessary. In this way it will be possible to clarify what is happening during the different phases in which the physical processes occur. These processes can be influenced by the rotation of the X-ray pulsars, the orbital period of the system, transit time of the matter from the primary to secondary star via wind, or disk, or intermediate, and so forth.

Moreover, all the results obtained can be scaled to extra-galactic distances and dimensions. The relative ease in performing multifrequency measurements of HMXBs suggests that they can be useful targets even for small satellites. In particular, in the X-ray and optical energy ranges, HMXBs are very suitable targets for experiments like SIXE (Spanish Italian X-ray Experiment) (Giovannelli *et al.*, 1993b, 1999b, 2001; Isern *et al.*, 2001) to be launched in small satellites such as the Spanish MINISAT-02.

### 7.7. Low mass X-ray binary systems

The LMXBs class is formed by systems where the optical star – the secondary – is later than type A with a mass $M \leq 1\ M_\odot$. They can be very old systems with age $\geq 10^9$ yr. In some very evolved systems such a companion can be even a white dwarf. Their spatial distribution is concentrated across the galactic plane, but mostly toward the galactic center: more than half of the known LMXBs are located within 20° of the galactic center (see Figure 70, Section 5.5.2).

The optical counterparts in LMXBs are intrinsically faint. Most of the spectra show a few characteristic emission lines superimposed on a roughly flat continuum, which is dominated by the emission of an accretion disk around the compact star – the primary. Such an emission is mainly the result of reprocessing of a fraction of the X-rays into optical photons in the disk. The contribution of the secondary star to the total luminosity is generally negligible.

The orbital periods of LMXBs range from 11.4 min for 4U 1820-30 (Stella, White & Priedhorsky, 1987) to 16.6 d for Cir X-1 (Kaluzienski *et al.*, 1976). The very short period systems ($P_{\rm orb} < 1$ hr) are expected to have degenerate dwarf mass companions and probably the transfer of mass is being driven by gravitational radiation.

The last updated catalog of LMXBs contains 150 systems (Liu, van Paradijs & van den Heuvel, 2001).

The importance of low mass X-ray binaries (LMXBs) resides mainly in the fact of them containing neutron stars (or black holes), which are objects of fundamental physical interest. Their study allows the derivation of information about the equation of state of high density matter and to test the general relativity in the presence of very strong field régime.

In LMXBs matter is transferred from a low mass ($\leq 1$ M$_\odot$) star (the secondary star) to a neutron star (the primary star) via an accretion disk. The X-ray radiation originates from the hot ($\sim 10^7$ K) plasma comprising the inner few tens km of the flow. This is very close to the neutron star, which has a radius $R_{\rm NS} \sim 10$ km and a mass $M_{\rm NS} \sim 1.4$ M$_\odot$. Therefore the study of the properties of such a flow can provide information about the star.

When the matter is descending into the neutron star's very deep gravitational potential well ($GM_{\rm NS}/R_{\rm NS} \sim 0.2c^2$), the temperature of the inner flow increases a great deal because of the release of a large amount of gravitational energy. The characteristic velocities near the NS are of order $(GM_{\rm NS}/R_{\rm NS})^{1/2} \sim 0.5c$. Therefore the time scale for the motion of the matter through the emitting region, namely, the dynamical time scale, is as short as $\tau_{\rm dyn} = (r^3/GM_{\rm NS})^{1/2} \sim 0.1$ ms for $r = 10$ km, and $\sim$2 ms for $r = 100$ km.

Milestones on the study of low mass X-ray binaries (LMXBs) have been the RXTE and BeppoSAX satellites, which rendered possible obtaining direct information about the properties of these flows at these time scales. Before the beginning of such missions in the middle 1980s, the neutron star (NS) LMXBs were deeply studied with previous experiments on board many satellites (i.e., Einstein, EXOSAT, ROSAT, Ginga). NS in LMXBs display a complex variety of quasi-periodic oscillation (QPO) modes in their X-ray flux. One important step in their knowledge was the introduction of X-ray colour–colour diagrams (CDs) (e.g Hasinger & van der Klis, 1989; Schulz, Hasinger & Trümper, 1989). Such diagrams resulted extremely useful in the study of correlations between the changes in X-ray spectrum and X-ray timing behaviour of LMXBs. These power spectral variations provided an observational handle of the bright LMXBs, which are galactic X-ray binary systems that in spite of being the brightest and longest known extra-solar cosmic sources had long defied interpretation owing to a lack of diagnostic properties. Several years ago van der Klis (1989; 1995) reviewed the quasi-periodic oscillations (QPOs) detected in LMXBs, before the discovery of the KHz QPOs, made with the RXTE satellite (Strohmayer *et al.*, 1996; van der Klis *et al.*, 1996). Such a discovery provided experimental evidence of the crude phenomenology, occurring in the vicinity of the neutron stars, as described before.

What is clear at this moment is that, for the first time, rapid X-ray variability phenomenon, directly linked with the NS's most distinguished characteristic, namely its compactness, have been detected. This is particularly evident if the phenomena are in some way related to the orbital motion. Indeed, a Keplerian orbital frequency $\nu_K = P_{orb}^{-1} = (GM_{NS}/4\pi^2 r_K^3)^{1/2}$ of 1200 Hz around a $1.4M_\odot$ NS as seen from infinity corresponds to an orbital radius $r_K = (GM_{NS}/4\pi^2 \nu_K^2)^{1/3}$ of 15 km directly constraining the equation of state of the bulk matter of nuclear density, and only just outside the general relativistically marginally stable orbit (van der Klis, 2001). Whatever the model, for the first time one has to seriously worry about general relativistic effects in describing the observable dynamics of the physical systems.

In the Ph.D. thesis of Wijnands (1999) and in his short review (Wijnands, 2001), and in van der Klis (2001), millisecond phenomena in X-ray binaries have been deeply discussed. Here we summarize some of the crucial information necessary for understanding what is going on in LMXBs.

On the basis of the correlations discovered, LMXBs were classified into the *Z sources* and *atoll* sources. The name derived by the shape traced in the CD by different sources (Figure 110):

- Z sources are those tracing out a Z shaped track with the branches labelled, from top to bottom, the *horizontal branch* (HB), the *normal branch* (NB), and the *flaring branch* (FB). Motion of the source across the Z track is owed to variations in the mass accretion rate, which is lowest on the horizontal branch, increasing on the normal branch onto the flaring branch, where it reaches the Eddington limit. Then, summarizing, Z sources are near-Eddington accretors and probably have somewhat stronger ($1$–$5 \times 10^9$ G) magnetic fields.
- Atoll sources are those tracing out a curved branch, which can be divided in the *island state* (IS) and the *banana branch*, subdivided into the *lower banana* (LB) and *upper banana* (UB) branch. The accretion rate is lowest in the IS and increasing on the banana branch from the LB onto the UB branch. Atoll sources are often X-ray bursts sources with luminosities between $10^{-3} L_{Edd}$ and a few $10^{-1} L_{Edd}$, and are thought to have somewhat weaker magnetic fields ($10^8$–$10^9$ G).

The shape of CDs and their different branches were deeply described by van der Klis (1996).

The power spectra on the HB are called horizontal branch oscillations (HBOs). Such spectra show band limited strong noise (called low frequency noise, LFN), which cuts off below several Hertz, simultaneous with 15–60 Hz QPOs, which are the HBOs. On the normal branch these QPOs can still be seen, often simultaneous with other 5–7 Hz QPOs seen on the FB, namely normal branch oscillations (NBOs). The NBOs smoothly merge with the 7–20 Hz QPOs seen on the FB,

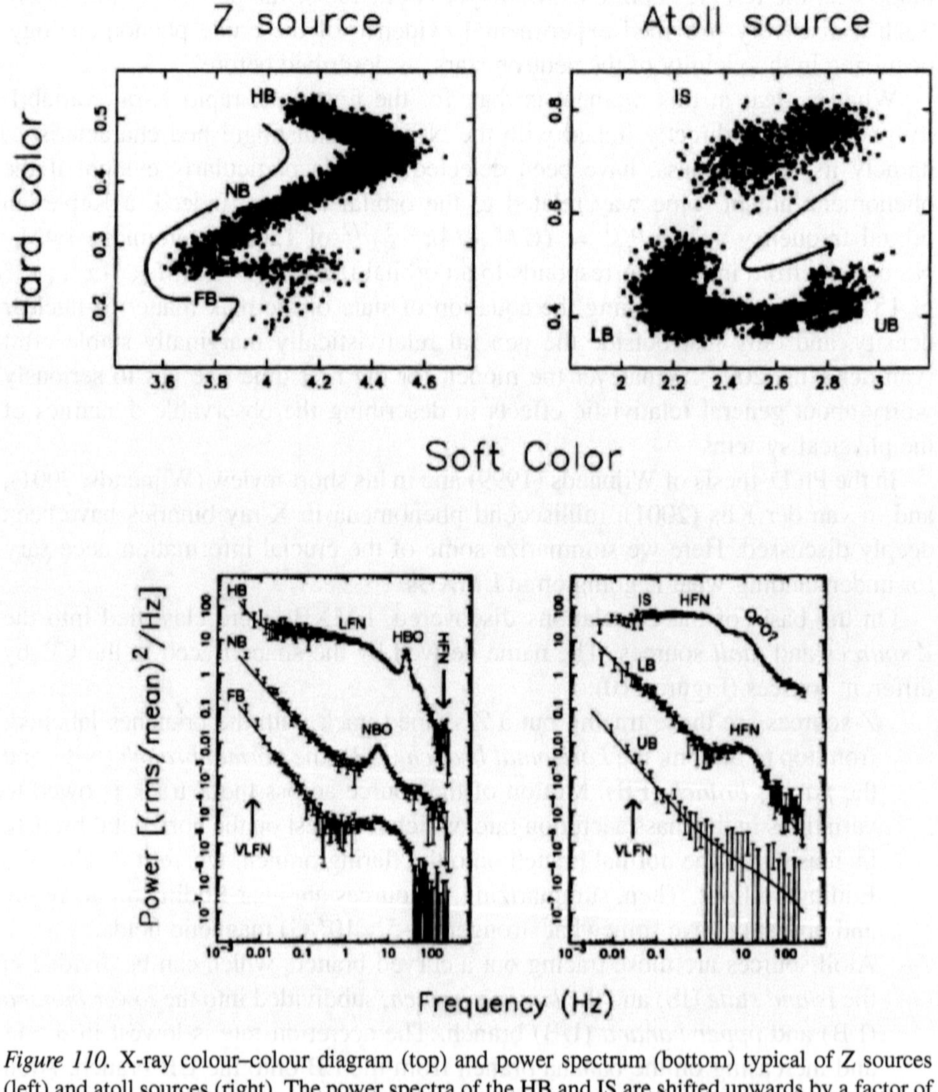


*Figure 110.* X-ray colour–colour diagram (top) and power spectrum (bottom) typical of Z sources (left) and atoll sources (right). The power spectra of the HB and IS are shifted upwards by a factor of $10^4$; the power spectra of the NB and LB by a factor of $10^2$ (Wijnands, 1999; 2001).

which are called flaring branch oscillations (FBOs). Two other noise components are found in all branches: one at very low frequencies (the very low frequency noise, VLFN), following a power law, and one at frequencies above 10 Hz (the high frequency noise, HFN), which cuts off between 50 and 100 Hz.

The power spectrum in the IS is dominated by band limited very strong noise, superimposed on which are broad bumps, sometimes called QPOs. The band limited noise is also called HFN but it is at much lower frequencies then the HFN observed in the Z sources, and it is most likely they are not related.

The differences between Z and atoll sources could be explained by assuming that the Z sources can reach higher mass accretion rates than the atoll sources (e.g., Hasinger & van der Klis, 1989 and references therein) and the neutron stars in Z sources have higher magnetic field intensity than in atoll sources.

However, with the advent of RXTE, new measurements demonstrated that the timing behaviour of the neutron star LMXBs is far more complex than previously thought, as discussed by Wijnands (1999; 2001). Indeed, all types of QPOs previously thought to be present only in Z sources have been observed also in the atoll sources. Therefore the Z and atoll sources are likely to be much more similar than previously suspected. The individual sources only differ in details from each other. The general behaviour of Z and atoll sources can then be summarized as follows (Wijnands, 2001):

- At low $\dot{M}$ the power spectra are dominated by band limited strong noise, which follows approximately a power law at high frequencies, but breaks at a certain frequency below which the power spectra are roughly flat. At frequencies above the break, either a QPO or a broad bump is present. At frequencies around ~100 Hz a broad noise component is present. KHz QPOs might be already be detectable in some sources bu not in all.
- If $\dot{M}$ increases the frequency of the break and of the QPO increases and the strength of both components decreases. The 100 Hz component does not change significantly (Wijnands & van der Klis, 1998; van Straaten *et al.*, 2000; di Salvo *et al.*, 2000), although it might become more coherent (van Straaten *et al.*, 2000). If not yet present, KHz QPOs appear and their frequencies increase with $\dot{M}$. The band limited noise, the low frequency QPOs, and the KHz QPOs gradually disappear as $\dot{M}$ increases further and the power spectra become increasingly dominated by a very low frequency noise component following a power law. QPOs near 5–7 Hz appear when the sources are at very high $\dot{M}$, but these high $\dot{M}$ levels might not be reached in all systems. If $\dot{M}$ increases even further either these QPOs disappear or first increase in frequency to about 20 Hz before disappearing.

Possible exceptions to this general picture are the X-ray dippers (e.g., Bałucińska-Church, 2001; Church *et al.*, 1998; Church, 2001) and the millisecond X-ray pulsars.

A fundamental question still open is why the Z and atoll sources resemble each other in the way they do, despite their significant different luminosities. It is still likely that $\dot{M}$ is higher in Z sources than in atoll sources, but it is unclear why the timing properties should be so similar. These similarities make it also increasingly unclear whether a difference in magnetic field strength is really present between Z and atoll sources.

Probably the source Cir X-1 could be the proper target for establishing the connections between the Z and atoll sources (e.g., Oosterbroek *et al.*, 1995).

In spite of the luminosity of Z sources being on average a factor of ~10 greater than that of atoll sources, both classes show remarkably similar properties derived from the detected KHz QPOs. A common phenomenon in LMXBs is the presence of a pair of KHz QPOs, centroid frequencies of $\nu_1$ and $\nu_2$, which drift in frequency while mantaining their frequency difference $\Delta\nu \equiv \nu_2 - \nu_1 \approx$ 250–360 Hz, roughly constant. However, in a few sources (Sco X-1: van der Klis *et al.*, 1997; 4U 1608-52: Mendez *et al.* 1998a,b; 4U 1735-44: Ford *et al.*, 1998; 4U 1728-34: Mendez & van der Klis, 1999) $\Delta\nu$ decreases significantly as $\nu_2$ increases.

Recently discussions either on the phenomenology and on models of LMXBs have been published by van der Klis (2000; 2001). The phenomenology of LMXBs strongly suggests that a beat frequency model of some kind is at work. Neutron star spin and disk Keplerian motion are periodic phenomena, which are present in the systems; therefore they are natural candidates for providing the basic frequencies. Stella (2002) discusses the beat frequency models and the relativistic precession models. We suggest to the interested reader to look at the latter reviews.

#### 7.7.1. *Soft X-Ray Transients*

Soft X-ray transients (SXTs) constitute a subclass of LMXBs and are characterized by low accretion rates. Non-steady mass transfer onto the compact object causes sporadic and sometimes repetitive large X-ray and optical outbursts, which usually last for several months, when the X-ray luminosity can increase up to a factor of $10^7$ (van Paradijs & McClintock, 1995). Therefore SXTs are typically quiescent and undetected for most of the time and undergo sporadic outbursts, during which they emit an intense X-ray flux. SXTs were initially classified on the basis of their spectral hardness, because of the lack of information on their nature (Cominsky *et al.*, 1978). The subsequent discovery of a number of phenomena observed in different classes of X-ray binaries, such as pulsations and bursts (e.g., White, 1989), showed clear relationships among transient an persistent accreting sources. The spectra of hard X-ray transients (HXTs) – associated with persistent X-ray pulsars orbiting around early type stars – and SXTs are characterized by $KT \geq 15$ keV and $KT \leq 15$ keV, respectively (White, Kaluzienski & Swank, 1984). The outbursts of SXTs, often accompanied by a significant increase in the luminosity of the optical companion and by the onset of type-I burst – thermonuclear flashes on the surface of the neutron star – activity. These properties associate SXTs with LMXBs, which contain old neutron stars.

SXTs are located in the galactic plane with a distribution similar to that of LMXBs (van Paradijs & White, 1995). The presence of SXTs in globular clusters is of particular interest owed to their possible evolutionary link with recycled millisecond pulsars (MSPs). It is possible that some of the dim ($L_X \leq 10^{34}$ erg s$^{-1}$) X-ray sources in globular clusters are SXTs in quiescence (e.g., Verbunt *et al.*, 1994). For an alternative explanation see e.g., Grindlay (1994).

A long review on SXTs is that by Campana *et al.* (1998) and a short updated review is contained in the Ph.D. thesis of Sánchez-Fernández (2003).

Briefly, contrary to persistent LMXBs, SXTs emit a significant flux of X-rays only during the outbursts, which are separated by long intervals of quiescence (much longer than the duration of outbursts). Their outbursts are very strong so that the systems in such conditions are the brightest sources in the X-ray sky. The decline from an outburst is slow: the duration can be of the order a few months. The luminosity during a typical outburst approaches the Eddington limit and the X-ray emission is dominated by thermal emission from the hot inner part of the accretion disk around the collapsed object. Most of the the optical contribution during outbursts is due to reprocessing of X-rays in the outer regions of the accretion disk. Typical luminosities at outbursts are of $\sim 10^{37}$–$10^{38}$ erg s$^{-1}$, but sometimes they can reach $\sim 10^{39}$ erg s$^{-1}$ (Tanaka & Lewin, 1995).

Therefore during outbursts the SXTs are similar to the persistent LMXBs both in X-ray and optical behaviour. The total luminosity is dominated by the emission of the accretion disk. When the SXT is returning to quiescence its optical emission reveals the presence of the optical companion. This is the unique possibility of directly detecting the characteristics of the secondary among the class of the LMXBs. However, for instance in cataclysmic variables, which can be considered LMXBs having as primary a white dwarf, such a detection is possible (e.g., Martinez-Pais *et al.*, 1994).

During quiescence the X-ray luminosity is strongly dependent on the nature of the compact star. For neutron star systems this luminosity is of $\sim 10^{38}$ erg s$^{-1}$, whilst for black hole systems it can be as small as $\leq 10^{31}$ erg s$^{-1}$ (e.g., van Paradijs, 1998 and references therein). Narayan, Garcia, & McClintock (1997) argued that this difference can be explained only if the accretion inner disk around the collapsed object of quiescent SXTs is advective. Bisnovatyi-Kogan, (1999) discussed the accretion disk theory from the standard model until advection. A general and exhaustive discussion on such a kind of disk is reported by Bisnovatyi-Kogan & Lovelace (2001).

Most of the SXTs contain black holes, whilst persistent LMXBs contain neutron stars. The orbital periods ($\sim$ hours) of most SXTs combined with the mass ratio between the compact and optical stars ($M_X/M_{opt} \geq 2$) imply that the secondary stars must have spectral types later than K. Table 15 shows the SXTs (Sánchez-Fernández, 2003). Systems with neutron stars or black holes, when firmly established, are distinguished by NS and BH, respectively (column 1); column 2 shows the year in which outbursts have been detected; the faintness of the optical counterparts of SXTs appears evident in column 3; in column 4, 5, and 6 orbital periods, mass functions of the systems, and spectral types are indicated, respectively, when known.

The X-ray spectra of SXTs when their luminosity is $>10^{37}$ erg s$^{-1}$ are dominated by an ultrasoft component (and this is the reason of the name given to this subclass of LMXBs), having roughly a bremsstrahlung shape, generally limited to photon energies below 10 keV. Such an ultrasoft component originates in the inner regions of a geometrically thin, optically thick accretion disk, from the sum of black bodies with surface temperature dependent on the distance, $r$, to the central compact source as $r^{-3/4}$.

TABLE 15
Soft X-Ray Transients with optical counterparts (Sánchez-Fernández, 2003).

| Source Name | Years of Outbursts | V mag (Quiesc.) | $P_{orb}$ (days) | f(M) | Secondary Spec. Type |
|---|---|---|---|---|---|
| A1455-314 (Cen X-4)$^{NS}$ | 1969, 79 | 18.3 | 0.63 | $0.21 \pm 0.08$ | K5-7IV |
| 4U1543-47$^{BH}$ | 1971, 83, 92 | 17 | 1.12 | $0.22 \pm 0.02$ | A2V |
| A1524-62 (KY TrA) | 1974, 90 | >21 | | | |
| A0620-00 (V616 Mon)$^{NS}$ | 1917, 75 | 18.3 | 0.32 | $2.72 \pm 0.06$ | K3-5V |
| X1908+005 (Aql X-1)$^{NS}$ | 1975, every 200–300 d | 21.6 | 0.79 | | |
| X1658-298 (V2134 Oph)$^{NS}$ | 1976, 78, 99 | >23 | 0.30 | | |
| H1705-25 (V2107 Oph)$^{BH}$ | 1977 | 21.5 | 0.52 | $4.65 \pm 0.21$ | K5V |
| H1608-522$^{NS}$ | 1978 | >20 | | | |
| EXO 0748-676 (UY Vol)$^{NS}$ | 1985 | >23 | 0.16 | | |
| GS1354-645 (BW Cir) | 1987, 97 | 22 | | | |
| GS2000+25 (QZ Vul)$^{BH}$ | 1988 | 22.5 | 0.34 | $5.01 \pm 0.12$ | K3-5V |
| GS1826-24 | 1988 | >21 | | | |
| GS2023+338 (V404 Cyg)$^{BH}$ | 1938, 56, 89 | 18.4 | 6.47 | $6.08 \pm 0.06$ | K0IV |
| GRS1124-68, (N Mus 1991)$^{BH}$ | 1991 | 20.4 | 0.43 | $3.01 \pm 0.15$ | K3-4V |
| GRO J0422+32 (V518 Per)$^{BH}$ | 1992, 93 | 22.4 | 0.21 | $1.13 \pm 0.09$ | M2V |
| GRS1009-45 (N Vel 1993)$^{BH}$ | 1993 | 22 | 0.29 | $3.17 \pm 0.12$ | K8V |
| GRS1716-249 (V2293 Oph) | 1993 | >22 | | | |
| GRO J1655-40 (N Sco 1994)$^{BH}$ | 1994 | 17.3 | 2.62 | $2.73 \pm 0.09$ | F6IV |
| XTE J1550-564$^{BH}$ | 1998, 99, 2000 | 21.6 | 1.54 | $6.86 \pm 0.71$ | G8IV-K4III |
| XTE J2012+381 | 1998 | 21.3 | | | |
| XTE J2123-058$^{NS}$ | 1998 | 22.5 | 0.25 | $0.61 \pm 0.08$ | K7V |
| XTE J1859+226 | 1999 | 23.2 | 0.31 | | |
| XTE J1118+480$^{BH}$ | 1999 | 18.8 | 0.17 | $6.0 \pm 0.4$ | M1V |
| XTE J1650-500 | 2001 | 21.6 | 0.212 | $0.64 \pm 0.03$ | M1V |

The soft components in the spectra of systems with BHs and NSs are qualitatively the same (i.e. same functional form), except for significantly lower characteristics temperatures in the BH systems, being typically the colour–temperature $KT \leq 1.2$ keV, when the X-ray luminosity is $L_X \sim 10^{38}$ erg s$^{-1}$ (Ebisawa *et al.*, 2001).

In BH SXTs, the thermal emission is accompanied by a hard power law component at high energies of the form $F_E \propto E^{-\alpha}$, which in some cases has been observed up to the MeV range (Gilfanov *et al.*, 1995). The power law component arises by Comptonization of soft seed photons in an optically thin region (Sunyaev

& Titarchuk, 1980). In NS SXTs, a variable isothermal component from a boundary layer between the disk and the neutron star arises in luminous states (Mitsuda *et al.*, 1984).

The soft component is not a general feature of SXT outbursts. Indeed, several transient events (e.g., in GRS 1716-249, GS 2023+338, and GRO J0422+32), as reported by Chen, Shrader & Livio (1997), have shown power law spectra without the soft excess, although they have light curves similar to other SXTs being the optical companion a low mass star. By the way, in their paper they prefer to call such soft X-ray transient systems X-ray novae. This term is also used by many other authors.

Accreting black holes show harder (20–500 keV) spectra than accreting low magnetic field neutron stars. A power law spectrum extending up to several hundreds keV can be considered as a black hole signature (Sunyaev *et al.*, 1991). However, SIGMA observations of low magnetic field neutron stars showed also that in such systems the spectra can be as hard as 20–200 keV (Barret & Vedrenne, 1994). The characteristic signatures for accreting BH and NS must then be searched in the slope of the observed hard energy tail, rather than simply in the extension of the spectra to high energies. Indeed, for BH systems the slope is ~1.5–2 (Sunyaev *et al.*, 1991; Grebenev *et al.*, 1993), while for NS systems the slope is ~3 (Barret & Vedrenne, 1994). A very clear example of such a difference is reported in Figure 111, where the GRANAT/SIGMA spectra of the bright NS LMXB GX5-1 and the BH LMXB GRS 1758-258 are shown (Gilfanov *et al.*, 1993; Churazov *et al.*, 1994).

Barret *et al.* (2000) reported on RXTE observations of four SXTs: 1E 1724-3045, GS 1826-238, SLX 1735-269, and KS 1731-260. The first three were in a low state with luminosities of $\sim (0.05\text{–}0.1)L_{\mathrm{Edd}}$ in the energy range 1–200 keV, whilst the fourth was in a high state with luminosity of $\sim^{0}.35L_{\mathrm{Edd}}$. The spectra of the low state sources are all consistent with resulting from thermal Comptonization with an electron temperature $KT_{\mathrm{e}} \approx 25\text{–}30$ keV. For KS 1731-260, in a high state, the spectrum is also dominated by thermal Comptonization, but with a much lower electron temperature ($KT_e \sim 3$ keV), and no significant hard X-ray emission. All the four sources have shown an iron $K\alpha$ line at 6.4 keV (marginal in 1E 1724-3045). The origin of the line and the reflection component is probably irradiation of the accretion disk by the X-ray source. They suggested a model in which the region of the main energy release, where hard X-rays originate, would be an optically thin boundary layer merged with an advection-dominant accretion flow (ADAF) and would be responsible for the rapid variability observed. Probably the observed soft component represents the unscattered emission from an optically thick accretion disk of variable inner radius. When the accretion rate increases, the inner disk radius shrinks and the strength of the reflected component and associated iron line increase. At the same time the Comptonization region cools off in response to an increased cooling flux from the accretion disk and from the reprocessed/reflected component, thus leading progressively to a quenching of the hard X-ray emission.

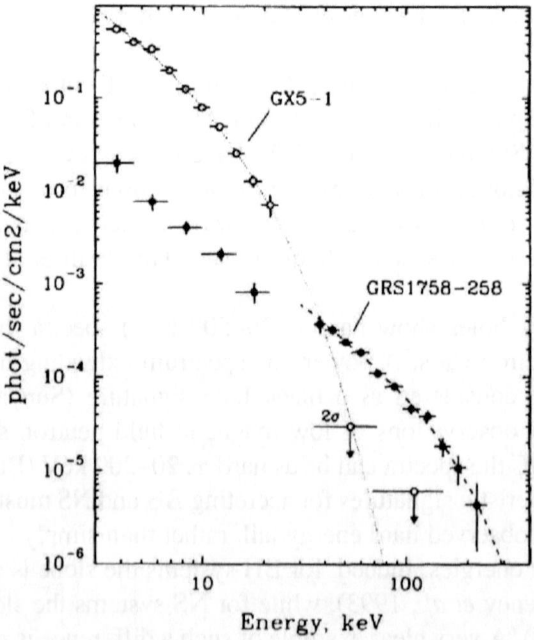


*Figure 111.* Spectrum of the BH LMXB GRS 1758-258 in comparison with that of the bright NS LMXB GX5-1, according to GRANAT/SIGMA observations (Gilfanov *et al.*, 1993).

If low state NSs accrete via ADAFs, the observation of X-ray burst, indicating that all the accreting matter actually accumulates onto the NS surface, argues against the existence of strong winds from such accretion flows. From their observations, Barret *et al.* (2000) concluded that the two criteria suggested for distinguishing among non-quiescent BHs and NSs are not contradicted: the first one states that, when thermal Comptonization is responsible for the hard X-ray emission, only BHs have $KT_e > 50$ keV; the second one states that only BHs are capable of emitting hard X-ray tails with luminosities $\geq 1.5 \times 10^{37}$ erg s$^{-1}$ in the energy range 20–200 keV. However, the first criterion is weakened by there being some NSs which do not display high energy cutoffs in their hard X-ray spectra.

Radio emission was observed during the outburst of A1742-289 (Davies *et al.*, 1976), Cen X-4 (Hjellming & Johnston, 1988) and Aql X-1 (Hjellming, Han & Roussel-Dupré, 1990). Now, it is possible to say that outbursts of SXTs are generally associated with radio outbursts (Hjellming *et al.*, 2000) which have been resolved on a small number of occasions into discrete ejections (Fender & Kuulkers, 2001), sometimes multiple, such as during the 1975 outburst of A0620-00 (Kuulkers *et al.*, 1999), of synchrotron emitting material. This manifests in a transient radio event, which becomes optically thin within a few days – presumably owing to decreasing self-absorption – as the ejected component expands and then fades monotonically.

These transient radio events could indicate that one of the side effects of the sudden appearance of a transient accretion disk environment in SXTs is a sudden production of a compact volume of mixed relativistic electrons and magnetic fields, which subsequently expands adiabatically. This is probably the result of the gas dynamical adjustments occurring during the formation of the hot X-ray emitting accretion disk (e.g., Hjellming, 1996 and references therein).

Recently, Belloni, Psaltis & van der Klis (2002) have studied an empirical model for a unified description of the power spectra of accreting NSs and BHs. This model is based on a superposition of multiple Lorentzians and offers the advantage that all QPOs and noise components are dealt with in the same way, without the need of deciding in advance the nature of each component. They have shown that the power spectra of low luminosity burst sources, of a BH candidate (BHC) in the low state, of a Z source, and of Cir X-1 can be decomposed into the sum of a small number (4–5) of Lorentzian components. Such components are characterized by a single characteristic frequency $\nu_{\max}$, which puts noise and QPO phenomena on equal footing.

Their power spectra decomposition suggests that the difference in the power spectra of BHC and NSs is caused by two effects. The first is that all characteristic frequencies are lower in the case of the BHC, probably reflecting the mass dependence of the dynamical timescale in the accretion flow, as was also noted by Sunyaev & Revnivtsev (2000). The second is more unexpected: in the power spectrum of one of their targets (XTE J1118+480), they have found the fractional rms of the Lorentzian with highest frequency much lower than in the other components. However, since these components have been detected in the latter BHC, albeit weaker than in NSs, does seem to suggest that it is not a feauture that requires a NS surface for its generation. Apparently a component with these characteristics can also be produced by the accretion disk around a BH. They then suggested that the underlying physical mechanisms producing similar components in all of the six RXTE sources analyzed (low luminosity X-ray bursters: 1E 1724-3045, SLX 1735-269, and GS 1826-24; the high latitude transient XTE J1118+480; the bright system Cir X-1; the Z source GX 17+2) may be the same and that the scatter in their different correlations is owed to a combination of differences in the analysis method and differences in the fundamental system parameters.

Sunyaev & Revnivtsev (2000) suggested a possible way of distinguishing between systems hosting a BH and a weakly magnetized NS based on their power spectra: NS systems can show significant power above ~300 Hz, while BHCs do not.

We believe that it is necessary to analyze wider samples of NSs and BHCs before to definitively state firm conclusions.

### 7.7.2. *Constraints on Soft X-Ray Transients' Parameters*

The discovery of an accreting NS with a 2.49 ms spin period in the soft X-ray transient SAX J1808.4-3658 (Wijnands & van der Klis, 1998), whose orbital period is $P_{\rm orb} = 2.01$ hr (Chakrabarty & Morgan, 1998) had important consequences for many aspects of our understanding of LMXBs and their evolution. One of these aspects is the tight limit one can place on the neutron star's radius, which in turn has significant implications for the equation of state and possible mass of this compact star.

Depending on which mechanism is operating during quiescence in SXTs, different spin period histories for the neutron stars are possible. If a radio pulsar is active during quiescence, the spin period evolution is determined by the competition between the disk spin up accretion during the outbursts (with a torque of $N = \dot{M}(GMr_{\rm m})^{1/2}$, where $r_{\rm m}$ is the magnetospheric radius) and spin down caused by radio pulsar losses in the quiescent phases. These two effects have a balance for

$$B_9 \simeq 2 \times 10^{-2} P_{-2}^{187/80} L_{37}^{41/80} M_{1.4}^{-91/320} R_6^{-199/80} \Theta_{100}^{-187/320} , \qquad (7.8)$$

where $L_{37}$ is the time averaged outburst luminosity in units of $10^{37}$ erg s$^{-1}$, $\Theta_{100}$ the ratio between the time spent in quiescence and the average outburst duration in units of 100, $P_{-2}$ the spin period in units of $10^{-2}$ s, $B = 10^9 B_9$, $M = 1.4 M_{1.4}$, and $R = 10^6 R_6$ (Campana *et al.*, 1998). The timescale for the period's evolution is

$$\tau \equiv P/\dot{P} \sim 10^{10} B_9^{-2/7} P_{-2}^2 \ \text{yr} \qquad (7.9)$$

in the case of a radio pulsar spin down, and

$$\tau \sim 10^9 B_9^{-2/7} M_{1.4}^{3/7} L_{37}^{-6/7} P_{-2}^2 \ \text{yr} \qquad (7.10)$$

in the case of disk-like accretion for the same set of parameters as above (e.g., Henrichs, 1983).

If the accreting matter is stopped at the magnetospheric radius during the quiescent phase, the neutron star enters the propeller régime (see Lipunov, 1987) and a pronounced spin down results that will affect its period evolution. In order to describe the action of the centrifugal barrier, different spin down mechanisms have been proposed: the spin down torque (Illarionov & Sunyaev, 1975) propeller mechanism is the same, in module, as for the spin up by a disk the torque $N = -\dot{M}(GMr_{\rm m})^{1/2}$.

However, the system spends a factor $\Theta$ more time in the propeller régime with an accretion rate much smaller. In this case the equilibrium does not depend on the NS parameters. NS remains in the propeller régime during quiescence if the values of $B$ and $P$ are proper. A net spin down will occur if

$$L_{37} \leq 0.6 (L_{33}^q)^{7/9} B_9^{4/9} M_{1.4}^{1/9} \Theta_{100}^{7/6} , \qquad (7.11)$$

where $L_{33}^q$ is the quiescent luminosity in units of $10^{33}$ erg s$^{-1}$. If the above conditions are not satisfied, the result will be a spin up of the NS. The spin up/down time scale in this régime is (e.g., Henrichs, 1983):

$$\tau \sim 10^7 B_9^{-2/7} M_{1.4}^{3/7} L_{37}^{-6/7} P_{-2}^{-1} \ \text{yr} , \qquad (7.12)$$

which is much shorter than in the case of a radio pulsar (Campana *et al.*, 1998).

The expected radio luminosity of the MSPs contained in SXTs can be determined through the empirical relation between the mean pulsed luminosity radio-pulsar magnetic field at 400 MHz and spin period for MSPs (Kulkarni, Narayan & Romani, 1990). Campana *et al.* (1998) derived a radio luminosity of

$$L_{\rm radio} \sim 40 B_9^{2/3} P_{-2}^{-4/3} \ {\rm mJy\ kpc^{-2}} \ . \tag{7.13}$$

The minimum detectable radio flux increases rapidly for decreasing periods. Moreover, the dispersion measure will heavily limit the observations: the mean dispersion measure of MSPs is $\sim$40 pc cm$^{-3}$, with only a few MSPs with a value >100 pc cm$^{-3}$.

Burderi *et al.* (2002) reported the results of BeppoSAX observations of the LMXB-atoll source KS 1731-260 during a quiescent phase. They derived for a 7 kpc distant source an X-ray luminosity $L_{\rm X} \sim 10^{33}$ erg s$^{-1}$. If the NS is spinning at a period of a few milliseconds, as inferred from the nearly coherent oscillations observed during type-I burst, the quiescent X-ray luminosity constrains the NS magnetic field strength. They considered all the mechanisms proposed for explaining the quiescent X-ray emission of a NS SXT and compared the corresponding expectations with the measured upper limits on the X-ray luminosity. With the upper limits on the radio pulsed emission from KS 1731-260, obtained during the quiescence of the system, they found that the most probable value of the magnetic field strength is less than $\sim 10^9$ G, and in the worst case is $<4 \times 10^9$ G.

Burderi & King (1998) from the observations of the SAX J1808.4-3658, placed a firm upper limit on the radius of the NS of $R_{\rm NS} < 13.8 \times M_{\rm NS}^{1/3}$ km. The limit is independent of distance or assumption about the magnetosphere geometry and could be significantly tightened by observations of the pulsations. They found also that the magnetospheric radius is $1 < r_{\rm m} <$ few times $R_{\rm NS}$. This fact could be simply a coincidence, but it may also suggest that whatever process induces field decay tends to stop once such conditions are achieved. This in turn supports the idea that the decay may be induced by accretion.

One strong method of diagnostic in X-ray binaries is the so-called *X-ray Dipping*. It consists of reduction in intensity which are generally at the orbital period owing to absorbing material in a bulge in the outer accretion disk or in blobbiness in a stellar wind. Dipping allows emission processes to be determined unambiguously since the requirement to fit non-dip plus several dip spectra, especially using high quality instruments aboard ASCA, RXTE, BeppoSAX, and Chandra, strongly constrains emission models. X-ray dipping provides information in two ways: via spectra and via ingress/egress times, which can give the size emission regions, and of particular interest is the size of Accretion Disk Corona (ADC) where Comptonization takes place. X-ray dipping is relatively common and it can appears in BHs, such as Cyg X-1, in galactic jet sources, such as GRO J1655-40, and in LMXBs, where $\sim$10% of sources are dippers (Church, 2001). He discussed X-ray dipping in Cyg X-1, GRO J1655-440, and in more than a dozen LMXBs. Measurement of dip ingress and egress times allowed emission regions sizes to be determined, and in particular revealed the very extended nature of the ADC in LMXBs with radii of $\sim (1\text{–}50) \times 10^9$ cm, suggesting the strong effect of the central object in both the formation and heating of the ADC. His measurements excluded the possibility that the Comptonization region can be a localized region such as in the vicinity of the NS. The survey of LMXBs based on ASCA and BeppoSAX data has demonstrated that the height of the blackbody emitting region on the NS is equal to the height of the radiatively supported inner disk. It has also shown that the blackbody component cannot originate in the accretion disk. These results will allow comparison with theory of flow between inner disk and star and of the state and geometry of the inner disk.

Church & Bałucińska-Church (1996) already used the X-ray dipping for determining parameters of the LMXBs X 1755-338 and X 1624-490, by using data from EXOSAT archive. Bałucińska-Church (2001) studied the spectral evolution of the continuum and disk line in the dipping source GRO J1655-40, analyzing data coming from the RXTE satellite. A highly shifted iron emission line was detected. The line originates in the inner part of the disk, which rotates with relativistic velocity $\sim$0.3 c producing the characteristic shape of a disk line with strong broadening and boosting. The region where the line is generated is most probably highly ionized (FeXII) although the ionization state is not well constrained. The intensity of the emission line decreases during dipping by the same factor as the emission from the accretion disk providing further evidence that the line originates in the disk. During deep saturated dipping the Comptonized component is totally absorbed but 10% of the emission is still seen, which must originate in the disk. It can be shown that either the outer and cooler parts of the disk are still visible during deep dipping or the absorber is blobby. In both cases the radius of the disk core, which makes the major contribution to the disk luminosity, is roughly in agreement with the radius of the line emitting region of $\sim$30 Schwarzschild radii ($r_S$). The radius of the X-ray emitting region is between 170 and 370 $r_S$, depending on the assumptions made about the absorber. The disk emission region is much larger than that previously reported by Kuulkers *et al.* (1998a,b). The difference is probably owed to an underestimation of the dip ingress/egress times made by the latter authors.

### 7.7.3. *Outburst Models*

Models of SXT outbursts are closely similar to those developed for dwarf nova (DN) outbursts. It is now generally accepted that SXT and DN outburst are owed to a sudden increase of the mass accretion onto the compact object. Indeed, LMXBs are very similar to CVs: one just has to replace the central white dwarf with a neutron star. The formalism of the accretion process is the same for the two cases; the only difference we must introduce is the dimension of the neutron star, whose magnetosphere is small enough to allow the disk to reach a much smaller inner radius and therefore much higher temperature. As a consequence radiation pressure can become important in the inner part of accretion disks around neutron stars, contrary to CVs.

One group of models involves accretion disk instabilities (e.g., Smak, 1985b, 2002 and references therein; Cannizzo, 1993 and references therein), and the other group involves instabilities in the mass transfer from the optical companion (e.g., Osaki, 1985, Osaki, 1996). In both approaches the compact star is continuously accreting from a disk-fed by an optical companion filling (or almost filling) its Roche lobe. In disk instability models the accretion disk can become locally thermally and viscously unstable owing to strong opacity variations caused by partial ionization of hydrogen. In contrast, the mechanism of mass transfer instability relies on the slow expansion of the superadiabatic convective layers of the optical companion star, which is irradiated by hard X-rays ($E \geq 10$ keV) supposedly produced by low level accretion onto the NS surface (Hameury, King & Lasota, 1986). After the onset of accretion at high rate the high energy radiation is intercepted by the evolving accretion disk and the companion contracts again within a few weeks, leaving a disk-fed at a low rate.

However, both classes of mechanisms triggering outbursts have several failures, as critically discussed by Lasota (1995): disk instability models fail in reproducing the observed recurrence time; mass instability models require a quiescent X-ray luminosity of $10^{34}$–$10^{35}$ erg s$^{-1}$ and a sub-giant or stripped giant companion in order to produce outburst recurrence in the observed range, and it is not.

In SXTs a modified disk instability model could probably be the most suitable. Indeed, van Paradijs (1996) remarked on the importance of X-ray heating of the accretion disk: if these effects are taken into account the modified disk instability models reproduce the recurrence times in agreement with those observed in SXTs. Such a kind of model has been invoked by Martinez-Pais *et al.* (1996) in order to explain several behaviour in the outburst spectra of the dwarf nova SS Cygni. This is a good example about the similarities between the outbursts in LMXBs and DN.

Now far a few critical words on the problem of accretion of matter and on the bursts in the case of binary systems containing a neutron star, where the accretion of matter is often disk-fed. The disk itself is rotating, so that the accreted material adds angular momentum to the neutron star. Thus, unlike radio pulsars – which are spinning down – accreting neutron stars in X-ray binaries should spin up as time passes. There is a strong funnelling of the accretion onto the poles and the local confinement of the accretion mound, at least until the ignition, most often results in stable burning. For this reason thermonuclear X-ray bursts are not seen from X-ray pulsars, with the exception of the SAX J1808.4-3658.

All of the observed regular X-ray bursting neutron stars accrete at $\dot{M} < 10^{-9}$ $M_\odot$ yr$^{-1}$ and show no evidence of magnetic fields strong enough to focus the accretion onto an area small enough to stabilize the burning. The thermonuclear instability forces the star to burn the fuel in a time-dependent manner (the fuel accumulated for a few hours until the instability set in, and then fuel is burned in about 10 seconds when $T > 10^9$ K).

The burning processes are quite complicated: most of the hydrogen is burned in a series of rapid proton captures followed by $\beta$-decays of the heavy nuclei – the rp process (Bildsten & Strohmayer, 1999).

Modelling the burning requires knowledge of the thermonuclear reaction rates for nuclei far on the proton-rich side of the valley of stability, but most of these rates have not yet been experimentally measured.

7.7.4. *Luminous X-Ray Sources in Globular Clusters*

Surveys of the dozen brightest ($L_X \geq 10^{36}$ erg s$^{-1}$) X-ray sources in globular clusters have shown that they are very similar to the LMXBs found in the rest of the Milky Way. They show X-ray bursts and lack of coherent pulsations in their persistent emission (Hertz & Wood, 1985; Parmar, Stella & Giommi, 1989; Verbunt *et al.*, 1995). The presence of type I bursts indicates the presence of accreting NSs rather than BHs in the luminous systems, which are considered similar to the galactic X-ray burst sources. However, their formation is more efficient by a factor of $\sim$200 in globular clusters than in the Galaxy itself.

As previously discussed, LMXBs show X-ray spectra generally modeled by the combination of one or more spectral components. At high luminosities, of $\sim 10^{37}$ erg s$^{-1}$, such spectral components consist of a modified blackbody and some form of power law like component, which is believed to come from the Comptonization of cooler photons. At lower luminosities the spectra are generally fitted with a power law with an exponential cutoff at high energies (White, Stella, & Parmar, 1988).

The importance of the study of X-ray sources in globular clusters reside in the fact that their distances are determined independently through main sequence modelling and other methods (e.g., Carretta *et al.*, 2000). Therefore more accurate values of luminosity than those typical for galactic LMXBs can be achieved. Moreover, the cluster reddening can be measured, so providing an independent estimate of low energy absorption. Finally, the mean cluster abundances can be determined, allowing any dependence of X-ray properties on abundance to be examined.

Sidoli *et al.* (2001) performed a detailed study of the broadband spectra of the luminous X-ray sources in 10 globular clusters (X 0512-401 in NGC 1851; X 1724-304 in Terzan 2; X 1730-335 in Liller 1; X 1732-304 in Terzan 1; X 1745-203 in NGC 6440; X 1746-370 in NGC 6441; X 1747-313 in Terzan 6; X 1820-303 in NGC 6624; X 1850-087 in NGC 6712; X 2127+119 in NGC 7078) by using the BeppoSAX measurements. With the sole exception of X 2127+119 in NGC 7078, all the spectra are well fitted by a two-component model, which is formed by a disk blackbody and Comptonized emission. The measured low energy absorptions are in agreement with those predicted from optical measurements of the host globular clusters. This implies that there is a little intrinsic X-ray absorption within the binary systems themselves and the aforementioned spectral model provides a good representation of the low energy continua. They found that the sources can be grouped into two blocks: the first composed by 3 ultra-compact sources ($P_{\rm orb} < 1$ hr), where the disk blackbody temperatures and inner radii appear physically realistic and the Comptonization seed photons temperatures and radii of the emission areas are consistent with the disk temperatures and inner radii. For the second block, which contains all the other sources, the disk blackbody parameters appear not to be physically realistic and the Comptonization parameters are unrelated to those of the disk blackbody emission.

This is an important result. Indeed, if such a spectral signature is characteristic of the ultra-compact binaries, this implies that amongst the sources studied by Sidoli *et al.* (2001) only the first block sources are ultra-compact. However, it is unclear why the difference between the two groups of binaries should exist. One possibility may be related to the mass ratio, which is similar in the ultra-compact systems and binaries containing black holes. In BH systems the soft components are also well fitted with disk blackbody models, which appear to have physically realistic parameters.

By using the total luminosity of the sources and the luminosity of the disk given by (e.g., Priedhorsky, 1986):

$$L_{\rm tot} = GM_{\rm NS}\dot{M}/R_{\rm NS} \ , \tag{7.14}$$

$$L_{\rm disk} = GM_{\rm NS}\dot{M}/2r_{\rm m} \ , \tag{7.15}$$

where the magnetospheric radius is:

$$r_{\rm m} = 20B_8^{4/7}R_6^{10/7}M_2^{1/7}L_{37}^{-2/7}\xi \ \ {\rm km} \tag{7.16}$$

being $\xi$ a parameter, which accounts for the disk geometry of the accretion, whose value is $\xi \sim 0.5$–1 (Ghosh & Lamb, 1979; 1992), $B = B_8 10^8$ G, $R_{\rm NS} = R_6 10$ km, $M_{\rm NS} = 2M_2{\rm M}_\odot$.

Sidoli *et al.* (2001) estimated the magnetic field strength by the total-luminosity/disk-luminosity ratio, which is a measure of the $R_{\rm NS}/2r_{\rm m}$ ratio (Priedhorsky, 1986):

$$L_{\rm disk}/L_{\rm tot} = 0.25B_8^{-4/7}R_6^{-3/7}M_2^{-1/7}L_{37}^{2/7}\xi \ . \tag{7.17}$$

Assuming a $2{\rm M}_\odot$ NS this ratio depends only on the magnetic field strength and total luminosity. The $L_{\rm disk}/L_{\rm tot}$ ratio reaches a maximum of 0.5 when $r_{\rm m} = R_{\rm NS}$. For the source X 1820-30 in NGC 6624, they derived $B = 2.1\times10^8\ G$ in agreement with the value derived from the saturation of the KHz QPO frequencies (Campana, 2000). In the case of the rapid burster X 1730-335 in Liller 1 they found $B \sim 8 \times 10^8$ G, which is consistent, within a factor of 2, with that estimated by Masetti *et al.* (2000) using the propeller effect model.

Sidoli *et al.* (2001) searched also for the correlations among parameters and metallicity of the host globular clusters. The temperature of the inner radius, ($KT_{\rm in}$, increases as the metallicity of the hosting cluster increases and the opposite is true for the inner disk radius, except for the ultra-compact source X 1820-303 in NGC 6624. However, this trend could be spurious since in the three ultra-compact binaries, located in NGC 1851, NGC 6712 and NGC 6624, the accreting matter might be only helium. In contrast, if this trend is real it suggests that matter accreting in the ultra-compact binaries in NGC 1851 and NGC 6712 is probably not pure helium. The total luminosity seems to increase with the metallicity, but it is necessary to be very careful with this, since the total sample is too small in order to paint general correlations. Indeed, the increase of the luminosity of the binary with the hosting cluster metallicity is mainly based on the behaviour of the three ultra-compact binaries.

Guainazzi, Parmar & Oosterbroek (2001), analyzing data from the BeppoSAX survey of bright NS systems in globular clusters, together with the results obtained on galactic LMXBs, noted that a two-component model is a reasonable description of the 0.1–200 keV spectra. At energies lower than a few keV, a thermal component dominates, either owing to the direct view of the NS surface, to the emission of the boundary layer, or to a multi-temperature blackbody originating in the accretion disk. At higher energies a power law with an high energy cutoff is the likely signature of thermal Comptonization of soft X-ray photons, probably produced by the same thermal mechanism as above. The optical depth correlates with the X-ray luminosity, which in turn anti-correlates with the electron temperature. This suggests that the properties of the Comptonizing plasma are driven by the X-ray energy output. They used the 'burster box' – invented by Barret *et al.* (1996) in order to better discriminate between NS and BH binaries – to show where the systems of their sample lie (see Figure 112).

The hatched box limits the areas where NS and BH systems can lie: inside the box there are NS systems, outside the box there are BH systems. Therefore the sources in globular clusters observed with the BeppoSAX contain NSs.

### 7.7.5. *Micro-Quasars*

We have already discussed micro-quasars, making the analogy with the macro-quasars. Here we want simply to spend a few words more in noting that micro-quasars constitute a small subclass of LMXBs, which show sporadic ejection of matter at apparently superluminal velocities. In these binaries of stellar mass the three basic ingredients of quasars are found: a black hole, an accretion disk heated by viscous dissipation, and collimated jets of high energy particles. But contrary to quasars, where the masses of the black holes are several million of solar masses, in micro-quasars the black holes are only a few solar masses; the accretion disks in micro-quasars reach thermal temperatures of several millions degrees instead of several thousands degrees in quasars, and the relativistic particle ejected can travel up to distances of a few light years only, instead of several million light years as in radio galaxies (e.g., Mirabel & Rodriguez, 1998). For a review see

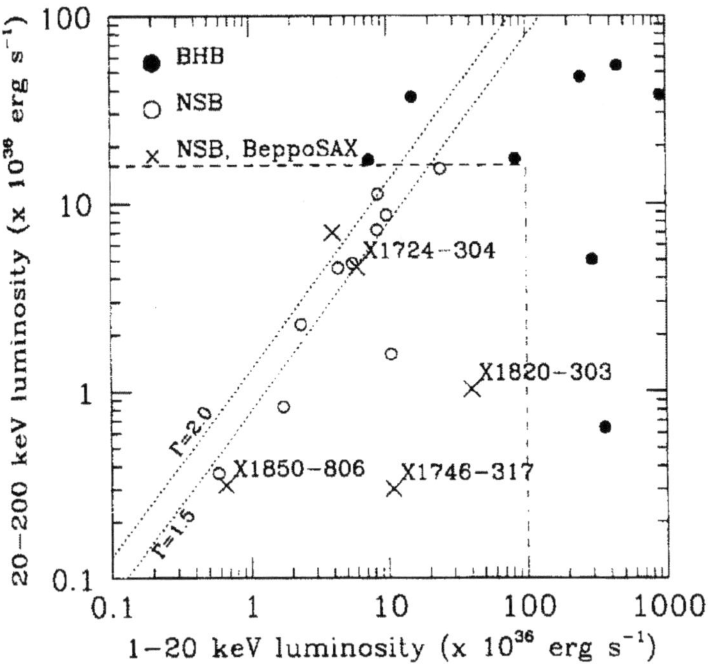


*Figure 112.* 1–20 keV versus 20–200 keV luminosity for black hole (filled circles) and neutron star (empty circles) binary systems. The sources in globular clusters observed by the BeppoSAX are plotted with *crosses*. The hatched box limits the areas where NS and BH systems can lie: inside the box there are NS systems, outside the box there are BH systems (Guainazzi, Parmar & Oosterbroek, 2001).

Mirabel & Rodriguez (1999) and for an update panorama the book *Micro-quasars* (Castro-Tirado, Greiner & Paredes, 2001).

Micro-quasars are the ideal laboratories for the comprehension of the mechanism of relativistic jets and its relation with the accretion flow onto a black hole, traced by the brightness and fast variability in X-ray energy regions. Since such binaries are galactic, then close by us, and bright at many wavelengths, multifrequency observations, possibly simultaneous, can allow a lot of information to be obtained about the entirety of processes occurring in accreting binaries with jets and the possibility of extrapolating the results, with some caution, to quasars. Indeed, the characteristic dynamical times in the flow of accreting matter onto a black hole are proportional to its mass, the events developing in time scales of minutes in micro-quasars could correspond to similar phenomena with duration of thousand years in a quasar of $\sim 10^9$ $M_\odot$, such as for instance 3C 273. Therefore phenomena which we cannot observe in quasars because of their long duration can now be observed in micro-quasars very easily, thanks to the new generation of experiments usually having high time resolution, high spectral and flux sensitivities, and high angular resolution.

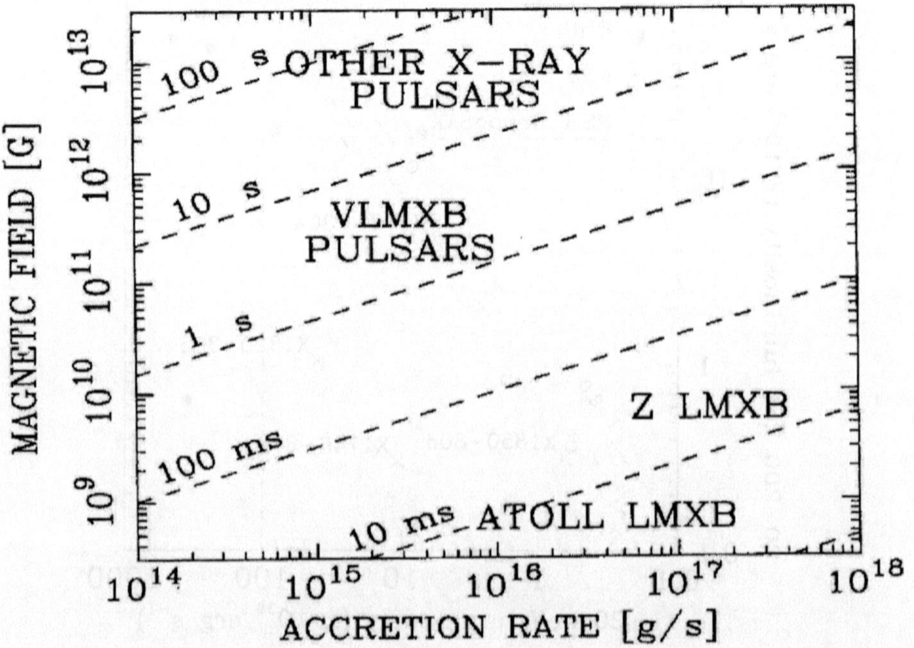


*Figure 113.* Schematic representation of the position occupied by different classes of X-ray binaries in the lg $B$–lg $\dot{M}$ diagram. Hatched lines represent different equilibrium spin periods (Stella, Mereghetti & Israel, 1996).

## 7.8. Anomalous X-ray pulsars

In the last few years a new class of X-ray binaries has been recognized. They are X-ray pulsars with properties clearly different from those of the common HMXBs. These pulsars were recognized as presenting several characteristics similar to those of LMXBs, and for this Stella, Mereghetti & Israel (1996), in the proceedings of the 1995 workshop on *Highlights of European Astrophysics*, suggested that this new group of pulsars constitutes a subclass of the LMXBs, characterized by lower luminosities, higher magnetic fields and smaller ages than non-pulsating LMXBs. They tentatively called the systems of this new group Very Low Mass X-ray Binaries (VLMXBs). Soon after these objects have been called Anomalous X-ray Pulsars (AXPs) (Mereghetti & Stella, 1995; van Paradijs, Taam & van den Heuvel, 1995) and this is now the current accepted name.

Figure 113 shows the magnetic field intensity versus the accretion rate of X-ray binaries. The positions occupied by LMXBs (Z and Atoll type), AXPs (VLMXBs), and other X-ray pulsars (HMXBs) are indicated (Stella, Mereghetti & Israel, 1996). The lines representing different equilibrium spin periods of the pulsars are indicated by the hatched lines.

Review papers on this topic are those by Mereghetti (2001a,b). A general review of accretion powered pulsars is that by Becker (2000). AXPs have spin periods ranging between ~6 and ~12 s, contrary to the larger spread of those of HMXBs (0.069–few $\times 10^3$ s). Spin periods of AXPs are monotonically increasing on timescales of $\sim 10^4$–$4 \times 10^5$ yr. Six AXPs are currently known and three of them are associated with SNRs (Mereghetti, 2001a). Association of X-ray pulsars and SNRs have been claimed since 1993 by Giovannelli's group (Giovannelli *et al.*, 1993a,c, 1994; Giovannelli & Sabau-Graziati, 2000b, 2001). The nature of AXP as neutron stars is supported by their $P_{\rm spin}$ and $\dot{P}_{\rm spin}$. However, the corresponding rotational energy loss ($\sim 10^{45} \Omega\dot{\Omega}$) erg s$^{-1}$) is not sufficient to power the luminosity of the AXP, typically $10^{34}$–$10^{36}$ erg s$^{-1}$. This poses problems in understanding the accretion processes in these objects: e.g., binary models assuming weakly magnetized neutron stars ($B \sim 10^{11}$ G) rotating close to their equilibrium period. This requires accretion rates of few $\times 10^{15}$ g s$^{-1}$, which is consistent with the luminosity of AXPs (Mereghetti & Stella, 1995). Alternatively, the AXPs could involve isolated neutron stars accreting matter from a residual accretion disk (Corbet *et al.*, 1995; van Paradijs, Taam & van den Heuvel, 1995; Ghosh, Angelini & White, 1997); AXPs result from the common envelope evolution of massive close X-ray binary systems. The formation of an accretion disk around an isolated neutron star could be possible through the falling material coming from the progenitor star after the supernova explosion (Chatterjee, Hernquist & Narayan, 2000).

Models based on strongly magnetized neutrons stars ($B \sim 10^{14}$–$10^{15}$ G) have been developed in the last decade just in order to explain the behaviour of the Soft Gamma ray Repeaters (SGRs) (Duncan & Thompson, 1992; Thompson & Duncan, 1995, 1996). SGRs are transient very short (<1 s) events, characterized by relatively soft bursts, peaked at ~20–30 keV, with super-Eddington luminosity. Up to now only four (perhaps five) SGRs are known (see reviews of Hurley, 2000a,b). Recent measurements of the spin down rates of SGRs and AXPs have been interpreted as evidence of very strong magnetic fields at the collapsed object poles, roughly two orders of magnitude greater than those of the 'normal' X-ray pulsars. However, this problem is one of the hottest in modern astrophysics. Indeed, for instance, Dar (2003a) argue that, instead, the observations support the hypothesis that SGRs and AXPs are neutron stars that have suffered a transition into a denser form of nuclear matter to become, presumably, strange stars or quark stars. Internal heat and slow gravitational contraction long after this transition can power both their quiescent X-ray emission and their star quakes, which produce 'soft' gamma ray bursts.

Reviews of observations of AXPs (Israel, Mereghetti & Stella, 2002), of the nature of SGRs and AXPs (Thompson, 2002), and of the general properties of recurrent bursts from SGRs (Aptekar *et al.*, 2002) have been published in the proceedings of the Rome 2000 mini-workshop on *Soft Gamma Ray Repeaters* (Feroci & Mereghetti, 2002).

The optical counterparts of AXPs are not known, with the possible exception of 4U 0142+61, for which Hulleman, van Kerkwijk & Kulkarni (2000) reported the discovery of a faint ($R \sim 25$ mag) blue object in its error box. If such an association should be true, this faintness would rule out the presence of an accretion disk, thus favoring the magnetar interpretation. On the basis of the limits in the optical and IR wavelength regions, the presence of a massive early type companion star, such as OB super giants or Be stars, can be excluded in AXPs. Moreover, no orbital motion signatures are present in their X-ray light curves.

Recently, the Newton-XMM observation of the AXP 1E 1048.1-5937 has provided the first clear evidence for a spectral variation (soft excess during the pulse minimum and no variations during the pulse rise and decay) as a function of the pulse phase in this source (Tiengo *et al.*, 2002). Such a behaviour is different from that of other AXPs which also show energy dependent pulse shapes like 1RXS J170849-4009 (Israel *et al.*, 2001; Gavriil & Kaspi, 2002) and 4U 0142+61 (Israel *et al.*, 1999).

The spectra of all AXPs are well described by the sum of a power law and a blackbody. There are no reasons to attribute this spectral shape to the presence of two physically distinct emitting components (e.g., Ghosh, Angelini & White, 1997). The contribution to the total flux is a strong function of the energy (Özel, Psaltis & Kaspi, 2001), whilst the pulsed fraction at different energies does not change significantly. As noted by Tiengo *et al.* (2002), in the case of two physically distinct spectral components, this would require an *ad hoc* coupling in the pulse profiles of the black body and power law components. The fact that the pulsed emission from 1E 1048.1-5937 can be fitted by the same spectral shape at all phases also suggests that it originates from a single physical component having a spectrum more complex than a pure blackbody.

Thompson, Lyutikov & Kulkarni (2002), in their paper on the *Electrodynamics of Magnetars*, commented on the strength of the surface magnetic field in the SGR and AXP sources, as inferred from their measured spin down rates, and the implications of their model for the narrow measured distribution of spin periods.

With more data on spin periods of AXPs and measurements of magnetic field intensity it will be possible to discriminate amongst different possibilities. This probably will occur with the data from XMM-Newton and AXAF-Chandra still under acquisition. However, this problem could be solved with a proper experiment, like SIXE (Spanish Italian X-ray Experiment) (e.g., Giovannelli *et al.*, 2001 and the references therein), dedicated to very long observations of selected sources ($A_{\rm eff} \times T_{\rm exp} \geq 10^{10}$ cm$^2$ s, where $A_{\rm eff}$ is the effective area of the X-ray detectors, and $T_{\rm exp}$ is the duration of the observation).

## 7.9. Cataclysmic Variables

### 7.9.1. *Historical Background and Classification*

In the 1950s it was recognized that the various phenomena displayed by the cataclysmic Variables (CVs) are all the consequence of accretion of matter onto a white dwarf from a low mass donor star (e.g., Warner, 1976; 1995). CVs are binary systems in which the primary component is a white dwarf ($M_{\mathrm{wd}} \sim 1\ \mathrm{M}_{\odot}$) and the secondary is a late type Main Sequence star ($M_{\mathrm{s}} \leq 1\ \mathrm{M}_{\odot}$) (e.g., Smak, 1985a).

More than 550 CVs are known, most of them discovered through optical observations, and some, especially those in which the magnetic field of the white dwarf is strong, discovered through X-ray observations, but with the detectors of the second and further generations, since CVs are in general not very bright in X-ray energy range. Those with known or suspected orbital period are listed by Ritter & Kolb (1998).

The first CV detected in the X-ray range, with rocket experiments, was the dwarf nova SS Cyg (Rappaport *et al.*, 1974; Heise *et al.*, 1978). The UHURU satellite detected two CVs, which were not recognized as such. Warner (1976) proposed the identification of 4U 1249-28 with EX Hya, and the variable AM Her, which on further optical studies was recognized as a CV (Forman *et al.*, 1978). The magnetic field in these two systems is strong ($\approx 10^7$–$10^8$ G). A few dozen CVs were detected in X-rays with HEAO-1 satellite, with EXOSAT, and with the Einstein satellite (e.g., reviews of Cordova & Mason, 1983; Cordova, 1995). Later Verbunt *et al.* (1997) recognized 91 CVs from a sample of 162 systems with known or suspected binary periods by using data of the ROSAT XRT-PSPC All Sky Survey.

Historically, because CVs were observed photometrically and without seeming to follow any regular pattern, they were given the term *cataclysmic* (from the Greek word kataklysmos = flood, storm; Hack & la Dous, 1993). As collecting of observational data progressed it became apparent that these objects were regular binary systems which for some reason changed in brightness; some of them also regularly (Recurrent Novae and Dwarf Novae) while some others only once (Classical Novae).

Therefore the classification of CVs was based on the optical outburst properties, by which one may distinguish four groups of CVs: (i) classical novae; (ii) recurrent novae; (iii) dwarf novae; (iv) nova-like objects (e.g., Giovannelli & Martinez-Pais, 1991 and references therein; Ulla, Giovannelli & Martinez-Pais, 1999).

This classification, however, is neither self-consistent nor adequate and it is much better to consider primarily the observed accretion behaviour (Smak 1985b). One obvious advantage of such an approach is connected with the time scales of various accretion phenomena, which are sufficiently short to avoid any major observational bias: the mass accretion rates in CVs usually range from $10^{-11}$ to $10^{-8}$ $\mathrm{M}_{\odot}\ \mathrm{yr}^{-1}$ (Patterson, 1984); the time scales are from tens of seconds (oscillations in dwarf novae at outbursts) to years (super-outbursts of SU UMa stars or long term variations in VY Scl stars).

Depending on the magnetic field intensity at the white dwarf, the accretion of matter from the secondary star onto the primary can occur either via an accretion disk (in the so-called Non-Magnetic CVs: NMCVs) or a channelling through the magnetic poles (in the case of Polars: PCVs) or in an intermediate way (in the case of Intermediate Polars: IPCVs).

CVs in a time scale of order between weeks and years flare up almost periodically, about few magnitudes in optical wavelengths; the duration of the outbursts is much shorter than the recurrence time.

In PCVs the white dwarf magnetic field is strong enough to make the Alfvén radius greater than the circularization radius, so no accretion disk is formed and the accretion structure is fully governed by the magnetic field, which canalize the accreting matter across the field lines. Owing to the intense magnetic field ($\sim$ 10–200 MG), the white dwarf rotation is synchronized with the binary orbital period (a few hours).

IPCV white dwarfs have moderate magnetic fields (order of a few MG); the Alfvén radius is smaller than the circularization radius but it is greater than the white dwarf radius. Therefore an accretion disk is formed in these systems but being disrupted at its inner region. In IPs matter follows again the magnetic field lines but just inside the Alfvén radius. The rotating white dwarf is asynchronous with the binary orbital period ($P_{\rm spin} \ll P_{\rm orb}$).

The last group defined by the accretion structure criterion, NMCVs, includes those systems whose white dwarf magnetic fields are not relevant in governing the accretion structure. In these systems the accretion disk extends down to the white dwarf surface and a boundary layer is formed. This family shows a great diversity of observational behaviour; for this reason the historical criterion of classification is, in this case, more appropriate for distinguishing their sub-classes. However, it is simply an attempt of classification for lack of a more general physical classification (e.g., Giovannelli, 1991 and references therein).

As recalled by Giovannelli & Sabau-Graziati (1999a), it is evident that the properties of an outburst in CVs depend crucially on the accretion rate, the mass of the white dwarf, and the chemical composition of its hydrogen rich envelope in which the thermonuclear runaway occurs. And the accretion process onto the white dwarf is strongly influenced by its magnetic field intensity. Indeed, the three kind of CVs (non-magnetic, polars, and intermediate polars) obey to relationships between the orbital period of the system and the spin period of the white dwarf (Warner & Wickramasinghe, 1991), where the magnetic field intensity plays a fundamental role. Then the investigation on the magnetic field intensities in white dwarfs is crucial in understanding the evolution of CVs systems. This means that it is necessary to study CVs belonging to different positions in the plane log $P_{\rm orb}$–log $P_{\rm spin}$, in order to find experimental indicators of their evolutionary history. But such a selection of CVs corresponds, in other words, to select systems having white dwarfs with different masses, magnetic moments and rotational velocities. And this is, in our opinion, the most powerful and general method to study CVs as gravimagnetic rotators, following Lipunov's idea (1987) on *The Ecology of Rotators*. In this way each system is completely described by those physical parameters. Thus through those parameters we could obtain the actual classification of CVs we are searching for.

The field strength distribution of MCVs differs from that of single white dwarfs, although both cluster around 30 MG. The distribution of isolated white dwarfs extend on a wide range of magnetic field strength ($\sim 10^5$–$10^9$ G), whilst in accreting white dwarfs of CVs, as far as is presently known, there is a lack of systems at both high and low field strengths. However, the apparent absence of low field MCVs might be explained by the IPs, which generally have unknown field strengths, and the lack of high field systems is still not understood (e.g., Beuermann, 1998).

Our opinion is that a more appropriate investigation of the class of the so-called IPs is necessary. Indeed, such systems could show surprises if deeply studied, as for instance occurred for the CV SS Cyg. This system, usually considered a NMCV because of a classification once made by Bath & van Paradijs (1983), whilst since 1984 it has been claimed as an IP by Giovannelli *et al.* (1985b) and later confirmed several times (e.g., Giovannelli & Martinez-Pais, 1991; Giovannelli, 1996a; Giovannelli & Sabau-Graziati, 1999a, Gaudenzi *et al.*, 2002). As explained by them, the behaviour of SS Cyg are sometimes close to those of a NMCV and sometimes to those of a PCV, being its magnetic field of $\sim 10^6$ G (Fabbiano *et al.*, 1981). This would teach a lesson: it is mandatory to observe CVs for long time in order to follow at least a whole period of the binary system between two successive outbursts. This is, of course, possible only for systems like dwarf novae where the almost periodical outbursts occur in time scales of weeks-months.

However, we can say that CVs form a broad stellar family of highly variable and dynamical members. When it comes to explaining particulars about, e.g., the detailed interaction between the transferred matter and the white dwarf's atmosphere; irregularities within regular photometric behaviour; turbulent transport in the disk; or the final fate of these objects, more is missing than what is known, rendering their study ever more challenging. At least, because CVs are natural multi-wavelength laboratories offering us the possibility of studying in detail the behaviour of plasma and radiation under extreme physical conditions. The understanding of stellar evolution, electromagnetism and polarization, mass and radiation transfer or 3-D geometrical effects, in a broad spectral range from hard X-rays to radio, is mandatory for improving the knowledge of the nature of CVs.

Variability, from milliseconds to hundreds of years, follows from different physical processes taking place in these systems and can be studied by means of several astronomical techniques. As our skills in developing further these techniques grow our understanding of the CVs insights also grows; and the more we learn about CVs the further techniques and theory develop. On the other hand, it is well known that conclusions obtained in the field of CVs have been extrapolated, upwards or downwards in scale, to other fields such as AGNs or LMXRBs, and vice versa.

From such exchanges of information and results astrophysical research in general always benefits.

For a recent review on NMCVs see the paper by Ulla, Giovannelli & Martinez-Pais (1999).

7.9.2. *Multifrequency Emissions from Non-Polar Cataclysmic Variables*

In CVs there are two stellar components that are responsible for minor but significant fractions of the total emission:

**i) the secondary stars** are cool main sequence stars with spectral type ranging from G8 to M6, corresponding to temperatures from 5,000 to 3,000 K. So they contribute substantially to the integrated radiation only in the red and IR regions;

**ii) the primary stars:** the temperatures of white dwarfs are known only in few cases: when they belong to high inclination systems, or when they accrete matter with a very low mass transfer rate. However, the white dwarf temperatures range between 10,000 and 50,000 K (Sion, 1986; 1991). Then statistically the white dwarfs in CVs seem to be hotter than single white dwarfs whose temperatures typically range between 10,000 and 20,000 K. Therefore white dwarfs are expected to radiate essentially in the UV, but they can be visible also in the optical range if they are not too hot. However, in CVs the geometrical dimensions of the white dwarfs are negligible with respect to those of the accretion disks; the consequence is that their emission can usually be compared with the background. The white dwarf is visible during the quiescent state of a low mass transfer system and manifests itself with the presence of the broad red wing of the $L_\alpha$ absorption line in the IUE spectrum, such as WZ Sge (Sion, Leckemby & Szkody, 1990) and reviewed by Szkody (1998).

Two other components responsible for the most conspicuous emission in CVs are: iii) the accretion disk; and iv) the boundary layer.

**iii) The accretion disk:** it does not have a homogeneous temperature, but spans a large range. Since the temperature distribution in disks is poorly known, in order to obtain a rough evaluation of their contribution to the total emission it is necessary to evaluate the contributions at different frequencies of a synthetic disk constituted of black bodies at different temperatures, the temperature distribution being that of a stationary accretion disk (e.g., la Dous, 1994). It then appears evident that the contribution of such an accretion disk is important in the whole range between EUV and IR, depending on the choice of the disk parameters. Furthermore, the UV radiation can be supplied from a zone in the vicinity of the white dwarf (some ten stellar radii), which could contain any optically thick material left there.

In a first approximation, in order to fit the observed flux distribution it is possible to assume that only the accretion disk contributes to the radiation, and that at each distance from the white dwarf it radiates like a black body with a radial temperature derived for a stationary accretion disk. For the theory of accretion disks see the book of Warner (1995) and references therein. With these assumptions la Dous (1989) computed the contribution function of a black body disk for a 1 $M_\odot$ white dwarf and accretion rate of $10^{-9}$ $M_{\odot}$ $yr^{-1}$. She divided the disk into annuli and computed the intensity emitted by each annulus, weighting with the area of the annulus. The UV radiation is dominated entirely by radiation from the central disk. The central and middle disk contribute to the optical. The IR radiation comes from the middle and outer disk. The contribution of the accretion disk to the total emission of a CV is shown in Figure 114.

**iv) The boundary layer:** a very important zone for the emission is that of the transition between the accretion disk and the white dwarf surface, namely the boundary layer. It is possible that all, or at least a significant fraction, of the kinetic energy of the material contained in the accretion disk must be radiated away within the geometrically very small boundary layer in order to have the possibility of the material accreting onto the white dwarf's surface. Then, whatever the situation, one can assume the presence of a strong X-ray source at the boundary layer, which will be visible also in the EUV and short-wavelength UV according to the choice of disposable theoretical parameters.

Most of the radiation then comes from the accretion disk and boundary layer, which contribute roughly 50% each. From the accretion disk the radiation is essentially emitted in the optical and UV, whilst from the boundary layer – optically thick (which occurs at high accretion rates) – the radiation is emitted in the soft X-ray range; when the accretion is at low rates the boundary layer is optically thin and appears as a hard thermal bremsstrahlung source.

These predictions have been tested experimentally, comparing the observations of CVs in optical, UV and soft X-ray ranges. The most promising technique has been used by Wood *et al.* (1989) and Horne (1994); this consists in observing eclipses at different frequencies in order to derive spectra at different radii of the disk surface. These experimental spectra are compared with those theoretically computed, by using models for deriving the physical parameters such as temperature, mass transfer rate, etc.. Alternatively, the image reconstruction is used to find the best map fitting the data best. The results from the high inclination dwarf novae Z Cha and OY Car have shown that the 'steady state accretion disk model' is correct during outburst, but not during quiescence, when the accretion disk becomes optically thin.

The picture of the boundary layer, coming from the X-ray, optical, and UV data, appears rather confusing, and in any case the emission from the boundary layer is much lower than predicted ($\leq 1/4$ of the disk's luminosity) (e.g., Mauche *et al.*, 1991; Belloni *et al.*, 1991). The explanation for the low emission from the boundary layer could be a fast rotational velocity of the white dwarf, close to the break up velocity; but this has been experimentally contradicted by the detection of the SiIV line from the photosphere of U Gem, which is consistent with a low rotational velocity (Sion *et al.*, 1994).

The ionization state of the winds from dwarf novae at outburst and in nova-like systems at high accretion rates can provide constraints on the boundary layer temperatures (Drew, 1990), but developed models (Vitello & Shlosman, 1993;

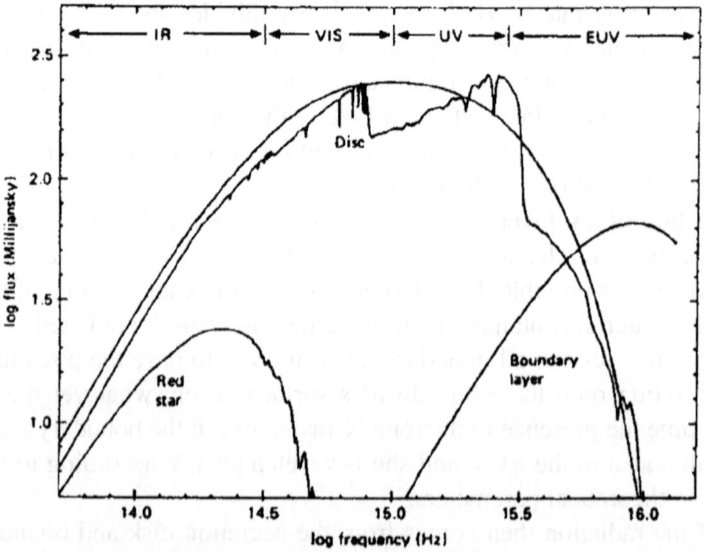


*Figure 114.* Contribution functions of the most important components of a cataclysmic system from IR to EUV (Pringle & Wade, 1985).

Shlosman & Vitello, 1993) using such a constraints (i.e. temperature of boundary layer less than 100,000 K) present many other difficulties not yet completely explained, such as a disk/wind asymmetry, as measured by Woods *et al.* (1992) with the IUE in low inclination systems. Probably, considering the presence of the white dwarf's magnetic field, the picture could be clarified. Indeed, the phase-dependent profiles imply an asymmetry which extends all the way to the inner disk. As shown by Shlosman, Vitello & Mauche (1996), the data are most consistent with an extensive moderately collimated rotating wind that originates from the surface of the accretion disk. A kinematic model and a rotating biconical disk wind (Knigge *et al.*, 1997) contain a high density transition region which can explain the FeII curtain which is evident in high resolution HST spectra of high inclination systems such as OY Car (Horne *et al.*, 1994).

Two more components of CVs as minor contributors to the total emission are: v) the gas stream; and vi) the hot spot.

**v) The gas stream:** it is definitively optically thin and cool and contains rather little material; so, probably, its contribution to the total emission of CVs is negligible at all frequencies as source of continuum, while it could contribute to the formation of lines in the red and IR regions;

**vi) The hot spot:** its structure and radiation characteristics are still an open problem; it is visible in many systems in optical photometry (less in the IR and never in the UV) as a periodically recurring hump in the orbital light curve. Its temperature must be $\leq 10,000$ K;

Finally, some observations of individual systems show the presence of a:

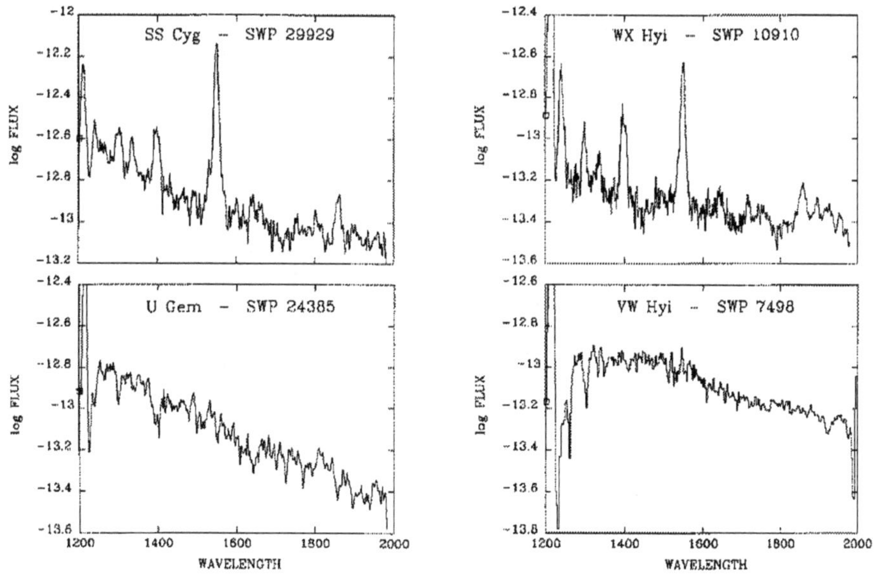


*Figure 115.* Typical UV spectra of dwarf novae in quiescence. Fluxes are in erg cm$^{-2}$ s$^{-1}$ Å$^{-1}$ (la Dous, 1993).

**vii) hot corona or chromosphere:** i.e., a shell of optically thin and rather hot gas, below and above the accretion disk. X-ray and UV line radiation are tentatively attributed to it, whilst it probably does not contribute to the UV, optical or IR continuum emission.

No radio emission from CVs has been measured. Only upper limits for individual systems of order of a few mJy are available (e.g $\leq$ 10 mJy in SS Cyg – Cordova, Mason & Hjellming 1983).

Figure 114 shows the contribution functions of the most important components of a cataclysmic system, previously discussed, as computed by Pringle & Wade (1985).

### 7.9.3. *Spectacular Spectral and Photometric Behaviour of Cataclysmic Variables*

Most dwarf novae in quiescence show strong emission lines in optical and UV ranges; however, some of them have only weak emissions in the optical and/or weak absorption lines in the UV. Figure 115 shows typical UV spectra of four dwarf novae in quiescence (la Dous, 1993).

Balmer emission lines show very often double peaked profiles, orbitally modulated, as shown in Figure 116 in the case of U Gem (Stover, 1981).

In some systems the absorption spectrum of the secondary cool main sequence star is visible in the red part of the optical spectrum. Figures 117a,b show the averaged quiescent spectra of SS Cygni after removing the primary motion: after (up) and before (down) secondary star subtraction. Figures 117c,d show the av-

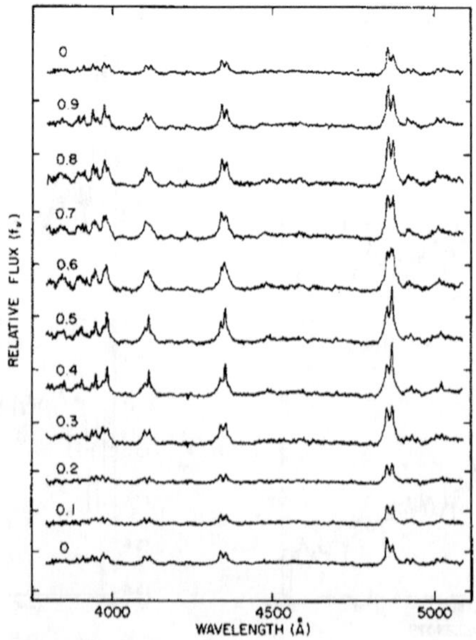


*Figure 116.* Double peaked Balmer lines orbitally modulated in U Gem. Phase zero is at mid-eclipse (Stover, 1981).

eraged quiescent spectra of SS Cygni after removing the secondary motion: after (up) and before (down) secondary star subtraction. In the bluer spectrum of Figure 117a double peaked $H_\delta$, $H_\gamma$ and $H_\beta$ are clearly visible, as well as in the redder spectrum of Figure 117b, taken in other epoch, double peaked $H_\alpha$ is clearly visible (Martinez-Pais *et al.*, 1994).

The radial velocity curves of both emission and absorption lines have the same periods as the hump or eclipse light curves, as shown in Figure 118 for SS Cygni in quiescence (Stover *et al.*, 1980). Emission and absorption lines are out of phase by $\sim$180°, being the phase lag of the emission with respect to the absorption of $\approx$5–10°. However, this lag varies from one system to another.

The variations in the profiles of double peaked emissions connected with the orbital phase are called S-wave. Figure 119 shows such an S-wave in the $H_\alpha$ emission line from SS Cygni in quiescence (Martinez-Pais *et al.*, 1994).

What do these spectacular spectral behaviours mean? As suggested from the theories of accretion disks and CVs (e.g., the book of Warner (1995) and the references therein) the generally observable emission line spectra of dwarf novae in quiescence are coming from an optically thin disk area, whilst the UV absorption spectra indicate that the inner accretion disk is optically thick. In optical and IR, the cool absorption spectrum is due to the secondary star. Periodic variations in line profiles and in radial velocities are correlated with the orbital phase, whilst random profile changes are owed to inhomogeneities in the accretion disk brightness distribution. The double peak profiles in systems with relatively high orbital

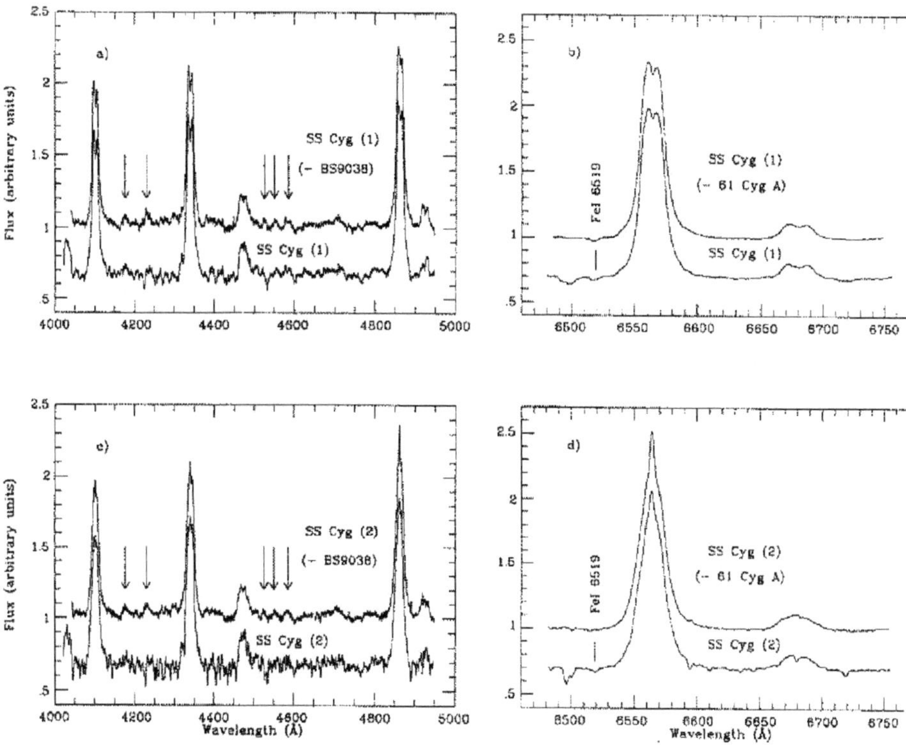


*Figure 117.* **a:** Spectra averaged after removing primary motion, after (up) and before (down) secondary star subtraction; **b:** Same as **a** but for spectra taken in another epoch and another range; **c:** Same as **a** but after removing secondary motion; **d:** Same as **c** but for spectra taken in another epoch and another range (Martinez-Pais *et al.*, 1994).

inclination are owed to the disk's rotation, as discussed for instance by Giovannelli *et al.* (1983) who derived the orbital parameters of SS Cygni starting from the measured separation of double peaks in the Balmer emission lines. The phase lag between the radial velocities determined from emission and absorption is an effect of the binary nature of the system, as the emission lines originate in the disk and the absorption spectrum in the secondary star.

During the outburst an absorption spectrum, of the resonance lines and of HeII (1640 Å) in the UV and of mostly the Balmer lines in the optical, appears, as computed by la Dous (1989). Figure 120 shows typical IUE SW spectra of four CVs in outburst (la Dous, 1993), in agreement with those computed in the UV range.

The UV resonance lines often display either strong asymmetric blue shifted line profiles or P Cygni profiles, like shown by Giovannelli *et al.* (1990a) for CIV (1548–1551 Å), indicating a process of mass loss from the system SS Cygni (Figure 121). The absorption spectrum owed to the late-type secondary star disappears because the hot continuum flux increases during outburst.

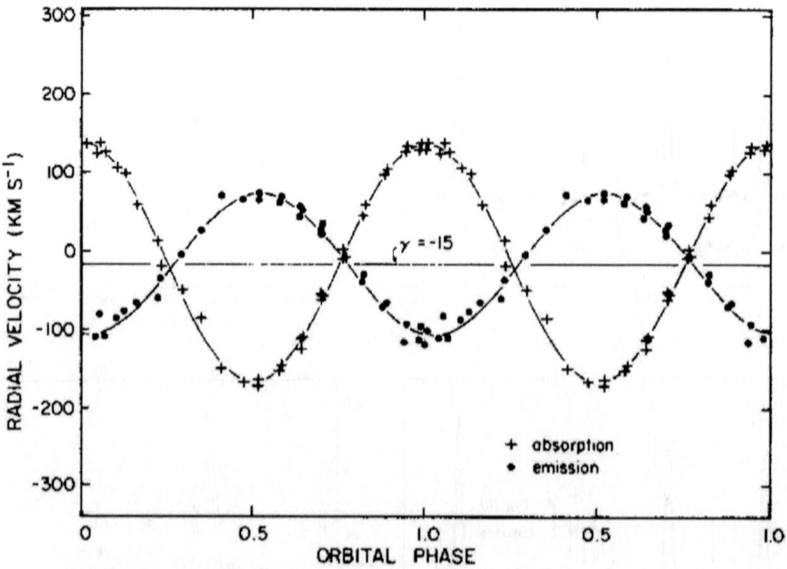


*Figure 118.* Radial velocity changes of SS Cygni during quiescence. The velocities of the Balmer emission and the cool absorptions are out of phase by ~180° (Stover *et al.*, 1980).

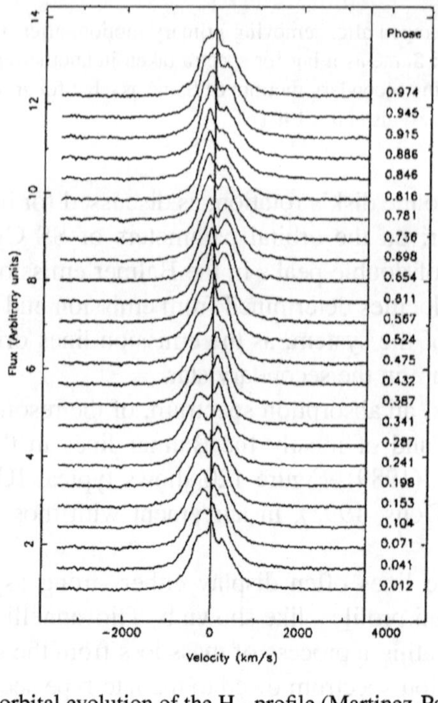


*Figure 119.* The orbital evolution of the $H_\alpha$ profile (Martinez-Pais *et al.*, 1994).

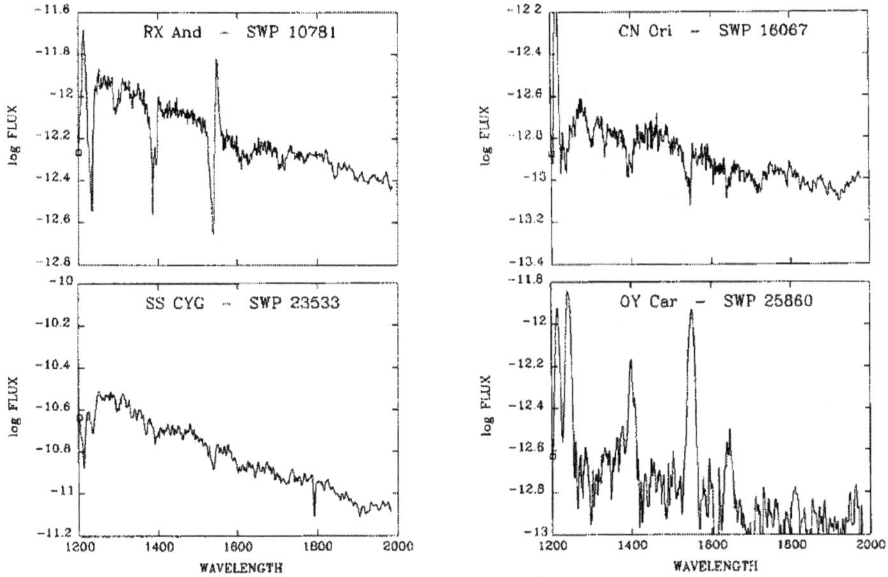


*Figure 120.* Typical UV spectra of dwarf novae in outburst. Fluxes are in erg $cm^{-2}$ $s^{-1}$ $Å^{-1}$ (la Dous, 1993).

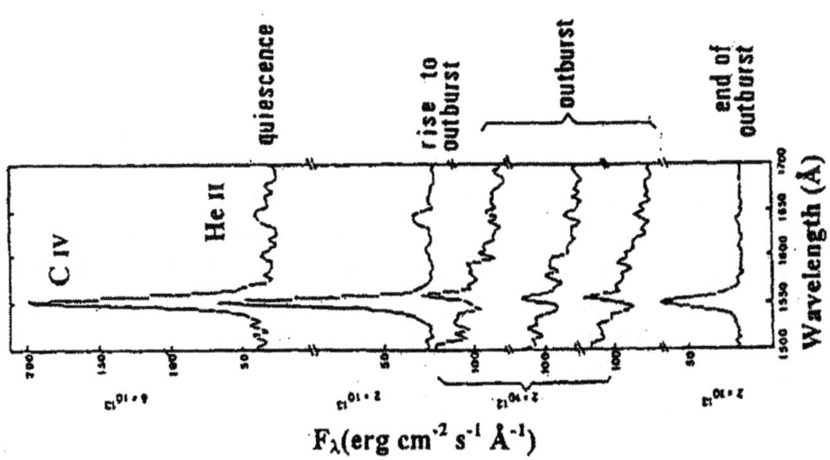


*Figure 121.* Quiescence–outburst–quiescence evolution of the CIV profile (Giovannelli *et al.*, 1990a).

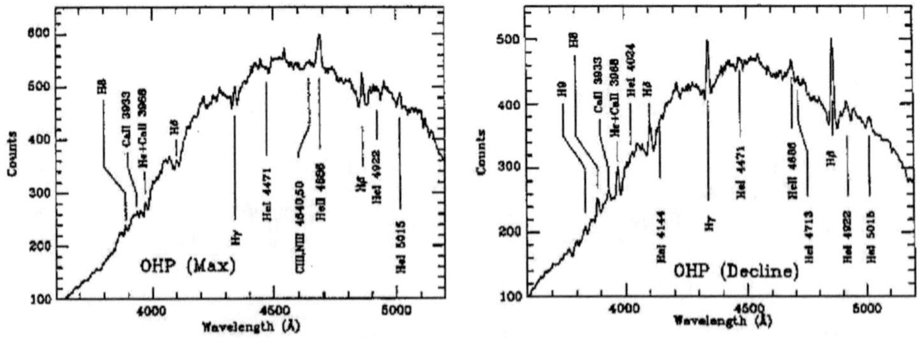


*Figure 122.* Mean of the OHP (Observatoire d'Haute Provence) spectra corresponding to: maximum (left) and the decline (right) (Martinez-Pais *et al.*, 1996).

At maximum and during decline, the optical lines sometimes have emission cores, as shown in Figure 122 for SS Cygni (Martinez-Pais *et al.*, 1996).

So, in general, during a quiescence–outburst–quiescence cycle, the emission lines gradually sink into broad absorption shells during the rise. The HeII (4686 Å) emission line and the CIII–NIII blend at 4650 Å, usually only marginally visible (or absent) in quiescence, essentially follow the development of the continuum flux during outburst. They are visible at late rise, fade during decline, and disappear again at a late stage of decline. At or shortly after maximum light emission cores begin to grow in the absorption lines, and during the decline the quiescent spectrum is restored (see Figure 122).

What do these spectacular spectral behaviours mean? As suggested from the theories of accretion disks and CVs, during outburst all the disk is assumed to be opaque; only absorption spectra are visible then. Blue shifted absorption profiles or P Cygni profiles are generated because of high velocity wind, which is expelled from the system during the outburst. The profile variations are owed to opacity changes in the accretion disk: the disk is partially optically thin during quiescence and becomes optically thick during outburst; the changes occur gradually. Profile changes are probably owed to brightness inhomogeneities in the disk.

What about the X-ray emission? Following the experimental results, during quiescence dwarf novae emit essentially hard X-rays (∼0.1–4.5 keV) and the flux distribution is rather well approximated by a thermal bremsstrahlung with $KT_{\rm brems} \approx$ 10 keV (Cordova & Mason, 1983). A direct correlation between the hard X-ray/optical fluxes ratio and $H_\beta$ equivalent width has been found by Patterson & Raymond (1985b), as shown in Figure 123.

During outburst dwarf novae emit soft X-rays (0.18–0.5 keV) with an increase of the flux of the order of 100 or more, although most of the radiation is hidden in the EUV range (Cordova & Mason, 1984). The soft X-ray spectra can be fitted either with black bodies at $KT_{\rm bb} \approx 25$–30 eV or, alternatively, with bremsstrahlung spectra at $KT_{\rm brems} \approx 30$–40 eV.

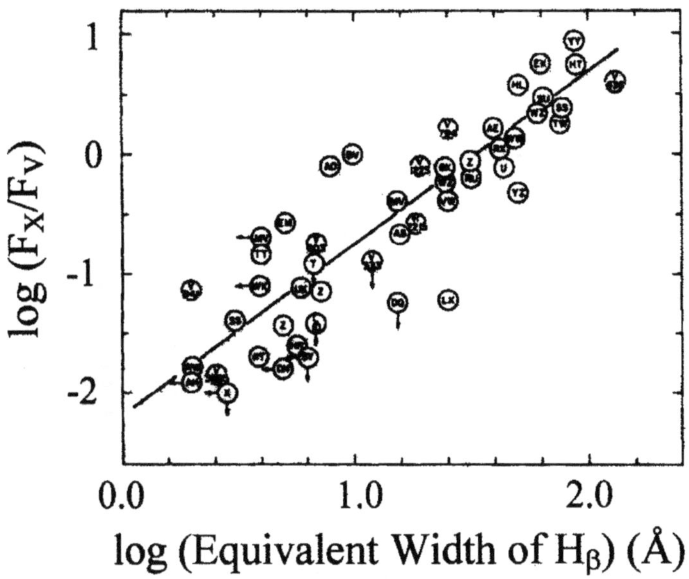


*Figure 123.* X-ray/optical flux ratio vs $H_\beta$-equivalent width in CVs (Patterson & Raymond, 1985a).

The most important features, shown in Figure 124, are the anti-correlation between the hard and soft X-ray emissions during the outburst cycle and the correlation between soft X-ray and optical emissions, as measured for SS Cyg (Watson, King & Heise, 1985), or – what is the same – anti-correlation between the hard X-ray and optical emission (Ricketts, King & Raine, 1979).

What does that mean? The UV flux and the bulk of optical flux in dwarf novae and nova-like stars originate in the accretion disk. The IR flux observed during quiescence and possibly some of the optical flux come from the secondary late-type star. The rise to an outburst either occurs simultaneously at all wavelengths when it is slow, or progressively starts later with decreasing wavelengths when it is fast, since ever more central hotter parts of the disk become involved. Indeed, several dwarf novae have been observed in the UV and optical during the rise to maximum outburst brightness and their behaviours are quite similar: the UV rise lags the optical rise by up to a day (e.g., VW Hyi: Hassall *et al.* (1983); CN Ori and RX And: Cordova, Ladd & Mason (1986); WX Hyi: Hassall, Pringle & Verbunt (1985). Figure 125 shows this lag for VW Hyi. With respect to the optical band, this lag is similar also in the EUV region covered by the Voyager (50–1200 Å) for SS Cyg (Polidan & Holberg, 1984) and VW Hyi (Polidan & Holberg, 1987).

This fact strongly supports the origin of the outburst being in the cooler outer part of the disk rather than in the hotter parts near the white dwarf; therefore mini-nova models for the outbursts are probably excluded (Cordova & Howarth, 1987). The two models for triggering the outbursts, compatible with the lag observations, are then:

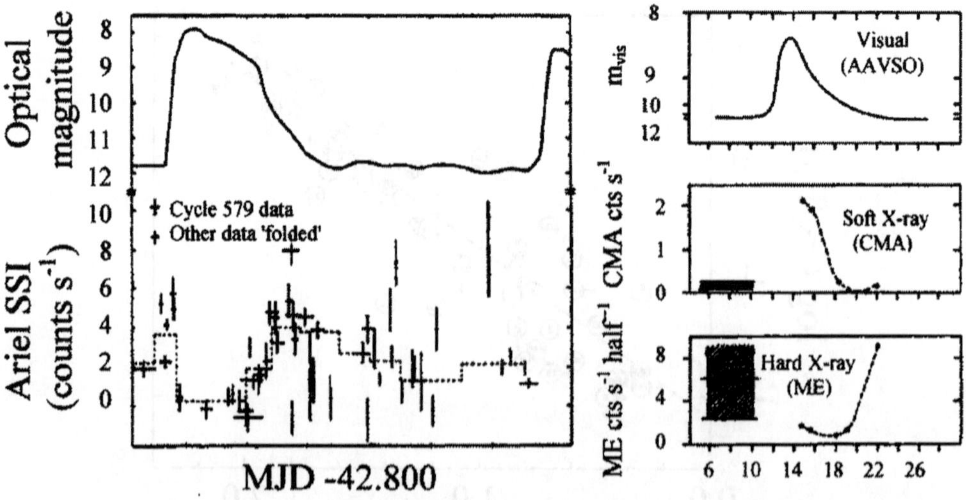


*Figure 124.* (Left) the anti-correlation between the optical and hard X-ray intensities (Ricketts, King & Raine, 1979). (Right) the anti-correlation between hard and soft X-ray intensities, which is correlated with the optical luminosity of SS Cygni (Watson, King & Heise, 1985).

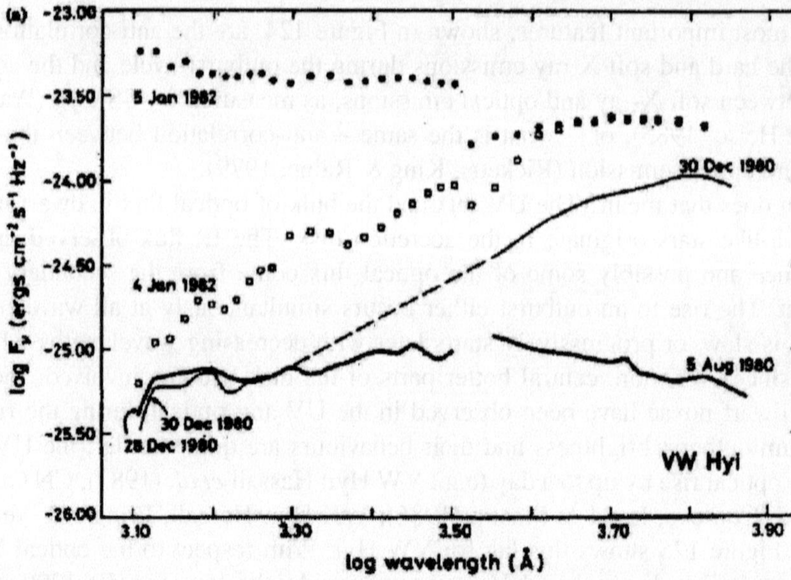


*Figure 125.* Various stages of the outburst evolution in VW Hyi. The quiescent spectrum (solid line) has been constructed with different observations. Two spectra taken on rises to outbursts on 1980 December 30 and 1982 January 4 show that the optical increases significantly while the UV is still low. The peak spectrum of 1982 January 5 is much flatter and resembles all the decline spectra of this system. The shift between the optical and UV probably results because the optical data were taken a few hours after the UV data when the star's brightness was decreasing (Schwarzenberg-Czerny *et al.*, 1985).

- an instability in the secondary star which allows the transfer of more mass to the disk;
- a thermal instability in the outer disk, which results in material stored there being suddenly transported through the disk. During the decline, the whole disk cools simultaneously. The contribution to the total emission from the boundary layer between the disk and the white dwarf surface is in UV and X-ray ranges: the boundary layer is optically thin during quiescence and then emits hard X-rays, but it is optically thick during outburst and then emits soft X-rays since the radiation is thermalized before escape (la Dous, 1993).

From the EUVE (Extreme Ultraviolet Explorer) Craig *et al.* (1997) have shown the EUV spectra of three NPCVs in outburst, namely VW Hyi, U Gem and SS CYG. VW Hyi shows the softest EUV spectrum peaked at ~250 Å. However, its boundary layer/disk luminosity ratio – $L_{\rm bl}/L_{\rm disk} \sim 0.2$ (Mauche *et al.*, 1991) – is in contrast with the boundary layer models. In U Gem the total size of the EUV emitting region is comparable to that of the white dwarf itself, which indicates that the outburst in mainly confined to the inner disk/boundary layer region (Long *et al.*, 1996). Its $L_{\rm bl}/L_{\rm disk} \sim 1$. SS Cyg is the hardest of the three EUV NMCVs. Its EUV spectrum is rather complex and changed by a factor 100 during the outburst with an almost constant spectral energy distribution. Quasi-coherent oscillations (~7–9 s) have been detected in the EUV emission (Mauche, 1996). SS Cyg shows no EUV emission longward ~130 Å. Mauche, Raymond & Mattei (1995) found that for SS Cyg the relation $L_{\rm bl}/L_{\rm disk} \sim 1$, valid for the boundary layer models is strongly violated, being this ratio $L_{\rm bl}/L_{\rm disk} \sim 0.07$. And this is one more proof that SS Cyg is not a NMCV, as already remarked, as claimed by Giovannelli's group.

### 7.9.4. *Some as Yet Open Questions*

Several fundamental questions concerning NMCVs (and CVs in general) still remain waiting for a proper answer. We will present briefly only some of them here.

One of them is the lack of a coherent classification, especially for NLs. On the other hand, in gross features and in most respects, DN and NLs, as well as quiescent novae, are almost indistinguishable, although, in addition to their different outbursts' behaviour, there appear to be some further minor differences which are not yet understood (see Hack & la Dous 1993). The question arises of whether the outburst behaviour, the current basis of almost all classification is really a suitable criterion for sorting CVs in physically related groups. There are also too many exceptions, either systems that do not fit in any particular group or that can be included in several of them, to be able to render the observational behaviour, at least as it is used at the present, suitable.

Could the classification of CVs as gravimagnetic rotators be the most suitable from a physical point of view?

Also it is necessary to explain the presence of a few objects inside the period gap. These are not so many, but their number increases proportionally to the number of new CVs discovered. Either a particular explanation has to be found for those objects, or a new general explanation is needed for the existence of the period gap.

Ultra-short period CVs (i.e. AM CVn systems) are still waiting for a general model. They are probably binary systems of two white dwarfs, but even this is still controversial.

Despite all the work developed during the last decades, the problem of modelling accretion disks in CVs is by no means closed, especially in quiescence. Closely related is the problem of the cause of outburts. We really do not know which of the present two families of models (Disc Instability Models or Secondary Instability Models) is responsible for the CVs outburst phenomenon, or in which system is each model valid, although Martinez-Pais *et al.* (1996) have partially solved this problem at least in the case of SS Cygni, where they found some evidence for an increase of the mass transfer rate from the secondary star as the mechanism responsible for symmetric outbursts.

Something similar can be said about the super-outburst phenomenon in SU UMa systems.

Another problem still open is connected with the three kinds of CVs, namely NMCVs, PCVs and IPCVs, which is, in our opinion, connected with the problem of a physical classification of CVs. Are these classes actually separated? Or some systems, usually placed in one of these classes, can behave in one or another way, typical for different classes, depending on their instantaneous physical conditions?

An example very clear is that of SS Cygni, usually classified as a non-magnetic dwarf nova. Several proofs have been shown and discussed many times by Giovannelli's group in order to demonstrate the Intermediate Polar nature of it (e.g., Giovannelli, 1996a, and references therein; Giovannelli & Sabau-Graziati, 1998); indeed, SS Cygni shows characteristics of a NMCV, as well as those of IP and sometimes even those of polars, although its position in the log $P_{\rm spin}$–log $P_{\rm orb}$ plane is very close to the line where IPs lie.

The investigation on the magnetic field intensities in white dwarfs is then crucial in understanding the evolution of CVs systems, by which it is possible to generate classical novae (e.g., Isern *et al.*, 1997) and type-Ia supernovae (e.g., Isern *et al.*, 1993).

In those catastrophic processes the production of light and heavy elements, and then the knowledge of their abundances provides strong direct inputs for cosmological models and cosmic ray generation problems.

### 7.9.5. *Multifrequency Emission from Magnetic Cataclysmic Variables*

The conclusions which we can derive from multifrequency observations of Non-Polar CVs, with emphasis on the IUE observations, can be found in the review papers by Ulla, Giovannelli & Martinez-Pais (1999), Cordova (1995, 1998) and Szkody (1998), and in the particular case of SS Cyg in the reviews by Giovannelli & Martinez-Pais (1991), Giovannelli (1996a) and Giovannelli & Sabau-Graziati (1998, 1999a). In the latter papers, the importance of the white dwarf magnetic field, as also in the so-called Non-Polar CVs, has been discussed many times, so that we are convinced that a general study of CVs could be more profitable by considering them as systems containing a gravi-magnetic rotator (white dwarf) with a mass $M$, angular velocity $\vec{\omega}$ and magnetic moment $\vec{\mu}$, whose axis is not necessarily coincident with the rotation axis. And the importance of the magnetic field appears, for instance, in the interpretation of data coming from the winds of CVs in outburst, or in the case of SS Cyg, currently still classified as NMCV, and, on the contrary, claimed to be an IPCV by Giovannelli and his group since many years (e.g., Giovannelli *et al.*, 1985b; Giovannelli & Martinez-Pais, 1991; Giovannelli, 1996a and the references therein).

Briefly, multifrequency observations of CVs by using also IUE made significant contributions on:

**i) knowledge of disk accreting and magnetic CVs**, as extensively discussed by Cordova (1995), and references therein. Indeed, crucial new information has been provided for understanding the dwarf nova outburst mechanism – extensively described by Giovannelli & Martinez-Pais (1991) – by means of the temporal behaviour of the various sources of emission during the progress of the outburst in optical, UV, soft and hard X-ray ranges; it has been possible to discriminate between the disk instability model and the Roche lobe overflow model for the mechanism underlying the dwarf nova outburst phenomenon.

**ii) nature of the high velocity winds.** Indeed, IUE revealed the presence of high velocity winds in CVs during outbursts. However, the ionization structure, mass loss, and geometry of the wind are still uncertain, as well as the relationship between the changes in the X-ray emission and the appearance and evolution of short period oscillations. A review on UV observations of magnetic cataclysmic variable is that by de Martino (1999).

**iii) boundary layer emission.** Multifrequency observations show that X-ray luminosity at all outburst phases is much lower (about at least a factor 10) than the UV/optical luminosity from the disk, as expected from the models (e.g., Mauche, 1998). This simply means that the boundary layer models are not correct.

In particular VW Hyi, whose EUV spectrum is peaked at ~250 Å, has $L_{\rm bl}/L_{\rm disk} \sim 0.2$ (Mauche *et al.*, 1991), which is in contrast with the predictions of the boundary layer models. In SS Cyg, Mauche, Raymond & Mattei (1995) find $L_{\rm bl}/L_{\rm disk} \sim 0.07$, which strongly violates the predictions of the boundary layer models. In contrast, in U Gem, $L_{\rm bl}/L_{\rm disk} \sim 1$, which indicates a good agreement with the predictions of the boundary layer models.

Physical models are constructed under specific hypothesis. If the hypotheses are correct, in the sense that they reproduce roughly the physical conditions occurring in nature, one can expect that models could fit nature rather well. If not, models are simply wrong or not sufficient. Boundary layer models are working i.e. for U Gem, but not in the cases of VW Hyi and SS Cyg. This means that the physical hypothesis used for their construction are encountered in U Gem but not in the other two systems. This could mean that in VW Hyi and SS Cyg the magnetic field of the white dwarf inhibits the inner part of the accretion disk to rich the white dwarf's surface; in other words the boundary layer (in its original meaning) does not exist. Indeed, as many times claimed, SS Cyg could be an intermediate polar (e.g., Giovannelli, 1996a and the references therein), or at least a frontier system between IPs and NMCVs (Giovannelli & Sabau-Graziati, 1999a).

**iv) underlying white dwarf and its photosphere.** IUE provided the first evidence that the white dwarf is heated by the dwarf nova outburst and subsequently cooled. A list of nine such systems has been reported by Szkody (1998). These measurements are very difficult because of the long quiescence–outburst–quiescence cycles (from weeks to years). The short outburst period dwarf nova VW Hyi cooled to 18,000 K (from 20,500 K) in the 14 days before the next outburst began (Verbunt *et al.*, 1987).

Time resolved spectroscopy with IUE and optical telescopes of CVs in quiescence provided measurements on the temperature of the white dwarf and its mass and rotational velocity. With higher time and spectral resolutions of HST, Sion *et al.* (1997) from the analysis of VW Hyi quiescent spectra have shown the first evidence of a thermonuclear runaway on the white dwarf for SS Cyg – as suggested by Gaudenzi *et al.* (1990), commented by Giovannelli & Sabau-Graziati (1999b), and evaluated by Gaudenzi *et al.* (2002) – as a possible mechanism for producing outbursts.

These observations are important for the relationship between dwarf novae and other types of CVs, and also for understanding the influence of outbursts on the boundary layer emission.

Verbunt (1987) presented and discussed on the UV observations of CVs by using the IUE archive. He presented a comparative study of the UV properties of CVs, determined the reddening of 51 systems from the 2200 Å feature. He noted that the spectral flux distribution of CVs does not depend on system type, orbital period, or (for dwarf novae) average length of interval between outburst maxima; it does not depend strongly on inclination. The exception is formed by the DQ Her systems, which may have relatively low fluxes at short wavelengths as compared to other systems. All systems for which observations are available show variability on a time scale of hours. In two nova-like variables in a low state and in five dwarf novae in quiescence there is evidence that most of the ultraviolet flux at short wavelengths is owed to the white dwarf. He calculated disk model spectra in order to show that the slope of the spectrum depends strongly on the white dwarf mass. Hence, determinations of the mass transfer rate from the spectral slope are subject to large errors in cases where the white dwarf mass is not known.

Paerels *et al.* (1996) presented high resolution spectroscopy of the stellar photospheric spectrum of the white dwarf in AM Her in the 75–120 Å band, obtained with the Short Wavelength Spectrometer on the EUVE satellite. They detected ionization edges and absorption lines from highly ionized neon (NeVI, NeVIII). Surprisingly they did not detect absorption at the OVI 2s and 2p edges, which are expected to be the strongest spectral features in this band in an atmosphere of solar composition at the density and temperature expected from the accretion region in this object. There was evidence for limb brightening of the spectrum in egress from eclipse, indicating the presence of a temperature inversion in the X-ray/EUV photosphere. These results are in agreement with the classical model for the soft X-ray/EUV emission from AM Her, namely hard X-ray irradiation of the atmosphere by the accretion shock.

**v) magnetic field of the white dwarf.** Indirect evaluations of magnetic field intensities in CV white dwarfs have been obtained through multifrequency observations (e.g., Fabbiano *et al.*, 1981). Our feeling is that the problem of magnetic fields in white dwarfs has been underestimated in the studies of CVs. Too many simplified models of disk accreting and magnetic CVs have been developed under the hypothesis that CVs can be sharply divided into three classes: Polar, IP, Non-Magnetic. Magnetic fields are smoothly varying in their intensities from one class to another, probably giving rise to misunderstandings in classifying CVs in the three classes mentioned, which probably contain systems which evolve from one class to another. Indeed, the discovery in some IPs of a circularly polarized optical emission suggests that these intermediate polars will evolve into polar systems (e.g., Mouchet, Bonnet-Bidaud & de Martino *et al.*, 1998).

Looking at the homogeneous set of data coming from the IUE for PCVs and IPCVs, it has been possible to obtain important information on common properties and peculiarities of these binaries (de Martino, 1999 and references therein), which render the two classes rather similar in some of their UV behaviour. MCVs are relatively bright UV sources ($L_{UV} \sim 2 \times 10^{30}$–$3 \times 10^{32}$ erg s$^{-1}$). Their IUE spectra show a blue continuum with strong resonance lines of NV (1240 Å), SiIV (1397, Å), CIV (1549 Å) as well as of HeII (1640 Å) in emission. The UV line luminosity is ~2–10% of the continuum (Bonnet-Bidaud & Mouchet, 1988; de Martino, 1995a,b). The UV line luminosities are comparable in both classes (PVCs and IPCVs), whilst the continuum luminosity in PCVs is ~10% that in IPCVs (Mouchet, Bonnet-Bidaud & de Martino, 1998). This means that the accretion rate is higher in IPCVs ($10^{16}$–$10^{17}$ g s$^{-1}$ than in PCVs because of the presence of an accretion annulus (it is a disk with an internal radius much greater than in the case of NMCVs), which is responsible of a brighter UV continuum in IP-CVs. Weak emission lines of HeII (2307, 2386, and 2733 Å), MgII (2800 Å) and AlIII (1857 Å) are also present in the spectra of both classes. An observational difference between PCVs and IPCVs is the presence of very weak emission lines of lower ionization states in polars, which are not in IPCVs. Line flux ratios of NV/CIV and NV/SiIV also indicate a higher ionization efficiency in IPCVs, being $L_{SiIV} < L_{NV} < L_{CIV}$, whilst in PCVs $L_{SiIV} > L_{NV} < L_{CIV}$ (de Martino, 1999). The effect of increasing ionization efficiency is the tendency to suppress SiIV and to increase NV and CIV. The higher ionization efficiency in IPCVs could be owed to an efficient absorption of the soft X-ray emission (de Martino, 1995b). These differences re-enforce the hypothesis that the two classes are evolutionary related (e.g., Mouchet, Bonnet-Bidaud & de Martino, 1998; de Martino, 1998).

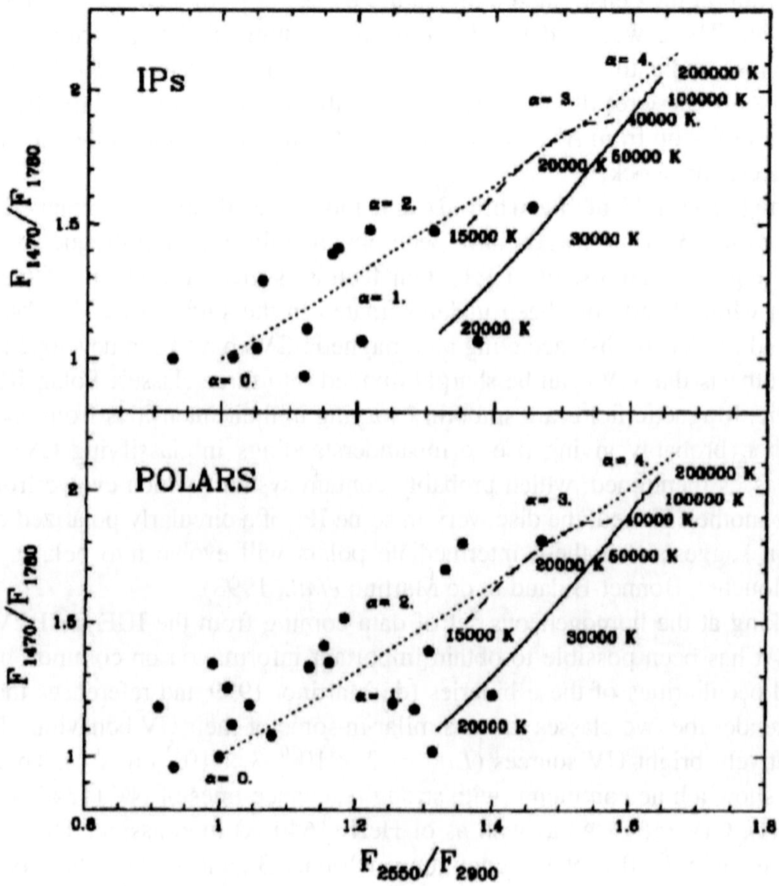


*Figure 126.* Far UV vs Near UV colour–colour diagrams for IPVCs (top) and PCVs (bottom). Power laws, $F_\lambda \propto \lambda^{-\alpha}$, (dotted lines), white dwarf (dashed lines), and black body (solid lines) distributions are reported too (de Martino, 1999).

The UV continuum properties can be inferred by colour–colour diagrams. Contrary to NMCVs, MCVs show a less steep UV energy distribution, since PCVs do not possess accretion disk and IPCVs have an accretion annulus (or truncated disk). Differences are also encountered in the two classes of MCVs and IPCVs.

Figure 126 shows the far UV vs near UV colour–colour diagram for MCVs (de Martino, 1999). Such a diagram was constructed measuring broad band continua in the IUE short wavelength range (1420–1520 Å and 1730–1830 Å) and in the long wavelength range (2500–2600 Å and 2850–2900 Å). Colours of blackbody and white dwarf distributions with different temperatures as well as power laws, $F_\lambda \propto \lambda^{-\alpha}$ are reported for comparison. In both classes the far UV and near UV colours cluster around the power law region with spectral index between 0 and 3. Typically PCVs have steeper far UV continua than in the near UV. Also their spectral shapes are steeper than those of IPCVs (de Martino, 1995a). The Rayleigh–Jeans tail of the hot soft X-ray/EUV reprocessed component searched for long time (e.g., Tanzi *et al.*, 1980) has not been observed in the time averaged spectra. Clearly the UV continua cannot be simply described by a single component but possess different contributions, as discussed by de Martino (1999) and already noted in the past, since 1984, by Giovannelli *et al.* (1985b).

Our suggestion is then in favor of studying CVs considering them as gravimagnetic rotators, avoiding in such a way any *a priori* classification. The fundamental parameters to be searched are the magnetic moment, the mass accretion rate and the orbital parameters of the systems. In this way it will be possible to fulfill the plane log $P_{\rm spin}$–log $P_{\rm orb}$ (Warner & Wickramasinghe, 1991), where in principle there are not restricted ranges of magnetic moment $|\vec{\mu}|$, or special correlations between $P_{\rm spin}$ and $P_{\rm orb}$ and $|\vec{\mu}|$. The distribution of objects in that diagram is owed to the interaction of braking torques and accretion torques, with the superposition of the observed or implied variations of the accretion rate on long time scale ($>10^2$ yr), acting on a continuum of magnetic moments.

## 7.10. Star Formation and the Evolution of Matter in the Universe

Following the interesting review of Saraceno & Orfei (2001), the understanding of the star formation process is critical in astrophysics because it is necessary for the comprehension of the evolution of stars, galaxies and finally of the Universe. Star formation is crucial for the energetics and chemistry of the interstellar medium, which may have strong implications on the origin and evolution of life. Eventually, life requires planets for its existence and the formation of planetary systems can only be studied as a part of the star forming process. The understanding of the star formation process will make clear the origin of the mass spectrum of stars, a crucial point of the evolution of the matter in the Universe. The cycle that matters follows, starting from the origin of the Universe, is schematically represented in Figure 127. Stars, once formed, are producing, through nuclear reactions, elements heavier than Hydrogen. At the end of their life part of the stellar mass will return to the interstellar medium (right side of the figure) to form new stars, part of it will be lost and will remain blocked in the remnants (White Dwarfs, Neutron Stars, Black Holes; left part of Figure 127). Figure 127 shows the importance of mass in the evolution of stars; in fact stars of less than $\sim$0.01 $M_\odot$ (D'Antona & Mazzitelli, 1994) will never reach the temperature necessary to start nuclear reactions, and they will undergo a continuous contraction (Brown Dwarfs: for a recent review see Persi, 2003).

More massive stars will produce energy through nucleosynthesis; their luminosity is a function of the mass as $L \propto M^4$, therefore the star life is $\propto M^{-3}$. This has a consequence for the search for extraterrestrial life: only stars of relatively low mass (less than $\sim$4 $M_\odot$) may host life in their planetary systems. Life needs a few billion years to develop and massive stars do not last so long. The final

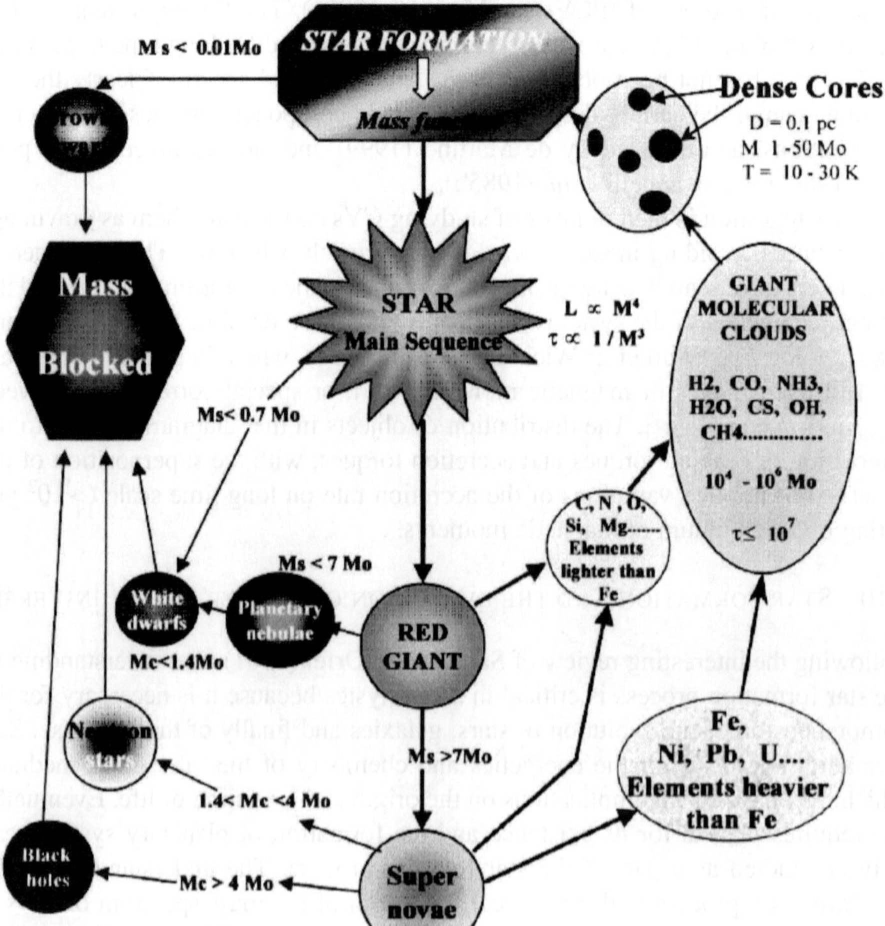


*Figure 127.* Star formation sketch. Clockwise from right part is shown the cycle that matter follows since the origin of the Universe, enriching at each cycle the interstellar medium of heavier elements. Ms indicate the original mass of the star and Mc the residual mass after the mass loss phase (courtesy of Saraceno & Orfei, 2001).

stages of stars also depend on stellar masses: stars of less than $\sim$7 $M_\odot$ will end their life as White Dwarfs. Amongst them, those of more than $\sim$0.7 $M_\odot$ will eject, during the post Main Sequence phase, up to 80% of their initial mass (the fraction increases with mass), passing through the stages of Red Giant and Planetary Nebulae. Star more massive than $\sim$7 $M_\odot$ will explode as supernovae producing a large amount of heavy elements (all the elements heavier than Fe can only be produced by Supernovae). The explosion ejects the largest part of the mass of the stars in the interstellar medium leaving a relatively small fraction (< 10%) of the initial mass in Black Holes or Neutron Stars. In conclusion: the less massive stars last for a longer time, but at the end of their life most of their masses is lost to the cycle. Increasing their masses, stars will last less (life $\propto M^{-3}$), but they increase the fraction of matter returned to the interstellar medium, enriched of heavy elements, to form new generations of stars. Therefore the chemical evolution of the Universe is mainly owed to massive stars. The evolution of the matter ejected by stars (see Evans, 1999 for a review) is schematically represented in the right hand side of Figure 127. The matter aggregates in Giant Molecular Clouds (GMCs), which show density enhancements in relatively small, cold regions called 'Dense Cores'. Stars form inside these cores and the mass of the dense cores settles the upper limit of the star masses. The masses of the Dense Cores have a lower limit given by the Jeans Mass, whiich depends on the temperature of the Cores, which depends on the cooling capacity of the gas. The cooling capacity of the gas of the Dense Cores depends on the abundance of molecules, such as CO, with a dipole momentum that can be excited at very low temperatures ($H_2$, the most abundant molecule, has only quadrupole momentum). The abundance of these molecules depends on the heavy elements produced in the previous generation of stars. So the mass spectrum influences the production of heavy elements and the production of heavy elements influences the mass spectrum. In the Early Universe, when heavy elements were not abundant, the gas of the clouds should have been much warmer; the Jeans Mass much higher; therefore stars much more massive than those we have nowadays should have been produced (Stahler, 1986). These first generations of massive stars have rapidly increased the abundance of the heavy elements of the Early Universe.

Massive stars, once formed, are able to destroy the cloud in a relatively short time (e.g., Blaauw (1991) noted three subgroups of sequentially formed stars in the Orion Complex).

The different phases of the evolution of a GMC are schematically represented in Figure 128 from Lada (1999). The upper panel represents the first stages: the cloud is very cold and only low mass stars are formed. They have not enough UV photons to create large ionized regions (HII regions). In the warmest regions of the cloud, around the newly formed stars, the gas remains molecular, but hot enough to excite $H_2$ in the near infrared, $H_2O$, and the high rotational transition of CO in the far infrared, as shown by the data of the ISO satellite (Saraceno *et al.* 1999).

An example is given in Figure 129, where the ISO spectrum of T Tauri is shown (Spinoglio *et al.*, 2000). At some point of the history of the cloud, massive stars (spectral type O, B) form. How they form it is not yet clear. Probably they form by coalescence of low mass stars. However, stars as massive as 120 $M_\odot$ (Massey & Hunter, 1998) are found in the OB association (Brown *et al.*, 1999). The enormous energy of these objects ($L \propto M^4$) can easily ionize (Lower panel of Figure 128) and dissipate the molecular cloud material, stopping the star formation process. It has been shown that ten O stars can ionize and disrupt a mass of $10^4$ $M_\odot$ (e.g., Franco, 1993, Whitworth, 1979) in a time of $\approx 10^6$ yr.

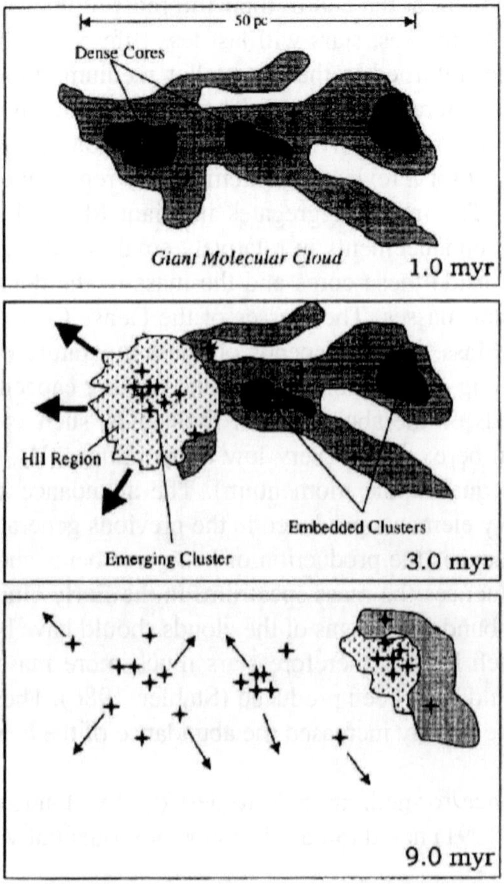


*Figure 128.* A schematic diagram showing the evolution of an OB association (from Lada, 1999) (cortesy of Saraceno & Orfei, 2001).

The small cold and dense fragments of molecular clouds where stars form are called the ‘Dense Core’. They appear as black spots in the optical wavelengths since their density does not allow us to see the background stars. They are strong emitters in CO and other molecules ($NH_3$, CS, $C^{18}O$) with density definitively smaller than that of $H_2$. Thus they can be detected only if high column densities are present.

Dense cores rotate very slowly (Benson & Myers, 1989), but, with a contraction of a factor $10^7$, the rotation can be so fast that the nucleus can fragment into several pieces (explaining why stars often belong to multiple systems), distributing the angular momentum on the different members of the cluster. Disks, which in a later phase will evolve in planetary systems, are also a consequence of this rotation (Jupiter has 98% of the angular momentum of the Solar System) and are formed by the material accreting on the star. The dissipation of angular momentum is

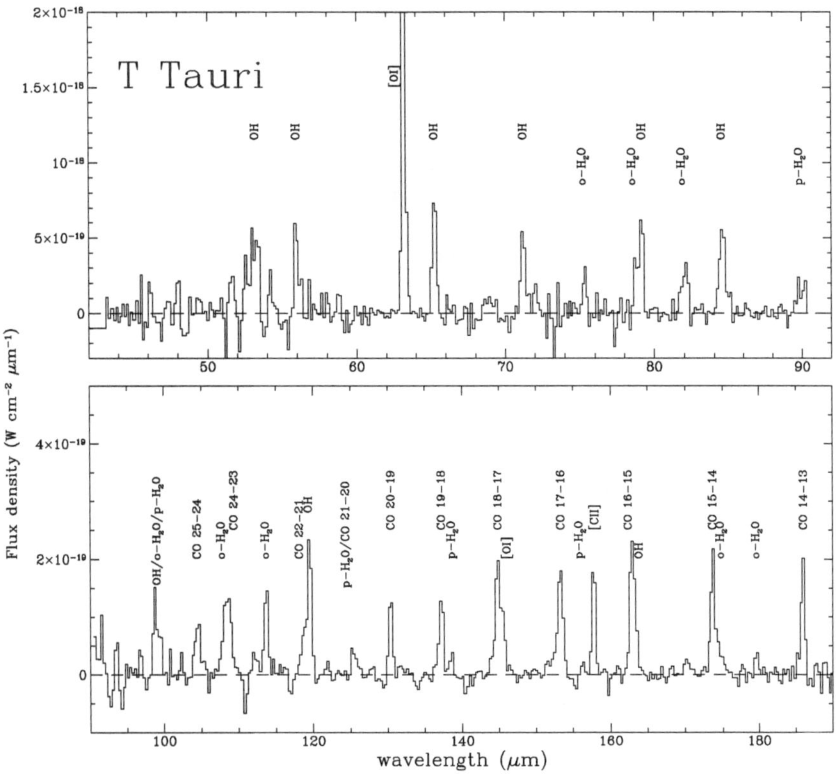


*Figure 129.* T Tauri FIR spectrum obtained with the ISO. It shows the major FIR coolants of the gas: CO lines with rotational quantum number ranging from 14 to 25, $H_2O$, and OH (Spinoglio *et al.*, 2000).

also one of the explanations for the energetic outflows that all protostars show. Protostars, being rotating systems, have also a relatively strong magnetic field, therefore outflows have a polar direction across the line of the magnetic field.

A protostar has the structure represented in Figure 130: a central object with a circumstellar disk. The matter coming from an envelope (the dense core) reaches the star dissipating across the disk both angular momentum and energy. Energetic outflows are observed across the polar axis. The Kuiper belt (100–200 AU) and the Oort cloud (0.1 pc) are the residuals of the protostellar disk and of the envelope from which the solar system was formed. As time proceeds, the circumstellar material of a protostar is expected to decrease in mass and the protostellar object appears more and more warm. This suggested to Lada & Wilking (1984) to give a first evolutionary sequence of the protostars found in the $\rho$ Ophiuchi cloud,

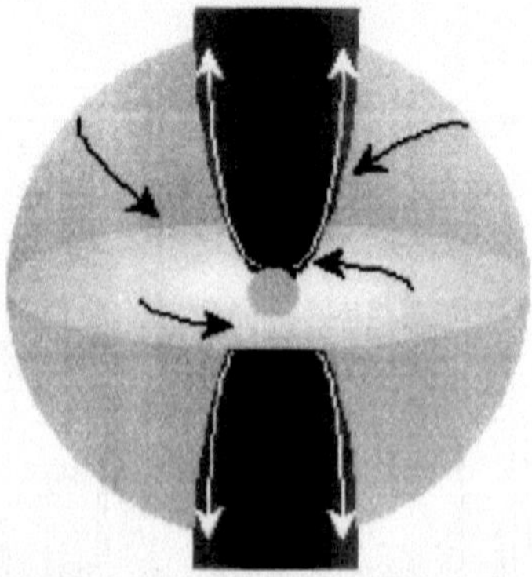

*Figure 130.* Structure of a protostellar object (courtesy of Saraceno & Orfei, 2001).

dividing the sources into three classes according to their spectral shapes. Later, the IRAS satellite extended the validity of this classification up to 100 $\mu$m (see Section 7.11.2) increasing a lot the number of protostars and pre-main sequence stars (e.g., Wilking, Lada & Young, 1989).

### 7.11. Pre-main sequence stars

The study of the Pre-Main Sequence Stars (PMSSs) is of great interest as it provides crucial information on stellar evolution and, particularly, on the role of magnetic fields, angular momenta, accretion of matter and mass loss processes, as well as indirect information on the formation processes of the Sun and Solar system.

Before the space era practically all observations of PMSSs were restricted to only the optical and near IR spectral regions. Consequently the related researches were devoted to study moderate and low temperature phenomena associated with dust, stellar photospheres, lower chromospheres, cool winds and low excitation shocks. With the advent of the IUE and of the new generations of more sensitive X-ray detectors on board satellites at the end of the seventies it was possible to obtain information about high energy processes in hotter temperature winds, high temperature shocked gas, on chromospheric and transition regions, on stellar coronae and flares. The current definition of different zones of the 'cool star' atmospheres comes from the analogy with the Sun, in which heating of the outer chromosphere is dominated by magnetic fields which are generated in the convective stellar envelope. The various temperature régimes are called: (i) chromosphere (1–5) $\times 10^4$ K; (ii) transition region ($5 \times 10^4$–$5 \times 10^5$ K); and (iii) corona ($T > 10^6$ K). Tight relations exist between the flux densities originating in these different temperature régimes (e.g., Rutten *et al.*, 1991 and references therein).

With the knowledge nowadays of PMSSs, we can say that they are sites of a variety of extraordinary astrophysical processes, including accretion disks and outflows. The former have been detected by means of observations of molecular lines in cold gas. The best known class of PMSSs is that of the TTSs, identified by strong $H_\alpha$ line in emission, which is believed to be generated in an extended circumstellar envelope, indicating a continuous mass accretion process. The presence of an accretion disk has been strengthened by the observation of nearly Keplerian differential rotation (Hartmann & Kenyon, 1987a,b) in the optical and near IR (NIR) spectra of FU Ori, a young stellar object (YSO) whose spectral energy distribution can be explained by the presence of a disk.

### 7.11.1. *Classification of Low Luminosity PMSSs*

PMS is an important evolutionary phase of all stars, which are evolving from their birth to the main sequence, and then to their ultimate fate. A first important difference between stars is owed to their masses (or luminosity). Indeed, as is well known, high mass stars evolve rapidly into the main sequence and therefore the experimental study of their evolution within the PMS phase is very difficult, owing to the lack of a good statistics. In contrast, low mass stars, because of their slower evolution towards the main sequence, can be studied in detail since it is possible to find large samples of objects belonging to different stages of the PMS evolutionary phase.

Owing to the large spread in the HR diagram, a well defined classification of low luminosity PMSSs is very difficult; however, these objects can be tentatively classified, depending on their evolutionary status, as follows (Imhoff & Appenzeller, 1987, and references therein):

- T Tauri Stars (TTSs): they are sub-giants of late spectral type, showing high variability, strong HI, CaII and FeII emission lines, and IR excess. The underlying photospheres are relatively cool, usually less than 6000 K, and typically ~4000 K. Masses and ages range from 0.5 $M_\odot$ to 2 $M_\odot$ and from $10^5$ to $10^7$ yr, respectively. They probably look like the Sun in its very early stages. A separation of TTSs in two classes is nowadays used, referring to the intensity of the optical emission lines of the CaII and Balmer lines. They are named Strong Emission Line Type (or Classical) and Weak Emission Line Type of TTSs: CTTSs and WTTSs, respectively.
- Herbig Be/Ae Stars: they are similar to TTSs, but with higher masses and higher photospheric temperatures.
- FU Orionis Like: this is a small group of PMSSs which have undergone to novalike flares standing for some years.
- Weak-Emission PMSSs: they are PMSSs with an activity even weaker than the WTTSs. This class contains both the ‘post T Tauri’ (sometimes older than the TTSs), and the ‘naked T Tauri’ which have a non relevant quantity of circumstellar matter (Walter *et al.*, 1988). Naked TTSs are very often classified as WTTSs (e.g., Bertout, 1989).

On the basis of IR spectral energy distributions of young stellar objects, another classification can be used in order to propose their evolutionary sequence (Adams, Lada & Shu, 1987) (see Section 7.11.2).

### 7.11.2. *Observational Properties of T Tauri Stars*

TTSs are newly formed low mass stars that have recently become visible in the optical range. Alfred H. Joy discovered these stars in the Taurus-Auriga dark cloud and named the class after its brightest member T Tauri. In this cloud a number of faint late spectral type stars, showing irregular and large light variations, with emission lines resembling the solar chromosphere were discovered (Joy, 1942; 1945; 1949). The physical nature of TTSs was first recognized by Ambartsumian (1947; 1952) after finding that they occur in groups that he named T associations and that these groups are often found in connection with O associations (groups of OB stars). He proposed that these objects are relatively low mass very young stars which have not yet reached the main sequence. The spectral peculiarities are therefore ascribed to certain atmospheric properties associated essentially with 'youth'. The arguments used to support this point of view include (Kuhi & Cram, 1989):

- the high space density of TTSs in the so-called T associations placed in regions of nebulosity;
- common radial velocities between TTSs and their associated molecular clouds (Herbig, 1977), indicating that there is a little differential motion between the stars and the cloud, which imply that the stars have not had enough time to diffuse away from the cloud; consequently, they must be very young;
- the location of TTSs on colour–magnitude diagrams (Walker, 1956) and HR diagrams of young clusters in the region predicted for low mass PMSSs from theoretical evolutionary tracks (Cohen & Kuhi, 1976, 1979);
- the very high abundance of lithium, which decreases as the inverse square root of the age for main sequence stars (Skumanich, 1972), and TTS lithium abundance Which seems to be consistent with interstellar abundances, which presumably are primeval (Zappala, 1972).

Magazzú & Rebolo (1989) derived the lithium abundance of 26 TTSs. They found a correlation between lithium abundance and mass of the stars: the higher is the mass the higher is the abundance. These results provide some support to the extra-mixing picture discussed by D'Antona & Mazzitelli (1984). However, as pointed out by Patterer *et al.* (1993), lithium line variations in WTTSs, which can be explained by magnetic activity, prevent the use of the derived equivalent widths in searching for relationships between lithium abundance and age, because they may be dependent on the phase at which the spectrum was taken. So that relations between Li I (6707 Å) abundance and age are probably useful only in statistical sense for a large sample of TTSs.

Optical spectroscopic criteria that define TTSs (Herbig, 1962) are the following:

- hydrogen Balmer lines and CaII H and K lines are in emission;
- anomalous fluorescent emission lines of Fe I (4063 and 4132 Å) are present;
- forbidden emission of [OI] and [SII] (4068 and 4076 Å) are observed in many TTSs;
- LiI (6707 Å) absorption is abundantly strong;

- after the hydrogen and CaII lines, the strongest emission lines are those of FeII, TiII, and HeI.

The emission line spectrum is superimposed on a continuum spectrum. Depending on this spectrum it is possible to recognize different sub-classes of TTSs, namely:

- extreme TTSs if the continuum spectrum is pure;
- veiled TTSs if the continuum spectrum is a late type of absorption spectrum with anomalous line strengths;
- moderate TTSs if the continuum spectrum is an almost normal absorption spectrum of type F through M (Bertout, 1989).

A separation of TTSs in two classes is used nowadays, referring to the intensity of the optical emission lines of the CaII and Balmer lines; the first and second classes have equivalent widths (EWs) of $H_\alpha$ greater and smaller than 10 Å, respectively (some authors use a value of 5 Å as limit in selecting the two classes, and in early objective prism surveys this value was often used in distinguishing TTS candidates ($H_\alpha$–EW >5 Å)). They are named Strong Emission (or Classical) and Weak Emission Line Type TTSs: CTTSs and WTTSs, respectively. However, according to the present knowledge, there is a gradual and smooth transition between the properties of CTTSs and WTTSs and furthermore the Balmer emission lines vary with time, often on short time scales.

Walter *et al.* (1988) define WTTSs as X-ray sources with an optical counterpart showing PMS characteristics, and in particular the LiI (6707 Å) absorption line is present with EW larger than 100 mÅ.

Figure 131 shows the medium resolution spectrograms in the spectral range 3200–8800 Å of; i) a WTTS (TAP 57 = 045251+3016) similar in many respect to a standard K7 dwarf (Feigelson *et al.*, 1987; Walter *et al.*, 1988); ii) a moderate M0 CTTS (DN Tau); iii) a veiled M0 CTTS (DF Tau); and iv) an extreme CTTS (DR Tau) with probable K5 spectral type (Bertout, 1989 and references therein).

Also in the UV range the spectra of the sub-classes of TTSs mentioned are different. Figure 132 shows a sequence of low dispersion long wavelength IUE spectra in the region of MgII lines. The observed fluxes are given in units of $10^{-13}$ erg $cm^{-2}$ $s^{-1}$ $Å^{-1}$ with offsets of $10^{-13}$ between the spectra. The range of spectral characteristics is depicted for a normal G dwarf (Mu Her), a weak emission PMS star (HD 283572), a WTTS (SU Aur), and a moderate emission TTS (GW Ori). It is possible to see a clear progression from a normal photospheric spectrum to spectra with increasing contributions from the continuous UV excess and emission lines (Imhoff & Appenzeller, 1987).

By comparing modern theoretical evolutionary calculations, we know today that all TTSs are PMSSs with $M < 3$ $M_\odot$ and most of them are younger than $3 \sim 10^6$ yr, following the conventional hydrostatic tracks. Figure 133 shows the position in the Hertzsprung–Russel diagram of all CTTSs and WTTSs with known $v \times \sin i$. Open circles and dark circles represent WTTSs and CTTSs, respectively. In both cases the circle area is proportional to the stellar $v \sin i$, which ranges from ~5 to

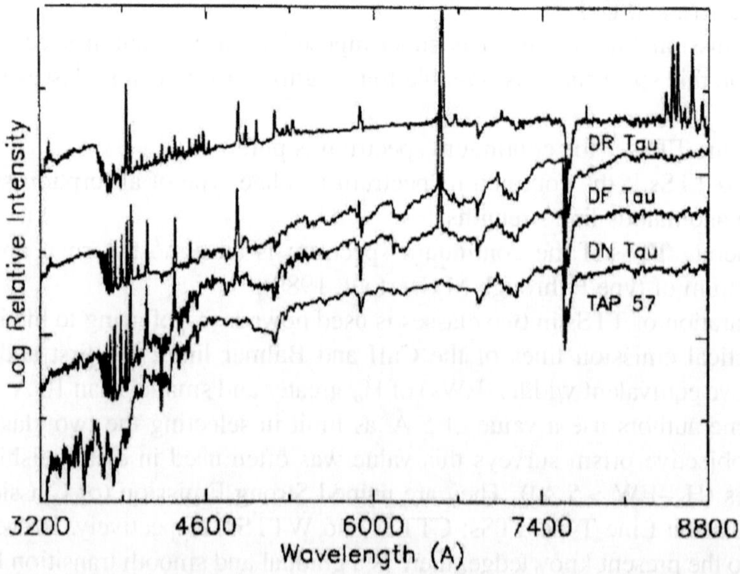


*Figure 131.* Medium resolution spectra of TAP 57 (K7 WTTS), DN Tau (moderate M0 CTTS), DF Tau (veiled M0 CTTS), and DR Tau (extreme TTS) (Bertout, 1989 and the references therein).

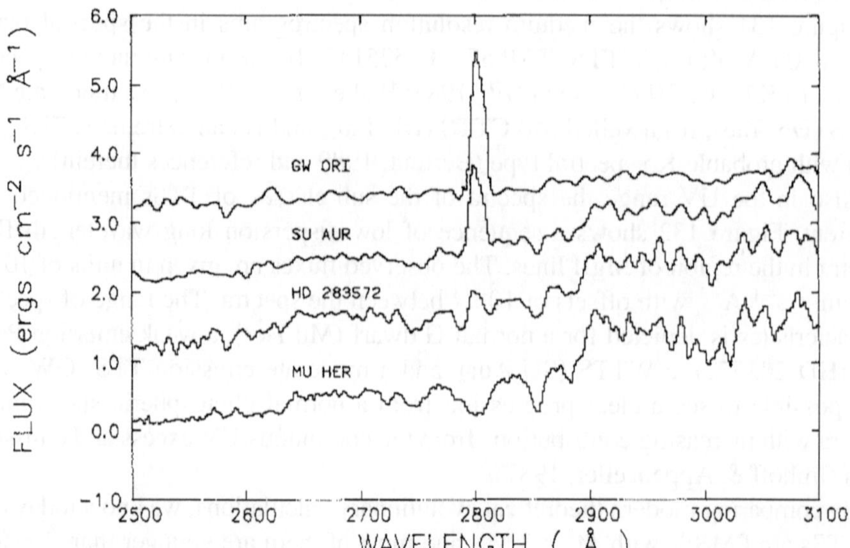


*Figure 132.* Long wavelength IUE spectra of MU Her (G dwarf), HD 283572 (weak emission line PMSS), SU Aur (WTTS), and GW Ori (moderate emission TTS) (Imhoff & Appenzeller, 1987 and the references therein).

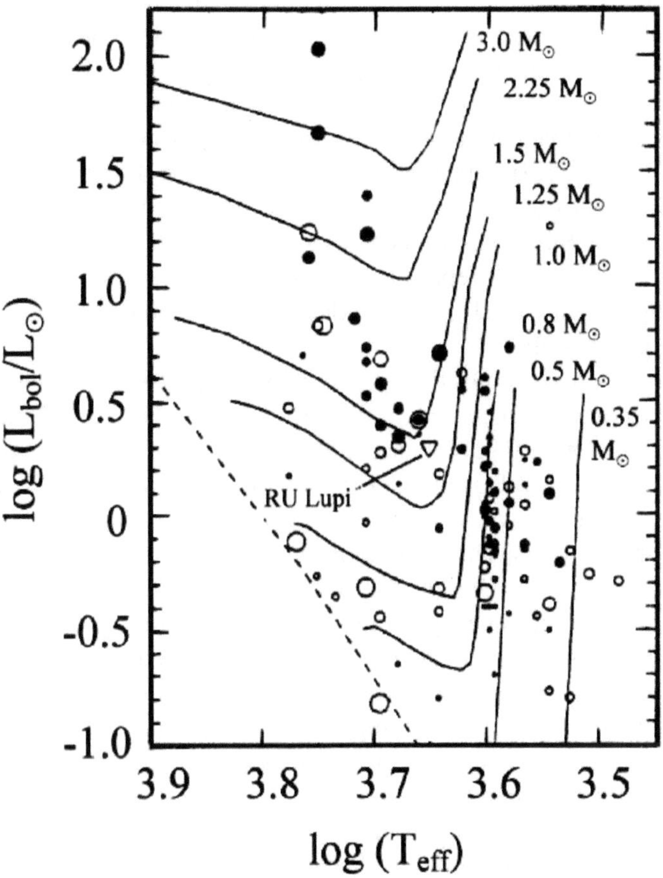


*Figure 133.* Position in the Hertzsprung–Russel diagram of all CTTSs and WTTSs with known $v \sin i$. Open and dark circles represent WTTSs and CTTSs, respectively. The position of RU Lupi, whose rotational velocity is still controversial, is marked with a triangle (Giovannelli, 1994, after Bertout, 1989).

100 km $s^{-1}$. Approximate PMS quasi-static evolutionary tracks for various masses are also plotted together with the zero age main sequence (ZAMS) represented by the dashed line (Bertout, 1989). In Figure 133, the position of RU Lupi that is marked with a triangle is reported too (Giovannelli, 1994).

However, the presence of disks and magnetic surface fields renders the determination of effective temperatures and intrinsic luminosities of TTSs very difficult, so the latter limits can be considered as order of magnitudes.

Indeed, direct imaging of the extreme TTS HL Tau (Grasdalen *et al.*, 1984) reveals structure with a disk-like appearance, which is, in general, theoretically supported by Beall (1986), and the detected circular polarization (~2%–4%) in the radio emission of two weak line TTSs (White, Pallavicini & Kundu, 1992a) is a direct confirmation of the presence of magnetic fields in the coronae of these stars.

Thanks to the advent of space experiments which opened new windows in the electromagnetic spectrum, T Tauri phenomenon now appears as one aspect of stellar formation that can no longer be treated independently of others aspects. So the term TTS (T Tauri Star) has become equivalent over the years to a PMS (Pre Main Sequence) low mass optical object, and the concept of young stellar object, first used by Strom (1972), is now used to designate all PMS stellar objects, embedded or visible. TTSs represent a cardinal class between deeply embedded low luminosity sources which can be studied only at IR and radio ranges, and solar type main sequence stars.

Most of present day knowledge on the physics of TTSs is coming from multi-frequency observations carried out in the past 25 years. The crucial measurements in UV – performed with IUE (e.g., see the review by Imhoff & Appenzeller, 1987) – and then in X-ray, as well as in other ranges allowed to substantially improve the knowledge of the physics of high energy processes in hot temperature winds, high temperature shocked gas, chromospheric and transition regions, stellar coronae and Sun-like flares (e.g., Haisch & Rodonó, 1989; Haisch, Strong & Rodonó, 1991). Moreover, multifrequency simultaneous observations provided the total energy distribution of some TTSs, and in particular, the energetic of flare-like events (FLEs) occurring in individual objects (e.g., Giovannelli *et al.*, 1989). In addition to the above-mentioned direct imaging of HL Tau (Grasdalen *et al.*, 1984), also other types of observations provide circumstantial but convincing evidence for the presence of cool disks around TTSs.Indeed:

- forbidden lines are blueshifted: if there is an optically thick circumstellar disk, the receding part of the flow is obscured and blueshifted lines are the result (e.g., Appenzeller, 1983);
- the observed polarization maps of several TTSs (e.g., Bastien, 1989 and references therein; Bastien & Menard, 1990) can be explained by multiple scattering in an extended disk and its bipolar lobes as suggested by Elässer & Staude (1978);
- the large degree of polarization of some TTSs combined with their bipolar outflows being often oriented perpendicular to the orientation of the electric polarization vector (e.g., Mundt & Fried, 1983; Hodapp, 1984) can also be explained by the presence of an optically thick disk (Bastien & Menard, 1988). Menard & Bastien (1992) performed measurements in a northern extension (59 objects) to about 15th mag of previous linear polarization surveys of TTSs and other YSOs. They found that 15 of the newly observed stars show large values of intrinsic polarization, in excess of 3%, and at least 43% and perhaps all TTSs are polarimetric variables. The statistics of these polarimetric data are compatible with a model of an optically thick circumstellar disk surrounding the TTSs of the sample;

– deep emission line studies of TTSs allow to interpret the permitted lines in terms of active stellar atmospheres, which emit narrow lines ($<$40 km $s^{-1}$), and extended turbulent envelopes, which produce the usual broad line profiles ($>$100 km $s^{-1}$) (Hamann & Persson, 1992a). A weak correlation between the $H_\alpha$ and narrow line strengths suggests that these two regions are somehow interacting. The narrow line widths constrain the velocities to $<$10 km $s^{-1}$ in the deep chromosphere and $<$22 km $s^{-1}$ higher up where the HeI and HeII form. This velocity structure is incompatible with models of TTS winds driven by the star. So the extended high velocity envelopes have a non-stellar origin and a non-spherical geometry, Therefore these results favor a model in which the envelope forms at a star accretion disk boundary layer and resides primarily near the disk plane.

For all these reasons, most of current models of TTSs assume that these objects are formed by a very active, but otherwise more or less normal, late type star surrounded by an accretion disk. The activity of the central star could be owed to a more rapid rotation with respect to late type main sequence stars. WTTSs are assumed to have less massive disks or at least lower accretion rates than the CTTSs. Whilst some WTTSs may be born without sizable circumstellar disk, CTTSs are expected to become WTTSs when their disks dissipate. The TTS evolutionary phase is expected to end when the effects of the accretion disks become negligible and the disk is dissipated by matter depletion, planetary systems formation, or dispersion by the TTS winds (Montmerle & André, 1989). Disk models have successfully reproduced the broad energy distributions and other spectral properties of CTTSs (e.g., Adams & Shu, 1986; Hartmann & Kenyon, 1987a,b; Kenyon & Hartmann, 1987; Weintraub, Masson & Zuckerman, 1987; Adams, Lada & Shu, 1988; Bertout, Basri & Bouvier, 1988; Bertout, 1989; Basri & Bertout, 1989; Bouvier & Bertout, 1992; Bertout *et al.*, 1993; Cameron & Campbell, 1993; Edwards *et al.*, 1993; Nakano, 1991, 1993; Terquem & Bertout, 1993). The presence of a disk around a moderate magnetic star is the most convincing explanation of the simultaneous multifrequency data (from UV to IR) in the case of RU Lupi (Giovannelli *et al.*, 1991a,b,c; 1995; Lamzin *et al.*, 1993). Many TTSs, and in particular WTTSs, have been detected as soft X-ray sources, at a level of $\sim 10^{30}$ erg $s^{-1}$ first by Ku & Chanan (1979). As in other late-type stars, the flux of the X-ray radiation depends on the rotational periods, roughly following the empirical relationship between rotational velocities and X-ray emission of cool stars (e.g., Bouvier, 1987; 1990). However, since the TTSs tend to rotate faster than the more evolved late-type stars, their X-ray emission is on average much higher than that of normal dwarfs. Therefore, X-ray mapping with the new generation of X-ray experiments, such as ROSAT, provides an efficient method of finding TTSs with weak emission lines, which are difficult to be detected by spectroscopic methods. The evolutionary status of WTTSs remains unclear. What is clear instead is that, in addition to their strong X-ray activity, they also show strong radio emission, which shows highly variability on timescales of hours (Cohen, Bieging & Schwartz, 1982;

Feigelson & Montmerle, 1985). A review on X-ray activity and magnetic fields of TTSs was published by Montmerle (1992). The UV spectrum of TTSs is dominated by chromosphere-like emission spectrum with strong lines of OI, MgII, FeII, SiII, SiIII, SiIV, CIV, and sometimes HeII and NV. A comparison with the Sun and other late-type-star chromospheres shows that the UV lines of low ionization stages are enhanced relative to the more highly ionized species, indicating a lower ionization temperature of the emitting gas (e.g., Imhoff & Giampapa, 1980). The FeII resonance lines and other metallic lines are sometimes absorbing and may change from emission into absorption on short time scales (e.g., Appenzeller *et al.*, 1980). TTSs show very often IR excess emission, which in some cases exceeds the optical one by a large factor (Mendoza, 1966, 1968). Most of the known TTSs in the nearby star forming regions have been detected by the IRAS satellite (at 12, 25, 60, and 100 $\mu$m) (Rucinski, 1985; Harris, Clegg & Hughes, 1988; Adams & Shu, 1988; Cohen, Emerson & Beichman, 1989). Moreover, a significant fraction of TTSs emit also in submm region (0.35–1.3 mm) (Weintraub, Sandell & Duncan, 1989; Adams, Emerson & Fuller, 1990; Beckwith *et al.*, 1990). From the observed energy distribution it is clear that the FIR and submm radiation is coming from cool circumstellar dust (e.g., Chini, 1989). Since this dust is very probably concentrated in thin circumstellar disk, the flux of that radiation can be used to evaluate disk masses, which are coming out of the order 0.001–1 $M_\odot$. The FIR excess occurs both in CTTSs and WTTSs and the estimated disk masses do not show obvious correlation with the Balmer emission line strength of the TTSs (Chini, 1989). Cohen, Emerson & Beichman (1989) have shown a correlation between the FIR excess and the [OI] strength. Hamann & Persson (1992b) have found a correlation between the IR excess, from R to N bands, and the CaII triplet line luminosity. Whatever excitation is involved, the existence of this correlation suggests that the line-emitting envelopes are somehow related to the presence of disks. The correlation, therefore, argues against the stellar driven (e.g., chromospheric) interpretation of the lines.

The presence of dust grains around TTSs is confirmed by the presence of strong 10 $\mu$m SiO dust emission or absorption features (Cohen & Witteborn, 1985). In most TTSs this feature is in emission; however, a deep absorption is observed in TTSs with very strong FIR excess. Sometimes this strong SiO signature is accompanied by the 3 $\mu$m water ice feature (e.g., Whitt *et al.*, 1988).

The NIR spectrum of the CTTSs, in addition to emission lines of hydrogen, helium and low ionization stages of metals, shows important molecular bands of $H_2$ and CO. This part of the spectrum has been described by many authors (e.g., Cohen, 1975; Evans *et al.*, 1987; Hamann, Simon & Ridgway, 1988; Hamann & Persson, 1992b).

The first detection of radio emission from TTSs was that of T Tauri itself performed by Spencer & Schwartz (1974). However, it was later discovered that this radio emission is coming from the IR companion of T Tau – named T Tau S, discovered with one-dimensional NIR speckle interferometry by Dyck, Simon &

Zuckermann (1982), much fainter than the optical component T Tau N – (Schwartz *et al.*, 1984; Schwartz, Simon & Campbell, 1986). This discovery opened a window in considering that also PMS stars lie very often in double systems, such as almost the 67% of all the stars (Lipunov, 1989).

A radio survey at 5 GHz of selected post TTSs and naked TTSs (White, Pallavicini & Kundu, 1992b) have shown a clear association between radio activity and youth and that those stars with high X-ray activity are likely to be detectable radio sources, even if the radio emission goes through high and low phases.

After measurements from the IRAS satellite (Beichman *et al.*, 1984) and from ground based observatories (e.g., Lada & Wilking, 1984) detection of IR sources in dense molecular clouds has been reported. Such observations can, in principle, represent detection of protostellar systems. Moreover, direct imaging of the extreme TTS HL Tau showing a disk-like structure around the star (Grasdalen *et al.*, 1984), and the existence of the solar system, convinced the community in believing on the possible presence of protoplanetary disks around young stars. Beall (1986) calculated the emitted spectrum from accretion disks in protostellar systems, successfully reproducing some of the spectra in the $\rho$ Oph cloud. He assumed an optically thick, finite disk which is truncated at its inner edge by the protostar which has not yet begun nuclear burning. The disk spectrum has a form which resembles a reddened blackbody with a higher temperature than that of the disk proper, since the disk spectrum flattens toward shorter wavelengths. Given the complexity of the spectra, it is also possible that circumstellar shells of dust can contribute to the observed radiation (Stahler, Shu & Taam, 1980). The actual physical situation is unlikely to be fitted by any single model.

Later on, attempts in developing hydrodynamic theory of equatorial disks have been made, but these are all very rudimentary and difficult to develop, since the flows generally involve a state intermediate between non-interacting Keplerian orbits and a fully collisional gas. Consequences of disks have been considered in theories designed to account for TTS spectra, namely:

- IR and optical continuum radiation from warm dust (Adams & Shu, 1986);
- emission from ionized gas at the disk–photosphere interface (Bertout, Basri & Bouvier, 1988);
- occultation of the receding part of a stellar wind, leading to blue shifted spectral lines (Edwards *et al.*, 1987; 1989).

A survey at 1.3 mm for circumstellar disks around 86 YSOs in Taurus-Auriga dark cloud (Beckwith *et al.*, 1990) have shown that 42% of them have detectable emission from small particles. The detected fraction is only slightly smaller for the weak line and naked TTSs than for the CTTSs, indicating that the former stars often have circumstellar material. In both categories, the column densities of particles are too large to be compatible with spherical distributions of circumstellar matter – the optical extinctions would be too large; the particles are almost certainly in spatially thin, circumstellar disk. Other measurements of TTSs, which were suspected of having circumstellar disks, performed at 0.8 and 1.1 mm (Adams, Emerson &

Fuller, 1990), have shown that the disk masses are larger than the 'minimum mass' solar nebula and suggest that planets can form in these systems. So that theories on accretion disks are urgent in explaining possibly all the multifrequency behaviour of TTSs. Some reviews on accretion disks around TTSs are those by Shu, Adams & Lizano (1987), Bertout (1989), Bertout & Bouvier (1989), Appenzeller & Mundt (1989), and Giovannelli (1996b). After that time many improvements in accretion disk models have been obtained, as we will comment underneath.

Current models of CTTSs (Kenyon & Hartmann, 1987; Bertout, Basri & Bouvier, 1988) search for accounting simultaneously for the spectrum all the way from the UV to the mid-IR. Although these models are unable to give a detailed quantitative interpretation for many emission line intensities and profiles, especially for $H_\alpha$, they describe the broad features of CTTSs spectra in terms of different components in addition to the central late-type star, namely:

– a circumstellar accretion disk;
– a boundary layer between the star and the inner part of the disk.

Indeed, in the immediate vicinity of the star an important phenomenon occurs, since at a distance $r = R^\star$, the Keplerian velocity in the disk around a 1 $M_\odot$ star is $\sim$250 km $s^{-1}$, which is by definition the break up velocity of this star; meanwhile the rotational velocities of TTSs are definitively smaller. On the average, for stars with $M < 1.25$ $M_\odot$ the equatorial velocities are <20 km $s^{-1}$ (Bouvier, 1990). Therefore near the equator the star rotates $\sim$10 times slower than the disk. Then a hot boundary layer must develop in order to ensure a continuous transition of the fluid velocities between the disk and the star. The main dissipative mechanism is owed to the viscosity which is not known, then the physical conditions within the boundary layer are poorly known (Basri & Bertout, 1989). According to some estimates (Montmerle, 1992), the reached temperatures are $\sim$8,000–10,000 K; therefore the bulk of the radiation is in the optical–near UV range. This is the reason because it is believed that the observed strong emission lines (in particular $H_\alpha$) and the UV excess originate in the boundary layer. The little correlation between disk mass and the $H_\alpha$ equivalent width found in the survey of Beckwith *et al.* (1990) supports this conclusion. Moreover, with the boundary layer, it is possible to explain qualitatively the correlation between the IR and UV excesses and the strength of the emission lines.

New models considering magnetic fields in accretion disks around protostars and the fragmentation of the disks have been developed by Nakano (1991). He investigated the densities of ions and the behaviour of magnetic field in disks around protostars by using the alpha model for the accretion disk. He found that it is highly possible that many protostars are surrounded by a swarm of fragments with mass of $\sim$0.01 $M_\odot$ at $r \sim 100$ AU. The spectrum of a protostar has a FIR excess compared with the composite spectrum of the central star and the steady accretion disk. The gaseous disk may become turbulent disturbed by the gravity of fragments, and heating by dissipation of the turbulence must contribute to the FIR spectrum. The fragments have an effective temperature of $\sim$100 K for a fairly long time and may have some contribution to the FIR spectrum. The turbulence caused by the fragments may also contribute to the angular momentum transport in the gaseous disk and then to the accretion. By mutual gravitational interaction some of the fragments may enhance the falling rate of matter to the star and cause the stellar activity such as the FU Orionis phenomenon. Repeating collisions and coalescence the fragments may grow to a companion star. It is to be remarked that the separation of binary components is mostly $<$100 AU (Abt & Levy, 1976), which is nearly equal to the radius at which fragmentation occurs. The fragments which have escaped capturing by the companion star will finally become brown dwarfs. The orbit of such small objects in binary systems are unstable and they will be ejected from the binary systems. Therefore various phenomena caused by the swarm of fragments can be expected. Nakano (1993) revises the latter model including the self-gravity of the disk and finds that there is a 'decoupling' radius inside which magnetic fields are decoupled from the gas. Fragmentation around this radius occurs and will be repeated as long as the gas continues to fall from the parent cloud and finally numerous fragments orbit the star at a distance roughly equal to the decoupling radius.

The evolution of disk accreting protostars is discussed also by Stringfellow (1991) via theoretical evolutionary calculations. He assumed that the boundary layer forms concurrently with the circumstellar disk, and both quickly become and remain geometrically thin compared to the radius of the protostar. The boundary layer itself is located slightly away and remains physically distinct from the surface of the protostar. The material being accreted from the disk passes through the boundary layer, where the accretion energy is liberated in its entirety, prior to settling onto the protostellar core. Thus the boundary layer and protostar contributions can be computed separately; the boundary layer being responsible for generating the accretion luminosity, and the protostar acting as the mass sink and defining the effective radius of the boundary layer. After calculations he made a comparison with CTTSs in the HR diagram. The total luminosity of the composite disk accreting system, given by the sum of the protostellar luminosity and the luminosity arising in the boundary layer, reproduces the stellar birthline when plotted against the core's $T_{\rm eff}$.

Bouvier & Bertout (1992) developed a disk model for a self-consistent explanation of both the UV and IR excesses, which strongly characterize TTS emission, extending to FIR, following the classical thin disk model successfully developed up to 5 $\mu$m (Bertout, Basri & Bouvier, 1988; Basri & Bertout, 1989).

Cameron & Campbell (1993) discussed the rotational evolution of magnetic TTSs with accretion disks and in particular the problem of the angular momentum transfer from the star to the disk. They found that for a stellar magnetic field of dipole form, with a polar strength of few hundreds Gauss, and an accretion rate of a few $\times 10^{-8}$ $M_\odot$ yr$^{-1}$, the star's rotation rate is quickly forced into a state where it evolves in quasi-static equilibrium for the remainder of the Hayashi phase.

On the contrary of the CTTSs, WTTSs show no UV or IR excess, and weak, if any, $H_\alpha$ in emission. The small equivalent width of this line suggests a chromospheric origin. The implication of current TTS models is that WTTSs should not be surrounded by circumstellar material, but simply represent evolved, active, solar-like PMSSs (e.g., Montmerle, 1997). The study of WTTSs in the mm band confirms this fact, with some exceptions of the possible existence of transition objects showing circumstellar disks with an inner cavity, hence deprived of excess $H_\alpha$ emission linked with the boundary layer (Montmerle & André, 1989; Montmerle, 1992). Therefore the difference between naked TTSs and WTTSs is owing to their NIR (~10 $\mu$m) emission: there is no excess in naked-TTSs, while this excess is small and optically thin in WTTSs. In CTTSs this excess is larger and optically thick (André, 1991).

On the basis of IR spectral energy distributions of YSOs, another classification can be used in order to propose their evolutionary sequence (Adams, Lada & Shu, 1987). In this scheme:

- Class I sources are deeply embedded very YSOs that are seen only in the IR and mm spectral ranges. Their IR spectral energy distributions are either flat or rising toward long wavelengths. A model including the star and its circumstellar disk both surrounded by a dusty envelope can successfully reproduce most of the observed spectral energy distributions. They therefore seem to point to young stellar objects (before the birthline) and often still in their protostellar accretion phase;
- Class II sources have energy distributions typical of CTTSs; they are either visible or IR sources and are probably somewhat more evolved than Class I objects. Their spectra originate from the star and a surrounding accretion disk;
- Class III sources show reddened blackbody-like energy distributions; they are often found at the cloud borders and correspond presumably to the more evolved cloud stars. They are either WTTSs or post-TTSs (PTTSs); indeed, objects detected primarily in the X-ray range are often belonging to Class III; their X-ray emission is probably caused by magnetic activity similar to that of the Sun, although stronger.
- Class IV sources are PTTSs and would correspond to the final stages of disk evolution, when the disk has completely disappeared. This fact can be experimentally proved only by mm observations, the material being very cold. These objects are indistinguishable from those of Class III, if observations are made only in the NIR range.

Class II and Class III objects do not simply represent CTTSs and WTTSs, but they may correspond to various stages between accretion disks, with their associated boundary layers, and hollow disks, clearing progressively 'inside out' (Montmerle & André, 1989). CTTSs would correspond to a (temporarily ?) stable keplerian accretion disk, not dominated by magnetic fields. Their mass loss may come predominantly from the disk. WTTSs would correspond to a more or less centrally cleared, hollow disk (typically up to distances ~1–10 AU). One pos-

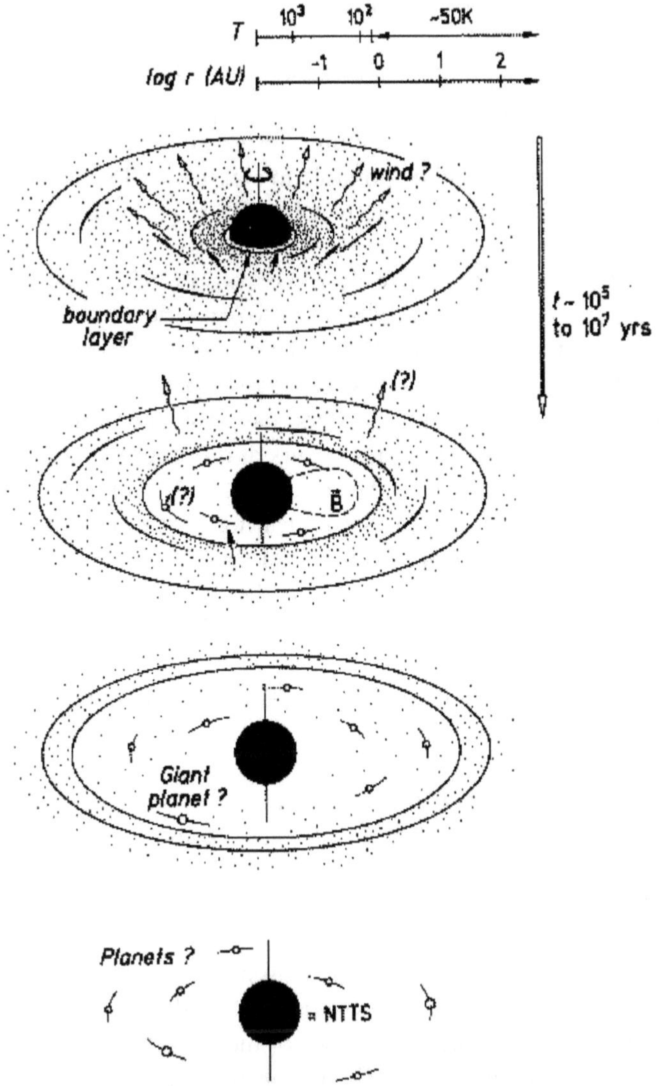


*Figure 134.* A possible evolutionary scenario for low mass PMSSs from the CTTS stage. The temperature and size scales are also indicated (Montmerle & André, 1989).

sible reason is the presence of hydromagnetic instabilities when magnetic fields close to the star dominate self-gravity. Montmerle & André (1989) suggested that an intriguing consequence of it could be the possibility of 'oscillations' between CTTS and WTTS states. In this framework some young stars, strongly magnetized, could become WTTSs directly, without passing through the CTTS phase. But at some point the disk evolves rapidly. In the case of planet formation the timescales may be very short: the order of $10^4$ yr for km-sized bodies (Cameron, 1988). Such hollow disks can be called Laplacian disks, even if they do not appear to shrink as Laplace originally suggested. In the above picture, PTTSs would correspond to the final stages of disk evolution, when the disk has completely disappeared. This fact can be experimentally proved only by mm observations, the material being very cold. In a certain sense it should be possible to classify PTTSs as Class IV objects, although indistinguishable from those of Class III, if observations are made only in the NIR range.

The proposed evolutionary scenario for low mass PMSSs, from the CTTS stage to the PTTS stage, is shown in Figure 134 (Montmerle & André, 1989). Panel (a) shows the current picture of a CTTS: a low mass star surrounded by a warm, optically thick, Keplerian accretion disk, separated from the star by a hot boundary layer. The mass loss observed in these stars may originate in the disk. The transition between CTTS and WTTS stages occurs at different times from star to star, somewhat between $\sim 3 \times 10^5$ and $\sim 3 \times 10^7$ yr. This transition occurs because of the evolution of the disk via a progressive clearing inside out, with a corresponding disappearance of the boundary layer (panel (b)). Dust grain condensation into planetesimals and later probably into planets then contribute to the development of a central cavity (panels (c) and (d)). Figure 134 also shows the temperature and size scales.

So disk models seem to be the most promising in explaining most of the behaviour of TTSs.

### 7.11.3. *Short Remarks on the Current Models*

Despite the impressive volume of observational information on TTSs, some basic questions have yet to be answered, namely:

- what are the energy source and mechanism of formation of the emission spectrum;
- the geometry and velocity field in the emission line formation region; and
- what is the cause of the brightness variability in all spectral ranges?

Generally speaking there are two classes of theoretical models: chromospheric and accretion models. In chromospheric models the region above the photosphere with $T > T_{\rm eff}$, which is responsible for the emission spectrum, is 'akin' to the solar chromosphere. In accretion models the hot circumstellar envelope is the result of the accretion of remnants of the protostellar cloud onto the young star (Lamzin, 1989). It is obvious that one can choose between the two classes of models by determining the geometry of the region responsible for the generation of the emission line spectrum.

For historical reasons a major early field of theoretical investigation on TTS phenomena was that of magnetic activity, since it is the main source of activity in solar type stars. Indeed, it is now popular to believe that several observed properties of CTTSs and WTTSs are likely caused by magnetic dynamo processes of solar type. However, this kind of models account for X-ray emission level, spot coverage and light variability in a day timescales, and $H_\alpha$ and CaII emission–line strength only for WTTSs. On the contrary, solar-type dynamo activity cannot account for the most intriguing properties of CTTSs, namely:

- strong $H_\alpha$, CaII and MgII emission lines;
- UV and IR continuum excess;
- large scale irregular variability.

The best explanation of the key properties of TTSs seems to come from disk models, including disk evolution processes and disk fragmentation. In this framework:

- WTTSs are young stars without disk or with a more or less centrally cleared, hollow disk whose properties reflect magnetic dynamo activity correlated with their rotation rates;
- CTTSs also display magnetic activity, but their more exotic properties are owed to the presence of a circumstellar disk interacting with the star (Bertout, 1993).

Therefore with the accretion disk models it is possible: (i) to explain a variety of apparently unrelated phenomena, such as UV and IR excesses, optical extinction, veiling, Balmer jump, forbidden line profiles, with a single physical process; (ii) to provide TTSs with a large cauldron of potential energy for covering their radiative burdens and for driving the powerful TTS winds. A correlation between the forbidden line flux – which is a mass loss rate indicator – and the IR excess of CTTSs – which is a measure of mass accretion via disk – has been found (Cabrit *et al.*, 1990). However, this connection between winds and accretion is not clarified yet.

However, results derived from the studies of the centrifugally driven winds from protostellar disks, discussed by Safier (1993a,b) – although they do not give the final answer on the origin of protostellar winds – strongly favor the disk driven picture. Indeed, in this way the large temperatures and ionization fractions, deduced from the forbidden-line emissions and the hydrogen recombination lines of TTSs, are very well justified, whereas they are not sustained by stellar winds.

Notwithstanding, it is obvious that many of the theoretical models are incomplete and oversimplified. Indeed great uncertainties exist about details of the early formation stages, which are characterized by the fragmentation processes and loss of angular momentum and magnetic flux. Although many arguments favor the accretion disk models of TTSs, even basic properties of such models – such as the viscosity and structure of the boundary layers – are highly uncertain or even unknown (Appenzeller, 1989). For these reasons the 'minority opinion' of Herbig (1989), following his own definition, can be take into account: *not every observed phenomenon requires disks to be invoked; sometimes more conventional explanations of certain observations could be as plausible or, in some cases, even more probable.*

### 7.12. Galactic Center

The Galactic Center (GC) is one of the most interesting places for testing theories in which frontier physics plays a fundamental role. This argument should be sufficient by itself for the writing of a long review. Indeed, there is an excellent review of Mezger, Duschl & Zylka (1996), which discusses the physical state of stars and interstellar matter in the Galactic Bulge ($R \sim 0.3$–3 kpc from the dynamic center of the Galaxy), in the Nuclear Bulge ($R < 0.3$ kpc) and in the Sgr A Radio and GMC Complex (the central ~50 pc of the Milky Way). This review reports also a list of review papers and conference proceedings related to the Galactic Center with the bibliographic details.

We will search briefly to point out the most important results derived by an immense amount of data spread across the whole electromagnetic spectrum.

First of all it is necessary to note the difficulty in observing the GC since the absorption is extremely high especially in the optical and UV ranges. The interstellar medium is transparent for energies below 2 $\mu$m and above ~2 keV. Therefore the X-ray band is a convenient window in exploring the GC. However, since around the GC the concentration of mass is very large, it is necessary to explore such a zone with high angular resolution instruments, in order to avoid blending of different sources.

A sketch of the Galaxy is reported in Figure 135 (Mezger, Duschl & Zylka, 1996).

The main components of the Milky Way are the Galactic Disk (GD), the Galactic Bulge (GB) and the Halo. The GD has a mass $M_{\rm GD} \sim 10^{11}$ $M_\odot$ and extends to a galactic radius $R_{\rm GD} \sim 14$ kpc. A reasonable supposition is that the GD is stabilized by a spherical halo of mass $M_{\rm Halo} \geq M_{\rm GD}$. A thin layer of ISM is located in the plane of symmetry of the GD. HII regions, which are tracers of massive star formation, indicate a two-armed spiral structure extending from 2 to 12 kpc (Georgelin & Georgelin, 1976, and Figure 4 of Mezger, Duschl & Zylka, 1996). The Central Region (CR) ($R \leq 3$ kpc) of the Galaxy is formed by the Nuclear Bulge (NB) ($R \leq 0.3$ kpc) and the GB (0.3 kpc $\leq R \leq 3$ kpc). Very often, the CR is currently called GB.

ASCA observations in the range 2–10 keV show a diffuse emission dominating the X-ray image. After removing the contributions of bright X-ray binaries, the spectrum of the diffuse X-ray emission from the central 1° of the Galaxy is shown in Figure 136. Major emission lines are identified with $K_\alpha$ lines from He and H like ions of Si, S, Ar, Ca, and Fe. A significant 6.4 keV Fe line, a fluorescent line from neutral atoms or ions below XVII is also present (Koyama *et al.*, 1996).

Massive and young associations such as the Arches and Quintuplet clusters (Figer *et al.*, 1999) and HeI emission–line stars (Najarro *et al.*, 1997), as well as Wolf–Rayet stars (Blum, Sellgren & Depoy, 1995; Figer, McLean & Morris, 1995) all represent clear evidence of episodes of star formation near the GC during the last $10^7$ yr (Krabbe *et al.*, 1995).

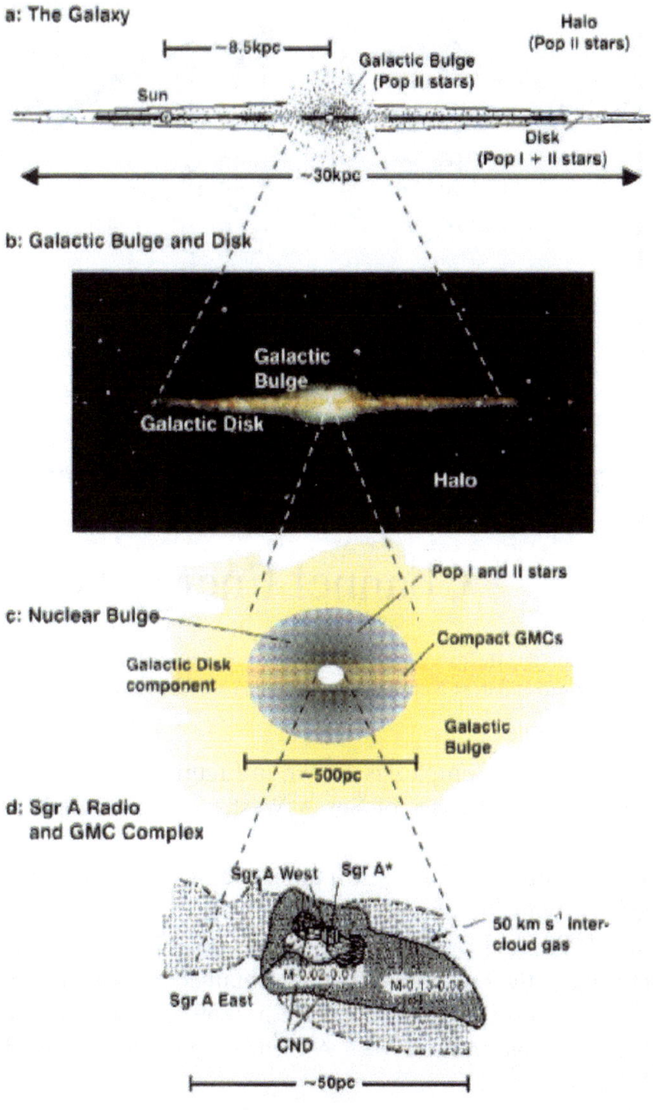


*Figure 135.* Sketch of the Galaxy: a) cross section; b) false color DIRBE image of the Galactic Bulge by using observations at 1.25, 2.2, and 3.5 $\mu$m (Weiland *et al.*, 1994); c) representation of Galactic and Nuclear Bulge and the thin layer of ISM; d) Sgr A and GMC Complex (Mezger, Duschl & Zylka, 1996).

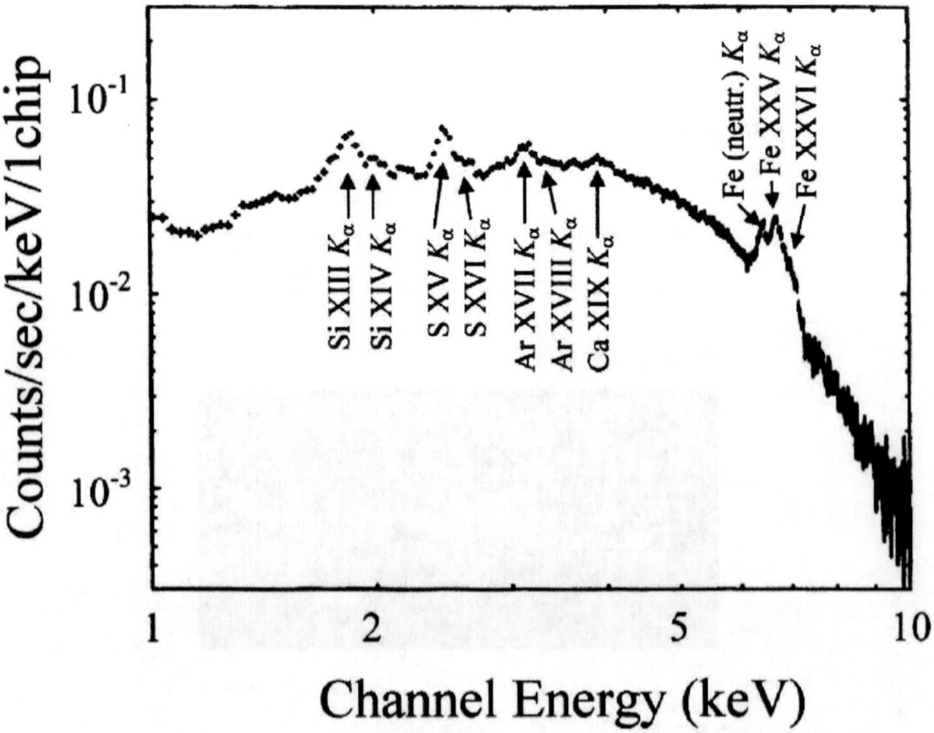


*Figure 136.* Spectrum of the diffuse X-ray emission from the central 1° Galactic Region, after removal of the bright X-ray binaries contributions (Koyama *et al.*, 1996).

A sketch of the Sgr A Complex is shown in Figure 137 (Sidoli, 2001). Sgr A* is embedded in the ionized gases of Sgr A West, and Sgr A East shell is located behind it.

The global spectrum of Sgr A* is still poorly known. There are no secure counterparts of Sgr A*, with the exception of the X-ray detections with CHANDRA (Baganoff *et al.*, 1999) and BeppoSAX (Sidoli *et al.*, 1999; Sidoli, 2000).

The first images of the GC (0.5–4.5 keV), obtained with the EINSTEIN Observatory, showed the point-like source 1E 1742.5-2859, whose position was coincident with Sgr A* (Watson *et al.*, 1981). A diffuse component around the GC was also detected.

Also GRANAT observatory produced the GC images in the 4–1000 keV energy range, with good angular resolution. It was shown that the dominant contribution to the hard X-ray flux from this region originates from the source 1E 1740.7-2942 located 50 arcmin apart from the dynamical center of our Galaxy. The temporal and spectral properties of this source in X-rays resemble that of most plausible galactic black hole candidate Cyg X-1 (Churazov *et al.*, 1993). In that paper the latter authors presented an overview of the GRANAT results.

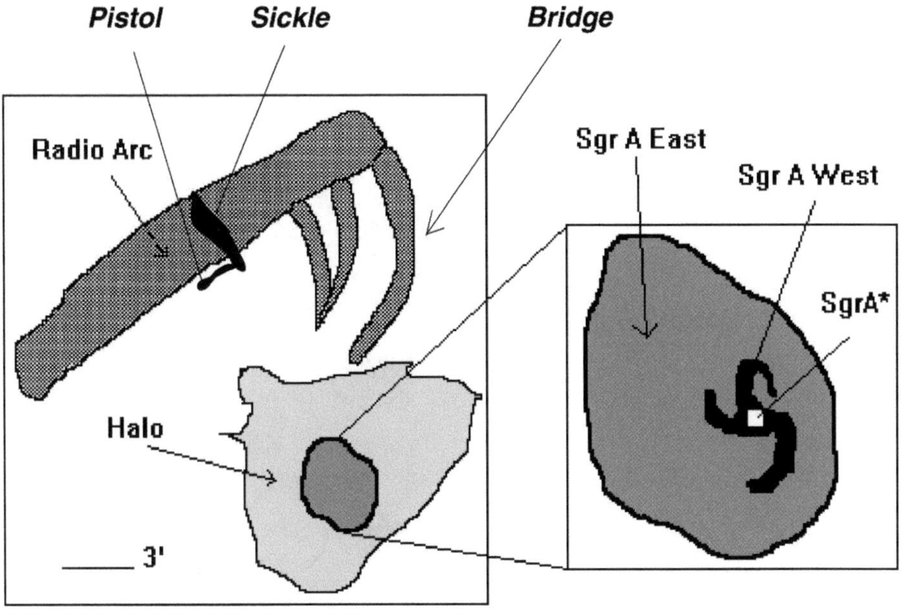


*Figure 137.* Sketch of the Sgr A Complex. North is to the top, East to the left. 1 arcmin corresponds to ~2.5 pc at 8.5 kpc. All these structures are contained within ~20 pc from the GC. However, their relative positions are still poorly known (Sidoli, 2001).

The ROSAT satellite observed in the range 0.1–2.4 keV the same region (Predehl & Trümper, 1994), detecting a diffuse emission and three point-like sources; one of them placed within 10 arcsec from Sgr A*. Within a few arcminutes from Sgr A* there is the transient X-ray source A1742-289, discovered by the ARIEL V satellite (Branduardi *et al.*, 1976) and the LMXB AX J1745.6-2901 (Maeda *et al.*, 1996). Sidoli (2001) derived the spectrum and the flux ascribed to the GC point source(s). A plasma model with temperature of ~4 keV, absorbed by a column density $N_{\rm H} = 7 \times 10^{22}$ cm$^{-2}$. A contamination of this spectrum from diffuse emission is shown by the presence of some emission lines, such as the sulfur line at ~2.4 keV and the iron line at 6.7 keV. The derived flux (2–10 keV), corrected for the absorption is $\sim 4 \times 10^{-11}$ erg cm$^{-2}$ s$^{-1}$ and the luminosity is $\sim 3 \times 10^{35}$ erg s$^{-1}$ (for an assumed distance of 8.5 kpc). This flux is compounded also by the diffuse emission component. Therefore only an upper limit to the 2–10 keV emission of Sgr A* can be estimated at a level of $10^{35}$ erg s$^{-1}$. From the observations of the GC performed with the CHANDRA satellite, Baganoff *et al.* (1999) detected one weak source in the range 0.1–10 keV, with luminosity $L_{\rm x} = 4 \times 10^{33}$ erg s$^{-1}$, probably the counterpart to Sgr A*.

Sidoli (2001) observed also the diffuse emission from the innermost 4° across the galactic plane. Figure 138 shows the regions observed and used for the spectral analysis. These regions include Sgr A (field 1), Sgr B2 (field 2), Sgr D (field 3), Sgr C (field 4), and the main GC molecular clouds complexes as derived by the

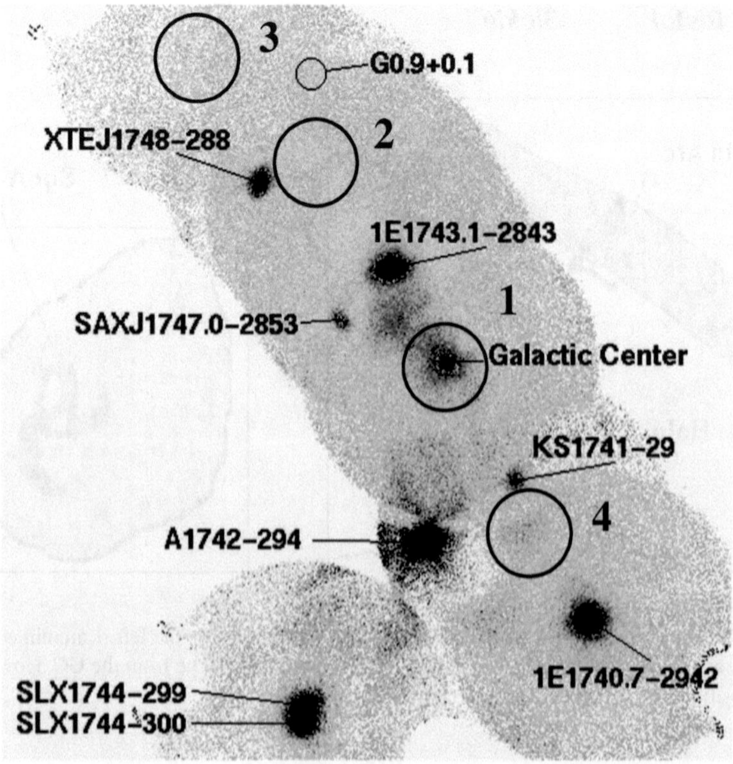


*Figure 138.* The four circles (8 arcmin radius) mosaic (MECS/BeppoSAX) indicate the extraction regions for the spectral analysis and the diffuse emission. They are the inner parts of four pointed observations on: Sgr A (1), Sgr B2 (2), Sgr D (3), and Sgr C (4) (Sidoli, 2001).

CO surveys (Oka *et al.*, 1998). These molecular clouds are part of a large layer of neutral gas distributed across the galactic plane: width ~50 pc, length ~500 pc.

Sgr A* is located close or at the dynamical center of the Milky Way. Upper limits for its size and mass are $\leq (2.5\text{–}4)\times 10^{13}$ cm and $\sim(2\text{–}3)\times 10^6$ $M_\odot$, respectively. It is embedded in the NB. Its radio-IR spectrum is best fitted by a model of an optically thin synchrotron radiation emitted by relativistic electrons with a quasi-monoenergetic distribution. Lynden-Bell & Rees (1971) made the hypothesis that most – if not all – normal galaxies have massive but dormant black holes at their center. Therefore a detailed study of our GC is crucial for understanding the phenomena related to the activity in the center of galaxies. Indeed, the angular resolution of the present day mm-VLBI comes close to resolving the Schwarzschild radii of the central black hole of a few $10^6$ $M_\odot$. A value of $2.5 \times 10^6$ $M_\odot$ of dark mass at the GC is supported by the measurements of the mass distribution and dynamics of IR stars in the central parsec of the Galaxy (Ghez *et al.*, 1998).

Miralda-Escudé & Gould (2000) predicted the existence of a cluster of black hole remnants around the GC. Integrating over the lifetime of the Galaxy and considering the rate of capture by Sgr A*, they concluded that ~20,000 of these black holes have migrated by dynamical friction to the GC and should still be there forming a compact cluster with radius of ~0.7 pc, for a distance to the GC of 8 kpc. It should be important to consider such a distance to the GC better that the canonic value of 8.5 kpc, since from the Optical Gravitational Lensing Experiment (OGLE), McNamara *et al.* (2000) derived the best distance to the GC equal to $7.9 \pm 0.3$ kpc.

How to detect the cluster black holes is still an open problem. Chanamé, Gould & Miralda-Escudé (2001) considered the possibility of detecting such a cluster by monitoring the pairs of images of background bulge stars, which are lensed by Sgr A*, looking for microlensing events induced by the surrounding black holes. On the other hand, Chanamé & Gould (2002) suggested millisecond pulsars as probe of mass segregation in the GC. Such a method is based on a small sample of any type of GC objects, provided they are significantly less massive than the black holes and constitute part of an old ($\geq$ 1 Gyr) population. In the presence of such a cluster and because of energy equipartition, lower mass objects would be expelled from the central regions and settle into a distribution very different from the cusp expected to be induced by the supermassive black hole alone. The authors showed that with a sample of 50 objects and using only their angular position in the sky relative to Sgr A*, it is possible to clearly differentiate between a distribution consistent with the presence of the cluster of black holes and a power law cusp distribution. Since the uncertainties involved in the age determination of less exotic objects is larger than those of the millisecond pulsars, these are the best candidates to perform such a test. In addition, by measuring their first and second period derivatives, millisecond pulsars offer the rare opportunity to determine the complete phase space information of the objects. This extra information improves the detection of mass segregation by ~30%.

Experimentally, this test will be possible by using data from the RXTE and INTEGRAL missions.

The luminosity contained in the radio-IR part of the spectrum is ~300 $L_\odot$. The optical to UV luminosity ratio for Sgr A* is $L_{\rm opt/UV} \leq 500\ L_\odot$. This is an upper limit if a standard accretion disk spectrum is fitted to the upper limit of the K-band flux density and a black hole mass of $\sim(2–3)\times10^6\ M_\odot$ is adopted. The X-ray luminosity of Sgr A* is less than a few $10^2\ L_\odot$ (Mezger, Duschl & Zylka, 1996).

Therefore, close to or at the dynamical center of the Milky Way there is a compact mass, which is probably a massive black hole (BH) or a cluster of black holes. Such a mass is much lower than that inferred for the BHs in most of the active galaxies, but well in the range of dark masses detected in the centers of the Seyfert and normal galaxies. Indeed, if M81 radio–IR spectrum is scaled to 8.5 kpc, the distance to the GC, it is possible to directly compare it with that of the Sgr A*. This comparison, shown in Figure 139, is in favor of the black hole hypothesis for the GC of the Milky Way (Mezger, Duschl & Zylka, 1996).

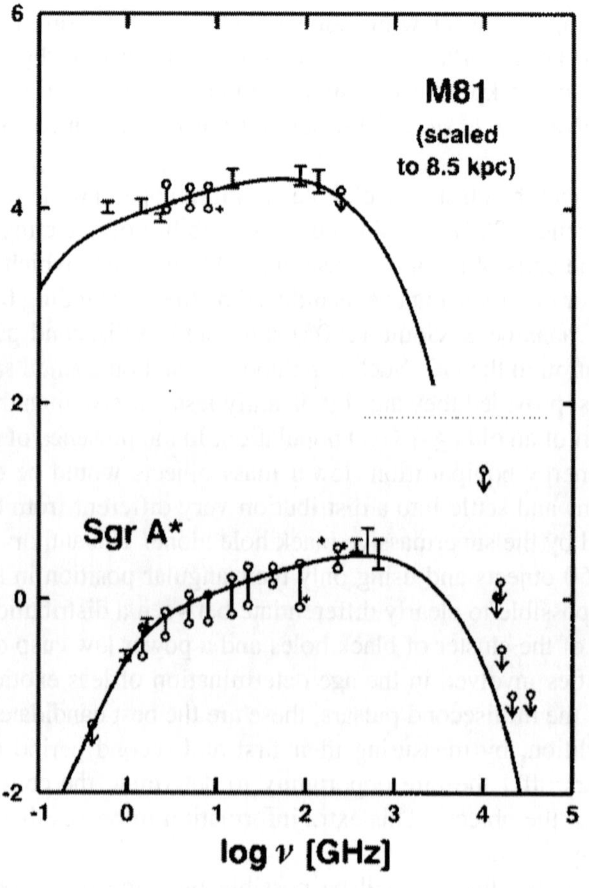


*Figure 139.* Comparison of the radio-IR spectrum of Sgr A* and the core spectrum of M81 (Reuter & Lesch, 1996), reduced to the distance of the GC. The full lines are models of spectra coming from synchrotron radiation of quasi-monoenergetic relativistic electrons. The spectra are self-absorbed for the lower frequencies, and optically thin for the higher frequencies (Mezger, Duschl & Zylka, 1996).

## 7.13. Supernova Remnants

We have already discussed on X-ray emission from SNRs in Section 5.5.4. Now we want to update this topic.

In his review paper Bandiera (2001) pointed out several of the numerous reasons, which push the astrophysicist and particle physicist communities to study SNRs.

There are two SNR compilations, which are accessible via WEB: the first is the catalog of David Green (2001) containing 231 SNRs and relative references. The second database is that of Sergei Trushkin (2002) and contains radio flux densities, spectra and references. A limited number of SNRs have been observed also in IR, optical, and X-ray energy ranges. The knowledge of SNRs in the X-ray energy range has been improved a lot with the advent of the ROSAT, mainly thanks to its high flux sensitivity and to its energy range, where SNRs mostly emit. Indeed, the majority of SNRs emits thermal radiation from a hot optically thin plasma with temperature in the range of (1–10) $\times 10^6$ K. The emission therefore peaks at energies between $\sim$0.1 keV and a few keV. A small minority of SNRs shows synchrotron emission with a power law energy spectrum decreasing with energy, so that the bulk of the emission is found at low energies. However, interstellar photoelectric absorption effectively limits the low energy flux. For an interstellar hydrogen column density $\leq 3 \times 10^{22}$ cm$^{-2}$, the band pass (0.1–2.4 keV) of the ROSAT X-ray Telescope (RXT) is quite suitable for an efficient detection of SNRs. Aschenbach (1999) reviewed on X-rays from SNRs.

There are 102 X-ray SNRs (Aschenbach, 1996). However, Busser (1998) discussed in his Ph. D. thesis the large number (350) of SNR candidates coming out from the ROSAT all sky survey (RASS). It seems possible that the fraction of SNRs showing up at X-rays will be increased. Unfortunately, the exposure times of the ROSAT sky survey were relatively short; moreover, because of its relatively soft energy range and significant absorption for objects at larger distances in the Galactic Plane, the limitations in detecting SNRs with the ROSAT have been rather strong. However, with the help of radio observations, it is possible to clarify the nature of most of the ROSAT SNR candidates (e.g., Schaudel *et al.*, 2002). Indeed, they reanalyzed the RASS data of these candidates and correlated the results with radio surveys like NVSS, ATNF, Molonglo and Effelsberg. A further correlation with SIMBAD and NED was performed for subsequent identification purposes. About 50 of the 350 candidates turned out to be likely galaxies or clusters of galaxies. On the contrary, they found 14 RASS sources as very probable SNR, but still candidates, since they are still searching for their real nature.

Reich (2002) briefly reviewed the origin and evolution of SNRs. He compared SNRs with supernova statistics and observational limitations. He discussed also the morphology and characteristics of the different types of SNRs, including recent results obtained with the Effelsberg 100 m telescope.

Most of the identified SNRs are of shell-type. This morphology is expected when releasing a strong blast wave, with an initial expansion velocity of $\approx$20,000 km s$^{-1}$, owed to a SN explosion, which releases $\sim 10^{51}$ erg into the ISM.

Woltjer (1972) discussed the following phases of the evolution of a shell-type SNR with an initial energy just of the latter value into a homogeneous ISM with density $n_0$ cm$^{-3}$. However, the hypothesis of homogeneity is not realistic. The effect of the stellar wind from the massive progenitor star modifies the ISM and the SNR evolves in a thin medium and only later will encounter the dense material previously swept up by the stellar wind (e.g., McKee, Van Buren & Larareff, 1984).

Not all SNRs show the shell-like structure. Some of them present instead a filled center structure: the prototype is the Crab Nebula, but other objects with similar properties have been discovered since then. They are called filled center SNRs, or Crab-like SNRs, or *plerions* (Weiler & Panagia, 1978).

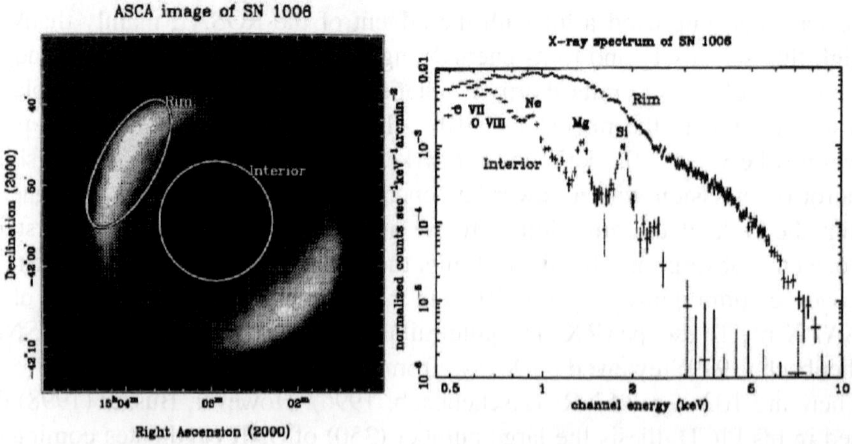


*Figure 140.* Left: the X-ray image obtained with the ASCA GIS shows a clear shell structure of SN 1006. Right: the X-ray spectra obtained with the ASCA SIS for the two regions marked in the image: the Rim and the Interior (Tsunemi *et al.*, 2000 after Koyama *et al.*, 1995).

Plerions are recognized not just on the basis of their morphology, but also of other properties, such as: i) a flat power law radio spectrum, with spectral index ranging from 0.0 to -0.3; ii) a high radio polarization, with a well organized pattern (not true for all plerions); iii) a power law X-ray spectrum, with photon index close to -2 (Asaoka & Koyama, 1990); iv) the detection of an associated pulsar (this is not the case for all plerions).

The details of the nature and structure of plerions are still unclear. However, there is a common agreement on the following points: i) a plerion is an expanding bubble, formed essentially by magnetic fields and relativistic electrons, and the observed synchrotron emission originates from these two components; ii) a continuous supply of magnetic flux and relativistic particle is required in order to explain the typical synchrotron emissivity, as well as the high frequency emission, where the synchrotron lifetime of the emitting particle is much shorter than the SNR age. A good review on Plerionic SNRs is that by Bandiera (2002).

Some SNRs produce both synchrotron emission and thin thermal emission. The first comes from the central part of the SNR whilst the second from the surrounding region, which is shock heated. However, ASCA discovered an enigmatic emission structure in SN 1006 (Koyama *et al.*, 1995), which was the brightest amongst historical SNe (Stephenson & Clark, 1978). It forms a typical shell structure.

Figure 140 shows the X-ray image of SN 1006 and its spectra. There is a thin thermal emission in its interior whilst there is a synchrotron emission in its rim. Koyama *et al.* (1997) also discovered a similar structure in the SNR RX J1713.7-3946. These may form a new group of SNRs, which are considered as the possible sites of particle acceleration for high energy cosmic rays. Allen *et al.* (1997) presented a wide band X-ray spectrum for Cas A, which is a typical young

SNR forming a shell structure. There are many emission lines below 10 keV, which comes from the thin thermal emission. Whereas there is a high energy trail component above 10 keV, which comes from the synchrotron emission from electrons having energy up to several TeV.

Tsunemi *et al.* (2000) presented observational results of Galactic SNRs focusing on their thin thermal emission. Their spectra provide a good diagnostic of the plasma. Generally such spectra cannot be fitted by using model spectra in collisional ionization equilibrium (CIE) condition. Some of them show gradients both in the temperature and in the abundance. The authors divide their discussion into two main parts: the firs devoted to Young SNRs and the second to Middle Aged SNRs. The X-ray spectrum of young SNRs, like Kepler' SNR, can be well fitted by the superposition of the fore shock component and the reverse shock component, both of which show gradients in various parameters: temperature, density, ionization parameter, etc. Cassiopeia-A, one of the typical young SNR, shows not point symmetric, but axially symmetric both in the intensity profile and in the line center of energy profile. This suggests the Doppler motion of the plasma of about several thousands of km $s^{-1}$. The Cygnus Loop, a middle aged SNR, also shows gradients: the shell regions show low metal abundance and low temperature, whilst the core region shows high metal abundance and high temperature. The plasma left in the core region must reflect the fossil of the supernova explosion. Based on the metal abundance, Tsunemi *et al.* (2000) conclude that the Cygnus Loop originated from the type II SN. The ROSAT observations of the Vela SNR, another typical middle aged SNR, revealed that there were several debris running over the shell, which was formed by the strong shock propagated in the surrounding medium. ASCA confirmed the abundance anomaly suggesting that they were the fossil of the progenitor star. The plasma both in the young and in the middle aged SNRs are in the mixing phase of the ejecta with the ISM.

An updated review on X-rays from SNRs is that by Aschenbach (2002). He remarks how the X-ray research on SNRs have been dominated in the 1990's by ROSAT and ASCA. ROSAT essentially produced well resolved images and coarse spectra, ASCA added high quality spectra and moreover extended the spectral range up to 7 keV and more. New scientific grounds were broken on the propagation of SNRs into the ISM, the physics of the SN explosions by the discovery of fragments of the progenitor stars, which shed new light on the question of element mixing in SN explosions, and on the production of several tens of TeV electrons in SNR shells, reviving the discussion on the origin of cosmic rays.

The today researches are performed with the Chandra and XMM-Newton observatories, and we are waiting for the results of the INTEGRAL mission. The former satellites produce excellent images and spectra with very high spatial and spectral resolutions and over a spectral bandpass never accomplished before. Aschenbach (2002) present images of various SNRs, which seem to show a distinction between the matter shocked by either the blast wave rushing into the ambient medium,

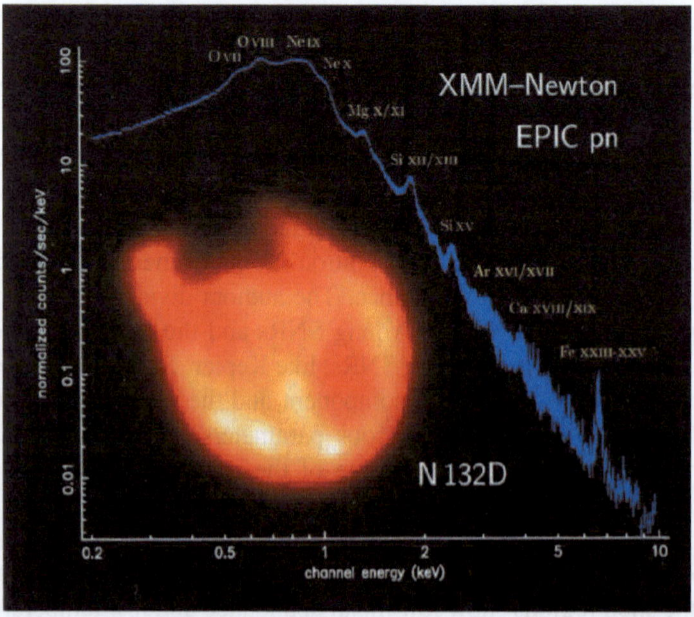


*Figure 141.* XMM-Newton EPIC pn camera image and total spectrum of the SNR N132D placed in the Large Magellanic Cloud (Behar *et al.*, 2001).

or to reverse shock heating the progenitor's ejecta, which includes differences in elemental abundances and their spatial distribution.

Just to show the quality of the images and spectra now available, Figure 141 shows the image of the SNR N132D in the LMC, ~3000 yr old, with superimposed the first high quality broad band X-ray spectrum of the whole remnant obtained with the XMM-Newton EPIC pn camera. The break out and the filamentary structure of the limb as well as a bright filament running across the middle of the image are clearly visible. Emission lines of Si, S, Ar, Ca and Fe are very prominent. The spectrum covers a flux range of 4.5 orders of magnitude in one exposure.

The Crab Nebula has been deeply studied thanks to the images of the Einstein HRI and the ROSAT HRI, with better resolution. However, only with the excellent Chandra image, it has been possible to reveal the detailed structure of the Crab Nebula, as shown in Figure 142 (Weisskopf *et al.*, 2000). At least three emitting regions can be recognized: the torus, the jet, and a more diffuse large scale region, extending far to the north east, for instance. In particular, the region around the pulsar and the main torus exhibit substantial ellipse-like substructure. The existence of the torus-like emission region as an equatorially confined pulsar wind nebula was suggested already in the very early times of the X-ray astronomy era by Aschenbach & Brinkmann (1975), interpreting lunar occultation measurements. The existence of the polar outflows (jets) was proposed by Brinkmann, Aschenbach & Langmeier (1985) on the base of a 2-dimension Fourier analysis of the Eistein HRI image, which were even more evident in the ROSAT HRI image if the analysis

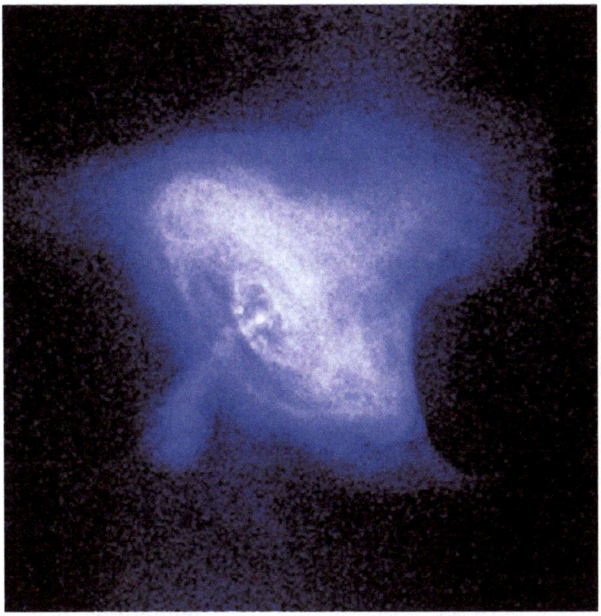

*Figure 142.* Chandra image of the Crab Nebula (Weisskopf *et al.*, 2000).

is restricted to just the spatial high frequency components (Aschenbach, 1992). For more details on the interpretation of this image see the review of Aschenbach (2002).

7.13.1. *Radio Pulsar–SNR and X-Ray Pulsar–SNR Associations*

A workshop on *The Relationship between Neutron Stars and Supernova Remnants* was held in Elba Island (Italy) on June 1998. The proceedings (Bandiera *et al.*, 1998) reported the status of the art, at that time, on this important problem. Indeed, how pulsars are born and if they are isolated or lie in binary systems, and what is the ratio between the number of isolated pulsars and the number of pulsars in binary systems still constitute open problems. Attempts in determining associations among pulsars and SNRs have been done by several authors in the cases of isolated pulsars (e.g., Kaspi, 1998; Helfand, 1998), AXPs (e.g., Mereghetti, 1998), and X-ray pulsars in binary systems, including AXPs (Giovannelli *et al.*, 1993a,c; 1994; Giovannelli & Sabau-Graziati, 2000a,b).

It has been demonstrated that such associations exist; this proves theories predicting formation of pulsars by SN explosions in collapsing isolated progenitors or SN symmetric or asymmetric explosions in binary system progenitors. Isolated pulsars are the remnants of a SN explosion of a collapsing isolated progenitor, or of a SN explosion in a binary system progenitor with disruption of the binary system itself. Pulsars in low and high eccentricity binary systems are the remnants of symmetric or asymmetric SN explosions, without the disruption of the binary nature of the systems.

In the latter two cases a neutron star is orbiting around an early type supergiant, giant or main sequence star.

This kind of association is very useful not only for the point of view of the evolution of binary systems, but also for calibrating the ages of the binary systems and associated SNRs: their ages must be of course the same. This allows making a check on the theories of the evolution of binary systems, the braking of the pulsars and the development of a SNR after the explosion. Evolutionary theories of SNRs are mainly based on their radio behavior, whilst evolutionary theories of binary systems are mainly based on their optical and X-ray behavior. Therefore such different methods of sounding such different cosmic objects are a powerful tool of cross-checking a lot of astrophysical problems.

Lipunov (1998) reviewed the problem of the population synthesis of binary radio pulsars (with white dwarfs, neutron stars, B-stars, and planets) including different assumptions about anisotropy during the formation of neutron stars. Binary pulsars' statistics strongly depends on the initial kick velocity ($w$) of newborn neutron stars. Results of his calculations demonstrated that the distribution of pulsars's high space velocities is in large contradiction with the binary pulsars zoo. Therefore the best coincidence between observations and calculations at $w \sim$ 100–200 km $s^{-1}$ indicates the need for a small natal kick velocity, as was previously shown by Kornilov & Lipunov (1984). This fact has been indirectly corroborated by Giovannelli *et al.* (1993a,c) who found a kick velocity of the X-ray pulsar A 0535+26 and its Be companion HDE 245770 (Flavia's star) $w \sim 160$ km $s^{-1}$, being such a system associated with the SNR S147.

However, for determining the veracity of most proposed associations, future proper motion and multifrequency studies are needed.

### 7.14. Gamma Ray Bursts

Gamma Ray Bursts (GRBs) constitute the hottest argument of modern astrophysics. Indeed, in spite of 2704 events recorded, their origin and nature is still controversial. Greiner (1999) presented and discussed the sky distribution of 1869 GRBs detected with CGRO-BATSE instrument, their localization and origin. Such an updated isotropic, but not homogeneous, distribution is shown in Figure 143, where 2704 GRBs are represented. The number of GRBs has further increased thanks to the detections of the BeppoSAX, RossiXTE, and HETE satellites and in the near future will increase too because of the measurements of the INTEGRAL mission.

GRBs are sudden events of a few seconds duration, different from each other both in intensity and duration. The same GRB manifests in different energy ranges with some delays. As an example, the light curves in different energy ranges of GRB 000615 are shown in Figure 144 (Nicastro *et al.*, 2001).

They emit an amount of energy, which can eclipse even a giant galaxy. Their origin is still a mystery. Are they galactic or extragalactic in origin?

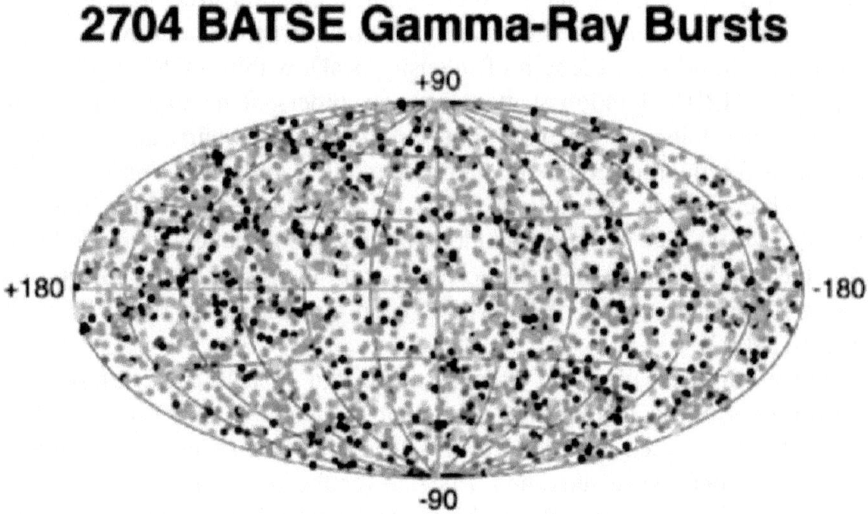


*Figure 143.* Distribution of 2704 GRBs detected by the BATSE experiment on board the CGRO from 1991 to 1999 (http://www.batse.msfc.nasa.gov/data/grb/skymap).

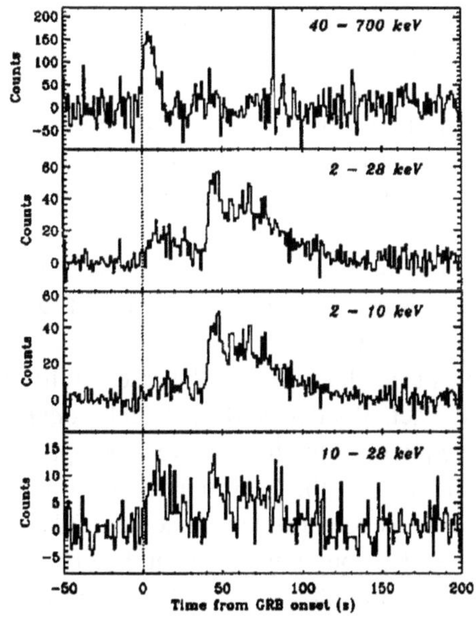


*Figure 144.* Example of GRB 000615 light curves in different energy ranges, from BeppoSAX (Nicastro *et al.*, 2001).

No one strong argument against the complete isotropic distribution has been given. The majority of GRBs has a rather complex temporal structure: in particular their variability time scale is significantly shorter than the duration. The typical ratio between duration and length of intensity peaks within a GRB is about 100. Note that the GRB's duration scatters over six orders of magnitude (1 ms to 1 Ks), and given additional spectral variations from burst to burst and even within bursts, one is faced with a confusing diversity of the *simply measurable quantities*. A statistical analysis of all available GRBs has been made by Belli (1996; 1997; 1998; 1999; 2001). The author concluded that they are galactic and divided in two classes: the first belong to the galactic plane and the second to the galactic halo. The possible galactic origin of GRBs has been also discussed by Kundt (1996; 1999; 2001; 2002).

Since their discovery in 1967 – thanks to the four VELA spacecraft, originally designed for verifying whether the Soviet Union abided the 1963 Limited Nuclear Test Ban Treaty – when 16 strong events were detected (Klebesadel, Strong & Olson, 1973), GRBs have remained a puzzle for the community of high energy astrophysicists. For this reason the problem of GRBs originated hundreds of papers most of them devoted to their physical interpretation. Mazets & Golenetskii (1988) reviewed on observations of GRBs until roughly the end of 1980s.

After the launch of the RXTE and BeppoSAX it has been possible to perform multifrequency observations of the probable counterparts associated with the GRBs just within a few hours of occurrence. Indeed, the BeppoSax measurements allowed the detection of the fading X-ray emission, which follows the higher energy photon emission associated with the GRB in its highest state. Such an emission has been called *afterglow* (Costa *et al.*, 1997) and extends at lower energy ranges, where the first Optical, IR, and Radio counterparts were detected since 1997 (e.g., reviews of Piran (1999a, 2000), Feroci (2001), Castro-Tirado (2002), Frontera (2003), and Pian (2002) and references therein). The first X-ray afterglow detected was that related to the GRB 970228 (Costa *et al.*, 1997). This was one of the most important measurements performed in the space. A previously unknown X-ray source was seen to vary by a factor of 20 on a 3 days timescale. Figure 145 shows such an event.

The precise X-ray position ($1'$) triggered the research for the eventual optical afterglow (OA), which was actually detected by Pedichini *et al.* (1997) and Guarnieri *et al.* (1997) in the rising phase of the light curve. The optical maximum ($V \sim 21.3$ mag was reached $\sim$20 hours after the GRB maximum emission (Groot *et al.*, 1997) and the power law decay was best fitted by $F \propto t^{-1.2}$ (Galama *et al.*, 1997; Bartolini *et al.*, 1998). An extended source was seen at the OA position by ground based and HST observations (van Paradijs *et al.*, 1997; Sahu *et al.*, 1997). Six months later HST detected in the position of the OA an object having $V = 28$ mag as well as the extended source with $V = 25.7$ mag (Fruchter *et al.*, 1997). The extended source surrounding the point source was interpreted as a galaxy. Later, the redshift of such a galaxy was determined as $z = 0.695$ (Djorgovski *et al.*, 1999). Reichart

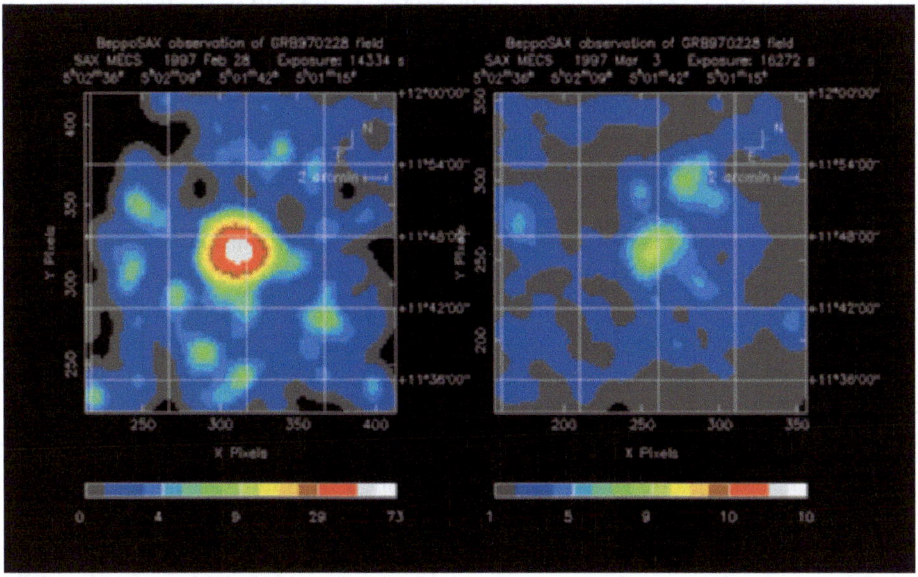

*Figure 145.* X-ray afterglow of the GRB 970228, from BeppoSAX (Costa *et al.*, 1997; courtesy of Enrico Costa and the BeppoSAX GRB Team).

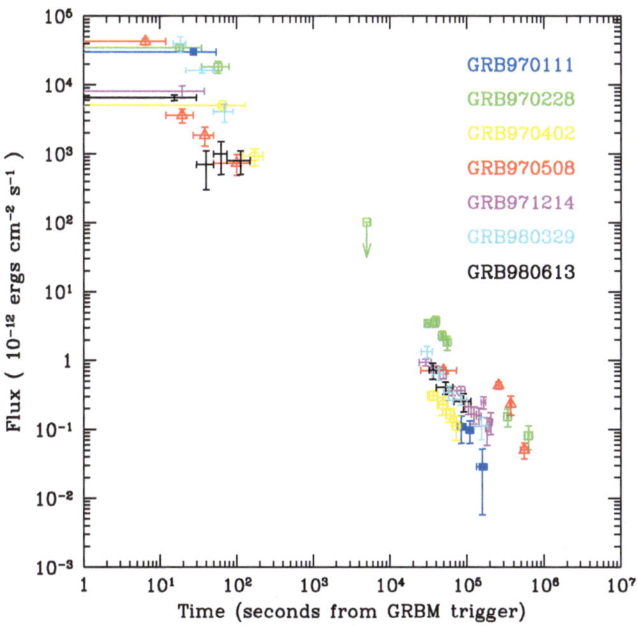


*Figure 146.* The decay of the flux of the most luminous BeppoSAX GRB X-ray afterglows (Costa, 1999) (courtesy of Enrico Costa and the BeppoSAX GRB Team).

(1999) proposed a type Ib/c supernova lies 'behind' the GRB, overtaking the light curve two weeks after. This has been confirmed by Galama *et al.* (2000).

After this important discovery the number of papers devoted to GRBs exploded, especially those proposing models for explaining the physics of the events. However, some review papers attempted to updated the situation on knowledge of GRBs both experimentally and theoretically (e.g., Costa, 1999; Feroci, 1999, 2001; Meszaros, 1999a,b; Piran, 1999b, 2000; Kulkarni *et al.*, 2001; Lamb, 2000; Kundt, 2001, 2002; Woosley, 2001; Castro-Tirado, 2002; Bisnovatyi-Kogan, 2002, 2003).

With the detection of the afterglow of the GRB 970228, the so called Afterglow Era for GRBs started. A confirmation that this Era was born arrived soon with the detection of the radio afterglow of the GRB 970508 (Frail *et al.*, 1997), whose distance was derived by the optical spectrum as $z = 0.835$ (Metzger *et al.*, 1997). So the 30 year old problem of fixing the distances to the GRB sources was apparently solved. With the discovery of the first radio afterglow and by scintillation measurements, the group of Dale Frail established its relative expansion, lending strong support for the fireball model.

For the GRB 990510, following the BeppoSAX/WFC detection, Vreeswijk *et al.* (1999a,b) found the optical counterpart placed at $z = 1.619$. This is the first GRB for which polarized optical emission was detected ($1.7 \pm 0.2\%$) $\sim$18.5 hr after the maximum emission (Covino *et al.*, 1999) and later on by Wijers *et al.* (1999). This confirms the synchrotron origin of the blast wave itself and represents the second case for jet-like outflow (Stanek *et al.*, 1999), being the first that of the GRB 970228.

Thus GRB counterparts are multifrequency emitters, contrary supernovae, which emit most of their energy at the optical frequencies. Recently the X-ray and optical afterglows of GRBs have been discussed in two excellent review by Frontera (2003) and Pian (2002), respectively.

Costa (1999) discussed on the X-ray afterglows of GRBs, detected with BeppoSAX. They pointed with the Narrow Field Instrument (NFI) at 12 directions of the 14 detected GRBs, at that time. In almost all cases they detected X-ray sources which, in the large majority they associate with the GRB, owing to their temporal behaviour. For a large fraction of these GRBs, transient sources in other wavelengths and host galaxies have been detected. They review the main spectral and temporal features of these afterglow sources, their energetics and the association with other features. Figure 146 shows the decay of most luminous afterglow sources. The lines on the left side of the plot indicate the average flux in the 2–10 keV band during the main burst itself.

Table 16 shows 28 GRBs detected by different satellites, for which the redshifts of the *host galaxies* have been determined (Djorgovski *et al.*, 2001; Greiner, 2002).

Bloom, Kulkarni & Djorgovski (2002) presented a study for measuring the location of GRBs relative to their host galaxies. They found the offsets of 20 long duration GRBs from their apparent host galaxy centers by means of images from the Palomar and Keck ground based telescopes and images from the HST space

TABLE 16

Gamma-ray bursts, detected by different satellites and the redshifts of the host galaxies (Djorgovski *et al.*, 2001; Greiner, 2002)

| GRB Name | Host Galaxy Redshift | Localization Source |
|---|---|---|
| 970228 | $0.625 \pm 0.002$ | BeppoSAX |
| 970508 | 0.835 | BeppoSAX |
| 970828 | 0.9579 | RXTE/ASM |
| 971214 | 3.418 | BeppoSAX |
| 980326 | 1 ? | BeppoSAX |
| 980329 | <3.9 | BeppoSAX |
| 980425 | 0.0085 | BeppoSAX |
| 980613 | $1.0964 \pm 0.0003$ | BeppoSAX |
| 980703 | $0.9660 \pm 0.0002$ | RXTE/ASM |
| 990123 | $1.6004 \pm 0.0005$ | BeppoSAX |
| 990506 | 1.3 | BATSE/PCA |
| 990510 | $1.619 \pm 0.002$ | BeppoSAX |
| 990705 | 0.86 | BeppoSAX |
| 990712 | $0.430 \pm 0.005$ | BeppoSAX |
| 991208 | $0.7055 \pm 0.0005$ | IPN |
| 991216 | 1.020 | RXTE/PCA |
| 000131 | 4.50 | Uly/KO/NE (IPN) |
| 000214 | 0.37-0.47 | BeppoSAX |
| 000301C | $2.0335 \pm 0.0003$ | RXTE/ASM+IPN |
| 000418 | $1.1185 \pm 0.0007$ | IPN |
| 000911 | 1.0585 | IPN |
| 000926 | 2.0369 | Uly/KO/NE (IPN) |
| 001109 | 0.37 | BeppoSAX |
| 010222 | 1.477 | BeppoSAX |
| 010921 | 0.45 | HETE/Uly/BeppoSAX |
| 011121 | 0.36 | BeppoSAX |
| 011211 | 2.14 | BeppoSAX |
| 020405 | 0.69 | Uly/MO/BeppoSAX |
| 020813 | 1.25 | HETE |
| 021004 | 2.3 | HETE |
| 021211 | 1.01 | HETE |
| 030226 | 1.98 | HETE |

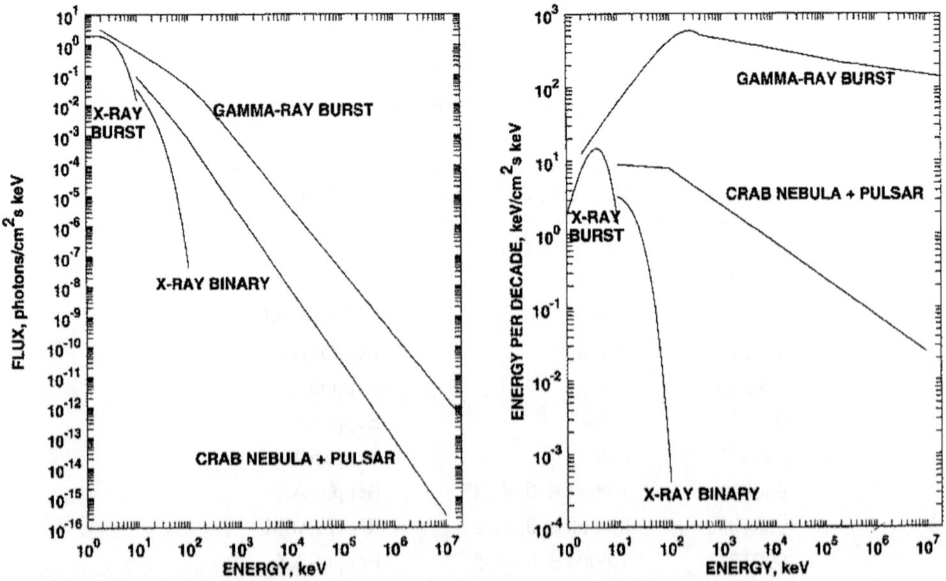


*Figure 147.* GRB energy spectrum compared with those of Crab Nebula & Pulsar, an X-ray burst and an X-ray binary (courtesy of Kevin Hurley, 2003).

based telescope. They discussed the robustness of the association between a GRB and its presumed host galaxy, which allows one to argue on the nature of GRB progenitors. They exclude the delayed merging remnants as the progenitors of long duration GRBs.

In the case of extragalactic origin of GRBs their energy, emitted during the burst, is an immense amount: e.g., the combination of the detection of the GRB 971214 by the BeppoSAX and the measurements of its X-ray afterglow (dal Fiume *et al.*, 2000) and the observations with the Keck Telescope on December 16, 1997 and January 10, 1998 in the optical R band of the afterglow source, allowed the determination of its distance at $z = 3.41$ (Kulkarni *et al.*, 1998). With such a red-shift the energy released during the burst was $\sim 10^{54}$ erg. The energetic afterglow of such a GRB was discussed by Ramaprakash *et al.* (1998).

Figure 147 shows the energy spectrum of an extragalactic GRB compared with the spectra of Crab Nebula & Pulsar, an X-ray burst and an X-ray binary (courtesy of Kevin Hurley, 2003): the left panel shows the flux vs energy; the right panel shows the energy per decade vs energy.

The spectra of most GRB above 25 keV are smooth (Schaefer *et al.*, 1998) and fitted well by the Band function (Band *et al.*, 1993). In BATSE data, no evidence for line features has been found (Briggs *et al.*, 1999; Preece *et al.*, 2000). Recently, evidence for a broad, transient emission feature (whose duration was $\sim$7 s) has been found in the PHEBUS data for GRB 910402. A time resolved spectrum of such a GRB, where the 2.4 MeV excess, with respect to the Band function, is

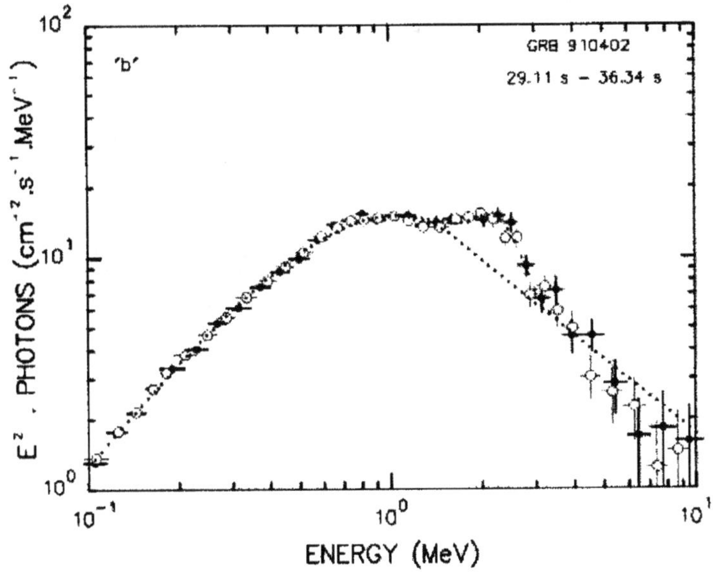


*Figure 148.* Time-resolved energy-spectrum of GRB 910402 detected by PHEBUS (Barat *et al.*, 2000). The fit to Band function (Band *et al.*, 1993) points out the 2.4 MeV excess (from Hurley, 2001).

shown in Figure 148 (Barat *et al.*, 2000). This GRB is one of the bursts studied for its X-ray afterglow by Tkachenko *et al.* (2000). The 7 s feature in excess of the Band function is best fit to a Gaussian centered at 2.4 MeV.

Therefore high resolution GRB spectroscopy may reveal more details, and consequently new surprises: This will be possible very soon with the INTEGRAL mission, as discussed by Hurley (2001). He commented also on the detection of emission from GRBs at energies greater and greater with steps occurring about every 5 years since 1980. Indeed, emission above 200 GeV coming from the Milagrito ground based experiment is probably associated with GRB 970417 (Atkins *et al.*, 2000). This energy is more than one order of magnitude greater than the previous record (Hurley, 1994) and it is close to or greater than the opacity limit for sources at $z > 0.3$ owing to pair production on extragalactic starlight. Thus, this suggests that if all GRBs are cosmological and their redshifts are $z \geq 1$, the emission record at 200 GeV may stand for some time. Such an idea will be tested by future experiments such as Milagro.

Important comments on trigger, energetics and beaming of GRBs have been discussed by Rees (1999). Starting from there now being consensus that GRBs involve extraordinary power outputs, and highly relativistic dynamics, he suggested that the trigger is probably a binary merger or collapse involving compact objects. The most plausible progenitors, ranging from the merging of two neutron stars to various hypernova-like scenarios, eventually lead to the formation of a black hole

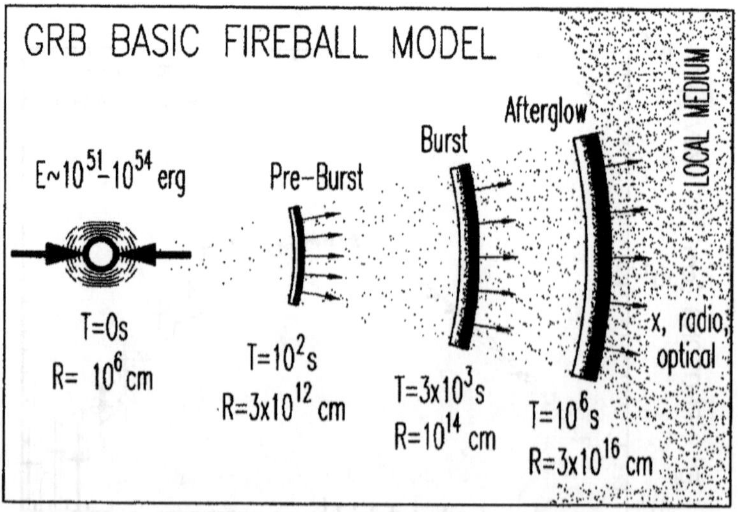


*Figure 149.* Sketch of the fireball model for GRBs (this work, after Feroci, 2001).

with a debris torus around it, the extractable energy being up to $10^{54}$ erg. In that paper, Rees discussed the various modes of energy extraction from such systems and gave some key theoretical issues that would discriminate among the various models.

The discovery of the GRB afterglow favored the *fireball model.* Such a model was proposed long time ago by Cavallo & Rees (1978) and later developed in detail by several authors (e.g., Meszaros, 1999a,b; Piran, 1999b; 2000). Such a model is schematically shown in Figure 149.

A *central engine* releases a large amount of energy – of the order of $10^{50}$–$10^{54}$ erg – in a very compact region and generates a photon pair fireball, which relativistically expands possibly in the form of repeated ejections of discrete shells. Different shells ejected at different velocities may catch up with each other causing an internal shock, which would cause the GRB prompt emission. After that the shells shock the interstellar medium, and this external shock front accelerate particles, which emit synchrotron radiation. Such an emission would be that we observe from the afterglow and it is consistent with what observed in the time average spectra of the GRB prompt emission (e.g., Tavani, 1996). This simple model has been further developed considering more realistic astrophysical conditions, such as self absorption of the synchrotron radiation, effects of Comptonization, varying density in the ISM, presence of pre-ejected winds, etc.. However, from the relatively simple fireball model, Sari & Piran (1997) derived a number of observational consequences, such as an energy distribution best fitted by a broken power law, separated by characteristic frequencies evolving with time. Such a prediction was later verified by Galama *et al.* (1998) in the case of the GRB 970508. The

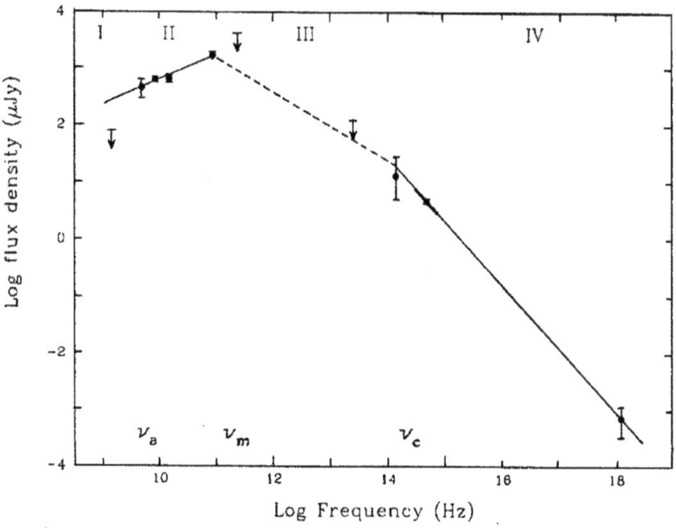


*Figure 150.* Emitted energy distribution as observed from GRB 970508 two weeks after the event. The best fit of data is consistent with the fireball model (Galama *et al.*, 1998).

multifrequency energy distribution obtained two weeks after the event is shown in Figure 150.

Later, such models have been improved. GRB afterglows are well described by synchrotron emission from relativistic blast waves expanding into an external medium. The blast wave is believed to amplify the magnetic field and accelerate the electrons into a power law distribution of energies promptly behind the shock. These electrons then cool both adiabatically and by emitting synchrotron and inverse Compton radiation. The resulting spectra are known to consist of several power law segments, which smoothly join at certain break frequencies, as discussed by Granot & Sari (2002).

Schaefer *et al.* (1998), using detections made with the BATSE/CGRO experiment, presented a composite spectrum of 19 bright bursts from 41 keV to 1.9 MeV, with no weak lines visible. The spectrum is relatively flat, as it is composed of some bursts with flat spectra and others with peaks over a wide range of energy. Such a composite spectrum is shown in Figure 151 up to ~1 MeV.

They used such a spectrum for calculating average redshift corrections in cosmological models. Expanding fireball models with shocked synchrotron emission are predicted to have low energy spectral slope ($\nu F_\lambda \propto \nu^\alpha$) that asymptotically approaches $\alpha = 4/3$ such that $\alpha$ should never exceed 4/3. This prediction was tested with more than 100 bright bursts with BATSE and Ginga data. Over 90% of the bursts have spectral slopes in agreement with this prediction. For the GRB 870303, which showed cyclotron lines, a strong case can be made that the slope violates the model limit, and then only from 2–5 keV.

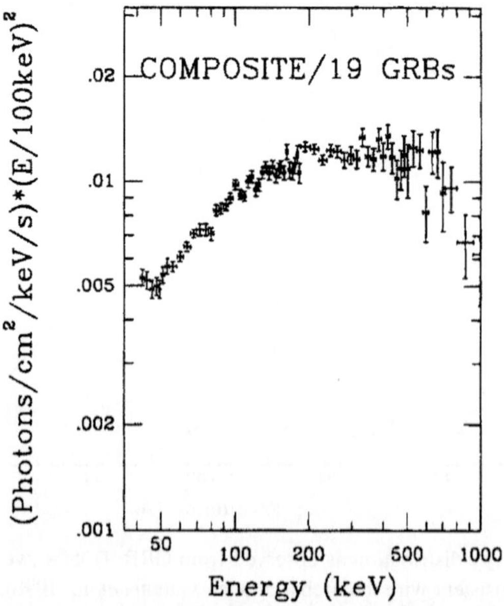


*Figure 151.* Composite spectrum of 19 bright BATSE bursts (Schaefer *et al.*, 1998).

Although the fireball model in its different variations, owing to refinements coming from many authors, is apparently the most suitable in explaining the behaviour of GRBs, many other models have been developed.

In the light of the BeppoSAX results (e.g., see the proceedings of the workshop on *Gamma Ray Bursts in the Afterglow Era* published in a special issue of the *Astronomy and Astrophysics Supplement Series: Vol. 138, No. 3, September 1999*), which favor the extragalactic origin of GRBs, Dar (1999a) discussed a model completely self-consistent in justifying the extragalactic origin of GRBs, apparently cutting in all the discussions on their origin. He used the argument that GRBs are produced in dense stellar regions by highly relativistic and narrowly collimated jets which happen to point in our direction. Such jets are also very powerful cosmic ray accelerators via Fermi processes (Dar, 1998). However, he also suggested the possibility of having galactic GRBs, most of which do not point in our direction. They could be the main source of Galactic cosmic rays at all energies.

In order to render the energetics of GRBs viable, Dar (2001; 2003b) discussed on the possible origin of GRBs and their afterglows from supernova highly relativistic jets. Within days after the explosion the parent star falls back to produce a thick accretion disk around the compact object. Instabilities in the disk induce sudden collapses with the ejection of highly relativistic *cannonballs* of plasma, similar to those ejected by micro-quasars. Each cannonball corresponds to an individual pulse in a GRB. He suggested that the afterglow of the GRB 970228 (located at $z = 0.695$) appears to be overtaken by a light curve akin to that of SN1998bw (located at $z_{\rm bw} = 0.0085$), when properly scaled by their differing redshifts (Dar, 1999b).

Panagia *et al.* (2001) discussed about the radio properties of supernovae and their possible association with GRBs. In particular, they discussed about the association between the interesting and unusual GRB 980425 with the radio SN199bbw.

The SN-GRB association in the case of GRB 970228 has been confirmed by Reichart (1999) and by Galama *et al.* (2000). Evidence of similar associations is found for a few other GRBs, namely: GRB 990712 (Hjorth *et al.*, 1999; Sahu *et al.*, 2000), GRB 970508 (Sokolov, 2001a,b), GRB 980703 (Holland *et al.*, 2001), and GRB 000418.

Then Dar (2001) forces a bit the possibility of such an association, but making reasonable assumptions. He says: '*It is quite possible that a good fraction of GRBs are associated with SNe, perhaps even all of the most frequent, long duration GRBs. The converse statement – that most SNe of certain types are associated with GRBs – appears at first sight to be untenable. The rate of Type Ib/Ic/II SNe has been estimated from their observed rate in the local Universe (e.g., van den Bergh & Tammann, 1991) and the star formation rate as function of redshift, to be* $10\,s^{-1}$ *in the observable Universe (Madau, 1998). The observed rate of GRBs is a mere* $1000\,yr^{-1}$. *Thus very few of these SNe produce visible GRBs. But if the SN-associated GRBs were beamed within an angle* $\theta \sim 3.6 \times 10^{-3}$, *only a fraction* $\pi\theta^2/4\pi \sim 3 \times 10^{-6}$ *would be visible, making the observed rates compatible and making possible a rough one to one SN–GRB association (or a ten to one association for* $\theta \sim 1 \times 10^{-2}$)'.

A recent confirmation for such an association is coming from the detection of a SN signature associated with GRB 011121 (Bloom, Kulkarni & Djorgovski, 2002). GRB 011121, detected by the BeppoSAX satellite (Piro *et al.*, 2001) is currently distinguished as the GRB with the lowest known redshift ($z = 0.36$ (Infante *et al.*, 2001)). The red 'bump', which shows a spectral rollover at ~7200 Å, is well described by a redshifted Type Ic SN which occurred at approximately the same time as the GRB event. The inferred luminosity is about half that of the bright SN 1998bw. These results serve as compelling evidence for a massive star origin of long duration GRBs. A discussion of such an origin is reported by Price *et al.* (2002) by using their near IR and radio observations of the afterglow, which was detected by the HST (Garnavich *et al.*, 2002).

Two emission features observed by the Chandra Observatory in the X-ray spectrum of the afterglow of the GRB 991216 were identified with the $Ly_\alpha$ line and narrow recombination continuum by hydrogenic ions of iron at a redshift $z = 1.00 \pm 0.02$ (Piro *et al.*, 2000b). Such an identification provided an unambiguous distance of a GRB. The line width and intensity measured imply that the progenitor of the GRB was a massive star system that ejected, before the GRB event, a quantity of iron of $\sim 0.01\,M_\odot$ at a velocity of ~0.1c. This could have been an event like that of a hypernova (Paczynski, 1998) or a collapsar (Woosley, 1993) where shortly before the GRB a substantial fraction of the progenitor mass is ejected. In particular, this event seems similar to a supernova explosion, as in the case described by the SupraNova model (Vietri & Stella, 1998).

An interesting model for GRBs has been proposed by Fargion during the years 1994–1998 (e.g., Fargion, 1999, 2001a,b and references therein) and Blackman, Yi & Field (1996). They suggested that GRBs, as well as Soft Gamma Repeaters (SGRs), are spinning and precessing Gamma Jets produced by a collimated $e^+e^-$ Jet via Inverse Compton Scattering in a very narrow (0.1°) angles, blazing and flashing the observer. The Jet arises in SN explosions; its energy decays slowly from earliest SN powers (corresponding to GRB) toward lower stable power as SGR régimes. GRBs and SGRs shared (sometimes) same spectra and time structure: then SGRs are low-power GRBs, but without SN relics (or GRB afterglows, signature of Jets in SN-GRBs). Moreover, weak isolated X-ray precursor signals, such as in GRB 980519, GRB 981226, GRB 000131, corresponding to huge isotropic $\sim 10^{47}$ erg $s^{-1}$, followed by the extreme GRB $\sim 10^{52}$ erg $s^{-1}$ powers, disagree with any Fireball explosive scenario. Fargion naturally interprets these X-ray precursors as rare marginal blazes of outlying X conical precessing Jet tails, surrounding the $\gamma$-Jet, later hitting in-axis as a GRB.

Vietri (1998) discussed the nature of GRBs. In particular, the author noted that the GRB 971214, whose duration was $\approx 30$ s, has released as much energy ($\sim 2 \times 10^{54}$ erg) as the observable universe (whose radius is $c/H_0$), if it were placed at $z = 3.41$ (Kulkarni *et al.*, 1998).

Woosley (2001) attempted a discussion of the many GRB models, which have been described by many authors in the proceedings of the workshop on *Gamma Ray Bursts in the Afterglow Era* (Costa, Frontera & Hjorth, 2001 and references therein).

Fenimore & Ramirez-Ruiz (2000) showed that the time variability of a GRB is correlated with the absolute magnitude of the burst: smooth bursts are intrinsically less luminous. This Cepheid-like relationship has been used to determine the redshift of a GRB from parameters measured only at $\gamma$-ray energies. They found the redshift of 220 GRBs. However, the relationship they used is based on only seven events. Thus such a relationship needs further confirmation with more events. Indeed, this work triggered that of Reichart *et al.* (2001). They found a possible Cepheid-like luminosity estimator for the long GRBs based on the variability of their light curves. For this purpose, they used BATSE/CGRO data for 13 bursts, Wind/Konus data for 5 bursts, Ulysses/GRB and NEAR/XGRS data for 1 burst from each experiment. They found a significant correlation between the variability of a burst's light curve and the burst's isotropic equivalent peak photon luminosity. Such a correlation is in agreement with those found by Fenimore & Ramirez-Ruiz (2000) and Stern, Poutanen & Svensson (1999).

The realization that GRBs are beamed with narrow opening angles, whilst the following afterglows can be observed over wider angles, led immediately to the search for orphan afterglows, which are defined as afterglows not associated with observed prompt GRB emission. Rhoads (1997) firstly suggested that observations of orphan afterglows would provide the possibility of estimating the opening angles and the true rate of GRBs. Nakar & Piran (2002) discussed the detectability of orphan afterglows under the hypothesis that the total energy is constant. Indeed, the observation that GRBs have a constant amount of total energy (Piran *et al.*, 2001; Panaitescu & Kumar, 2001; Frail *et al.*, 2001) and that the observed variability in the apparent luminosity arises mostly from variation in the jet opening angles leads to a remarkable result, namely: the post-jet break afterglow light curve is universal (Granot *et al.*, 2002). In the study of energetics of GRBs, Jimenez, Band & Piran (2001) – analyzing a sample of eight bursts with BATSE spectra and spectroscopic redshifts plus four bursts with BATSE spectra and host galaxy $R$ magnitudes – found that the burst rate is proportional to the galaxy luminosity at the epoch of the burst. Assuming that the total energy emitted has a log normal distribution, they found that the average emitted energy (under the assumption that is radiated isotropically) is $E_{\gamma,\mathrm{iso}} \sim 1.3 \times 10^{53}$ erg (for $H_0 = 65$ km s$^{-1}$ Mpc$^{-1}$, $\Omega_\mathrm{m} = 0.3$, and $\Omega_\Lambda = 0.7$). The distribution has a logarithmic width of $\sigma_\gamma \sim 1.7$. The corresponding distribution of X-ray afterglow energy (for seven bursts) has $E_{\mathrm{X,iso}} = 3.5 \times 10^{51}$ erg and $\sigma_\mathrm{X} \sim 1.3$. They used their results for successfully fitting all bursts with BATSE spectra, for which there were afterglow searches.

We do not want to enter into more details in the forest of models. The interested reader can refer to a recent review on GRB theories published by Meszaros (2002). He reviewed the GRB phenomenon from a theoretical point of view, with emphasis on the fireball shock scenario of the prompt emission and the longer wavelength afterglow. A lighter and pleasant article on GRB has been published by Gehrels, Piro & Leonard (2002).

In spite of the enthusiasm in accepting the cosmological nature of GRBs, our opinion is, at the moment, that the extragalactic origin of GRBs has not yet been definitively demonstrated, at least for most of them. The problem in explaining the origin of the immense amount of energy associated with extragalactic GRBs is still under discussion. Such an enormous energy coming out from single events would be justified in a general way, avoiding *ad hoc* models. Physical restrictions on the models of GRBs have been discussed by e.g., Bisnovatyi-Kogan (2002, 2003) and Kundt (1996, 1999, 2001, 2002).

Indeed, for instance, Band *et al.* (1993) in their study of time averaged spectra accumulated by the BATSE/CGRO found that the spectra are well described at low energy by a power law continuum with an exponential cutoff, $N_\mathrm{E}(E) \propto E^\alpha e^{-E/E_0}$, and by a steeper power law, $N_\mathrm{E}(E) \propto E^\beta$ with $\alpha > \beta$, at high energy. However, the spectral parameters $\alpha$, $\beta$, and $E_0$ vary from burst to burst with no universal values. The break in the spectrum, $E_0$, ranges from below 100 keV to more than 1 MeV, but peaks below 200 keV with only a small fraction of the spectra breaking above 400 keV. Consequently it is unlikely that a majority of the burst spectra are shaped directly by pair processes, unless bursts originate from a broad redshift range. They found also that the correlations amongst burst parameters do not fulfill the predictions of the cosmological models of burst origin. However, their sample may not be appropriate for such a test. They did not find correlations with burst morphology or the spatial distribution.

Given enough statistics it should be possible to segregate between the local and cosmological hypothesis. If GRBs are at cosmological distance, they should show a dipole in their distribution showing the reflection of the local group motion of $\sim$650 km $s^{-1}$, as noted by Roberto Terlevich in the discussion after Mitrofanov's presentation of his paper on the statistics of GRBs at the Frascati Workshop 1995 (Mitrofanov, 1996).

### 7.15. Cosmic Counterparts of Gamma Ray Sources

After the EGRET experiment on board CGRO satellite, more than 200 $\gamma$-ray sources have been discovered to populate the sky. The number is similar to that we had after UHURU for the X-ray sources, at the beginning of 1970s. Now the X-ray sources detected are of the order of hundreds of thousands, thanks to the strong increase of sensitivities. Then, probably, increasing the sensitivity of the $\gamma$-ray detectors the number of $\gamma$ ray sources could follow the same trend of the X-ray ones.

The second COS B catalog (Swanenburg *et al.*, 1981) was formed by 25 $\gamma$-ray sources, most of them not recognized with known cosmic sources. Also most of the 271 $\gamma$-ray sources of the $3^{rd}$ EGRET Catalog (Hartman, *et al.*, 1999) are still not associated with known cosmic sources (see Section 4.3.3.3). Many of these sources concentrate towards the galactic plane and correlate with the spiral arms of the Milky Way, which indicates a significant contribution from population-I objects (e.g., Romero, Benaglia & Torres, 1999; Romero, 2001).

What kind of objects could be the counterparts of the $\gamma$-ray sources? For sure, the following: EGRET AGNs, EGRET pulsars, EGRET LMC, COMPTEL sources (750 keV – 30 MeV), OSSE blazars (50 keV – 10 MeV), TeV sources (300 GeV – 3 TeV), and Young Open Clusters (YOCs) (still under discussion). Romero (2001 and references therein) suggested a list of $\gamma$-ray counterpart candidates, such as the early-type stars, accreting neutron stars, radio quiet pulsars, interacting SNRs, and black hole candidates. Paredes *et al.* (2000), Grenier (2001) and Romero (2001) have suggested the possibility of having micro-quasars as $\gamma$-ray counterparts of the $3^{rd}$ EGRET Catalog sources.

From a statistical point of view two populations can be identified: one associated with the Gould belt, a nearby star forming region, and the other formed by higher luminosity sources at lower latitudes. Indeed, it has been suggested (Grenier, 2000) that a number of unidentified $\gamma$-ray sources could lie in Gould's belt: a lane of massive stars, most of which are less that 20–30 million years of age, arches across the sky. This starburst disk dominates our cosmic neighborhood up to a thousand light years away, and includes many of the brightest, most conspicuous stars in the sky. Most of the stable unidentified EGRET $\gamma$ ray sources, at mid-latitudes closely follow the curve and apparent width of the belt. It has been shown (Gehrels *et al.*, 2000) that these mid-latitude sources are distinct from the population of unidentified sources in the Galactic Plane, and are likely to be associated with Gould's belt.

The micro-quasar LS5039, which is a massive X-ray binary with persistent non-thermal radio emission, has been physically associated with the $\gamma$-ray source 3EG J1824-1514 (Paredes *et al.*, 2000). Romero *et al.* (2001) suggested the association of the 3EG J0542+2610 with the transient X-ray source A0535+26/HDE245770 – the best studied system of such a class (Giovannelli & Sabau-Graziati, 1992) – suggesting that this Be/accreting pulsar can produce variable hadronic $\gamma$-ray emission through the mechanism originally proposed by Cheng & Ruderman (1989), where a proton beam accelerated in a magnetospheric electrostatic gap impacts the transient accretion disk. Giovannelli & Ziółkowski (1990) discussed on the possibility of the formation of a temporary accretion disk at the periastron passage of the neutron star around the Be star and Finger, Wilson & Harmon (1996) observed an accretion disk around the neutron star during a giant outburst of the system. So the association could be really true. However, more direct measurements, especially simultaneous in different energy ranges, are necessary in order to definitively clarify such an association.

The possibility of having high energy $\gamma$-ray emission from close binary systems has been already discussed in Section 7.6.3.

Recently, Kaufman-Bernadó, Romero & Mirabel (2002) suggested a model for galactic variable $\gamma$-ray sources based on the idea of precessing microblazars, which have been proposed by Mirabel & Rodriguez (1999) as micro-quasars with jets forming a small angle with the line of sight, by analogy with the unified model for AGNs. Thus microblazars are sources with highly variable and enhanced non-thermal flux owed to Doppler boosting. Taking into account that more than 130 HMXRBs have been detected (Liu, van Paradijs, J. & van den Heuvel, 2000) and that this number should be a small fraction of the total number of these systems in the Galaxy, Kaufman-Bernadó, Romero & Mirabel (2002) concluded that '*it is not unreasonable to expect the existence of a few tens of microblazars at mid and low galactic latitudes that could be responsible of the variable galactic $\gamma$-ray sources detected by EGRET*'. The recent discovery of the X-ray transient V 4641 Sgr, which has a $\sim$9 $M_\odot$ black hole and shows extreme superluminal velocities (Orosz *et al.*, 2001), could be a microblazar. The best way of detecting microblazars is to measure the electron–positron annihilation feature in their spectra. This will be possible with the INTEGRAL-IBIS.

**Which kind of open problems still survive?**

One is the problem of knowing the behavior of $\gamma$ ray sources in the energy range $\approx$10–300 GeV; indeed, such a range is above that measurable with the space based detectors and below that measurable with the ground based detectors. So this is essentially a technological and/or physical problem. The technological problem is self-evident; the physical problem could be related to the eventuality of discovering a new method of detection of photons of such energies.

Another problem is that connected with the cutoffs at high energies of the $\gamma$-ray sources, not yet completely known, and moreover not yet measured but few exceptions.

Finally, the last problem is related to the origin of $\gamma$ rays, once they are detected from a cosmic source; in other words, from which physical part of the cosmic source $\gamma$-rays are coming? i.e.: (i) from jets, where electrons and positrons are interacting with the surrounding matter and/or radiation, giving rise to X-rays and $\gamma$-rays via bremsstrahlung or to a formation of neutral and charged pions, which decay into $\gamma$-rays and neutrinos; (ii) from spherical accretion of matter onto collapsed objects.

## 7.16. Young Open Clusters: Galactic Gamma Ray and Cosmic Ray Contributors

The importance of the multifrequency study of Young Open Clusters (YOCs) is crucial in order to demonstrate that they are sites of acceleration of particles and how they can contribute to the production of $\gamma$-rays and cosmic rays.

It has been largely demonstrated that YOCs are very exciting sites of high energy processes, thanks to multifrequency (from IR to $\gamma$-ray energy range) observations started with the particularly interesting YOC Berkeley 87. A review on YOCs as $\gamma$-ray emitters has been written by Giovannelli (1996c). Briefly, here we would like to summarize in order to note the importance of YOCs as possible counterparts of a part of the unknown EGRET sources and sites of production of 10–15% of galactic cosmic rays.

### 7.16.1. *Nuclear Gamma Rays from Energetic Particle Interactions*

Gamma-ray line emission from nuclear de-excitation following energetic particle reactions was deeply discussed by Ramaty, Kozlowsky, & Lingenfelter (1979). The calculated $\gamma$-ray spectra and intensities can be used for the study of astrophysical sites which contain large fluxes of energetic protons and nuclei.

The interaction of energetic particles with ambient matter produces nuclear $\gamma$-rays of energies ranging from tens of keV to about 20 MeV. These $\gamma$-rays exhibit a great wealth of spectral structure, ranging from very narrow lines to broad features, depending on the composition and energy spectrum of the energetic particles, and on the composition and physical state of the ambient medium.

Observable $\gamma$-ray line emission is coming from many astrophysical sites including Solar Flares, Interstellar Medium, Neutron Stars and Black Holes, Novae, X-Ray Novae, Supernovae, Supernova Remnants, OB associations, HII regions, Nuclei of Galaxies, Blazars, and Young Open Clusters, as demonstrated by Giovannelli (2002 and the references therein).

TABLE 17

Main $\gamma$-ray lines from carbon and oxygen (adapted from Ramaty, Kozlowsky, & Lingenfelter (1979).

| Photon Energy (MeV) | Emission Mechanism | Production Processes | Mean Life $\times 10^{-14}$ (s) |
|---|---|---|---|
| 4.438 | $^{12}C^{*4.439} \rightarrow$ g.s. | $^{12}C(p,p')^{12}C^{*}$ | 5.62 |
| | | $^{12}C(\alpha,\alpha')^{12}C^{*}$ | 5.62 |
| | | $^{14}N(p,x)^{12}C^{*}$ | 5.62 |
| | | $^{14}N(\alpha,x)^{12}C^{*}$ | 5.62 |
| | | $^{16}O(p,x)^{12}C^{*}$ | 5.62 |
| | | $^{16}O(\alpha,x)^{12}C^{*}$ | 5.62 |
| 6.129 | $^{16}O^{*6.131} \rightarrow$ g.s. | $^{16}O(p,p')^{16}O^{*}$ | 2400 |
| | | $^{16}O(\alpha,\alpha')^{16}O^{*}$ | 2400 |
| | | $^{20}Ne(p,x)^{16}O^{*}$ | 2400 |

TABLE 18

The abundances of $^{12}C$ and $^{16}O$ and energy of the most significant direct de-excitation lines (adapted from Ramaty, Kozlowsky, & Lingenfelter, 1979).

| Nucleus | Abundance | Direct De-excitation Lines (MeV) |
|---|---|---|
| $^{12}C$ | $4.2 \times 10^{-4}$ | 4.438, 15.11 |
| $^{16}O$ | $6.9 \times 10^{-4}$ | 2.741, 6.129, 6.917, 7.117 |

Energetic particle interactions with ambient matter can lead to $\gamma$-ray line emission in a variety of ways. These include: direct excitation of nuclear levels, production of excited secondary nuclei, production of radioactive species – which decay into excited levels of other nuclei –, production of neutrons and positrons.

Many excited levels decay by photon emission and these $\gamma$ rays can be detected. The strongest de-excitation line in $^{12}C$ at 4.438 MeV results from the de-excitation of its first excited state at 4.439 MeV. Such a level in $^{12}C$ can also be populated by spallation reactions on $^{16}O$ and $^{14}N$ (see Table 17).

Excited states of $^{12}C$ above the 4.439 MeV level decay primarily by particle emission, and hence they are not important sources of $\gamma$-ray lines. An exception is the 15.11 MeV level, which, because of conservation of isotopic spin, cannot decay by emitting $\alpha$-particles and hence de-excites only by $\gamma$-ray emission.

Nuclear excitation of $^{16}O$ can produce several $\gamma$-ray lines. The first excited state of $^{16}O$ at 6.05 MeV decays by $e^+e^-$ pair emission. The second, third, and fourth excited states at 6.131, 6.919, and 7.119 MeV decay almost exclusively to the ground state, producing $\gamma$-rays at 6.129, 6.917, and 7.117 MeV, respectively (see Table 18).

7.16.2. *The Young Open Cluster Berk 87*

The YOC Berk 87 lies within the error box of the COS B source 2CG 075+00 and its center lies within the error box of the unidentified X-ray source 1H 2018+366. Close to this position an X-ray enhancement was detected by EXOSAT (Warwick *et al.*, 1988).

Berk 87 is located at the boundary of the Cygnus X region; its distance is ~0.9 kpc from radial velocity measurements (Turner & Forbes, 1982) and from the depth of the interstellar lines of the cluster members (Polcaro *et al.*, 1989). The cluster is very dense and about 100 member stars are listed by Turner & Forbes (1982) within a 16 arcmin circle centered on the apparent cluster center, corresponding to ~5 pc radius at a distance of 0.9 kpc. A much higher star density (5 stars $arcmin^{-2}$) has been derived from the deep CCD $V$ and $R$ images of the central part of the cluster (Polcaro *et al.*, 1991a). Berk 87 is probably part of the star formation region ON2, where many compact HII regions, strong OH masers and CO and ammonia molecular clouds have been detected (e.g., Dent, Macdonald & Andersson, 1988 and references therein).

Berk 87 catched the attention of many groups in the last decades because it contains several peculiar objects, namely:

- the rare WR star ST3 (WO-type) (Stephenson, 1966), also named WR 142 in the catalog of van der Hucht *et al.* (1981) and in the recent one of van der Hucht (2001), and Sand 5 (Sanduleak, 1971). This star shows strong emission of the OVI (3811–34 Å) doublet (e.g., Barlow & Hummer, 1982), a terminal wind velocity of $\sim$5200 km $s^{-1}$ (Polcaro et al, 1991b) and mass loss rate of $3 \times 10^{-5}$ $M_\odot$ $yr^{-1}$ (Barlow, 1991). The WO stars are believed to represent the final short lived stage of high mass ($M > 40$ $M_\odot$) stars;
- the variable star V439 Cyg whose characteristics are reported by Polcaro *et al.* (1989);
- the emission–line supergiant HDE 229059 and the red supergiant irregular variable BC Cyg, whose ages look incompatible with the estimated cluster age: $\sim 4 \times 10^6$ yr (Turner & Forbes, 1982);
- a cluster of compact HII regions with an extremely dense member, G 75.77+ 0.34 (Mattews *et al.*, 1973; Fanti, Mantovani & Tomasi, 1981), having an estimated density of $\sim 10^5$ $cm^{-3}$ (Dent, Macdonald & Andersson, 1988) and no visible exciting star.

The optical observations of the sky region close to ST3 and other nearby stars of the cluster indicate the presence of diffuse emission in the CIV (5802–12 Å)

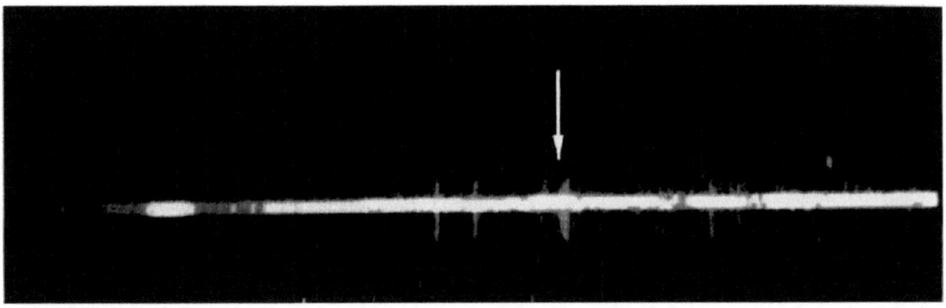

*Figure 152.* Image of ST3 and surrounding sky spectra. The CIV diffuse emission is marked by the arrow (Polcaro *et al.*, 1991b).

doublet. Figure 152 shows the image of ST3 spectrum with such a diffuse emission (Polcaro *et al.*, 1991b).

By the analysis of the shape of this diffuse emission, they suggested the presence of shock waves which are produced by the interactions of the strong WR stellar wind with the other cluster members. This scenario can easily explain the observations of X-ray and $\gamma$-ray photons from the entire region. Indeed, the existence of shock fronts, producing a non-radiative heating of the inter-cluster medium gives rise to a diffuse hard X-ray emission detected from the cluster region by EXOSAT satellite (Warwick *et al.*, 1988). The location of Berk 87 within the error box of the COS B source 2CG 075+00, labelled as diffuse one (Pollock, 1987) with an observed flux of $\sim 1.3 \times 10^{-6}$ photons cm$^{-2}$ s$^{-1}$ for energies greater than 100 MeV and the aforementioned behaviour of the cluster in optical and X-ray ranges enabled Polcaro *et al.* (1991b) to argue that a system of shocks is present in the central region of the cluster. Then an enhanced energy density of relativistic cosmic rays can be produced by the acceleration of cosmic ray particles emitted in the winds of the young stars of Berk 87.

EGRET collaboration reported that six $\gamma$-ray sources ($E > 100$ MeV) are coincident with SNRs (Dingus 1994). However, the observations of five of them by CYGNUS extensive air shower detector do not show evidence of $\gamma$-ray emission at energies $\sim 10^{14}$ eV (Allen *et al.* 1995).

Assuming that the observed $\gamma$-rays are coming from $\pi^0$ decay, the SNRs spatially coincident with OB associations or HII regions should also be important sources of galactic cosmic rays. The COMPTEL instrument on board CGRO detected from the Orion complex an excessive 3–7 MeV emission, which was attributed to cosmic ray induced lines of $^{12}C^*$ and $^{16}O^*$ at 4.44 MeV and 6.13 MeV, respectively, at a level of $\sim 10^{-4}$ photons cm$^{-2}$ s$^{-1}$ (Bloemen *et al.* 1994a). Later a more detailed analysis of the data with a better understanding of the instrumental background have reduced the signal to a $<3\sigma$ result (Bloemen *et al.* 1999b). So, combining all available data they set a $2\sigma$ upper limit on the 3–7 MeV emission of Orion of $3 \times 10^{-5}$ photons cm$^{-2}$ s$^{-1}$. However, from the direction of Orion complex, the EGRET experiment detected the source 2EG J0617-0652 with a flux of $\sim(2.5 \pm 0.6) \times 10^{-6}$ photons cm$^{-2}$ s$^{-1}$ (Thompson *et al.* 1995).

An alternative scenario for the production of high energy $\gamma$ rays for the COS B diffuse sources was proposed by Polcaro *et al.* (1991a,b,c). In that model the interactions of strong stellar winds (thousands km s$^{-1}$) from peculiar stars – the WR star ST3 in the case of the YOC Berk 87 has a wind with a velocity of $\sim$5200 km s$^{-1}$ – with the dense inter-cluster medium produce shock waves. The high velocities of the wind particles ($\sim$1000–5200 km s$^{-1}$: Manchanda *et al.* 1996) coming from the numerous early-type stars seem to provide good conditions for the injection of particles into the system of shock fronts. There shock waves can accelerate the ambient cosmic ray particles up to the maximum energy of 10–1000 PeV. In the case of protons the maximum energy is $\sim 10^{16}$ eV (Cesarsky & Montmerle, 1983):

$$E_{\rm Max} = 4 \times 10^{15} Z(B/10^{-5})(w_\infty/v_\infty)^2 \ \text{eV}\,. \tag{7.18}$$

The association of the COS B source 2CG 075+00 with Berk 87 was considered virtually certain by Polcaro *et al.* (1991b). They estimated a local enhancement of the energy density of cosmic rays of $\sim$100 eV cm$^{-3}$ in the central part of Berk 87, by using the COS B detected $\gamma$-ray flux from 2CG 075+00, $F_{\gamma(\rm COSB)}(E > 100\rm MeV) \simeq 1.3 \times 10^{-6}$ photons cm$^{-2}$ s$^{-1}$. The observation of Berk 87 by EGRET-CGRO instrument, in the energy range 20 MeV – 30 GeV, confirmed this identification. EGRET detected a $\gamma$-ray point source (CGRO J 2021+37), practically coincident with Berk 87 cluster. The measured flux from this EGRET point source was (Polcaro *et al.* 1995) $F_{\gamma(\rm EGRET)}(E > 100\rm MeV) = (9.14 \pm 0.94) \times 10^{-7}$ photons cm$^{-2}$ s$^{-1}$. Such a value is consistent with that measured by COS B. This identification suggests that Berk87-like YOCs can form one important class of objects producing relativistic cosmic rays (Manchanda *et al.* 1996; Giovannelli, Bednarek & Karakuła: 1996).

### 7.16.3. *Cosmic Ray Energy Density in Berk 87*

The EGRET-team resolved the source 1EG J2021+37 into two sources, namely 2EG J2019+3719 and 2EG J2026+3610. The error box of the latter source does not contain Berk 87, while the former source is still fully compatible with Berk 87 and its flux is the same as that reported for the source in the first EGRET catalog; the slope of its spectrum is $\gamma = 1.9 \pm 0.1$. Such a slope (at that time unknown) was used by Giovannelli, Bednarek & Karakuła (1996) in order to extrapolate at higher energies the emission of Berk 87. Figure 153 shows such an extrapolation.

In the energy range 20 MeV – 30 GeV, the flux from Berk 87 placed at a distance to the Earth of $\sim$0.9 kpc (Turner & Forbes 1982) corresponds to a total $\gamma$-ray emissivity of $N_\gamma \simeq 10^{38}$ ph s$^{-1}$. With the measured spectral index $\sim$2, Giovannelli, Bednarek & Karakuła (1996) obtained $L_\gamma \simeq 9 \times 10^{34}$ erg s$^{-1}$.

They estimated the total energy in relativistic protons as $E_{\rm p} \simeq 9 \times 10^{45}$ erg, which gives a cosmic ray energy density $\rho_{\rm CR} \simeq 97$ eV cm$^{-3}$ in this region. This value is two orders of magnitude greater than the average density of cosmic rays in the vicinity of the Earth.

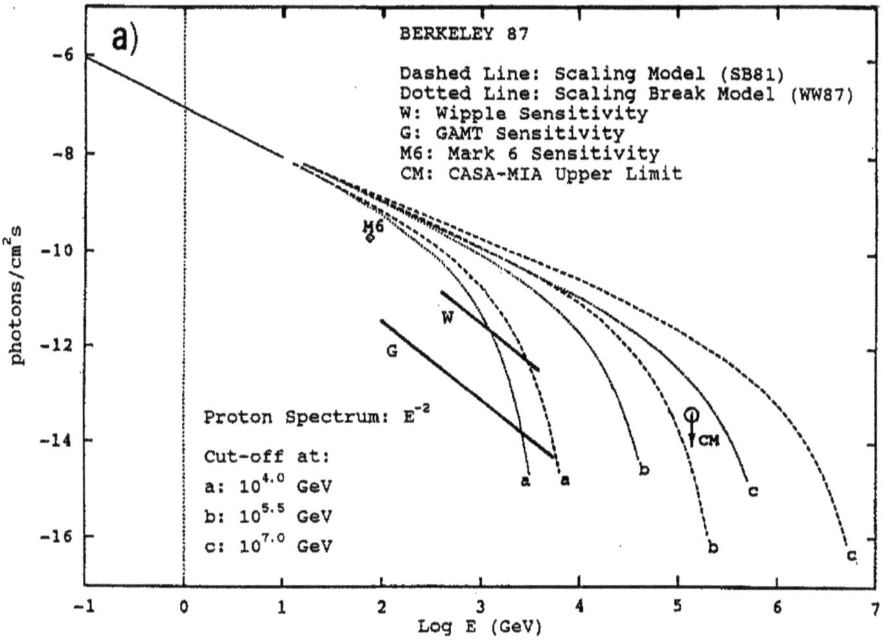


*Figure 153.* (a) Extrapolation of the EGRET $\gamma$ ray flux of Berk 87 to higher energies, assuming an input power law proton spectrum $\propto E^{-2}$. Cutoffs of the proton spectrum are: (a) $10^4$ GeV; (b) $10^{5.5}$ GeV; (c) $10^7$ GeV. The full line represents the $\gamma$-ray spectrum in the EGRET energy range with a spectral index of 2. The dashed and dotted lines show the $\gamma$-ray spectra computed with the scaling (Stephens & Badhwar (1981) (SB81)) and scaling break (Wdowczyk & Wolfendale (1987) (WW87)) models of the p–p interactions, respectively. The sensitivities of the Mark-6, Whipple and GAMT telescopes are also reported. CM represents the upper limit of the CASA-MIA experiment (from Giovannelli, Bednarek, & Karakuła, 1996).

The maximum energy for protons is (Polcaro *et al.* 1991b): $E_{\rm Max} \simeq 1.7 \times 10^{16}$ eV. The protons will lose only a part of their primary energy in a single interaction with the ambient matter of Berk 87. Since their Larmor radius in the Berk 87 magnetic field ($\sim$2 pc) is smaller than the dimensions of the cluster itself ($\approx$5 pc), protons are confined in the cluster and can further cool in collisions with the ambient matter. The $\pi^{\pm}$ produced in proton–proton interactions decay into secondary electrons (positrons) and neutrinos. The energy carried out by neutrinos is of the order of that carried out by $\gamma$-rays (Giovannelli, Karakuła, & Tkaczyk 1982a,b, 1983).

The secondary electrons (positrons) cool predominantly via bremsstrahlung process on a time scale of a few hundreds years (for the parameters of Berk 87). However, since electrons are secondary products of $\pi$-decay, their energy density is lower than that of the primary protons. For this reason the secondary electrons give a negligible contribution to the $\gamma$-ray spectrum above the energy of the half rest mass of $\pi^0$ ($\simeq$67.5 MeV), which is into the lowest energy range of the EGRET experiment.

TABLE 19

Surface abundances for some nuclear species at the WC stage for an evolving star of 120 $M_\odot$ (adapted from Ødegaard, 1996).

| Element | Initial Abundance | WC Abundance | Accumulated Mass Loss ($\dot{M}/M_\odot$) | Mean Mass Fraction |
|---|---|---|---|---|
| $^4$He | 0.26 | 0.217 | 70.4 | 0.61 |
| $^{12}$C | $3.2 \times 10^{-5}$ | 0.499 | 10.8 | 0.094 |
| $^{16}$O | 0.01 | 0.216 | 1.58 | 0.014 |
| $^{20}$Ne | $1.7 \times 10^{-3}$ | $8.8 \times 10^{-4}$ | 0.157 | $1.4 \times 10^{-3}$ |
| $^{71}$Ga | $2.9 \times 10^{-8}$ | $9.37 \times 10^{-4}$ | $4.17 \times 10^{-2}$ | $3.6 \times 10^{-4}$ |
| $^{70}$Ge | $4.6 \times 10^{-8}$ | $4.83 \times 10^{-5}$ | $2.27 \times 10^{-3}$ | $2.0 \times 10^{-5}$ |

In contrast, the $\gamma$-rays from bremsstrahlung of the secondary electrons could give a significant contribution below that energy.

#### 7.16.4. *Isotopes from Wolf–Rayet Winds*

According to evolutionary models the abundances of $^{12}$C, $^{16}$O, and $^{22}$Ne increment of >100 times at the transition from the WN to the WC phase. The abundances of $^{26}$Mg and $^{25}$Mg are equal to that of $^{24}$Mg in WC stars. Anomalous abundances found in cosmic rays and meteorites could indicate a WR origin. IR spectroscopy should provide proper Ne abundances for many WR stars.

Wolf–Rayet stars are producers and emitters of $^{26}$Al and then responsible of the observed $^{26}$Al–1.809 MeV $\gamma$-ray line in the direction of the Galactic Center (Blake & Dearborn 1989; Diehl *et al.* 1995). IR observations show the signature of evolved hot massive stars in the Galactic Center (Krabbe *et al.* 1991). However, supernovae and novae are likely competing contributors to the $^{26}$Al in the Galactic Center (Meynet 1994; Diehl 1999).

Table 19 shows the surface abundances of some elements at the WC stage for an evolving star of initial mass of 120 $M_\odot$ (Ødegaard 1996).

As far as the energy production and nucleosynthesis are concerned, the H burning used by Meynet *et al.* (1994) is expanded by duly taking the NeNa and MgAl burning modes into account, and in particular all the channels leading to the production and destruction of $^{26}$Al (Arnould, Paulus & Meynet 1997).

The s processes developing during the core He burning phase (Prantzos, Hashimoto, & Nomoto 1990) includes in particular the following radionuclides with half-lives between $10^5$ and $10^8$ yr that are of cosmological interest: $^{26}$Al, $^{36}$Cl, $^{41}$Ca, $^{60}$Fe, $^{81}$Kr, $^{93}$Zr, $^{97}$Tc, $^{98}$Tc, $^{99}$Tc, $^{107}$Pd, $^{129}$I, $^{135}$Cs, $^{146}$Sm, $^{182}$Hf, and $^{205}$Pb. Therefore we can expect heavy ions accelerated to energies of $\approx$ a few $\times 10^{18}$ eV.

So in Berk 87-like YOCs, containing WR stars, one can expect heavy ions accelerated to energies up to $E_{\rm Max} \approx$ a few $\times 10^{18}$ eV. Moreover, energetic protons from early-type stars interacting with the YOC ambient matter – reach in high Z isotopes – can produce observable $\gamma$-ray line in emission.

An open problem is connected with the consistency of the predicted fluxes of such heavy ions with the measurements. In other words the problem is strictly related to a better knowledge of the cross sections of such reactions.

Our suggestion is the following: knowing the primary composition (as in the case of WRs) one can try to fit the cosmic ray spectrum. And this method could be used also for the other classes of cosmic ray emitters, for which the composition can be theoretically evaluated and experimentally tested for instance by means of new generation satellites. This is just the contrary we are doing now: from the measured spectrum we go back to the primary composition.

### 7.16.5. *Possible Cosmic Ray Induced Lines from Young Open Clusters and Their Future Detection*

The Orion complex belongs to the giant molecular clouds Orion OMC1 ($M \approx 10^5$ $M_\odot$) whose density is $\sim10^3$–$10^5$ cm$^{-3}$ (Lang 1992). Its core – roughly coincident with the ORI OB1 association, is placed at a distance of $\sim$460 pc, with a mass of $\sim2 \times 10^3$ $M_\odot$ (Bruch & Sanders 1983), where the density is $\sim10^4$–$10^6$ cm$^{-3}$ – contains the young (few $\times 10^6$ yr) open cluster NGC 1976 (Lang 1992 and references therein). Its position is compatible with that of the peak of the $\gamma$-ray emission measured by COMPTEL, centered between ORI A and ORI B (Bloemen *et al.* 1994b), which was attributed to cosmic ray induced lines of $^{12}C^*$ and $^{16}O^*$ at 4.44 MeV and 6.13 MeV, respectively at a level of $\sim10^{-4}$ photons cm$^{-2}$ s$^{-1}$. But a further analysis of the COMPTEL data brought to the conclusions that the 'emissivity' for the Orion complex in the 3–7 MeV range is actually a $2\sigma$ upper limit of $3 \times 10^{-5}$ photons cm$^{-2}$ s$^{-1}$ (Bloemen *et al.*, 1999b). However, from the direction of the Orion complex, the EGRET experiment detected the source 2EG J0617-0652 with a flux of $\sim(2.5\pm0.6) \times 10^{-6}$ photons cm$^{-2}$ s$^{-1}$, which is consistent with a mass of the cloud of $\sim$(1–2) $\times 10^5$ $M_\odot$. This value is in agreement with that ($\sim1.2 \times 10^5$ $M_\odot$) derived by the COS B measured flux ($\sim2 \times 10^{-6}$ photons cm$^{-2}$ s$^{-1}$ in the range 70–5000 MeV; Bignami & Hermsen 1983) and with the determination from radio measurements ($\sim1.5 \times 10^5$ $M_\odot$; Blitz 1980).

The physical conditions in Berk 87 are very similar to those at the center of the ORI OB1 association, where the density is $\approx10^5$ cm$^{-3}$ (Lang 1992) and where the YOC NGC 1976 is placed. Then it is reasonable to expect nuclear de-excitation lines of $^{12}C^*$ and $^{16}O^*$ from Berk 87. Therefore the expected flux in the range 3–7 MeV from Berk 87 can be simply derived by the flux measured in Orion complex scaled with the distance. This expected flux is $\leq 6.6 \times 10^{-6}$ photons cm$^{-2}$ s$^{-1}$. These lines, if present, can be easily detected by the INTEGRAL-SPI spectrometer (Vedrenne *et al.* 1999).

We want to remark that Berk87-like YOCs can form a class of cosmic sources (Manchanda *et al.* 1996), which is responsible of the generation of a significant fraction (10–15%) of cosmic rays (Giovannelli 1997). Since they emit photons in a broad band of the electromagnetic spectrum from IR to VHE $\gamma$-ray regions, they are good targets for the INTEGRAL mission.

Then summarizing:

- The CGRO observations have shown the association of 2EG J2019+3719 (2CG 075+00) with Berk 87 YOC;
- In the inner part of Berk 87 ($\sim$0.8 pc radius) the mean energy density of cosmic rays can be as high as 97 eV cm$^{-3}$. This means that acceleration of cosmic rays does occur;
- The angular resolution of EGRET did not allow a definitive association of Berk 87 with the source 2EG J2019+3719, whose error box is more than twice of the apparent dimensions of Berk 87 cluster. The INTEGRAL-IBIS imager (Lebrun *et al.* 1999b) could instead resolve definitively this enigma, being its angular resolution ($\sim$12 arcmin FWHM) compatible with the dimensions of Berk 87 ($\sim$16 arcmin diameter) and of its central part ($\sim$6 arcmin diameter), where the cosmic ray density is $\sim$97 eV cm$^{-3}$;
- The $\gamma$-ray spectrum expected from $\pi^0$ decay generated by p–p interactions and the upper limit of the $\gamma$-ray flux at 140 TeV of the CASA-MIA experiment suggest a cutoff around $10^{15}$ eV for a proton spectrum with spectral index 2, like that of 2EG J2019+3719 $\equiv$ Berk 87 measured by the EGRET experiment (Giovannelli, Bednarek, & Karakuła 1996).

All this renders the class of Berk87-like YOCs a good target for high energy experiments, such as INTEGRAL and ground based VHE experiments (e.g., Mark 6 and Whipple).

The physical conditions in Berk 87, very similar to those at the center of the ORI OB1 association suggest the possibility of measuring nuclear de-excitation lines of $^{12}C^*$ and $^{16}O^*$ with the INTEGRAL-SPI spectrometer at an expected flux of $\leq 6.6 \times 10^{-6}$ photons cm$^{-2}$ s$^{-1}$ in the range 3–7 MeV. Moreover, the positrons resulting from energetic particle interactions should produce 0.511 MeV annihilation radiation from Berk 87 at a flux level of $\leq 10^{-5}$ photons cm$^{-2}$ s$^{-1}$ (using the emissivity from Kozlovsky, Lingenfelter & Ramaty, 1987), which could be easily measured by the INTEGRAL-SPI spectrometer.

Therefore we can remark that Berk-87 itself and Berk-87-like YOCs are good targets for detecting nuclear de-excitation lines, and $e^+e^-$ annihilation radiation. The INTEGRAL mission will probably solve such a problem.

### 7.17. THE HARDEST WINDOWS TO THE UNIVERSE

Since 1945, with the advent of the rockets used for scientific purposes, astrophysical experimentation has seen an enormous expansion of the wavebands available for study. New disciplines have been developed, making it possible to study astronomical objects over a continuous range of wavelengths spanning 20 orders of magnitude. About half of this electromagnetic spectrum comprises the $\gamma$-ray range and its study is considered an important part of the high energy astrophysics. The major advances in the development of new techniques for observations from the ground and from space have led to an intense experimental activity in this field.

One of the motivations of $\gamma$-ray astronomy is the identification of the origin and acceleration mechanisms of the high energy cosmic rays. The results of $\gamma$-ray observations at energies greater than $10^{11}$ eV have already provided important evidences about the discrete nature of the sources.

In the 1980s and 1990s the $\gamma$-ray astronomy has been the subject of several reviews (e.g., Watson, 1984; Protheroe, 1986; Ramana Murthy & Wolfendale, 1986; Weekes, 1988, 1992, 1996, 1997; Cawley, 1989: Gaisser, 1990a,b,c; Lamb, 1996, 1997; Lorenz, 1999; Krennrich, 1999).

Gamma-rays may be generated by many production mechanisms, such as matter–antimatter annihilation, radioactive decay, energetic particle collisions and by hot plasmas, as well as by interactions of electrons with matter (synchrotron radiation), magnetic fields (bremsstrahlung) and light (inverse Compton effect) and decay of neutral pions generated in cosmic ray nuclear interactions with matter. Corresponding to this variety of emission phenomena there are different $\gamma$-ray energy loss processes of importance with respect to the detection of $\gamma$-rays in the >1 keV energy domain. Then the detectors used are based on the dominant process of energy loss for the photons. In a first approximation they can be divided into different categories: detectors for energies where the photoelectric absorption is predominant, detectors for energies where the Compton scattering is the dominant loss process and those for energies where electron pair production is most important (e.g., Fichtel & Trombka, 1981). So it is possible to summarize as follows:

- In the low energy (LE) region (~0.1–0.5 MeV) the photoelectric absorption is dominant and the types of detectors are scintillator counters and solid state detectors;
- in the medium energy (ME) range (0.5 – 30 MeV), the Compton scattering is the dominant loss process and the designed detector, the Compton telescope, consists of two arrays of detectors (converter and absorber) each surrounded by an anticoincidence system and separated by about 1 m;
- in the high energy (HE) range (30 MeV – 10 GeV) the pair production process becomes dominant and the type of detector used is the spark chamber and photons with energies up to 40 GeV can be stopped. For energies below 10 GeV $\gamma$-rays cannot penetrate the atmosphere and observations must be performed by satellites and high flying balloons;
- in the very high energy (VHE) region (10 GeV – 100 TeV) the Cherenkov radiation created by the secondary charged particles of the shower can be measured on clear nights using the atmospheric Cherenkov technique (ACT). At energies above $\sim 10^{11}$ eV the intensity of astrophysical fluxes becomes too low for satellites detectors. Indeed, the integral flux at the top of the atmosphere for all particles is $\geq 1$ m$^{-2}$ h$^{-1}$ sr$^{-1}$ at energies $\leq 10^{14}$ eV, and only $\sim$4 m$^{-2}$ yr$^{-1}$ sr$^{-1}$ at energies $\leq 10^{17}$ eV. However, the $\gamma$-rays are so energetic that they initiate electromagnetic cascades in the upper atmosphere.

The VHE $\gamma$-ray astronomy is also known as TeV $\gamma$-ray astronomy. Here, as well as in the next two regions, $\gamma$-rays cannot be uniquely identified because the showers initiated by $\gamma$-ray must be separated from those produced by the charged cosmic rays and the distinguishing features between both types of showers are difficult to use from the detection point of view. This kind of detection technique was extensively discussed by Auriemma *et al.* (1995). They demonstrated the full feasibility of an experiment in the VHE range with resolution enough good to render their proposed Cherenkov telescope the most attractive in this energy range;

- at energies greater than $10^{14}$ eV it is possible to detect the charged particles of the shower on the ground using the air shower technique (AST). The ultra high energy (UHE) region (100 TeV – 100 PeV) is also referred as the field of the *PeV* $\gamma$-ray astronomy. Above $10^{16}$ eV the charged particles of the shower excite the atmospheric nitrogen and oxygen, emitting fluorescence light by de-excitation, which can be detected also on the ground;
- the extremely high energy (EHE) region (100 PeV – 100 EeV) is the highest range in which a detector is looking for emission from the space. This is the Fly's Eye detector consisting of more than 100 small mirrors directed so that the entire sky is viewed all the time.

Although each detection technique may be applied, in principle, in the aforementioned entire energy regions, in practice it is difficult to carry out observations between the highest energy observed with satellites (~30 GeV) and the energy threshold of the ACT (~300 GeV) (e.g., see the review of Saggion & Bastieri, 2002). The great areas of the detectors based on the AST have made it possible, at present, to extend their energy thresholds downward almost reaching the upper limit of the VHE range (~100 TeV), but the detection technique is still not optimized at these energies. A new generation of detectors must be developed to cover the observations in these two gaps.

### 7.18. NEUTRINOS

In the latter years there was large progress in the investigation of the problem of neutrino mixing. Different indications in favor of non-zero neutrino masses and mixing angles have been found. The problem of the neutrino masses and mixing is the most important of today's neutrino physics. There are different indications that neutrinos have small masses and that there is neutrino mixing. Such indications were obtained in solar neutrino experiments (Cleveland *et al.*, 1995; Anselmann *et al.*, 1995; Abdurashitov *et al.*, 1995; Suzuki, 1998), in atmospheric neutrino experiments (Fukuda *et al.*, 1994; Becker-Szendy *et al.*, 1995; Allison *et al.*, 1997; Fukuda *et al.*, 1998a,b,c), and in Liquid Scintillator Neutrino Detector (LSND) experiment (Athanassopoulos *et al.*, 1996).

The results of the four experiments (Homestake, GALLEX, SAGE, Kamiokande) on solar neutrino presently running show not only discrepancies with the expectations of Standard Solar Models (SSMs) but, if constrained by the solar luminosity, also unsolvable contradictions within the sequences of nuclear reactions in the same way as a negative or zero flux for the neutrino from $^7$Be. These problems could be solved by hypothesizing new properties for neutrinos as oscillations in vacuum or in matter, or others. If the indications in favor of neutrino oscillations will be confirmed, they will represent the first observation of processes in which the lepton numbers are not conserved. It is popular believing the investigation of such processes crucial in understanding the physics at a scale much larger than the electroweak scale (Bilenky, Giunti & Grimus, 1998).

The important problem for the next generation of experiments is a deep investigation of neutrino oscillations mainly in the region of $\Delta m^2 > 10^{-4}$ eV$^2$, in which at present there are indications in favor of oscillations. Many neutrino experiments are running, or going to start. A list of them, with the proper references is reported in Bilenky, Giunti & Grimus (1998).

With such experiments, probably in a few years from now, we will know much more about the fundamental properties of neutrinos, such as masses, mixing, and their nature (Dirac neutrino or Majorana neutrino).

Results about the 'solar neutrino' have been recently published by Frame (2002), Ishii (2002), and about 'solar and atmospheric neutrinos' by Bettini (2002).

Reviews of 'solar neutrino puzzle: present situation and future scenario' and 'neutrino mixing and oscillations' have been recently published by Aliani *et al.* (2002) and by Bilenky (2002), respectively. Auriemma (2002) discussed about 'the quest for the mass of neutrino' making an excursus about direct evidence for neutrino mass, the solar neutrino and atmospheric neutrino anomalies through a short presentation of the results from experiments. He then discussed the scenario of neutrino oscillations combining both SuperKamiokande and MACRO data, from which standard oscillations $\nu_\mu \longrightarrow \nu_\tau$, and non-standard oscillations $\nu_\mu \longrightarrow \nu_s$ – being $\nu_s$ a new singlet neutrino state not coupled to $W^\pm$ – remain available for an interpretation.

From an experimental point of view the two types of scenario are indistinguishable in the low energy range explored by SuperKamiokande, whilst they could produce a difference in the zenith angle modulation of MACRO. The analysis of MACRO data gives a better fit for the vacuum $\nu_\mu \longrightarrow \nu_\tau$ oscillations (Auriemma, 2002 and the references therein).

**Which is the hottest problem still surviving?**

Star evolution models need the knowledge of the velocities of the reactions inside the stars. Therefore the cross-sections of such reactions must be known as well as possible.

As is well known, a p–p reaction chain is responsible of the production of 98.5% of the energy in the Sun. The net result is the transformation of 4 protons in a nucleus of helium and the associated nuclear energy production.

Modern experimental techniques allow us to approach energies typical for stellar fusions. LUNA (Laboratory for Underground Nuclear Astrophysics) experiment under Gran Sasso measured the cross section of the reaction $^{3}He + ^{3}He \longrightarrow ^{4}He + 2p$ at energies (typical for the stars) between 16.5 and 25 keV, corresponding to Gamov's peak, i.e. the energy at which the reactions in the interior of the Sun occur (e.g., Bettini, 2002). This is one of the fundamental reactions for the solar neutrino problem.

The first direct measure of the cross section of the above reaction down to the lower edge of the solar Gamov's peak (Bonetti *et al.*, 1999) exclude definitively an hypothetical resonance in such a channel contrary to predictions in several theoretical papers.

Then once more **the knowledge of the cross-sections of nuclear reactions occurring in the stars appears as one of the most crucial points of all astroparticle physics.** Future direct measurements of the cross sections of the $^{3}He(^{4}He,\gamma)^{7}Be$ and $^{7}Be(p,\gamma)^{8}Be$ reactions of the p–p chain and $^{14}N(p,\gamma)^{15}O$ reaction of the CNO-cycle will allow a substantial improvement in our knowledge on star evolution.

## 8. Some Ideas on Next Generation Experiments

In the study of cosmic sources three resolutions are crucial for a better comprehension of the physics governing their behavior, namely (e.g., Giovannelli & Sabau-Graziati, 1997):

**i) Resolution in Energy:** better resolution allows the detailed studies of astrophysical plasmas and line profiles (e.g., the X-ray missions ASCA, BeppoSAX, CHANDRA, XMM, the X and $\gamma$-ray missions INTEGRAL, and the future SPECTRUM X-$\gamma$, if it will be launched);

**ii) Resolution in Position:** increasing this resolution it is possible to distinguish close sources in a crowded field (i.e. Galactic Center X-ray sources) (e.g., CHANDRA, XMM, INTEGRAL). Future X-ray experiments could discern protoclusters and protogalaxies, and finally, with interferometric techniques, it will be probably possible to look at accretion disks and stellar surfaces;

**iii) Resolution in Sensitivity:** increasing this resolution it is possible: a) to detect weaker sources (i.e. ROSAT has detected more than 100,000 soft X-ray sources, most of them before ROSAT confused in the background radiation); b) to detect either flux variations at lower levels in a fixed $\Delta t$, or flux variations at higher levels but in shorter time scales (e.g., CHANDRA, XMM, RXTE, ASCA and INTEGRAL).

Since most of the energy ranges are forbidden by ground based observations, space based experiments must be developed, following the general trend in increasing their sensitivities, compatibly at reasonable cost. But the spectral resolution in most of the energy bands is very close to the theoretical resolution. So, only by in-

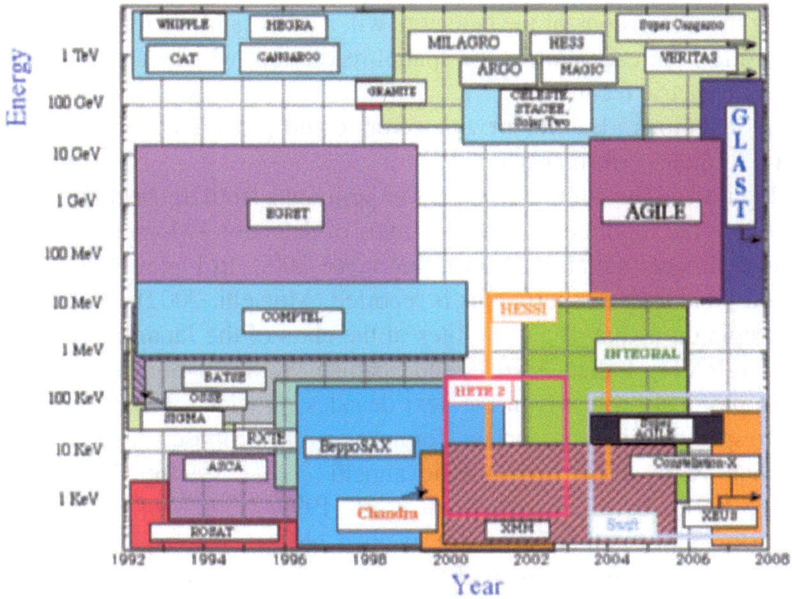


*Figure 154.* Main high energy experiments and their ranges of detection from 1992 to 2008 (courtesy of Aldo Morselli, 2002).

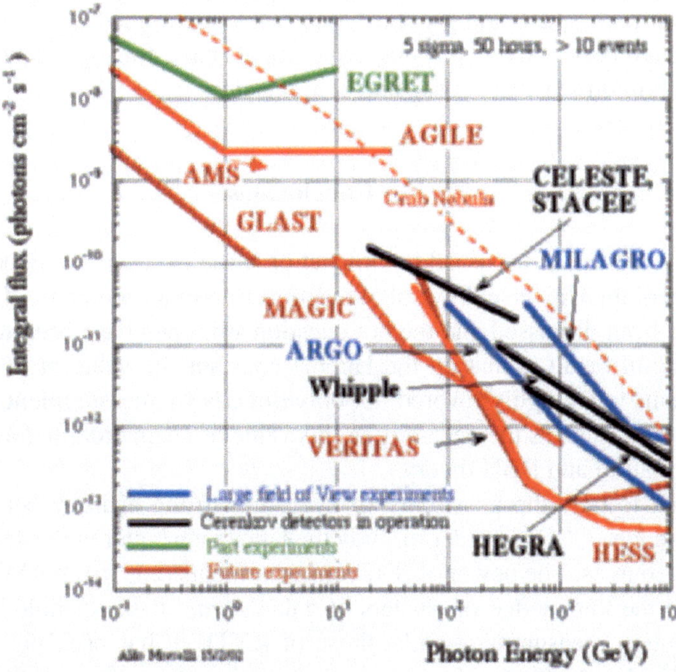


*Figure 155.* Sensitivity of the past and future experiments. Most of them will be operating in 2002. All sensitivities are at $5\sigma$. Cerenkov telescopes' sensitivities (Veritas, MAGIC, Whipple, Hess, Celeste, Stacee, Hegra) are for 50 hours of observations. Large field of view detectors sensitivities (AGILE, GLAST, Milagro, ARGO) are for 1 year's observation. MAGIC sensitivity is based on the availability of high efficiency PMT's (courtesy of Aldo Morselli, 2002).

creasing the dimensions of the detectors will it be possible to reach lower threshold in intensities. An exception is the spatial resolution, which can be improved a lot by interferometric techniques.

Only ingenious ideas for new detectors could provide a significant step in the increase of their performances.

In Figure 154 the main high energy experiments from the beginning of 1990s to next decade are reported with their ranges of detection (Morselli, 2002). Most of these experiments are already operating since 2002. In Figure 155 the sensitivity of the past and future experiments is reported (Morselli, 2002).

Our opinion is that small satellites of the class of the Japanese ASTRO-Series, American-RXTE, Italian-AGILE, and Spanish MINISAT-Series can solve important astrophysical problems in the next decades, provided that they could be dedicated to specific missions, such as, for instance, the proposed payload SIXE (Spanish Italian X-ray Experiment) (Giovannelli *et al.*, 1993b; Giovannelli *et al.*, 1999a,b, 2001, 2002; Gómez-Gomar *et al.*, 1999; Isern *et al.*, 2001). Of course, this line of mini-satellites does not exclude bigger projects, provided that these could also allow different individual countries and Agencies to support also the smaller programs.

In general we believe that in the next decades the advances in the knowledge of the physics will necessarily pass through an improvement of 'passive physics' experiments both space and ground based. This is enforced also by the fact that the big ground based 'active physics' experiments have already reached reasonable upper limits in dimensions, costs, and complication.

## 9. Conclusions

In this paper we have discussed the impact of space experiments on knowledge of the physics of the Universe. Several of the most important problems of astrophysics today have been discussed and the hot question still open have been remarked.

Meaningful result is that on the Hubble constant. Its value of $56 \pm 6$ km s$^{-1}$ Mpc$^{-1}$ seems to be highly favored not only directly by measurements, but also by indirect proofs, such as the necessity of this value of $H_0$ in order to have agreement with the inflation and BBN theory.

We want to note the possibility of using the strong analogy between macro-QSOs and micro-QSOs in order to study the latter for a better understanding also of the former sources. The power of X-ray measurements (e.g., from ASCA) appeared evident for the knowledge of clusters of galaxies and then cosmology. With high time resolution measurements, like those of RXTE, it has been pointed out how it is possible to investigate degenerate matter of neutron stars. With Beppo-SAX measurements a substantial improvement on knowledge of the GRB origin has been reached. With simultaneous multifrequency measurements it is possible to improve our knowledge of the accretion processes onto collapsed objects and then

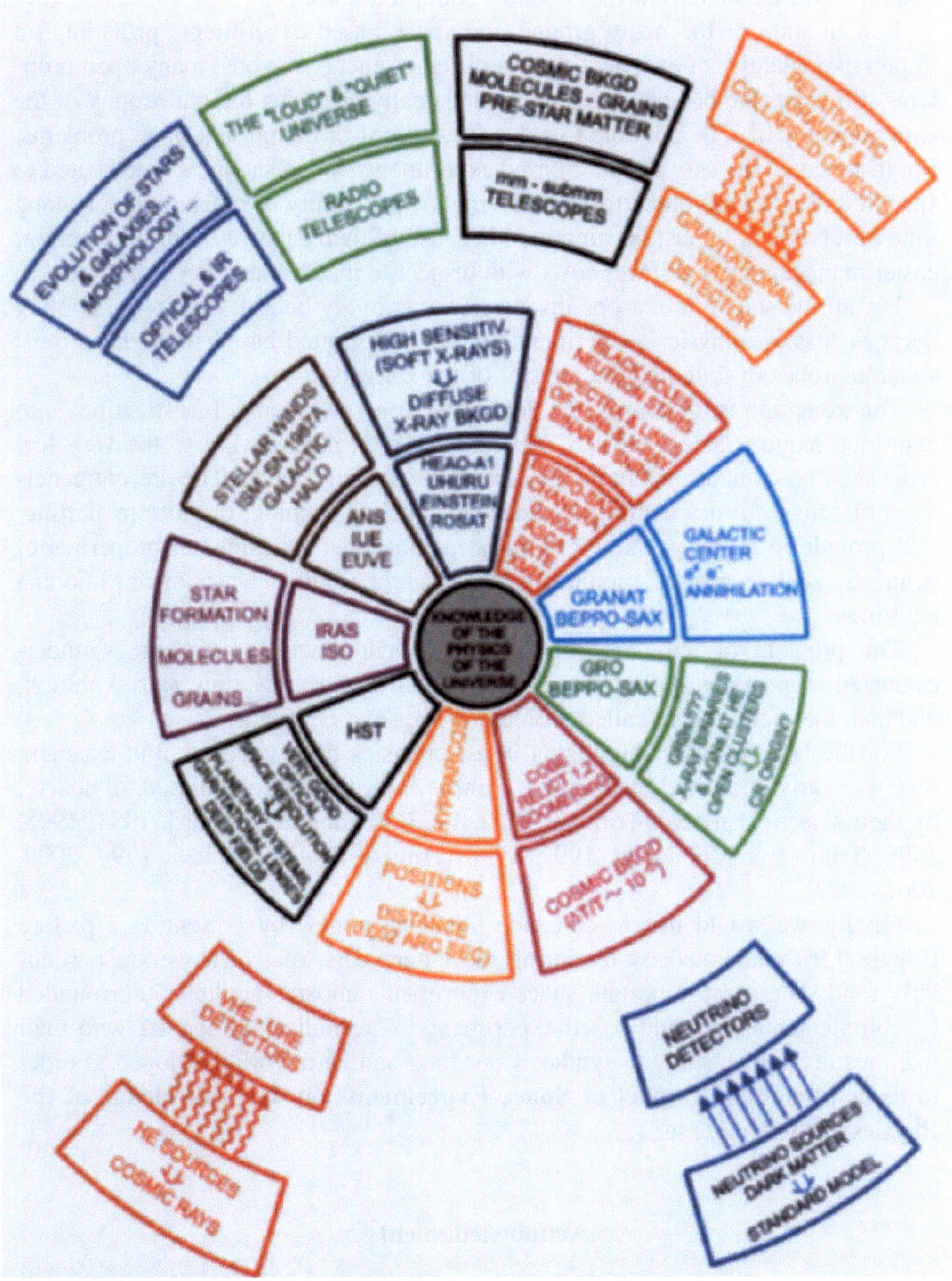


*Figure 156.* Sketch of the most important space and ground based passive physics experiments with their main contributions to the knowledge of the physics of the Universe (updated from Giovannelli & Sabau-Graziati, 1997).

of plasma physics. The importance of magnetic fields has been largely pointed out in all the cosmic sources having inside a collapsed object.

But, in spite of the many ground and space based experiments providing an impressive quantity of excellent data in different energy regions, many open problems still exist. We believe that only by drastically changing the philosophy of the experiments will it be possible to solve faster most of the present open problems. For instance, in the case of space based experiments, small satellites – dedicated to specific missions and problems, and having the possibility of scheduling very long time observations – must be supported because of their relative faster preparation, easier management and lower costs with respect to medium and large satellites.

For all the scientific reasons discussed we strongly believe that in the coming decades 'passive physics' experiments space and ground based will be the most suitable probes in sounding the physics of the Universe.

The exception is, of course, the field of neutrino astronomy. Investigations into neutrinos require 'active physics' very large experiments owing to the very low cross sections of neutrinos in their interactions with the matter. The present generation of active physics neutrino experiments already running, or close to starting, will provide in a few years crucial information about fundamental properties of neutrinos, such as masses, mixing, and their nature as Dirac neutrinos or Majorana neutrinos.

The problem of the cross-sections in nuclear reactions, sometimes under-evaluated, appears as probably the most crucial in understanding star evolution, and then indirectly the so-called problem of the solar neutrino.

For the last ten years' highlights in astrophysics the reader can find excellent reviews yearly published by Virginia Trimble and collaborators, biased, of course, by their selection effects (Trimble, 1992, 1993; Trimble & Leonard, 1994, 1995, 1996; Trimble & McFadden, 1997, 1998; Trimble & Aschwanden, 1999, 2000, 2001, 2002).

Finally, we would like to conclude this long review by presenting a picture (Figure 156) elaborated by us during the latter years, of which we are particularly fond. The most important space experiments aboard satellites – surrounded by complementary ground based experiments – are indicated together with their main results, which show in synthesis our long and hard work developed in order to demonstrate **The Impact of Space Experiments on our Knowledge of the Physics of the Universe**.

## Acknowledgments

We are deeply indebted to our dear friend, Szczepan Karakuła, with whom we had numerous discussions regarding the subject of this review, both in terms of its logical arguments and its contents. His contributions live on in our memories and in this text in spite of his sudden and untimely death on July 16th, 1996.

We would like to thank Dr Corinne Rossi, Department of Physics, University of Rome 'La Sapienza', who had the patience to read this paper carefully and critically. Her comments have been taken into account.

We are pleased to thank the many colleagues who gave permission for publishing some of their figures, which rendered this paper more suitable for faster comprehension of the arguments discussed. Particular thanks are due to Mr Juan Antonio Balsalobre (DCETE-INTA) and Mr Massimo Frutti (IASF-CNR) for helping with the graphics. One of us (FG) is indebted to Mr Jose Torres Riéra, Director of the Departamento de Ciencias del Espacio y Tecnología Electrónicas of INTA for hospitality, to Professor Francisco Sánchez Martínez, Director of the Instituto de Astrofísica de Canarias (IAC) for the permission of using their magnificent library, and to Mrs Lourdes Abellán García and Mr Antonio J. Bacallado Abreu, librarians of the IAC, for their help in searching particular unusual publications.

This work has been partially supported by the Spanish PNIE (Plan Nacional de Investigaciones Espaciales: National Plan for Space Investigations) and the Italian ASI (Agenzia Spaziale Italiana: Italian Space Agency).

This research has made use of NASA's Astrophysics Data System.

## References

Abbott, D.C.: 1982, *Astrophys. J.* **263**, 723.

Abdurashitov, J.N., Faizov, E.L., Gavrin, V.N., Gusev, A.O., Kalikhov, A.V., Knodel, T.V., *et al.*: 1995, *Nucl. Phys. B Proc. Suppl.* **38**, 60.

Abell, G.O.: 1958, *Astrophys. J. Suppl. Ser.* **3**, 211.

Abell, G.O.: 1965, *Ann. Rev. Astron. Astrophys.* **3**, 1.

Abell, G.O.: 1975, in *Stars and Stellar Systems IX: Galaxies and the Universe*, A. Sandage, M. Sandage, and J. Kristian (eds.), University of Chicago, Chicago, USA, p. 601.

Abramopoulos, F., Ku, W.: 1983, *Astrophys. J.* **271**, 446.

Abt, H.A., Levy, S.G.: 1976, *Astrophys. J. Suppl. Ser.* **30**, 273.

Adams, W.S.: 1941, *Astrophys. J.* **93**, 11.

Adams, F.C., Shu, F.H.: 1986, *Astrophys. J.* **308**, 836.

Adams, F.C., Lada, C.J., Shu, F.H.: 1987, *Astrophys. J.* **312**, 788.

Adams, F.C., Lada, C.J., Shu, F.H.: 1988, *Astrophys. J.* **326**, 865.

Adams, F.C., Shu, F.H.: 1988, in *Comets to Cosmology*, A. Lawrence (ed.), Springer *Lett. Not. Phys.* **297**, p. 164.

Adams, F.C., Emerson, J.P., Fuller, G.A.: 1990, *Astrophys. J.* **357**, 606.

Akimoto, F., Tawara, Y., Furuzawa, A., Kumada, A., Yamashita, K.: 1996, in *UV and X-ray Spectroscopy of Astrophysical and Laboratory Plasmas*, K. Yamashita & T. Watanabe (eds.), Universal Academy Press, Frontiers Science Series **15**, 399.

Albert, C.E., White, R.A., Morgan, W.W.: 1977, *Astrophys. J.* **211**, 309.

Alessandro, B.: 2001, in *Frontier Objects in Astrophysics and Particle Physics*, F. Giovannelli & G. Mannocchi (eds.), Italian Physical Society, Editrice Compositori, Bologna, Italy **73**, 363.

Alexander, D.: 1992, *Vistas in Astronomy* **35**, 11.

Aliani, P., Antonelli, V., Ferrari, R., Picariello, M.,Torrente-Lujan, E.: 2002, in *Le Rencontres de Physique de la Vallée d'Aoste: Results and Perspectives in Particle Physics*, M. Greco (ed.), *Frascati Physics Series* **XXVII**, 151.

Allen, G.E., Berley, D., Biller, S., Burman, R.L., Cavalli-Sforza, M., Chang, C.Y., *et al.*: 1995, *Astrophys. J. Lett.* **448**, L25.

Allen, G.E., Keohane, J.W., Gotthelf, E.V., Petre, R., Jahoda, K., Rothschild, R.E., *et al.*: 1997, *Astrophys. J. Lett.* **487**, L97.

Allen, S.W.: 1999, in *Diffuse Thermal and Relativistic Plasma in Galaxy Clusters*, Böhringer, H., Feretti, L., Schuecker, P. (eds.), MPE Report 271B, p. 129.

Allen, S. W., Fabian, A. C., Edge, A. C., Bautz, M. W., Furuzawa, A., Tawara, Y.: 1996, *Mon. Not. Roy. Astr. Soc.* **283**, 263.

Allison, W.W.M., Alner, G.J., Ayres, D.S., Barrett, W.L., Bode, C., Border, P.M., *et al.*: 1997, *Phys. Lett. B* **391**, 491.

Alpar, M.A., Shaham, J.: 1985, *Nature* **316**, 239.

Anders, E., Grevasse, N.: 1989, *Geochim. Cosmochim. Acta* **53**, 197.

Anderson, R.C., Brune, W.H., Henry, R.C., Feldman, P.D. Fastie, W. G.: 1979, *Astrophys. J. Lett.* **233**, L39.

André, P: 1991, in *Young Star Clusters and Early Stellar Evolution*, F. Palla, P. Persi & H. Zinnecker (eds.), *Mem. S.A.It.* **62**, 829.

Angel, J.R.P., Stockman, H.S.: 1980, *Ann. Rev. Astron. Astrophys.* **18**, 321.

Angelini, L., Stella, L., Parmar, A.N.: 1989, *Astrophys. J.* **346**, 906.

Anselmann, P., Hampel, W., Heusser, G., Kiko, J., Kirsten, T., Laubenstein, M., *et al.*: 1995, *Phys. Lett. B* **357**, 237.

Antonucci, R.R.: 1993, *Ann. Rev. Astron. Astrophys.* **31**, 473.

Apparao, K.M.V., Tarafdar, S.P.: 1994, *Astrophys. J.* **420**, 803.

Appenzeller, I.: 1983, *Rev. Mexicana Astron. Astrophys.* **7**, 151.

Appenzeller, I.: 1989, in *Low Mass Star Formation and Pre-Main Sequence Objects*, Bo Reipurth (ed.), ESO Conference and Workshop Proceedings **33**, 491.

Appenzeller, I., Krautter, J., Mundt, R., Chavarria, C., Wolf, B.: 1980, *Astron. Astrophys.* **90**, 184.

Appenzeller, I., Mundt, R.: 1989, *Astron. Astrophys. Rev.* **1**, 291.

Aptekar, R.L., Butterworth, P.S., Cline, T.L., Frederiks, D.D., Golenetskii, S.V., Il'Inskii, V.N., *et al.*: 2002, in *Soft Gamma Ray Repeaters*, M. Feroci, S. Mereghetti (eds.), *Mem. S.A.It.* **73**, 485.

Arbatsumian, V.A.: 1947, in *Stellar Evolution and Astrophysics*, Acad. Sci. Armenian SSR, Erevan.

Arbatsumian, V.A.: 1952, *Trans. Int. Astron. Union* **VIII**, 665.

Arnaud, K.A., Branduardi-Raymont, G., Culhane, J.L., Fabian, A.C., Hazard, C., McGlyn, T.A., *et al.*: 1985, *Mon. Not. Roy. Astr. Soc.* **217**, 105.

Arnaud, M., Rothenflug, R., Böhringer, H., Neumann, D., Yamashita, K.: 1996, in *UV and X-Ray Spectroscopy of Astrophysical and Laboratory Plasmas*, K. Yamashita, T. Watanabe (eds.), Universal Academy Press, Inc., Tokyo, Japan, p. 163.

Arnaud, M., Neumann, D.M., Aghanim, N., Gastaud, R., Majerowicz, S., Highes, J.P.: 2001, *Astron. Astrophys. Lett.* **365**, L80.

Arnet, W.D., Bahcall, J.N., Kirshner, R.P.: 1989, *Ann. Rev. Astron. Astrophys.* **27**, 629.

Arnould, M., Paulus, G., Meynet, G.: 1997, *Astron. Astrophys.* **321**, 452.

Arp, H.C., Burbidge, G. Hoyle, F., Narlikar, J.V.: 1992, *Nature* **356**, p. 287.

Arp, H.C., Keys, C.R., Rudnicki, K.: 1993 *Progress in New Cosmologies*, Plenum Press, New York, USA.

Asaoka, I., Koyama, K.: 1990, *Publ. Astron. Soc. Japan* **42**, 625.

Aschenbach, B.: 1992, *Zeiss Information with Jena Review* **1**, 6.

Aschenbach, B.: 1996, MPE-Report **263**, 213.

Aschenbach, B.: 1999, in *Multifrequency Behaviour of High Energy Cosmic Sources*, F. Giovannelli & L. Sabau-Graziati (eds.), *Mem. S.A.It.* **70**, 1045.

Aschenbach, B.: 2002, in Proc. of the 270. WE-Heraeus Seminar on Neutron Stars, Pulsars, and Supernova Remnants, W. Becker, H. Lesch, & J. Trümper (eds.), MPE Report **278**, 13.

Aschenbach, B., Brinkmann, W.: 1975, *Astron. Astrophys.* **41**, 147.

Athanassopoulos, C., Auerbach, L.B., Burman, R.L., Cohen, I., Caldwell, D.O., Dieterle, B.D., *et al.*: 1996, *Phys. Rev. Lett.* **77**, 3082.

Atkins, R., Benbow, W., Berley, D., Chen, M.L., Coyne, D.G., Dingus, B.L., *et al.*: 2000, *Astrophys. J. Lett.* **533**, L119.

Auman, H.H.: 1985, *Publ. Astron. Soc. Pac.* **97**, 885.

Auman, H.H., Beichman, C.A., Gillet, F.C., de Jong, T., Houck, J.R., Low, F.J., *et al.*: 1984, *Astrophys. J. Lett.* **278**, L23.

Auriemma, G.: 1993, in *Frontier Objects in Astrophysics and Particle Physics*, F. Giovannelli & G. Mannocchi (eds.), Italian Physical Society, Editrice Compositori, Bologna, Italy **40**, 423.

Auriemma, G.: 2002, in *Multifrequency Behaviour of High Energy Cosmic Sources*, F. Giovannelli & L. Sabau-Graziati (eds.), *Mem. S.A.It.* **73**, 43.

Auriemma, G., Angeloni, L., Belli, B.M., Bernardi, A., Cardini, D., Costa, E., *et al.*: 1978, *Astrophys. J. Lett.* **221**, L7.

Auriemma, G., Capaccioli, M., Costa, E., Dalí, G., Errico, L., Frenkel, A., *et al.*: 1995, *Astroparticle Physics* **3**, 215.

Avni, Y., Bahcall, J.N.: 1974, *Astrophys. J. Lett.* **192**, L139.

Avni, Y., Bahcall, J.N.: 1975, *Astrophys. J. Lett.* **202**, L131.

Awaki, H., Koyama, K., Kunieda, H., Takano, S., Tawara, Y., Ohashi, T.: 1991, *Astrophys. J.* **366**, 88.

Awaki, H., Mushotzky, R., Tsuru, T., Fabian, A.C., Fukazawa, Y., Loewenstein, M., *et al.*: 1994, *Publ. Astron. Soc. Japan* **46**, L65.

Axford, I. W.: 1981a, *17th Intern. Cosmic Ray Conf.*, Paris, France **12**, 155.

Axford, I.W.: 1981b, *Texas Symp. on Relativistic Astrophysics*, New York Academy of Sciences, Annals **375**, 297.

Ayres, T. R., Linsky, J. L.: 1980, *Astrophys. J.* **241**, 279.

Baade, W., Zwicky, F.: 1934, *Phys. Rev.* **45**, 138.

Bååth , L.B.: 1990, in *Parsec-Scale Radio Jets*, J.A. Zensus & T.J. Pearson (eds.), Cambridge University Press, Cambridge, UK, p. 91.

Backer, D.C., Kulkarni, S.R., Heiles, C.E.,Davis, M.M., Goss, W.M.: 1982, *Nature* **300**, 615.

Baganoff, F., Angelini, L., Bautz, M., Brandt, N., Cui, W., Doty, J., *et al.*: 1999, *Astron. Astrophys. Suppl. Ser.* **195**, 6201.

Bahcall, J.N.: 1978a, *Ann. Rev. Astron. Astrophys.* **16**, 241.

Bahcall, J.N.: 1978b, in *Physics and Astrophysics of Neutron Stars and Black Holes*, R. Giacconi & R. Ruffini (eds.), Bologna, Società Italiana di Fisica; North Holland Publishing Co., Amsterdam, Holland, p. 63.

Bahcall, J., Ostriker, J.: 1975, *Nature* **256**, 23.

Bahcall, N.A.: 1977a, *Ann. Rev. Astron. Astrophys.* **15**, 505.

Bahcall, N.A.: 1977b, *Astrophys. J. Lett.* **218**, L93.

Bahcall, N.A., Ostriker, J.P., Perlmutter, S., Steinhardt, P.J.: 1999, *Science* **284**, 1481.

Bai, T., Sturrock, P.A.: 1989, *Ann. Rev. Astron. Astrophys.* **27**, 421.

Bailes, M., Lorimer, D.: 1995, in *Millisecond Pulsars: A Decade of Surprise*, A.S. Fruchter, M. Tavani & D.C. Backer (eds.), *ASP Conf. Ser.* **72**, p. 17.

Bally, J., Stark, A.A., Wilson, R.W., Henkel, C.: 1988, *Astrophys. J.* **324**, 223.

Baltrusaitis, R.M., Cady, R., Cassiday, G.L., Cooper, R., Elbert, J.W., Gerhardy, P.R., *et al.*: 1985, *Phys. Rev. D* **54**, 1875.

Bałucińska-Church, M.: 2001, *Adv. Space Res.* **28**, 349.

Band, D., Matteson, J., Ford, L., Schaefer, B., Palmer, D., Teegarden, B., *et al.*: 1993, *Astrophys. J.* **413**, 281.

Bandiera, R.: 2001, in *Frontier Objects in Astrophysics and Particle Physics*, F. Giovannelli & G. Mannocchi (eds.), Italian Physical Society, Editrice Compositori, Bologna, Italy **73**, 225.

Bandiera, R.: 2002, in *Multifrequency Behaviour of High Energy Cosmic Sources*, F. Giovannelli & L. Sabau-Graziati (eds.), *Mem. S.A.It.* **73**, 107.

Bandiera, R., Masini, E., Pacini, F., Salvati, M. & Woltjer, L. (eds.): 1998, *The Relationship between Neutron Stars and Supernova Remnants*, *Mem. S.A.It.* **69 N. 4**.

Barat, C., Lestrade, J.P., Dezalay, J.-P., Sunyaev, R., Terekhov, O., Kuznetsov, A.: 2000, *Astrophys. J.* **538**, 152.

Barcons, X.: 1996, in *Multifrequency Behaviour of High Energy Cosmic Sources*, F. Giovannelli & L. Sabau-Graziati (eds.), *Mem. S.A.It.* **67**, 481.

Baring, M.G., Harding, A.K.: 1997, in *Proceedings Fourth Compton Symposium*, C.D. Dermer, M.S. Strickman & J.D. Kurfess (eds.), *AIP Conf. Proc.* **410**, p. 638.

Barker, E.S., Lugger, P.M., Weiler, E.J., York, D.G.: 1984, *Astrophys. J.* **280**, 600.

Barlow, M.J.: 1991, in *Wolf–Rayet Stars and Interrelation with Other Massive Stars in Galaxies*, K.A.van der Hucht & B. Hidayat (eds.), Kluwer Academic Publishers, IAU Symp. **143**, 281.

Barlow, M.J., Hammer, D.C.: 1982, in *Wolf–Rayet Stars: Observations, Physics, Evolution*, C. de Loore, & A. Willis (eds.), Reidel Pub. Co., IAU Symp. **99**, 387.

Barrett, A.H., Meeks, M.L., Weinreb, S.: 1964, *Astron. J.* **69**, 134.

Barret, D., Vedrenne, G.: 1994, *Astrophys. J. Suppl. Ser.* **92**, 505.

Barret, D., Grindlay, J.E., Bloser, P.F., Zhang, S.N., Fishman, G.J., Harmon, B.A., *et al.*: 1996, *Astron. Astrophys. Suppl. Ser.* **120**, 121.

Barret, D., Olive, J.F., Boirin, L., Done, C., Skinner, G.K., Grindlay, J.E.: 2000, *Astrophys. J.* **533**, 329.

Bartolini, C., Bonifazi, A., Guarnieri, A., Piccioni, A., Giangrande, A., Giovannelli, F.: 1981, *IAU Circ.* **N. 3655**.

Bartolini, C., Bianco, G., Guarnieri, A., Piccioni, A., Giovannelli, F.: 1983, *Hvar Obs. Bull.* **7(1)**, 159.

Bartolini, C., Beskin, G.M., Guarnieri, A., Masetti, N., Piccioni, A.: 1998, in *Gamma Ray Bursts*, C.A. Meegan, R.D. Preece & T.M. Koshut (eds.), *AIP Conference Proceeedings* **428**, 540.

Basko, M.M.: 1978, *Astrophys. J.* **223**, 268.

Basri, G., Bertout, C.: 1989, *Astrophys. J.* **341**, 340.

Bassani, L., Dean, A.J.: 1981, *Nature* **294**, 332.

Bastien, P.: 1989, in *Polarized Radiation of Circumstellar Origin*, G.V. Coyne *et al.* (eds.), Vaticano Observatory, Vatican City, p. 541.

Bastien, P., Menard, F: 1988, *Astrophys. J.* **326**, 334.

Bastien, P., Menard, F.: 1990, *Astrophys. J.* **364**, 232.

Bath, G.T., van Paradijs, J.: 1983, *Nature* **305**, 33.

Battrick, B. (ed.): 1997, *Hipparcos Venice '97*, ESA Publication Division, ESTEC, Nordwijk, The Netherlands, ESA **SP-402**.

Beall, J.H.: 1986, *Astrophys. J.* **316**, 227.

Becker, W.: 2000, *Adv. Space Res.* **25**, 647.

Becker, W., Trümper, J.: 1997, *Astron. Astrophys.* **326**, 682.

Becker, W., Trümper, J.: 1999, *Astron. Astrophys.* **341**, 803.

Becker-Szendy, R., Bratton, C.B., Casper, D., Dye, S.T., Gajewski, W., Ganezer, K., *et al.*: 1995, *Nucl. Phys. B (Proc. Suppl.)* **38**, 331.

Beckman, J.: 2001, in *The Bridge between the Big Bang and Biology: Stars, Planetary Systems, Atmospheres, Volcanoes: Their Link to Life*, F. Giovannelli (ed.), CNR President Bureau, Special Volume, Roma, Italy, p. 161.

Beckwith, S.V.W., Sargent, A.I., Chini, R., Güsten, R.: 1990, *Astron. J.* **99**, 924.

Bedijn, P.J.: 1986, in *Light on Dark Matter*, F.P. Israel (ed.), D. Reidel Publ. Co., Dordrecht, Holland, p. 119.

Bednarek, W.: 1997, *Astron. Astrophys.* **322**, 523.

Bednarek, W., Calvani, M.: 1991, in *Frontier Objects in Astrophysics and Particle Physics*, F. Giovannelli & G. Mannocchi (eds.), Italian Physical Society, Editrice Compositori, Bologna, Italy **28**, p. 175.

Bednarek, W., Giovannelli, F., Karakuła, S., Tkaczyk, W.: 1990a, *Astron. Astrophys.* **236**, 268.

Bednarek, W., Giovannelli, F., Karakuła, S., Tkaczyk, W.: 1990b, *Astron. Astrophys.* **236**, 175.

Bednarek, W., Giovannelli, F.: 1999, in *Multifrequency Behaviour of High Energy Cosmic Sources*, F. Giovannelli & L. Sabau-Graziati (eds.), *Mem. S.A.It.* **70**, 1071.

Bednarek, W., Giovannelli, F., Sabau Graziati, L.: 1999, in *3rd INTEGRAL Conference Proc.*, A. Bazzano, G.C. Palumbo & C. Winkler (eds.), *Astro. Lett. & Communication* **38**, 185.

Behar, E., Rasmussen, A.P., Griffiths, R.G., Dennerl, K., Audard, M., Aschenbach, B., *et al.*: 2001, *Astron. Astrophys. Lett.* **365**, L242.

Beichman, C.A.: 1987, *Ann. Rev. Astron. Astrophys.* **25**, 521.

Beichman, C.A., Jennings, R.E., Emerson, J.P., Baud, B., Harris, S., Rowan-Robinson, M., *et al.*: 1984, *Astrophys. J. Lett.* **278**, L45.

Beichman, C., Boulanger, F., Rice, W., Persson, C.J., Viallefond, F: 1987, in *Star Formation in Galaxies*, C. Persson (ed.), NASA, Washington, DC, US Govt. Print. Off., p. 297.

Belli, B.M.: 1996, in *Multifrequency Behaviour of High Energy Cosmic Sources*, F. Giovannelli & L. Sabau-Graziati (eds.), *Mem. S.A.It.* **67**, 445.

Belli, B.M.: 1997, *Astrophys. J. Lett.* **479**, L31.

Belli, B.M.: 1998, *Astron. Astrophys. Suppl. Ser.* **138**, 415.

Belli, B.M.: 1999, in *Multifrequency Behaviour of High Energy Cosmic Sources*, F. Giovannelli & L. Sabau-Graziati (eds.), *Mem. S.A.It.* **70**, 941.

Belli, B.M.: 2001, in *Gamma Ray Bursts in the Afterglow Era*, E. Costa, F. Frontera & J. Hjorth (eds.), Springer Verlag, Heidelberg, Germany, p. 50.

Belloni, T., Verbunt, F., Beuermann, K., Bunk., W., Izzo, C., Kley, W., *et al.*: 1991, *Astron. Astrophys. Lett.* **246**, 44.

Belloni, T., Klein-Volt, M., Méndez, M., van der Klis, M., van Paradijs, J.: 2000, *Astron. Astrophys.* **355**, 271.

Belloni, T., Psaltis, D., van der Klis, M.: 2002, *Astrophys. J.* **572**, 392.

Bennett, C.L., Smoot, G.F., Hinshaw, G., Wright, E.L., Kogut, A., De Amici, G., *et al.*: 1992, *Astrophys. J. Lett.* **396**, L7.

Bennett, D.P., Stebbins, A., Bouchet, F.R.: 1992, *Astrophys. J. Lett.* **399**, L5.

Bennett, D.P., Kogut, A., Hinshaw, G., Banday, A.J., Wright, E.L., Gorski, K.M., *et al.*: 1994, *Astrophys. J.* **436**, 423.

Benson, P.J., Myers, P.C.: 1989, *Astrophys. J. Suppl. Ser.* **71**, 89.

Benvenuti, P., F.D. Macchetto, E.J. Schreier (eds.): 1996, in *Science with Hubble Space Telescope - II*, Space Telescope Science Institute, Baltimore, MD, USA.

van den Bergh, S.: 1989, *Astron. Astrophys. Rev.* **1**, 111. 63.

van den Bergh, S., Tammann, G.A.: 1991, *Ann. Rev. Astron. Astrophys.* **29**, 363.

de Bernardis, P., Ade, P.A.R., Bock, J.J., Bond, J.R., Borrill, J., Boscaleri, A., *et al.*: 2000, *Nature* **404**, 955.

Bernstein, R.A.: 1998, *PhD Thesis*, California Institute of Technology, CA, USA.

Bertaux, J.L.: 1984, *Local Interstellar Medium*, Y. Kondo, F.C. Bruhweiler & B.D. Savage (eds.), IAU Coll. N. 81, p. 3.

Bertout, C,: 1989, *Ann. Rev. Astron. Astrophys.* **27**, 351.

Bertout, C.: 1993, in *Cataclysmic Variables and Related Physics*, O. Regev & G. Shaviv (eds.), *Annals of the Israel Physical Society* **10**, 241.

Bertout, C., Basri, G., Bouvier, J.: 1988, *Astrophys. J.* **330**, 350.

Bertout, C., Bouvier, J.: 1989, in *Low Mass Star Formation and Pre-Main Sequence Objects*, Bo Reipurth (ed.), ESO Conference and Workshop Proceedings **33**, 215.

Bertout, C., Bouvier, J., Duschl, W.J., Tscharnuter, W.M.: 1993, *Astron. Astrophys.* **275**, 236.

Bettini, A.: 2002, in *Le Rencontres de Physique de la Vallée d'Aoste: Results and Perspectives in Particle Physics*, M. Greco (ed.), *Frascati Physics Series* **XXVII**, 95.

Beuermann, K.: 1998, in *High Energy Astronomy and Astrophysics*, P.C. Agrawal & P.R. Visvanathan (eds.), India Univ. Press, p. 100.
Bhat, C.L., Mayer, C.J., Wolfendale, A.W.: 1984, *Astron. Astrophys.* **140**, 284.
Bhattacharya, D.: 1995, in *X-ray Binaries*, W.H.G. Lewin, J. van Paradijs & E.P.J. van den Heuvel (eds.), Cambridge University Press, p. 223.
Bielefeld, M.J., Reedy, R.C., Metzger, A.E., Trombka, J.T., Arnold, J.R.: 1976, in *Proc. Lunar Sci. Conf. 7th*, Pergamon Press, New York, USA, p. 2661.
Biermann, P.L.: 1993, *Astron. Astrophys.* **271**, 649.
Biermann, P.L.: 1994, in *HE Astrophysics. Models and Observations from MeV to EeV*, J. M. Matthews (ed.), World Scientific, Singapore, p. 217.
Biermann, P. L.: 1995, in *Frontier Objects in Astrophysics and Particle Physics*, F. Giovannelli & G. Mannocchi (eds.), Italian Physical Society, Editrice Compositori, Bologna, Italy **47**, 469.
Biermann, P. L.: 1999, *Astrophys. and Space Sci.* **264**, 423.
Biermann, P.L., Cassinelli, J.P.: 1993 *Astron. Astrophys.* **277**, 691.
Biermann, P.L., Strom, R.G.: 1993, *Astron. Astrophys.* **275**, 659.
Bignami, G.F., Hermesen, W.: 1983, *Ann. Rev. Astron. Astrophys.* **21**, 67.
Bildsten, L., Strohmayer, T.: 1999, in *Physics Today* **52, N. 2**, 40.
Bilenky, S.M.: 2002, in *Le Rencontres de Physique de la Vallée d'Aoste: Results and Perspectives in Particle Physics*, M. Greco (ed.), *Frascati Physics Series* **XXVII**, 179.
Bilenky, S.M., Giunti, C., Grimus, W.: 1998, in *Le Rencontres de Physique de la Vallée d'Aoste: Results and Perspectives in Particle Physics*, M. Greco (ed.), *Frascati Physics Series* **XII**, 147.
Binggeli, B., Sandage, A., Tammann, G.A.: 1985, *Astron. J.* **90**, 1681.
Bird, D.J., Corbato, S.C., Dai, H.Y., Dawson, B.R., Elbert, J.W., Emerson, B.L., *et al.*: 1994, *Astrophys. J.* **424**, 491.
Bisnovatyi-Kogan, G.S.: 1999, in *Observational Evidence for Black Holes in the Universe*, S.K. Chakrabarti (ed.), Kluwer Academic Publisher, the Netherlands, p. 1.
Bisnovatyi-Kogan, G.: 2002, in *Multifrequency Behaviour of High Energy Cosmic Sources*, F. Giovannelli & L. Sabau-Graziati (eds.), *Mem. S.A.It.* **73**, 357.
Bisnovatyi-Kogan, G.: 2003, in *Frontier Objects in Astrophysics and Particle Physics*, F. Giovannelli & G. Mannocchi (eds.), Italian Physical Society, Editrice Compositori, Bologna, Italy **85**, 291.
Bisnovatyi-Kogan, G.S., Lovelace, R.V.E.: 2001, *New Astro. Rev.* **45**, 663.
Blaauw, A.: 1991, in *The Physics of Star Formation and Early Stellar Evolution*, C.J. Lada & N.D. Kylafis (eds.), Kluwer Academic Publ., *NATO ASI Ser.* **C342**, 125.
Black, J.H., Fazio, G.G.: 1973, *Astrophys. J. Lett.* **185**, L7.
Blackman, E.G., Yi, I., Field, G.B.: 1996, *Astrophys. J. Lett.* **479**, L79.
Blake, J.B., Dearborn, D.S.P.: 1989, *Astrophys. J. Lett.* **338**, L17.
Blandford, R.D., Rees, M.J.: 1978, in *BL Lac Objects*, A.N. Wolfe (ed.), University of Pittsburg Press, Pittsburg, U.K., p. 328.
Blandford, R.D., Königl, A.: 1979, *Astrophys. J.* **232**, 34.
Blandford, R.D., Levinson, A.: 1995, *Astrophys. J.* **441**, 79.
Blandford, R., Gehrels, N.: 1999, *Physics Today* **June**, 2.
Blasi, P.: 2001a, in *Frontier Objects in Astrophysics and Particle Physics*, F. Giovannelli & G. Mannocchi (eds.), Italian Physical Society, Editrice Compositori, Bologna, Italy **73**, 97.
Blasi, P.: 2001b, in *Frontier Objects in Astrophysics and Particle Physics*, F. Giovannelli & G. Mannocchi (eds.), Italian Physical Society, Editrice Compositori, Bologna, Italy **73**, 383.
Blinnikov, S.I., Chugai, N.N., Lozinskaya, T.A.: 1988, *Sov. Sci. Rev. E. Astrophys. Space Phys.* **6**, 195.
Blitz, L.: 1980, in *Giant Molecular Clouds in the Galaxy*, P.M. Solomon & M.G. Edmunds (eds.)(Oxford: Pergamon), p. 1.
Blitz, L., Magnani, L., Mundy, L.: 1984, *Astrophys. J. Lett.* **282**, L9.
Bloemen, H.: 1989, *Ann. Rev. Astron. Astrophys.* **27**, 469.

Bloemen, H., Hermsen, W., Swanenburg, B.N., de Vries, C.P., Diehl, R., Schönfelder, V., *et al.*: 1994a, *Astrophys. J. Suppl. Ser.* **92**, 419.
Bloemen, H., Wijnands, R., Bennett, K., Diehl, R., Hermsen, W., Lichti, G., *et al.*: 1994b, *Astron. Astrophys. Lett.* **281**, L5.
Bloemen, H., Diehl, R., Hermsen, W., Knödleseder, J., Schönfelder, V.: 1999a, in *Proceedings Third Integral Workshop: The Extreme Universe*, A. Bazzano, G.G.C. Palumbo & C. Winkler (eds.), Astro. Lett. & Communications **38**, 391.
Bloemen, H., Morris, D., Knödleseder, J., Bennet, K., Diehl, R., Hermsen, W., *et al.*: 1999b, Astro. Lett. & Commun. **38**, 349.
Bloemen, J.B.G.M.: 1985, *Ph. D. Thesis*, Leiden University, Holland.
Bloemen, J.B.G.M., Strong, A.W., Mayer-Hasselwander, H.A., Blitz, L., Cohen, R.S., Dame, T.M., *et al.*: 1986, *Astron. Astrophys.* **154**, 25.
Bloom, J.S., Kulkarni, S.R., Djorgovski, S.G.: 2002, *Astron. J.* **123**, 1111.
Blum, R.D., Sellgren, K., DePoy, D.L.: 1995, *Astrophys. J. Letter* **440**, L17.
de Boer, K.S., Koorneef, J., Savage, B.D.: 1978, *Bull. Am. Astron. Soc.* **10**, 726.
de Boer, K.S., Koorneef, J., Savage, B.D.: 1980, *Astrophys. J. Lett.* **136**, L7.
Boesgaard, A.M., Steigman, G.: 1985, *Ann. Rev. Astron. Astrophys.* **23**, 319.
Bogun, S., Lemke, D., Klaas, U., Herbstmeier, U., Assendorp, R., Richter, G., *et al.*: 1996, *Astron. Astrophys. Lett.* **315**, L71.
Bohlin, R.C., Savage, B.D., Drake, J.F.: 1978, *Astrophys. J.* **224**, 132.
Bohlin, R.C., Hill, J.K., Jenkins, E.B., Savage, B.D., Snow, T.P., Spitzer, L., *et al.*: 1983, *Astrophys. J. Suppl. Ser.* **51**, 277.
Böhringer, H., Briel, U.G., Schwarz, R.A., Voges, W., Hartner, G., Trümper, J.: 1994, *Nature* **368**, 828.
Böhringer, H., Feretti, L., Schuecker, P. (eds.): 1999, *Diffuse Thermal and Relativistic Plasma in Galaxy Clusters*, MPE-Report **271**.
Boldt, E.: 1987, *Phys. Rep.* **146**, 215.
Bolton. C.T.: 1972, *Nature* **235**, 271.
Bond, J.R., Carr, B.J., Hogan, C.J.: 1986, *Astrophys. J.* **306**, 28.
Bonetti, R., Broggini, C., Campajola, L., Corvisiero, P., D'Alessandro, A., Dessalvi, M., *et al.*: 1999, *Phys. Rev. Lett.* **82**, 5205.
Bonnet, R.M.: 1992, in *The Astronomy and Astrophysics Encyclopedia*, S.P. Maran (ed.), Cambridge University Press, p. 621.
Bonnet-Bidaud, J.-M., Mouchet, M.: 1988, *A Decade of UV Astronomy with the IUE Satellite* ESA **SP-281, Vol. 2**, 271.
Bookbinder, J., Cowie, L.L., Ostriker, J.P., Krolik, J.H., Rees, M.: 1980, *Astrophys. J.* **237**, 647.
Boughn, S.P., Kuhn, J.R.: 1986, *Astrophys. J.* **309**, 33.
Boulanger, C., Baud, B., van Albada, G.D.: 1985, *Astron. Astrophys. Lett.* **144**, L9.
Boulanger, F., Maddalena, R.J., Thaddeus, P.: 1986, in *Light on Dark Matter*, F.P. Israel (ed.), D. Reidel Publ. Co., Dordrecht, Holland, p. 203.
Boulanger, F., Perault, M.: 1988, *Astrophys. J.* **330**, 964.
Bouvier, J.: 1987, in *Protostars and Molecular Clouds*, Th. Montmerle & C. Bertout (eds.), CNES, Gyf-sur-Yvette, France, p. 189.
Bouvier, J.: 1990, *Astron. J.* **99**, 946.
Bouvier, J., Bertout, C.: 1992, *Astron. Astrophys.* **263**, 113.
Bowyer, S.: 1991, *Ann. Rev. Astron. Astrophys.* **29**, 59.
Bowyer, S., Edelstein, J., Lampton, M.: 1997, *Astrophys. J.* **485**, 52.
Bowyer, S., Korpela, E.J., Edelstein, J., Lampton, M., Morales, C., Pérez-Mercader, J., *et al.*: 1999, *Astrophys. J.* **526**, 10.
Bradt, H., Mayer, W., Naranan, S., Rappaport, S., Spada, G.: 1967, *Astrophys. J. Lett.* **161**, L1.

Bradt, H., Giacconi, R. (eds.): 1973, *X-Ray and Gamma Ray Astronomy*, IAU Symposium N. 55, D. Reidel Publ. Co., Dordrecht, Holland.
Bradt, H.V.D., McClintock, J.E.: 1983, *Ann. Rev. Astron. Astrophys.* **21**, 13.
Bradt, H., Rothschild, R., Swank, J: 1996, in *Multifrequency Behaviour of High Energy Cosmic Sources*, F. Giovannelli & L. Sabau-Graziati (eds.), *Mem. S.A.It.* **67**, 593.
Braes, L.L.E., Miley, G.K.:, 1971, *Astron. Astrophys.* **14**, 160.
Branduardi, G., Ives, F.C., Sanford, P.W., Brinkman, A.C., Maraschi, L.: 1976, *Mon. Not. Roy. Astr. Soc.* **175**, 47p.
Braun, R.: 1985, *Ph.D. Thesis, University of Leiden, Holland.*
Bregman, J.N.: 1980, *Astrophys. J.* **236**, 577.
Bregman, J.N., Glassgold, A.E., Huggins, P.J., Neugebauer, G., Soifer, B.T., Matthews, K., *et al.*: 1986, *Astrophys. J.* **301**, 708.
Bridle, A.H., Perley, R.A.: 1984, *Ann. Rev. Astron. Astrophys.* **22**, 319.
Briel, U.G., Henry, J.P., Schwarz, R.A., Boehringer, H., Ebeling, H., Edge, A.C., Hartner, G.D., Schindler, S., Truemper, J., Voges, W.: 1991, *Astron. Astrophys. Lett.* **246**, L10.
Briel, U.G., Henry, J.P.: 1994, *Nature* **372**, 439.
Briggs, M.S., Band, D.L., Preece, R.D., Paciesas, W.S., Pendleton, G.N.: 1999, *Astrophys. Lett. & Communications* **39(1-6)**, 237.
Brinkmann, W., Aschenbach, B., Langmeier, A.: 1985, *Nature* **313**, 662.
Brinkmann, W., Fabian, A.C., Giovannelli, F. (eds.): 1990, *Physical Processes in Hot Cosmic Plasmas*, Kluwer Academic Publisher, Dordrecht, Holland.
Brown, A., Jordan, C,: 1981, *Mon. Not. Roy. Astr. Soc* **196**, 757.
Brown, A.G.A., Blaauw, A., Hoogerwerf, R., de Bruijne, J.H.J., de Zeeuw, P.T.: 1999, in *The Origin of Stars and Planetary Systems*, C.J. Lada & N.D. Kylafis (eds.), Kluwer Academic Publ., *NATO ASI Ser.* **C342**, 411.
Brownlee, D.E.: 1985, *Ann. Rev. Earth Planet. Sci.* **13**, 147.
Bruch, A., Sanders, W.L.: 1983, *Astron. Astrophys.* **121**, 237.
Bruhweiler, F.C., Kondo, Y.: 1981, *Astrophys. J. Lett.* **248**, L123.
Bruhweiler, F.C., Kondo, Y.: 1982a, *Astrophys. J* **259**, 232.
Bruhweiler, F.C., Kondo, Y.: 1982b, *Astrophys. J. Lett.* **260**, L91.
Bruhweiler, F., Oegerle, W., Weiler, E., Stencel, R., Kondo, Y.: 1984a, in *Future of Ultraviolet Astronomy Based on Six Years of IUE Research*, J.M. Mead, R.D. Chapman, Y. Kondo (eds.), NASA Conference Publication **2349**, 200.
Bruhweiler, F., Oegerle, W., Weiler, E., Stencel, R., Kondo, Y.: 1984b, in *The Local Interstellar Medium*, F. Bruhweiler, Y. Kondo, B. Savage (eds.), NASA Conference Publication **2345**, 64.
Burderi, L., King, A.R.: 1998, *Astrophys. J. Lett.* **505**, L135.
Burderi, L., Di Salvo, T., Stella, L., Fiore, F., Robba, N.R., van der Klis, M., *et al.*: 2002, *Astrophys. J.* **574**, 930.
Burger, M., van Dessel, E., Giovannelli, F., Sabau-Graziati, L., Bartolini, C., Guarnieri, A., *et al.*: 1996, in *Multifrequency Behaviour of High Energy Cosmic Sources*, F. Giovannelli & L. Sabau-Graziati (eds.), *Mem. S.A.It.* **67**, 365.
Burns, J.O., White, R.A., Hanish, R.J.: 1980, *Astron. J.* **85**, 191.
Burns, J.O., White, R.A., Haynes, M.P.: 1981a, *Astron. J.* **86**, 1120.
Burns, J.O., White, R.A., Hough, D.H.: 1981b, *Astron. J.* **86**, 1.
Burrows, D.: 1996, *The CUBIC Handbook*, (publication available from http://www.astro.psu.edu/xray/cubic/papers/handbook/).
Bussard, R.W., Ramaty, R., Drachman, R.J.: 1979, *Astrophys. J.* **228**, 928.
Busser, J.-U.: 1998, *Ph. D. Thesis*, University of Münich, Germany.
Byram, E.T., Chubb, T.A., Friedman, H.: 1966, *Science* **152**, 66.
Cabrit, S., Edwards, S., Strom, S.E., Strom, K.M.: 1990, *Astrophys. J.* **354**, 687.
Calvetti, M.: 2000, *INFN Notizie*, **February**, p. 1.

Cameron, A.C., Campbell, C.G.: 1993, *Astron. Astrophys.* **274**, 309.
Cameron, A.G.W.: 1968, in *Ahrens*, p. 125.
Cameron, A.G.W.: 1973, *Space Sci. Rev.* **15**, 121.
Cameron, A.G.W.: 1982, in *Essays in Nuclear Astrophysics*, C.A. Barnes, D.D. Clayton, D.N. Schramm (eds.), Cambridge University Press, Cambridge, UK, p. 23.
Cameron, A.G.W.: 1988, *Ann. Rev. Astron. Astrophys.* **26**, 441.
Cameron, R.A., Kurfess, J.D., Johnson, W.N., Kinzer, R.L., Kroeger, R.A., Leising, M.D., *et al.*: 1992, in *The Compton Observatory Science Workshop*, NASA-GSFC, p. 3.
Campana, S.: 2000, *Astrophys. J. Lett.* **534**, L79.
Campana, S., Colpi, M., Mereghetti, S., Stella, L., Tavani, M.: 1998, *Astron. Astrophys. Rev.* **8**, 279.
Canizares, C.R.: 1975, *Astrophys. J.* **201**, 589.
Canizares, C.R., Buote, D.A.: 1997, in *X-Ray Imaging and Spectroscopy of Cosmic Hot Plasmas*, F. Makino & K. Mitsuda (eds.), Waseda University, Tokyo, Japan, p. 43.
Cannizzo, J.K.: 1993, *Astrophys. J.* **419**, 318.
Carr, B.J.: 1988, in *Comets to Cosmology, Proceedings Third IRAS Conference (London)*, A. Lawrence (ed.), Springer-Verlag, Berlin, p. 265.
Carretta, E., Gratton, R.G., Clementini, G., Fusi Pecci, F.: 2000, *Astrophys. J.* **533**, 215.
Carruthers, G.P.: 1970, *Astrophys. J. Lett.* **161**, L81.
Carter, D.: 1984, *Astron. Express* **1**, 61.
Casares, J.: 2001, in *Binary Stars: Selected Topics on Observations and Physical Processes*, F.C. Lazaro & M.J. Arévalo (eds.), Lect. Not. in Physics **563**, 277.
Cassinelli, J.P.: 1979, *Ann. Rev. Astron. Astrophys.* **17**, 275.
Cassinelli, J.P., Hartmann, L.: 1977, *Astrophys. J.* **212**, 488.
Cassinelli, J. P., Lamers, H. J. G. L. M.: 1987, in *Exploring the Universe with the IUE Satellite*, Y. Kondo (ed.), D. Reidel Publ. Co., Dordrecht, Holland, p. 139.
Castellani, V., Cassatella, A.: 1987, in *Exploring the Universe with the IUE Satellite*, Y. Kondo (ed.), D. Reidel Publ. Co., Dordrecht, Holland, p. 637.
Castro-Tirado, A.J.: 2002, in *Multifrequency Behaviour of High Energy Cosmic Sources*, F. Giovannelli & L. Sabau-Graziati (eds.), *Mem. S.A.It.* **73**, 337.
Castro-Tirado, A.J., Brandt, S., Lund, N.: 1992, *IAU Circ.* **N. 5590**.
Castro-Tirado, A.J., Greiner, J., Paredes, J.M.. (eds.): 2001, *Micro-quasars*, Supplement to Vol. **276** of *Astrophysics Space Science.*
Catala, C., Foing, B.H.: 1988, in *Comptes-Rendus du Premier Atelier MUSICOS*, C. Catala & B.H. Foing (eds.), Paris Observatory Press.
Catala, C., Foing, B.H.: 1990, *Proc. Second MUSICOS Workshop*, C. Catala & B.H. Foing (eds.), Paris Observatory Press.
Catala, C., Foing, B.H., Baudrand, J., Cao, H., Char, S., Chatzichristou, H., *et al.*: 1993, *Astron. Astrophys.* **275**, 245.
Caux, E., Serra, G.: 1987, in *Star Formation in Galaxies*, C.J. Lonsdale (ed.), NASA PRINT OFF, Washington DC, USA, p. 93.
Cavaliere, A., Gursky, H., Tucker, W.H.: 1971, *Nature* **231**, 437.
Cavallo, G., Rees, M.J.: 1978, *Mon. Not. Roy. Astr. Soc.* **183**, 359.
Cawley, M.F.: 1989, in *Frontier Objects in Astrophysics and Particle Physics*, F. Giovannelli & G. Mannocchi (eds.), Italian Physical Society, Editrice Compositori, Bologna, Italy **19**, 235.
Celotti, A. (ed.): 2001, *AGN in the Year 2000 Mem. S.A.It.* **72 N. 1**.
Cerruti-Sola, M., Perinotto, M.: 1985, *Astrophys. J.* **291**, 237.
Cesarsky, C.J., Montmerle, T.: 1983, *Space Sci. Rev.* **36**, 173.
Chadwick, P.M., Dickinson, M.R., Dipper, N.A., Kendall, T.R., McComb, T.J.L., Orford, K.J., *et al.*: 1998, *Astrophys. J.* **503**, 391.
Chakrabarty, D., Morgan, E.H.: 1998, *IAU Circ.* **N. 6877**.
Chanamé, J., Gould, A., Miralda-Escudé, J.: 2001, *Astrophys. J.* **563**, 793.

Chanamé, J., Gould, A.: 2002, *Astrophys. J.* **571**, 320.
Charles, P.A., Booth, L., Densham, R.H., Bath, G.T., Howarth, I.D., Willis, A.J., *et al.*: 1983, *Mon. Not. Roy. Astr. Soc.* **202**, 657.
Charles, P.A., Seward, F.D.: 1995, in *Exploring the Universe*, Cambridge University Press, UK, p. 21.
Charnley, S.: 2001, in *The Bridge between the Big Bang and Biology: Stars, Planetary Systems, Atmospheres, Volcanoes: Their Link to Life*, F. Giovannelli (ed)., CNR President Bureau, Special Volume, Roma, Italy, p. 139.
Chatterjee, P., Hernquist, L., Narayan, R.: 2000, *Astrophys. J.* **534**, 373.
Chen, W., Shrader, C.R., Livio, M.: 1997, *Astrophys. J.* **491**, 312.
Cheng, K.S., Ruderman, M.: 1989, *Astrophys. J. Letter* **337**, L77.
Cheng, L.,-X., Leventhal, M., Smith, D. M., Purcell, W. R., Tueller, J., Connors, A., *et al.*: 1997, *Astrophys. J. Lett.* **481**, L43.
Chester, T.J.: 1986, in *Light on Dark Matter*, F.P. Israel (ed.), D. Reidel Publ. Co., Dordrecht, Holland, p. 3.
Chevalier, R.A.: 1982, *Astrophys. J.* **258**, 790.
Chevalier, R.A., Fransson, C.: 1984, *Astrophys. J. Lett.* **279**, L31.
Chi, X., Wolfendale, A.W.: 1989, *J. Phys. G* **15**, 1509.
Chi, X., Szabelski, J., Vahiar, M.N., Wdowczyk, J., Wolfendale, A.W.: 1992, J. Phys. G. Nucl. Part. Phys. **18**, 539.
Chi, X., Wdowczyk, J., Wolfendale, A.W.: 1993, in *Frontier Objects in Astrophysics and Particle Physics*, F. Giovannelli & G. Mannocchi (eds.), Italian Physical Society, Editrice Compositori, Bologna, Italy **40**, 383.
Chiang, J., Fichtel, C. E., von Montigny, C., Nolan, P. L., Petrosian, V.: 1995, *Astrophys. J.* **452**, 156.
Chini, R.: 1989, in *Low Mass Star Formation and Pre-Main Sequence Objects*, Bo Reipurth (ed.), ESO Conference and Workshop Proceedings **33**, 173.
Christiansen, W.N., Hindman, J.V.: 1951, *Nature* **168**, 356.
Christiansen, W.N., Hindman, J.V.: 1952, *Aust. J. Sci. Res., A* **5**, 437.
Chupp, E.L.: 1982, in *Gamma Ray Transient and Related Astrophysical Phenomena*, R.E. Lingenfelter, R.E. Hudson & H.S. Worral (eds.), *American Institute of Physics*, New York, USA, p. 363.
Chupp, E.L., Forrest, D.J., Hibie, P.R., Suri, A.N., Tsai, C., Dunphy, P.P.: 1973, *Nature* **241**, 332.
Churazov, E., Gilfanov, M., Sunyaev, R., Pavlinsky, M., Grebenev, S., Dyachkov, A., *et al.*: 1993, in *Frontier Objects in Astrophysics and Particle Physics*, F. Giovannelli & G. Mannocchi (eds.), Italian Physical Society, Editrice Compositori, Bologna, Italy **40**, 175.
Churazov, E., Gilfanov, M., Sunyaev, R., Khavenson, N., Novikov, B., Dyachkov, A., *et al.*: 1994, *Astrophys. J. Suppl. Ser.* **92**, 381.
Church, M.J.: 2001, *Adv. Space Res.* **28, Ns. 2-3**, 323.
Church, M.J., Bałucińska-Church, M.: 1996, in *Multifrequency Behaviour of High Energy Cosmic Sources*, F. Giovannelli & L. Sabau-Graziati (eds.), *Mem. S.A.It.* **67**, 395.
Church, M.J., Balucinska-Church, M., Dotani, T., Mitsuda, K., Inoue, H., Takahashi, T.: 1998, *Adv. Space Res.* **22**, 935.
Claas, J.: 1990, *Ph.D. Thesis*, University of Leiden, Holland.
Clark, G.W.: 1962, unpublished memorandum.
Clark, G.W.: 1975, *Astrophys. J. Lett.* **207**, L105.
Clark, G.W., Garmire, G.P., Kraushaar, W.L.: 1968, *Astrophys. J. Lett.* **153**, L203.
Clarke, T.E., Kronberg, P.P., Böhringer, H.: 1999, in *Diffuse Thermal and Relativistic Plasma in Galaxy Clusters*, Böhringer, H., Feretti, L., Schuecker, P. (eds.), MPE-Report **271B**, 82.
Clay, J.: 1927, *Proc. Nederlandsche Akad. v. Wet.* **30**, 1115.
Clayton, D.D.: 1982, in *Essays in Nuclear Astrophysics*, C.A. Barnes, D.D. Clayton & D.N. Schramm (eds.), Cambridge University Press, Cambridge, UK, p. 73.

Clayton, D.D.: 1984, *Astrophys. J.* **280**, 144.
Clearly, M.N., Heiles, C., Haslam, C.G.T.: 1979, *Astron. Astrophys. Suppl. Ser.* **36**, 95.
Cleveland, B.T., Daily, T., Davis, R., Distel, J., Lande, K., Lee, C.K., *et al.*: 1995, *Nucl. Phys. B Proc. Suppl.* **38**, 47.
Coburn, W., Heindl, W.A., Rothschild, R.E., Gruber, D.E., Kreykenbohm, I., Wiòms, J., *et al.*: 2002, *Astrophys. J.* **580**, 394.
Coe, M.J., Roche, P., Everall, C., Fishman, G.J., Hagedon, K., Finger, M., *et al.*: 1994, *Astron. Astrophys.* **289**, 784.
Cohen, M.: 1975, *Mon. Not. Roy. Astr. Soc.* **173**, 279.
Cohen, M., Kuhi, L.V.: 1976, *Astrophys. J.* **210**, 365.
Cohen, M., Kuhi, L.V.: 1979, *Astrophys. J. Suppl.* **41**, 743.
Cohen, M., Bieging, J.H., Schwartz, P.R.: 1982, *Astrophys. J.* **253**, 707.
Cohen, M.H., Urwin, S.C.: 1984, in *Proc. IAU Symp. N. 110*, p. 95.
Cohen, M., Witteborn, F.C.: 1985, *Astrophys. J.* **294**, 345.
Cohen, M., Emerson, J.P., Beichman, C.A.: 1989, *Astrophys. J.* **339**, 455.
Colafrancesco, S.: 1995, in *Frontier Objects in Astrophysics and Particle Physics*, F. Giovannelli & G. Mannocchi (eds.), Italian Physical Society, Editrice Compositori, Bologna, Italy **47**, 51.
Colafrancesco, S.: 1996, in *Multifrequency Behaviour of High Energy Cosmic Sources*, F. Giovannelli & L. Sabau-Graziati (eds.), *Mem. S.A.It.* **67**, 57.
Colafrancesco, S.: 1997, in *Frontier Objects in Astrophysics and Particle Physics*, F. Giovannelli & G. Mannocchi (eds.), Italian Physical Society, Editrice Compositori, Bologna, Italy **57**, 71.
Colafrancesco, S.: 1999, in *Multifrequency Behaviour of High Energy Cosmic Sources*, F. Giovannelli & L. Sabau-Graziati (eds.), *Mem. S.A.It.* **70**, 851.
Colafrancesco, S.: 2002a, in *Multifrequency Behaviour of High Energy Cosmic Sources*, F. Giovannelli & L. Sabau-Graziati (eds.), *Mem. S.A.It.* **73**, 141.
Colafrancesco, S.: 2002b, in *Multifrequency Behaviour of High Energy Cosmic Sources*, F. Giovannelli & L. Sabau-Graziati (eds.), *Mem. S.A.It.* **73**, 930.
Colafrancesco, S.: 2002c, in *Multifrequency Behaviour of High Energy Cosmic Sources*, F. Giovannelli & L. Sabau-Graziati (eds.), *Mem. S.A.It.* **73**, 981.
Coleman, G.D., Worden, S.P.: 1976, *Astrophys. J.* **205**, 475.
Coleman, G.D., Wu, C.-C., Weedman, D.W.: 1980, *Astrophys. J. Suppl. Ser.* **43**, 393.
Collmar, W., Bennett, K., Bloemen, H., Blom, J.J., Hermsen, W., Lichti, G.G., *et al.*: 1997, *Astron. Astrophys.* **328**, 33.
Comastri, A., Setti, G., Zamorani, G., Hasinger, G.: 1995, *Astron. Astrophys.* **296**, 1.
Combes, F.: 1991, *Ann. Rev. Astron. Astrophys.* **29**, 195.
Cominsky, L., Jones, C., Forman, W., Tananbaum, H.: 1978, *Astrophys. J.* **224**, 46.
Corbet, R.H.D.: 1984, *Astron. Astrophys.* **141**, 91.
Corbet, R.H.D.: 1986, *Mon. Not. Roy. Astr. Soc.* **220**, 1047.
Corbet, R.H.D., Smale, A.P., Ozaki, M., Koyama, K., Iwasawa, K.: 1995, *Astrophys. J.* **443**, 786.
Cordova, F.A.: 1986, in *The Physics of Accretion onto Compact Objects*, K.P. Mason, M.G. Watson & N.E. White (eds.), *Lecture Notes in Physics* **266**, 339.
Cordova, F.A.: 1995, in *X-Ray Binaries*, W. Lewin, J. van Paradijis & E. van den Heuvel (eds.), Cambridge University Press, p. 331.
Cordova, F.A.: 1998, in *Ultraviolet Astrophysics, Beyond the IUE Final Archive*, W. Wamstecker & R. González-Riestra, (eds.), ESA Publication Division, ESTEC, Nordwijk, The Netherlands, ESA **SP-413**, 375.
Cordova, F.A., Chester, T.J., Tuohy, I.R., Garmire, G.P.: 1980, *Astrophys. J.* **235**, 163.
Cordova, F.A., Mason, K.O.: 1983, in *Accretion-Driven Stellar X-Ray Sources*, Cambridge University Press, p. 154.
Cordova, F.A., Mason, K.O., Hjellming, R.M.: 1983, *Pub. Astron. Soc. Pacific* **95**, 69.
Cordova, F.A., Mason, K.O.: 1984, *Mon. Not. Roy. Astr. Soc.* **206**, 879.

Cordova, F.A., Ladd, E.F., Mason, K.O.: 1986, in *Magnetospheric Phenomena in Astrophysics*, R.I. Epstein & W.C. Feldman (eds.), AIP **144**, 250.
Cordova, F.A., Howarth, I.D.: 1987, in Exploring the Universe with the IUE Satellite, Y. Kondo (ed.), D. Reidel Publ. Co., Dordrecht, Holland, p. 395.
Costa, E., Frontera, F., Heise, J., Feroci, M., in't Zand, J., Fiore, F., *et al.*: 1997, *Nature* **387**, 783.
Costa, E. *on behalf of the BeppoSAX GRB Team*: 1999, *Astron. Astrophys. Suppl. Ser.* **138**, 425.
Costa, E., Frontera, F., Hjorth, J. (eds.): 2001, *Gamma Ray Bursts in the Afterglow Era*, Springer Verlag, Heidelberg, Germany.
Coté, J., Waters, L.B.F.M.: 1986, in *Light on Dark Matter*, F.P. Israel (ed.), D. Reidel Publ. Co., Dordrecht, Holland, p. 77.
Covino, S., Lazzati, D., Ghisellini, G., Saracco, P., Campana, S., Chincarini, G., *et al.*: 1999, *Astron. Astrophys. Lett.* **348**, L1.
Cox, D.P., Anderson, P.R.: 1982, *Astrophys. J.* **253**, 268.
Cox, P., Mezger, P.G.: 1986, in *Light on Dark Matter*, F.P. Israel (ed.), D. Reidel Publ. Co., Dordrecht, Holland, p. 23.
Cox, D.P., Reynolds, R.J.: 1987, *Ann. Rev. Astron. Astrophys.* **25**, 303.
Cox, P., Kessler, M.F. (eds.): 1999, *The Universe as Seen by ISO*, ESA SP-427 (two volumes).
Cowie, L., Songaila, A.: 1986, *Ann. Rev. Astron. Astrophys.* **24**, 499.
Craig, N., Abbott, M., Finley, D., Jessop, H., Howell, S.B., Mathioudakis, M., *et al.*: 1997, *Astrophys. J. Suppl. Ser.* **113**, 131.
Cruddace, R.G., Dupree, A.K.: 1984, *Astrophys. J.* **277**, 263.
Czerny, B.(ed.): 2001, *Adv. Space Res.* **28, Issue 2-3**.
Dalgarno, A.: 1985, in *Molecular Astrophysics: State of the Art and Future Directions, NATO ASI Ser.* N. C157, 3.
Dalgarno, A.: 1992, in *The Astronomy and Astrophysics Encyclopedia*, Maran, S.P (ed.), Cambridge University Press, UK, p. 343.
Daly, R.A.: 1988, *Astrophys. J. Lett.* **333**, L1.
Dame, T.M., Ungerechts, H., Cohen, R.S., de Geus, E.J., Grenier, I.A., May, J., *et al.*: 1987, *Astrophys. J.* **322**, 706.
D'Amico, N.: 2000, in *Pulsar Astronomy and Beyond*, M. Kramer, N. Wex & N. Wielebinski (eds.), *ASP Conf. Ser.* **202**, 27.
D'Antona, F., Mazzitelli, I: 1984, *Astron. Astrophys.* **138**, 431.
D'Antona, F., Mazzitelli, I: 1994, *Astrophys. J. Suppl. Ser.* **90**, 467.
Dar, A.: 1998, *Astrophys. J. Lett.* **500**, L93.
Dar, A.: 1999a, *Astron. Astrophys. Suppl. Ser.* **138**, 505.
Dar, A.: 1999b, in *Frontier Objects in Astrophysics and Particle Physics*, F. Giovannelli & G. Mannocchi (eds.), Italian Physical Society, Editrice Compositori, Bologna, Italy **65**, 279.
Dar, A.: 2001, in *Frontier Objects in Astrophysics and Particle Physics*, F. Giovannelli & G. Mannocchi (eds.), Italian Physical Society, Editrice Compositori, Bologna, Italy **73**, 313.
Dar, A.: 2003a, talk (unpublished) presented at the Vulcano Workshop 2002: *Frontier Objects in Astrophysics and Particle Physics.*
Dar, A.: 2003b, in *Frontier Objects in Astrophysics and Particle Physics*, F. Giovannelli & G. Mannocchi (eds.), Italian Physical Society, Editrice Compositori, Bologna, Italy **85**, 253.
Dar, A., Loeb, A., Nussinov, S.: 1989, in *Results and Perspectives in Particle Physics*, M. Greco (ed.), Edition Frontières, Gif-sur-Yvette Cedex, France, p. 29.
David, L., Forman, W., Jones, C.: 1990, *Astrophys. J.* **359**, 29.
Davidsen, A., Bowyer, S., Lampton, M., Cruddace, R.: 1975, *Astrophys. J.* **198**, 1.
Davidsen, A. F., Hartig, G.F., Fastie, W.G.: 1977, *Nature* **269**, 203.
Davies, R.D.: 1981, *Phil. Trans. Series A* **303, N. 1480**, 581.
Davies, R.D., Walsh, D., Browne, I., Edwards, M.R., Noble, R.G.: 1976, *Nature* **261**, 476.

De Amici, G., Bersanelli, M., Kogut, A., Levin, S., Limon, M., Smoot, G.F.: 1991, *Astrophys. J.* **381**, 341.

Dean, A.J., Ramsden, D.: 1981, *Phil. Trans. Roy. Soc*, London, UK **A 301**, 577.

Dent, W.R.F., Macdonald, G.H., Andersson, M.: 1988, *Mon. Not. Roy. Astr. Soc.* **235**, 1397.

DePoy, D.L.: 1987, in *Star Formation in Galaxies*, C. Persson (ed.), NASA, Washington, DC, US Govt. Print. Off., p. 701.

Dermer, C.D., Schlickeiser, R.: 1992, *Science* **257, N. 5077**, 1642.

Dermer, C.D., Schlickeiser, R.: 1993, *Astrophys. J.* **416**, 458.

Deutsch, L.K., Willner, S.P.: 1986, *Astrophys. J. Lett.* **306**, L11.

Devereux, N.: 1987, *Astrophys. J.* **323**, 91.

De Zotti, G., Granato, G.L., Baccigalupi, C., Magliocchetti, M., Mazzei, P., Perrotta, F., *et al.*: 2002, in *Multifrequency Behaviour of High Energy Cosmic Sources*, F. Giovannelli & L. Sabau-Graziati (eds.), *Mem. S.A.It.* **73**, 876.

Dickman, R.L.: 1978, *Astrophys. J. Suppl. Ser.* **37**, 407.

Dickman, R.L., Snell, R.L., Schloerb, F.P.: 1986, *Astrophys. J.* **309**, 326.

Diehl, R.: 1996, in *Multifrequency Behaviour of High Energy Cosmic Sources*, F. Giovannelli & L. Sabau-Graziati (eds.), *Mem. S.A.It.* **67**, 303.

Diehl, R.: 1999, in *Wolf–Rayet Phenomena in Massive Stars and Starburst Galaxies*, K.A. van der Hucht, G. Koenisberger & P.R.J. Eenens (eds.), IAU Symp. **193**, 205.

Diehl, R.: 2000, in *The Fifth Compton Symposium*, M.L. McConnell & J.M. Ryan (eds.), *AIP Conf. Proc.* **510**, 13.

Diehl, R., Dupraz, C., Bennet, K., Blömen, H., Hermsen, W., Knödleser, J., *et al.*: 1995, *Astron. Astrophys.* **298**, 445.

Diercksen, G.H.F., Huebner, W.F., Langhoff, P.W. (eds.): 1985, *Molecular Astrophysics: State of the Art and Future Directions*, *NATO ASI Ser.* N. C157.

Dinerstein, H.L.: 1986, *Astron. J.* **92**, 1381.

Dinerstein, H.L., Robinson, E.L.: 1986, in *Light on Dark Matter*, F.P. Israel (ed.), D. Reidel Publ. Co., Dordrecht, Holland, p. 145.

Dingus, B.L.: 1994, *Presentation at Snowmass, June 29-July 14*, Snowmass, CO, USA.

van Dishoeck, E.F., Black, J.H.: 1988, *Astrophys. J.* **334**, 771.

Djorgovski, S.G., Kulkarni, S.R., Bloom, J.S., Frail, D.A.: 1999, *GCN Circ.* **N. 289**.

Djorgovski, S.G., Kulkarni, S.R., Bloom, J.S., Frail, D.A., Harrison, F.A., Galama, T.J., *et al.*: 2001, in *Gamma Ray Bursts in the Afterglow Era*, E. Costa, F. Frontera & J. Hjorth (eds.), Springer Verlag, Heidelberg, Germany, p. 218.

Donati, J.F.: 1993, in *Inside the Stars*, W.W. Weiss & A. Baglin (eds.), *ASP Conf. Ser.* **40**, 136.

la Dous, C.: 1989, *Astron. Astrophys.* **211**, 131.

la Dous, C.: 1993, in *Cataclysmic Variables and Related Objects*, M. Hack & C. la Dous (eds.), NASA **SP-507**, 15.

la Dous , C.: 1994, *Space Sci. Rev.* **67**, 1.

Doyle, S., Balick, B., Corradi, R.L.M., Schwarz, H.E.: 2000, *Astron. J.* **119**, 1339.

Draine, B.T.: 1981, *Astrophys. J.* **245**, 880.

Draine, B.T., Lee, H.M.: 1984, *Astrophys. J.* **285**, 89.

Drechsel, H., Rahe, J., Kondo, Y. (eds.): 1987, *Cataclysmic Variables. Recent Multi-Frequency Observations and Theoretical Developments*, *Astrophys. Space Sci.* **130**.

Dressler, A.: 1980, *Astrophys. J.* **236**, 351.

Drew, J.: 1990, in *Physics of Classical Novae*, A. Cassatella & R. Viotti (eds.), Springer-Verlag, *Lecture Notes in Physics* **369**, 228.

Drury, L.: 1983a, *Rep. Progr. Phys.* **46**, 973.

Drury, L.: 1983b, *Space Sci. Rev.* **36**, 57.

Drury, L.O'c.: 1990, *Proc. 21st Intern. Cosmic Ray Conf. (Adelaide)* **12**, 85.

Drury, L.O'c., Aharonian, F.A., Voelk, H.J.: 1994, *Astron. Astrophys.* **287**, 959.

Dube, R.R., Wickes, W.C., Wilkinson, D.T.: 1979, *Astrophys. J.* **32**, 333.
Dubinski, J., Carberg, R.G.: 1991, *Astrophys. J.* **378**, 496.
Dufour, R.J., Shields, G.A., Talbot, R.J. Jr.: 1982, *Astrophys. J.* **252**, 461.
Duncan, R.C., Thompson, C.: 1992, *Astrophys. J. Lett.* **392**, L9.
Dunham, T. Jr.: 1937, *Pub. Astron. Soc. Pacific* **49**, 26.
Dunham, T. Jr., Adams, W.S.: 1937, *Pub. Am. Astron. Soc.* **9**, 5.
Dupree, A.K., Raymond, J.: 1982, *Astrophys. J. Lett.* **263**, L63.
Dupree, A.K., Baliunas, S.L., Guinan, E.T., Hartmann, L., Nassiopoulos, G.E., Sonneborn, G.: 1987, *Astrophys. J. Lett.* **317**, L85.
Dwek, E.: 1981, *Astrophys. J.* **247**, 614.
Dwek, E.: 1986, *Astrophys. J.* **302**, 363.
Dyck, H.M., Simon, T., Zuckermann, B.: 1982, *Astrophys. J. Lett.* **255**, L103.
Dzikowski, T., Gawin, J., Grochalska, B., Korejwo, J., Wdowczyk, J.: 1985, in *Proc. 19th Internat. Cosmic Ray Conf. (La Jolla)* **1**, 238.
Ebisawa, K., Mukai, K., Kotani, T., Asai, K., Dotani, T., Nagase, F., *et al.*: 2001, *Astrophys. J.* **550**, 1007.
Eddington, A.S.: 1926, *Proc. Roy. Soc.* **A111**, 424.
Edelson, R.A., Malkan, M.A.: 1986, *Astrophys. J.* **308**, 59.
Edelson, R., Krolik, J., Madejski, G., Maraschi, L., Pike, G., Urry, C.M., *et al.*: 1995, *Astrophys. J.* **438**, 120.
Edwards, S., Cabrit, S., Strom, S.E., Heyer, I., Strom, K.M., Anderson, E.: 1987, *Astrophys. J.* **321**, 473.
Edwards, S., Cabrit, S., Ghandour, L.O., Strom, S.E.: 1989, in *Low Mass Star Formation and Pre-Main Sequence Objects*, Bo Reipurth (ed.), in *ESO Conference and Workshop Proceedings* **33**, 385.
Edwards, S., Strom, S.E., Hartigan, P., Strom, K.M., Hillebrand, L.A., Herbst, W., *et al.*: 1993, *Astron. J.* **106**, 372.
Efstathiou, G., Frenk, C.S., White, S.D.M., Davis, M.: 1988, *Mon. Not. Roy. Astr. Soc.* **235**, 715.
Egger, R.J., Freyberg, M.J., Morfill, G.E.: 1996, *Space Sci. Rev.* **75**, 511.
Eggleton, P. (ed.): 1976, Structure and evolution of Close Binary Systems, D. Reidel Publ. Co., Dordrecht, Holland.
Ehrenfreund, P., O'Tuairisg, S., Foing, B.H., Sonnentrucker, P., Cami, J.: 2001, in *The Bridge between the Big Bang and Biology: Stars, Planetary Systems, Atmospheres, Volcanoes: Their Link to Life*, F. Giovannelli (ed)., CNR President Bureau, Special Volume, Roma, Italy, p. 150.
Eichhorn, H.: 1992, in *The Astronomy and Astrophysics Encyclopedia*, Maran, S.P (ed.), Cambridge University Press, UK, p. 661.
Eilek, J.: 1999, in *Diffuse Thermal and Relativistic Plasma in Galaxy Clusters*, Böhringer, H., Feretti, L., Schuecker, P. (eds.), MPE Report **271B**, 71.
Eilek, J., Owen, F.N., Zhou, F.: 1999, *Lect. Notes in Physics* **530**, 136.
Elässer, H., Staude, H.J.: 1978, *Astron. Astrophys. Lett.* **70**, L3.
Elbaz, D., Aussel, H., Cesarsky, C.J., Desert, F.X., Fadda, D., Franceschini, A., *et al.*: 1999, *The Universe as Seen by ISO*, P. Cox & Kessler, M.F. (eds.), ESA **SP-427, Vol. 2**, 999.
Ellis, G.F.R.: 1984, Ann. Rev. Astron. Astrophys. **22**, 157.
Elvis, M.: 1976, *Mon. Not. Roy. Astr. Soc.* **177**, 7p.
Elvis, M., Soltan, A., Keel, W.C.: 1984, *Astrophys. J.* **283**, 479.
Erlykin, A.D.: 1993, in *Frontier Objects in Astrophysics and Particle Physics*, F. Giovannelli & G. Mannocchi (eds.), Italian Physical Society, Editrice Compositori, Bologna, Italy **40**, 405.
Erlykin, A.D.: 1995, in *Frontier Objects in Astrophysics and Particle Physics*, F. Giovannelli & G. Mannocchi (eds.), Italian Physical Society, Editrice Compositori, Bologna, Italy **47**, 483.
Erlykin, A.D.: 1997, in *Frontier Objects in Astrophysics and Particle Physics*, F. Giovannelli & G. Mannocchi (eds.), Italian Physical Society, Editrice Compositori, Bologna, Italy **57**, 363.

Erlykin, A.D., Osborne, J.L., Wolfendale, A.W., Zhang, L.: 1996, *Astron. Astrophys. Suppl. Ser.* **120**, 623.

Erlykin, A.D., Wolfendale, A,W.: 1999, in *Frontier Objects in Astrophysics and Particle Physics*, F. Giovannelli & G. Mannocchi (eds.), Italian Physical Society, Editrice Compositori, Bologna, Italy **65**, 467.

Erlykin, A.D., Wolfendale, A,W., Vahia, M.N.: 1999, in *Frontier Objects in Astrophysics and Particle Physics*, F. Giovannelli & G. Mannocchi (eds.), Italian Physical Society, Editrice Compositori, Bologna, Italy **65**, 455.

Evans, N.J.II: 1999, *Ann. Rev. Astron. Astrophys.* **37**, 311.

Evans, R.D.: 1985, *The Atomic Nucleous*, McGraw-Hill, New York, USA, p. 713.

Evans, N.J.II, Levreault, R.M., Beckwith, S., Skrutskie, M.: 1987, *Astrophys. J.* **320**, 364.

Ewen, H.I.: 1951, *Ph. D. Thesis*, Harvard University, MA, USA.

Fabbiano, G.: 1984, in *X-Ray Astronomy '84*, M. Oda & R. Giacconi (eds.), ISAS, Tokyo, Japan, p. 333.

Fabbiano, G.: 1988, *Astrophys. J.* **325**, 544.

Fabbiano, G.: 1989, *Ann. Rev. Astron. Astrophys.* **27**, 87.

Fabbiano, G., Hartmann, L., Raymond, J. Steiner, J., Branduardi-Raymond, G., Matilsky, T.: 1981, *Astrophys. J.* **243**, 911.

Fabbiano, G., Feigelson, E., Zamorani, C.: 1982, *Astrophys. J.* **256**, 397.

Fabbiano, G., Panagia, N.: 1983, *Astrophys. J.* **266**, 568.

Fabbiano, G., Trinchieri, G.: 1985, *Astrophys. J.* **296**, 430.

Fabbiano, G., Trinchieri, G.: 1987, *Astrophys. J.* **315**, 46.

Faber, S., Gallagher, J.: 1976, *Astrophys. J.* bf 204, 365.

Fabian, A.C.: 1979, *Proc. Roy. Soc. London* **A 366**, 449

Fabian, A.C.: 1981, in *The Structure and Evolution of Normal Galaxies*, S.M. Fall & D. Lynden-Bell (eds.) Cambridge University Press, UK, p. 181.

Fabian, A.C.: 1999, in *Diffuse Thermal and Relativistic Plasma in Galaxy Clusters*, Böhringer, H., Feretti, L., Schuecker, P. (eds.), MPE Report **271B**, p. 123.

Fabian, A.C., Rees, M.J., Stella, L., White, N.E.: 1989, *Mon. Not. Roy. Astr. Soc.* **238**, 729.

Fabian, A.C., George, I.M., Miyoshi, S., Rees, M.J.: 1990, *Mon. Not. Roy. Astr. Soc.* **242**, 14.

Fabian, A.C., Barcons, X.: 1992, *Ann. Rev. Astron. Astrophys.* **30**, 429.

Fabricant, D., Lecar, M., Gorenstein, P.: 1980, *Astrophys. J.* **241**, 552.

Fabricant, D., Gorenstein, P.: 1983, *Astrophys. J.* **267**, 535.

Falcke, H., Biermann, P.L.: 1996, *Astron. Astrophys.* **308**, 321.

Fanti, R., Mantovani, F., Tomasi, P.: 1981, *Astron. Astrophys. Suppl. Ser.* **43**, 1.

Fargion, D.: 1999, *Astron. Astrophys. Suppl. Ser.* **138**, 507.

Fargion, D.: 2001a, in *Frontier Objects in Astrophysics and Particle Physics*, F. Giovannelli & G. Mannocchi (eds.), Italian Physical Society, Editrice Compositori, Bologna, Italy **73**, 347.

Fargion, D.: 2001b, in *Gamma Ray Bursts in the Afterglow Era*, E. Costa, F. Frontera & J. Hjorth (eds.), Springer Verlag, Heidelberg, Germany, p. 315.

Feldman, P.D., Brune, W.H., Henry, R.C.: 1981, *Astrophys. J. Lett.* **249**, L51.

Feenberg, E., Primakoff, H.: 1948, *Phys. Rev.* **73**, 449.

Feigelson, E.D.: 1984, in *Cool Stars, Stellar System, and the Sun*, S. Baliunas & L. Hartmann (eds.), Springer-Verlag, New York, USA, p. 27.

Feigelson, E.D., Montmerle, T.: 1985, *Astrophys. J. Lett.* **289**, L19.

Feigelson, E.D., Jackson, J.M., Mathiew, R.D., Myers, P.C., Walter, F.D.: 1987, *Astron. J.* **94**, 1251.

Felten, J.E., Gould, R.J., Stein, W.A., Woolf, N.J.: 1966, *Astrophys. J.* **146**, 955.

Fender, R.P., Garrington, S.T., McKay, D.J., Muxlow, T.W.B., Pooley, G.G., Spencer, R.E., Stirling, A.M., Waltman, E.B.: 1999, *Mon. Not. Roy. Astr. Soc.* **304**, 865.

Fender, R.P., Kuulkers, E.: 2001, *Mon. Not. Roy. Astr. Soc.* **324**, 923.

Fenimore, E.E., Ramirez-Ruiz, E.: 2000, *astro-ph/0004176*.

Feretti, L., Dallacasa, D., Govoni, F., Giovannini, G., Taylor, G.B., Klein, U.: 1999, *Astron. Astrophys.* **344**, 472.

Ferland, G.J., Persson, S.E.: 1989, *Astrophys. J.* **347**, 656.

Ferlet, R.: 1999, *The Astron. Astrophys. Rev.* **9**, 153.

Fermi, E.: 1949, *Phys. Rev. 2nd Ser.* **75**, N. 8, 1169.

Fermi, E.: 1954, *Astrophys. J.* **119**, 1.

Feroci, M.: 1999, in *Multifrequency Behaviour of High Energy Cosmic Sources*, F. Giovannelli & L. Sabau-Graziati (eds.), *Mem. S.A.It.* **70**, 905.

Feroci, M.: 2001, in *Frontier Objects in Astrophysics and Particle Physics*, F. Giovannelli & G. Mannocchi (eds.), Italian Physical Society, Editrice Compositori, Bologna, Italy **73**, 287.

Feroci, M., Mereghetti, S. (eds.): 2002, *Soft Gamma Ray Repeaters, Mem. S.A.It.* **73 N. 2**, 455-593.

Fichtel, C.E., Hartman, R.C., Kniffen, D.A., Thompson, D.J., Ogelman, H., Ozel, M.E., *et al.*: 1975, *Astrophys. J.* **198**, 163.

Fichtel, C.E., Simpson, G.A., Thompson, D.J.: 1978, *Astrophys. J.* **222**, 833.

Fichtel, C.E., Trombka, J.I.: 1981, *Gamma Ray Astrophysics: New Insight into the Universe*, NASA **SP-453**.

Fichtel, C.E., Bertsch, D.H., Chiang, J., Dingus, B.L., Esposito, J.A., Fierro, J.M., *et al.*: 1994, *Astrophys. J. Suppl. Ser.* **94**, 551.

Fierro, J.M., Bertsch, D.L., Brazier, K.T., Chiang, J., D'Amico, N., Fichtel, C.E., *et al.*: 1993, *Astrophys. J. Lett.* **4137**, L27.

Figer, D., Kim, S.S., Morris, M., Serabyn, E., Rich, R.M., McLean, I.S.: 1999, *Astrophys. J.* **525**, 750.

Figer, D., McLean, I., Morris, M.: 1995, *Astrophys. J. Lett.* **447**, L29.

Filippenko, A.V., Leonard, D.C., Matheson, T., Li, W., Moran, E.C., Riess, A.G.: 1999, *Pub. Astron. Soc. Pacific* **111**, 969.

Finger, M.H.: 1998, *Adv. Space Res.* **22, Issue 7**, 1007.

Finger, M.H., Cominsky, L. R., Wilson, R.B., Harmon, B.A., Fishman, G.J.: 1994, in *The Evolution of X-Ray Binaries*, S.S. Holt & C.S. Day (eds.), *AIP Conf. Proc.*, **308**, 459.

Finger, M.H., Wilson, R.B., Harmon, B.A.: 1996, *Astrophys. J.* **459**, 288.

Finger, M.H., Prince, T.A.: 1997, in *Proc. Fourth Compton Symposium*, C.D. Dermer, M.S. Strickman & J.D. Kurfess (eds.), *AIP Conf. Proc.* **410**, 57.

Fishman, G.J., Meegan, C.A., Wilson, R.B., Paciesas, W.S., Pendleton, G.N.: 1992, in *The Compton Observatory Science Workshop*, NASA-GSFC, p. 26.

Fitzpatrick, E.L., Savage, B.D.: 1983, *Astrophys. J.* **267**, 93.

dal Fiume, D., Orlandini, M., Frontera, F., del Sordo, S., Piraino, S., Santangelo, A., *et al.*: 1998, *Nuclear Physics B (Proc. Suppl.)* **69/1-3**, 145.

dal Fiume, D., Amati, L., Antonelli, L.A., Fiore, F., Muller, J.M., Parmar, A., *et al.*: 2000, *Astron. Astrophys.* **355**, 454.

Fixsen, D.J., Cheng, E.S., Cottingham, D.A., Eplee, R.E., Jr., Isaacman, R.B., Mather, J.C., *et al.*: 1994, *Astrophys. J.* **420**, 445.

Fixsen, D. J., Dwek, E., Mather, J. C., Bennett, C. L., Shafer, R. A.: 1998, *Astrophys. J.* **508**, 123.

Foing, B.H., Char, S.: 1988, in *Comptes-Rendus du Premier Atelier MUSICOS*, C. Catala & B.H. Foing (eds.), Paris Observatory Press, p. 31.

Ford, Eric C., van der Klis, M., van Paradijs, J., Méndez, M., Wijnands, R., Kaaret, P.: 1998, *Astrophys. J. Lett.* **508**, L155.

Forman, W., Kellog, E., Gursky, H., Tananbaum, H., Giacconi, R.: 1972, *Astrophys. J.* **178**, 309.

Forman, W., Jones, C., Cominsky, L., Julien, P., Murray, S., Peters, G., *et al.*: 1978, *Astrophys. J. Suppl.* **38**, 357.

Forman, W., Schwarz, J., Jones, C., Liller, W., Fabian, A.: 1979, *Astrophys. J. Lett.* **234**, L27.

Forman, W., Jones, C., Tucker, W.: 1985, *Astrophys. J.* **293**, 102.

Forman, W., Jones, C.: 1990, in *Clusters of Galaxies*, W.R. Oegerle, M.J. Fitchett, L. Danly (eds.), Cambridge University Press, Cambridge, Great Britain, p. 257.

Forman, W., Jones, C.: 1994, in *Cosmological Aspects of X-Ray Clusters of Galaxies*, W.C. Seitter (ed.), Kluwer Academic Publishers, Dordrecht, Holland, p. 39.

Forman, W., Jones, C., Tucker, W.: 1994, *Astrophys. J.* **429**, 77.

Frail, D., Kulkarni, S., Nicastro, L., Feroci, M., Taylor, G.: 1997, *Nature* **389**, 261.

Frail, D.A., Kulkarni, S.R., Sari, R., Djorgovski, S.G., Bloom, J.S., Galama, T.J., *et al.*: 2001, *Astrophys. J. Lett.* **562**, L55.

Frame, K.C. (for the SNO Collaboration): 2002, in *Le Rencontres de Physique de la Vallée d'Aoste: Results and Perspectives in Particle Physics*, M. Greco (ed.), *Frascati Physics Series* **XXVII**, 65.

Franco, J.: 1993, *Rev. Mex. Astron. Astrofis.* **26**, 13.

Frank, J., King, A.R., Raine, D.J.: 1985, *Accretion Power in Astrophysics*, Cambridge University Press, UK.

Fransson, C., Cassatella, A., Gilmozzi, R., Kirshner, R.P., Panagia, N., Sonneborn, G., *et al.*: 1989, *Astrophys. J.* **336**, 429.

Friberg, P., Hjalmarson, Å.: 1990, in *Molecular Astrophysics: A Volume Honouring Alexander Dalgarno*, T.W. Hartquist (ed.), Cambridge University Press, p. 3.

Friedlander, M., Gehrels, N., Macomb, D.J. (eds.): 1992, *Compton Gamma Ray Observatory*, M. , *AIP Conf. Proc.* **280**.

Fritz, G., Davidsen, A., Meekins, J.F., Friedman, H.: 1971, *Astrophys. J. Lett.* **164**, L81.

Frontera, F.: 2003, in *Supernovae and Gamma Ray Bursts*, K.W. Weiler (ed.), Springer, Berlin (in press).

Fruchter, A., Livio, M., Macchetto, D., Petro, L., Sahu, K., Thorsett, S., *et al.*: 1997, *IAU Circ.* **N. 6747**.

Frye, G.M., Albats, P., Zych, A., Staib, J., Hopper, V., Rawlinson, W., *et al.*: 1972a, *Bull. Am. Phys. Soc.* **17**, 524.

Frye, G.M., Albats, P., Zych, A., Staib, J., Hopper, V., Rawlinson, W., *et al.*: 1972b, *Nature* **233**, 466.

Fukazawa, Y.: 1998, Ph.D. Thesis, University of Tokyo, Japan.

Fukazawa, Y., Makishima, K., Ebisawa, K., Fabian, A.C., Gendreau, K.C., Ikebe, Y., *et al.*: 1994a, *Publ. Astr. Soc. Pacific* **46**, 141.

Fukazawa, Y., Ohashi, T., Fabian, A.C., Canizares, C.R., Ikebe, Y., Makishima, K., *et al.*: 1994b, *Publ. Astron. Soc. Japan Lett.* **46**, 55.

Fukazawa, Y., Ohashi, T., Tamura, T., Ikebe, Y., White, R.E., Makishima, K.: 1996, in *UV and X-ray Spectroscopy of Astrophysical and Laboratory Plasmas*, K. Yamashita & T. Watanabe (eds.), Universal Academy Press, Frontiers Science Series **15**, 383.

Fukazawa, Y., Makishima, K., Tamura, T., Nakazawa, K., Ezawa, H., Ikebe, Y., *et al.*: 2000, *Mon. Not. Roy. Astr. Soc.* **313**, 21.

Fukuda, Y., Hayakawa, T., Inoue, K., Ishida, T., Kajita, T., Koshio, Y., *et al.*: 1994, *Astrophys. J.* **435**, 225.

Fukuda, Y., et Super-KAMIOKANDE Collaboration: 1998a, *Phys. Rev. Lett.* **81, Issue 6**, 1158.

Fukuda, Y., et Super-KAMIOKANDE Collaboration: 1998b, *Phys. Rev. Lett.* **81, Issue 8**, 1562.

Fukuda, Y., et Super-KAMIOKANDE Collaboration: 1998c, *Phys. Lett. B* **436, Issue 1-2**, 33.

Fukugita, M.: 1988, *Phys. Rev. Lett.* **61**, 1046.

Fukugita, M., Yanagida, T.: 1985, *Phys. Rev. Lett.* **55**, 2645.

Fukugita, M., Hogan, C.J., Peebles, P.J.E.: 1998, *Astrophys. J.* **503**, 518.

Fusco-Femiano, R., Dal Fiume, D., Feretti, L., Giovannini, G., Grandi, P., Matt, G., *et al.*: 1999, *Astrophys. J. Lett.* **513**, L21.

Gaisser, T.K.: 1990a, in *Astrophysical Aspects of the Most Energetic Cosmic Rays*, M. Nagano & F. Takahara (eds.), World Scientific, p. 146.

Gaisser, T.K.: 1990b, *Science* **247**, 1049.

Gaisser, T.K.: 1990c, *Cosmic Rays and Particle Physics*, Cambridge and New York, Cambridge University Press.
Galama, T.J., Groot, P.J., van Paradijs, J., Kouveliotou, C., Robinson, C.R., Fishman, G.J., *et al.*: 1997, *Nature* **387**, 479.
Galama, T.J., Wijers, R.A.M.J., Bremer, M., Groot, P.J., Strom, R.G., Kouveliotou, C., *et al.*: 1998, *Astrophys. J. Lett.* **500**, L97.
Galama, T.J., Tanvir, N., Vreeswijk, P.M., Wijers, R.A.M.J., Groot, P.J., Rol, E., *et al.*: 2000, *Astrophys. J.* **536**, 185.
Gallagher, J., Hunter, D.: 1987, in *Star Formation in Galaxies*, C. Persson (ed.), NASA, Washington, DC, US Govt. Print. Off., p. 166.
Gao, Yi-T., Stecker, F.W., Cline, D.B.: 1991, *Astron. Astrophys.* **249**, 1.
Garcia-Primo, M.A.: 2001, *Astrophys. Space Sci.* **276**, 3.
Garmany, C.D., Conti, P.S.: 1984, *Astrophys. J.* **284**, 705.
Garnavich, P.M., Stanek, K.Z., Wyrzykowski, L., Infante, L., Bendek, E., Holland, S.T., *et al.*: 2002, *AAS* **200**, 2505.
Gaudenzi, S., Giovannelli, F., Lombardi, R., Claudi, R.: 1986, in *New Insights in Astrophysics*, E. Rolfe (ed.), ESA **SP-263**, 455.
Gaudenzi, S., Giovannelli, F., Lombardi, R., Claudi, R.: 1990, *Acta Astron.*, **40**, 105.
Gaudenzi, S., Claudi, R.U., Giovannelli, F., Lombardi, R., Pelosi, M., Strappolini, M.: 2002, in *Multifrequency Behaviour of High Energy Cosmic Sources*, F. Giovannelli & L. Sabau-Graziati (eds.), *Mem. S.A.It.* **73 N. 1**, 213.
Gavriil, F.P., Kaspi, V.M.: 2002, *Astrophys. J.* **567**, 1067.
Gear, W.K., Gee, G., Robson, E.I., Nolt, I.J.: 1985, *Mon. Not. Roy. Astr. Soc.* **217**, 281.
Gehrels, N., Share, G.H. (eds.): 1988, *Nuclear Spectroscopy of Astrophysical Sources*, *AIP Conf. Proc.* **170**.
Gehrels, N., Tueller, J.: 1992, in *The Compton Observatory Science Workshop*, C.R. Shrader, N. Gehrels and B. Dennis (eds.), NASA Conference Publication **3137**, 446.
Gehrels, N., Chipman, E., Kniffen, D.A.: 1993, *Astron. Astrophys. Suppl. Ser.* **97, N. 1**, 5.
Gehrels, N., Macomb, D.J., Bertsch, D.L., Thompson, D.J., Hartman, R.C.: 2000, *Nature* **404**, 363.
Gehrels, N., Piro, L., Leonard, J.T.: 2002, *Scientific American* **287, N. 6**, 53.
Geiss, J.: 1993, in *The Origin and Evolution of the Elements*, N. Prantzos, E. Flam & M. Cassé (eds.), Cambridge University Press, p. 89.
Gendreau, K.C.: 1995, *PhD Thesis*, Massachusetts Institute of Technology, Cambridge, MA, USA.
Gendreau, K.C., Mushotzky, R., Fabian, A.C., Holt, S.S., Kii, T., Serlemitsos, P.J., *et al.*: 1995, *Publ. Astr. Soc. Japan* **47**, 5.
Genzel, R.: 1999, *The Universe as Seen by ISO*, P. Cox & Kessler, M.F. (eds.), ESA **SP-427, Vol. 1**, 13.
George, I.M., Fabian, A.C.: 1991, *Mon. Not. Roy. Astr. Soc.* **248**, 760.
Georgelin, Y.M., Georgelin, Y.P.: 1976, *Astron. Astrophys.* **49**, 57.
Ghez, A.M., Klein, B.L., Morris, M., Becklin, E.E.: 1998, *Astrophys. J.* **509**, 678.
Ghosh, P.: 1994, in *The Evolution of X-Ray Binaries*, S.S. Holt & C.S. Day (eds.), *AIP Conf. Proc.*, **308**, 439.
Ghosh, P.: 1995, *J. Astrophys. & Astron.* **16 N. 2**, 289.
Ghosh, P.: 1998, *Adv. Space Res.* **22**, 1017.
Ghosh, P., Lamb, F.K.: 1978, *Astrophys. J. Lett.* **223**, L83.
Ghosh, P., Lamb, F.K.: 1979, *Astrophys. J.* **234**, 296.
Ghosh, P., Lamb, F.K.: 1992, in *X-Ray Binaries and Recycled Pulsars*, E.P.J. van den Heuvel & S. Rappaport (eds.), Kluwer Academic Publ., Dordrecht, Holland, p. 487.
Ghosh, P., Angelini, L., White, N.E.: 1997, *Astrophys. J.* **478**, 713.

Giacconi, R.: 1978, in *Physics and Astrophysics of Neutron Stars and Black Holes*, R. Giacconi & R. Ruffini (eds.), Bologna, Società Italiana di Fisica; North Holland Publishing Co., Amsterdam, Holland, p. 17.

Giacconi, R. (ed.): 1981, *X-Ray Astronomy with Einstein satellite*, D. Reidel Publ. Co., Dordrecht, Holland.

Giacconi, R., Gursky, H., Paolini, F.R., Rossi, B.: 1962, *Phys. Rev. Lett.* **9**, 439.

Giacconi, R., Gursky, H., Van Speybroeck, L.P.: 1968, *Ann. Rev. Astron. Astrophys.* **6**, 373.

Giacconi, R., Gursky, H., Kellog, E., Schreier, E., Tananbaum, H.: 1971, *Astrophys. J. Lett.* **167**, L67.

Giacconi, R., Murray, S., Gursky, H., Kellog, E., Schreier, E., Tananbaum, H.: 1972, *Astrophys. J.* **178**, 281.

Giacconi, R., Ruffini, R.: 1978, *Physics and Astrophysics of Neutron Stars and Black Holes*, Bologna, Società Italiana di Fisica; North Holland Publishing Co., Amsterdam, Holland.

Giacconi, R., Branduardi, G., Briel, U., Epstein, A., Fabricant, D. Feigelson, E., *et al.*: 1979, *Astrophys. J.* **230**, 540.

Giacconi, R., Zamorani, G.: 1987, *Astrophys. J.* **313**, 20.

Giangrande, A., Giovannelli, F., Bartolini, C., Guarnieri, A., Piccioni, A.: 1980, *Astron. Astrophys. Suppl. Ser.* **40**, 289.

Gilfanov, M., Churazov, E., Sunyaev, R., Khavenson, N., Novikov, B., Dyachkov, A., *et al.*: 1993, *Astrophys. J.* **418**, 844.

Gilfanov, M., Churazov, E., Sunyaev, R., Vikhlinin, A., Finoguenov, A., Sitdikov, A., *et al.*: 1995, in *The Lives of the Neutron Stars*, M.A. Alpar, U. Kiziloglu & J. van Paradijs (eds.), Kluwer Academic Publ., Dordrecht, The Netherlands, p. 331.

Gillet, F.C.: 1986, in *Light on Dark Matter*, F.P. Israel (ed.), D. Reidel Publ. Co., Dordrecht, Holland, p. 61.

Gillet, F.C., Backman, D.E., Beichman, C., Neugebauer, G.: 1986, *Astrophys. J.* **310**, 842.

Gilmozzi, R., Cassatella, A., Clavel, J., Fransson, C., Gonzales, R.: 1987, *Nature* **328**, 318.

Giménez, A.: 1999, in *Multifrequency Behaviour of High Energy Cosmic Sources*, Giovannelli, F. & Sabau-Graziati, L. (eds.), *Mem. S.A.It.* **70**, 1281.

Giménez, A., Sabau-Graziati, L.: 1996, in *Multifrequency Behaviour of High Energy Cosmic Sources*, Giovannelli, F. & Sabau-Graziati, L. (eds.), *Mem. S.A.It.* **67**, 563.

Giommi, P., Capalbi, M., Fiocchi, M., Memola, E., Perri, M., Piranomonte, S., *et al.*: 2002, in Blazar Astrophysics with BeppoSAx and Other Observatories, P. Giommi, E. Massaro & G. Palumbo (eds.), ASI Special Publication, printed at ESA-ESRIN, p. 63.

Giommi, P., Massaro, E., Palumbo, G. (eds.): 2002, Blazar Astrophysics with BeppoSAx and Other Observatories, ASI Special Publication, printed at ESA-ESRIN.

Giovannelli, F.: 1981, *Space Sci. Rev.* **30**, 213.

Giovannelli, F. (ed.): 1985, *Multifrequency Behaviour of Galactic Accreting Sources*, Editrice SIDEREA, Roma, Italy.

Giovannelli, F.: 1991, *Frontier Objects in Astrophysics and Particle Physics*, F. Giovannelli & G. Mannocchi (eds.), Italian Physical Society, Editrice Compositori, Bologna, Italy **28**, 3.

Giovannelli, F.: 1994, *Space Sci. Rev.* **69**, 1.

Giovannelli, F.: 1996a, in *Multifrequency Behaviour of High Energy Cosmic Sources*, F. Giovannelli & L. Sabau-Graziati (eds.), *Mem. S.A.It.* **67**, 401.

Giovannelli, F.: 1996b, in *Multifrequency Behaviour of High Energy Cosmic Sources*, F. Giovannelli & L. Sabau-Graziati (eds.), *Mem. S.A.It.* **67**, 209.

Giovannelli, F.: 1996c, in *Multifrequency Behaviour of High Energy Cosmic Sources*, F. Giovannelli & L. Sabau-Graziati (eds.), *Mem. S.A.It.* **67**, 289.

Giovannelli, F.: 1997, in *Frontier Objects in Astrophysics and Particle Physics*, F. Giovannelli & G. Mannocchi (eds.), Italian Physical Society, Editrice Compositori, Bologna, Italy **57**, 337.

Giovannelli, F.: 2002, in *Chemical Enrichment of Intra-cluster and Intergalactic Medium*, R. Fusco-Femiano & F. Matteucci (eds.), *ASP Conf. Proc.* **253**, 251.

Giovannelli, F., Karakuła, S., Tkaczyk, W.: 1982a, *Astron. Astrophys.* **107**, 376.

Giovannelli, F., Karakuła, S., Tkaczyk, W.: 1982b, *Acta Astron.* **32**, 121.

Giovannelli, F., de Loore, C., Bartolini, C., Burger, M., Ferrari-Toniolo, M., Giangrande, A., *et al.*: 1982c, in *Proc. Third European IUE Conference*, E. Rolfe (ed.), ESA **SP-176**, 233.

Giovannelli, F., Gaudenzi, S., Piccioni, A., Rossi, C.: 1983, *Acta Astron.* **33**, 319.

Giovannelli, F., Karakuła, S., Tkaczyk, W.: 1983, *Astron. Astrophys.* **125**, 121.

Giovannelli, F., Ferrari-Toniolo, M., Persi, P., Bartolini, C., Guarnieri, A., Piccioni, A., *et al.*: 1984a, in *Proc. Fourth European IUE Conference*, E. Rolfe (ed.), ESA **SP-218**, 439.

Giovannelli, F., Bartolini, C., Candy, P., Gaudenzi, S., Guarnieri, A., Lombardi, R., *et al.*: 1984b, in *Proc. of the Fourth European IUE Conference*, E. Rolfe (ed.), ESA **SP-218**, 391.

Giovannelli, F., Bisnovatyi-Kogan, G.S., Golynskaya, I.M., Kurt, V.G., Lamzin, S.A., Mizyakina, T.A., *et al.*: 1984c, in *Proc. Fourth European IUE Conference*, E. Rolfe (ed.), ESA **SP-218**, 359.

Giovannelli, F., Ferrari-Toniolo, M., Persi, P., Golynskaya, I.M., Kurt, V.G., Mizyakina, T.A., *et al.*: 1985a, in *Multifrequency Behaviour of Galactic Accreting Sources*, F. Giovannelli (ed.), Edizioni Scientifiche SIDEREA, Roma, p. 284.

Giovannelli, F., Gaudenzi, S., Lombardi, R., Bartolini, C., Guarnieri, A., Piccioni, A.: 1985b, in *Multifrequency Behaviour of Galactic Accreting Sources*, F. Giovannelli (ed.), SIDEREA, Roma, Italy, p. 37.

Giovannelli, F., Karakuła, S., Tkaczyk, W.: 1985, in *Multifrequency Behaviour of Galactic Accreting Sources*, F. Giovannelli (ed.), SIDEREA, Roma, Italy, p. 185.

Giovannelli, F., Polcaro, V.F.: 1986, *Mon. Not. Roy. Astr. Soc.* **222**, 619.

Giovannelli, F., Mannocchi, G. (eds.): 1987, *HE-UHE Behaviour of Accreting X-Ray Sources*, Italian Physical Society, Editrice Compositori, Bologna, Italy, **8**.

Giovannelli, F., Rossi, C., Covino, E., Errico, L., Vittone, A.A., Bisnovatyi-Kogan, G.S., *et al.*: 1988, *A Decade of UV Astronomy with the IUE Satellite* ESA **SP-281, Vol. 2**, 101.

Giovannelli, F., Castaldo, D., Covino, E., Vittone, A.A., Rossi, C.: 1989, in *Proc. of the IAU Coll. N. 104 on Solar and Stellar Flares: Poster Papers*, B.M. Haisch & M. Rodonó (eds.), Kluwer Academic Publishers, Dordrecht, The Netherlands, p. 139.

Giovannelli, F., Mannocchi, G. (eds.): 1989, *Frontier Objects in Astrophysics and Particle Physics*, Italian Physical Society, Editrice Compositori, Bologna, Italy, **19**.

Giovannelli, F., Sheffer, E.K., Kopaeva, I.F., Averintsev, M.B., Bisnovatyi-Kogan, G.S., Golynskaya, I.M., *et al.*: 1989, in *Frontier Objects in Astrophysics and Particle Physics*, F. Giovannelli & G. Mannocchi (eds.), Italian Physical Society, Editrice Compositori, Bologna, Italy **19**, 197.

Giovannelli, F., Gonzalez Martinez-Pais, I., Gaudenzi, S., Lombardi, R., Claudi, R.U.: 1990a, *Astrophys. Space Sci.* **169**, 125.

Giovannelli, F., Burger, M., van Dessel. E.L., de Martino, D., Waters, R., Bartolini, C., *et al.*: 1990b, *Astrophys. Space Sci.* **169**, 139.

Giovannelli, F., Bednarek, W., Karakuła, S., Tkaczyk, W.: 1990c, in *Physical Processes in Hot Cosmic Plasmas*, W. Brinkmann, A.C. Fabian & F. Giovannelli (eds.), Kluwer Academic Publisher, p. 425.

Giovannelli, F., Ziółkowski, J.: 1990, *Acta astron.* **40**, 95.

Giovannelli, F., Mannocchi, G. (eds.): 1991, *Frontier Objects in Astrophysics and Particle Physics*, Italian Physical Society, Editrice Compositori, Bologna, Italy, **28**.

Giovannelli, F., Martinez-Pais, I.G.: 1991, *Space Sci. Rev.* **56**, p. 313.

Giovannelli, F., Errico, L., Vittone, A.A., Rossi, C.: 1991a, *Astron. Astrophys. Suppl. Ser.* **87**, 89.

Giovannelli, F., Errico, L., Vittone, A.A., Bisnovatyi-Kogan, G.S., Kurt, V.G., Lamzin, S.A., *et al.*: 1991b, *Mem. S.A.It.* **62**, 867.

Giovannelli, F., Rossi, C., Errico, L., Vittone, A.A., Bisnovatyi-Kogan, G.S., Kurt, V.G., *et al.*: 1991c, in *Angular Momentum Evolution of Young Stars*, S. Catalano & J.R. Stauffer (eds.), Kluwer Academic Publ., Dordrecht, The Netherlands, p. 97.

Giovannelli, F., Sabau-Graziati, L.: 1992, *Space Sci. Rev.* **59**, 1.

Giovannelli, F., Mannocchi, G. (eds.): 1993, *Frontier Objects in Astrophysics and Particle Physics*, Italian Physical Society, Editrice Compositori, Bologna, Italy, **40**.

Giovannelli, F., Sabau-Graziati, L.: 1993, in *Frontier Objects in Astrophysics and Particle Physics*, F. Giovannelli & G. Mannocchi (eds), Italian Physical Society, Editrice Compositori, Bologna, Italy, **40**, 15.

Giovannelli, F., Polcaro, V.F., Sabau Graziati, L.,Norci, L.: 1993a, in *Frontier Objects in Astrophysics and Particle Physics*, F. Giovannelli & G. Mannocchi (eds.), Italian Physical Society, Editrice Compositori, Bologna, Italy **40**, 323.

Giovannelli, F., Sabau-Graziati, L., La Padula, C., Errico, L., Frutti, M., Inarta, S., *et al.*: 1993b, *Astron. Astrophys. Suppl. Ser.* **97**, 395.

Giovannelli, F.,Sabau Graziati, L., Norci, L., Polcaro, V.F.: 1993c, *Proc. 23rd Intern. Cosmic Ray Conf. (Calgary)* **1**, 33.

Giovannelli, F., Sabau-Graziati, L.: 1994, in *Proc. 11th ESA Symp. on European Rocket and Balloon Programmes and Related Research*, ed. E. Rolfe, ESA **SP-335**, 263.

Giovannelli, F., Norci, L., Polcaro,V.F., Sabau Graziati, L.: 1994, in *Evolutionary Links in the Zoo of Interacting Binaries*, F. D'Antona, V. Caloi,C. Maceroni & F. Giovannelli (eds.), *Mem. S.A.It.* **65**, 373.

Giovannelli, F., Mannocchi, G. (eds.): 1995, *Frontier Objects in Astrophysics and Particle Physics*, Italian Physical Society, Editrice Compositori, Bologna, Italy, **47**.

Giovannelli, F., Sabau-Graziati, L.: 1995, in *Frontier Objects in Astrophysics and Particle Physics*, F. Giovannelli & G. Mannocchi (eds.), **47**, 15.

Giovannelli, F., Vittone, A.A., Rossi, C., Errico, L., Bisnovatyi-Kogan, G.S., Kurt, V.G., *et al.*: 1995, *Astron. Astrophys. Suppl. Ser.* **114**, 341.

Giovannelli, F., Bednarek, W., Karakuła, S.: 1996, *J. Phys. G: Nucl. Part. Phys.* **22**, 1223.

Giovannelli, F., Sabau-Graziati, L. (eds.): 1996a, *Multifrequency Behaviour of High Energy Cosmic Sources*, Mem. S.A.It. **67**.

Giovannelli, F., Sabau-Graziati, L.: 1996b, in *Multifrequency Behaviour of High Energy Cosmic Sources*, F. Giovannelli & L. Sabau-Graziati (eds.), *Mem. S.A.It.* **67**, 17.

Giovannelli, F., Karakuła, S., Tkaczyk, W.: 1996c, in *Multifrequency Behaviour of High Energy Cosmic Sources*, F. Giovannelli & L. Sabau-Graziati (eds.), *Mem. S.A.It.* **67**, 515.

Giovannelli, F., Mannocchi, G. (eds.): 1997, *Frontier Objects in Astrophysics and Particle Physics*, Italian Physical Society, Editrice Compositori, Bologna, Italy, **57**.

Giovannelli, F., Bednarek, W., Karakuła, S., Sabau-Graziati, L., Tkaczyk, W.: 1997, in *Frontier Objects in Astrophysics and Particle Physics*, F. Giovannelli & G. Mannocchi (eds.), Italian Physical Society, Editrice Compositori, Bologna, Italy **57**, 279.

Giovannelli, F., Sabau-Graziati, L.: 1997, in *Frontier Objects in Astrophysics and Particle Physics*, F. Giovannelli & G. Mannocchi (eds.), Italian Physical Society, Editrice Compositori, Bologna, Italy, **57**, 3.

Giovannelli, F., Sabau-Graziati, L.: 1998, in *Ultraviolet Astrophysics Beyond the IUE Final Archive*, W. Wamstecker & R.Gonzalez-Riestra (eds.), ESA **SP-413**, 419.

Giovannelli, F., Mannocchi, G. (eds.): 1999, *Frontier Objects in Astrophysics and Particle Physics*, Italian Physical Society, Editrice Compositori, Bologna, Italy, **65**.

Giovannelli, F., Sabau-Graziati, L. (eds.): 1999, *Multifrequency Behaviour of High Energy Cosmic Sources*, Mem. S.A.It. **70**.

Giovannelli, F., Sabau-Graziati, L.: 1999a, in *Multifrequency Behaviour of High Energy Cosmic Sources*, F. Giovannelli & L. Sabau-Graziati (eds.), *Mem. S.A.It.* **70**, 987.

Giovannelli, F., Sabau-Graziati, L.: 1999b, in *Multifrequency Behaviour of High Energy Cosmic Sources*, F. Giovannelli & L. Sabau-Graziati (eds.), *Mem. S.A.It.* **70**, 773.

Giovannelli, F., Bednarek, W., Polcaro, V.F., Sabau Graziati, L.: 1999, in *3rd INTEGRAL Conference Proc.*, A. Bazzano, G.C. Palumbo & C. Winkler (eds.), *Astro. Lett. & Communication* **38**, 399.

Giovannelli, F., Sabau-Graziati, L., La Padula, C., Isern, J., Bravo, E., Garcia-Berro, E., *et al.*: 1999a, in *Multifrequency Behaviour of High Energy Cosmic Sources*, Giovannelli, F., Sabau-Graziati, L. (eds.), *Mem. S.A.It.* **70**, 1317.

Giovannelli, F., Sabau-Graziati, L., La Padula, C., Isern, J., Bravo, E., Garcia-Berro, E., *et al.*: 1999b, in *Frontier Objects in Astrophysics and Particle Physics*, F. Giovannelli & G. Mannocchi (eds.), Italian Physical Society, Editrice Compositori Bologna, Italy **65**, 623.

Giovannelli, F., Sabau-Graziati, L.: 2000a, in *The Chemical Evolution of the Milky Way: Stars versus Clusters*, F. Matteucci & F. Giovannelli (eds.), Kluwer Publ. Co., ASSL **255**, 151.

Giovannelli, F., Sabau-Graziati, L.: 2000b, in *Hot Points in Astrophysics*, Joint Institute for Nuclear Research (JINR), Dubna, Russia, p. 224.

Giovannelli, F., Bednarek, W., Sabau-Graziati, L.: 2001, in *4th INTEGRAL Workshop on Exploring the Gamma Ray Universe*, A. Gímenez, V. Reglero & C. Winkler (eds.), **ESA SP-459**, p. 285.

Giovannelli, F., Mannocchi, G. (eds.): 2001, *Frontier Objects in Astrophysics and Particle Physics*, Italian Physical Society, Editrice Compositori, Bologna, Italy, **73**.

Giovannelli, F., Sabau-Graziati, L.: 2001, *Astrophys. Space Sci.* **276**, 67.

Giovannelli, F., Sabau-Graziati, L., Isern, J., La Padula, C.D., Angulo, M., Baussells, J., *et al.*: 2001, in *Frontier Objects in Astrophysics and Particle Physics*, F. Giovannelli & G. Mannocchi (eds.), Italian Physical Society, Editrice Compositori Bologna, Italy **73**, 565.

Giovannelli, F., Sabau-Graziati, L. (eds.): 2002a, *Multifrequency Behaviour of High Energy Cosmic Sources*, Mem. S.A.It. **73 N. 1**.

Giovannelli, F., Sabau-Graziati, L. (eds.): 2002b, *Multifrequency Behaviour of High Energy Cosmic Sources*, Mem. S.A.It. **73 N. 4**.

Giovannelli, F., Sabau-Graziati, L.: 2002c, in *Multifrequency Behaviour of High Energy Cosmic Sources*, F. Giovannelli & L. Sabau-Graziati (eds.), *Mem. S.A.It.* **73 N. 4**, 791.

Giovannelli, F., Sabau-Graziati, L., Isern, J., La Padula, C.D., Angúlo, M., Bravo, E., *et al.*: 2002, in *Multifrequency Behaviour of High Energy Cosmic Sources*, F. Giovannelli & L. Sabau-Graziati (eds.), *Mem. S.A.It.* **73, N. 1**, 402.

Giovannelli, F., Mannocchi, G. (eds.): 2003, *Frontier Objects in Astrophysics and Particle Physics*, Italian Physical Society, Editrice Compositori, Bologna, Italy **85**.

Glushkov, A.V., Efimov, N.N., Efremov, N.N., Makarov, I.T., Pravdin, M.I., Dedenko, L.I.: 1985, in *Proc. 19th Intern. Cosmic Ray Conf. (La Jolla)* **2**, 186.

Gnedin, Yu.N., Khozov, G.V., Larionov, V.M.: 1981, *Soviet Astron. Lett.* **7**, 256.

Gnedin, Yu.N., Pavlov, G.G., Apparao, K.M.V., Chitre, S.M.: 1981, *Astrophys. Space Sci.* **76 no. 1**, 105.

Gnedin, Yu.N., Khozov, G.V., Larionov, V.M.: 1983, *Astrophys. Space Sci.* **93**, 207.

Gnedin, Yu.N., Larionov, V.M., Lyutyi, V., Zaytzeva, G.: 1988, *Adv. Space Res.* **8, Ns. 2-3**, 157.

Gomez-Gomar, J., Isern, J., Bravo, E., Giovannelli, F., La Padula, C.D.: 1999, *Astrophys. Space Sci.* **263**, 389.

Gorenstein, P., Tucker, W.A.: 1976, *Ann. Rev. Astron. Astrophys.* **14**, 373.

Gould, R.J.: 1965, *Phys. Rev. Lett.* **12**, 511.

Gould, R.J., Schreder, G.: 1966, *Phys. Rev. Lett.* **16, Issue 6**, 252.

Granot, J., Panaitescu, A., Kumar, P., Woosleyy, S.E.: 2002, *Astrophys. J. Lett.* **570**, L61.

Granot, J., Sari, R.: 2002, *Astrophys. J.* **568**, 820.

Grasdalen, G.L., Strom, S.E., Strom, K.M., Capps, R.W., Thompson, D., Castelaz, M: 1984, *Astrophys. J. Lett.* **283**, L57.

Gratton, L.: 1990, *Unpublished Reflexions* (preprint).

Grebenev, S., Sunyaev, R., Pavlinsky, M., Churazov, E., Gilfanov, M., Dyachkov, A., *et al.*: 1993, *Astron. Astrophys. Suppl. Ser.* **97, N. 1**, 281.

Green, D.A.: 1984, *Mon. Not. Roy. Astr. Soc.* **209**, 449.

Green, D.A.: 2001, *SNR Catalog*, (http://www.mrao.cam.ac.uk/surveys/snrs)

Greiner, J.: 1999, in *Multifrequency Behaviour of High Energy Cosmic Sources*, eds. F. Giovannelli & L. Sabau-Graziati, *Mem. S.A.It.* **70**, 891.

Greiner, J.: 2002, *GRBs Table*, (http://www.mpe.mpg.de/~jgc/grbgen.html)

Greiner, J., Morgan, E.H., Remillard, R.A.: 1996, *Astrophys. J. Lett.* **473**, L107.

Grenier, I.A.: 2000, *Astron. Astrophys. Letter* **364**, L93.

Grenier, I.A.: 2001, in *The Nature of Unidentified Galactic Gamma Ray Sources*, A. Carraminana, O. Reimer & D. Thompson (eds.), Kluwer Academic publishers, Dordrecht, Holland, p. 51.

Grenier, I.A.: 2003, *astro-ph/0303498.*

de Grijp, M.H.K., Miley, G.K., Lub, J., de Jong, T.: 1985, *Nature* **314**, 240.

Grimm, H.-J., Gilfanov, M., Sunyaev, R.: 2002, *Astron. Astrophys.* **391**, 923.

Grimm, H.-J., Gilfanov, M., Sunyaev, R.: 2003, *Mon. Not. R. astr. Soc.* **339**, 793.

Grindlay, J.E.: 1978, *Astrophys. J.* **221**, 234.

Grindlay, J.E.: 1981, in *X-Ray Astronomy with the Einstein Satellite*, R. Giacconi (ed.), D. Reidel Publ. Co., Dordrecht, Holland, p. 79.

Grindlay, J.E.: 1994, *Astrophys. J. Suppl. Ser.* **92**, 465.

Grindlay, J.E., Helmken, H.F., Hambury-Brown, R., Davis, J, Allen, L.R.: 1975, *Astrophys. J. Lett.* **197**, L9.

Grindlay, J.E., Gursky, H., Schnopper, H., Parsignault, D., Hesie, J., Brinkman, A., *et al.*: 1976, *Astrophys. J. Lett.* **201**, L127.

Groot, P.J., Galama, T.J., van Paradijs, J., Strom, S., Telting, J., Rutten, R.G.M., *et al.*: 1997, *IAU Circ.* **N. 6584**.

Grove, J.E.: 1996, in *Multifrequency Behaviour of High Energy Cosmic Sources*, eds. F. Giovannelli & L. Sabau-Graziati, *Mem. S.A.It.* **67**, 127.

Gruber, D.A.: 1992, in *The X-Ray background*, X. Barcons, A.C. Fabian (eds.), Cambridge University Press, Cambridge, UK, p. 44.

Gruber, D.A., Matteson, J.L., Jung, G.V., Kinzer, R.L.: 1985, in *Proc. 19th Intern. Cosmic Ray Conf. (La Jolla)* **1**, 349.

Grün, E., Gustafson, B., Mann, I., Baguhl, M., Morfill, G.E., Staubach, P., *et al.*: 1994, *Astron. Astrophys.* **286**, 915.

Guainazzi, M., Parmar, A.N., Oosterbroek, T.: 2001, in *X-Ray Astronomy: Stellar Endpoints, AGN, and the Diffuse X-ray Background*, N.E. White, G. Malaguti & G.G.C. Palumbo (eds.), *AIP Conf. Proc.* **599**, 638.

Guarnieri, A., Bartolini, C., Piccioni, A., Giovannelli, F.: 1985, in *Multifrequency Behaviour of Galactic Accreting Sources*, F. Giovannelli (ed.), Edizioni Scientifiche SIDEREA, Roma, p. 310.

Guarnieri, A., Bartolini, C., Masetti, N., Piccioni, A., Costa, E., Feroci, M., *et al.*: 1997, *Astron. Astrophys. Lett.* **328**, L13.

Guidoboni, E., Marmo, C., Polcaro, V.F.: 1993, in *Frontier Objects in Astrophysics and Particle Physics*, eds. F. Giovannelli & G. Mannocchi, Italian Physical Society, Editrice Compositori, Bologna, Italy **40**, 289.

Guilbert, P.W., Rees, M.J.: 1988, *Mon. Not. Roy. Astr. Soc.* **233**, 475.

Guinan, E.F., Wacher, S.W., Baliunas, S.L., Loeser, J.G., Raymond, J.C.: 1986, in *New Insight in Astrophysics (8 Years of UV Astronomy with IUE)*, E. Rolfe (ed.), ESA **SP-263**, 197.

Gull, S.F.: 1973, *Mon. Not. Roy. Astr. Soc.* **161**, 47.

Gum, C.S., Kerr, F.J., Westerhout, G.: 1960, *Mon. Not. Roy. Astr. Soc.* **121**, 132.

Gunn, J.E., Ostriker, J.P.: 1970, *Astrophys. J.* **160**, 979.

Gunn, J.E., Gott, J.R.: 1972, *Astrophys. J.* **176**, 1.

Gursky, H., Kellog, E.M., Leong, C., Tananbaum, H., Giacconi, R.: 1971a, *Astrophys. J. Lett.* **165**, L43.

Gursky, H., Kellog, E.M., Murray, S., Leong, C., Tananbaum, H., Giacconi, R.: 1971b, *Astrophys. J. Lett.* **167**, L81.

Gursky, H., Ruffini, R. (eds.): 1975, *Neutron Stars, Black Holes and Binary X-Ray Sources*, D. Reidel Publ. Co., Dordrecht, Holland.

Gursky, H., Ruffini, R., Stella, L. (eds.): 2000, *Exploring the Universe: A Festschrift in Honor of Riccardo Giacconi, Advanced Series in Astrophysics and Cosmology*, Series Editors: F. Li Zhi & R. Ruffini **Vol. 13**, World Scientific, Singapore, New Jersey, London, Hong Kong.

Haas, M., Lemke, D., Stickel, M., Hippelein, H., Kunkel, M., Herbstmeier, U., *et al.*: 1999, *The Universe as Seen by ISO*, P. Cox & Kessler, M.F. (eds.), ESA **SP-427, Vol. 2**, 885.

Hack, M., Hutchings, J.B., Kondo, Y., McCluskei, G.E., Plavec, M., Polidan, R.S.: 1975, *Astrophys. J.* **198**, 453.

Hack, M., la Dous, C.: 1993, in *Cataclysmic Variables and Related Objects*, M. Hack & C. la Dous (eds.), NASA **SP-507**, 1.

Haisch, B.M.: 1983, in *Activity in Red Dwarf Stars*, P.B.Byrne & M. Rodono (eds.), D. Reidel Publ. Co., Dordrecht, Holland, p. 255.

Haisch, B.M., Rodonó, M. (eds.): 1989, Proc. of the IAU Coll. N. 104 on *Solar and Stellar Flares*, Kluwer Academic Publishers, Dordrecht, The Netherlands.

Haisch, B.M., Strong, K.T., Rodonó, M.: 1991, *Ann. Rev. Astron. Astrophys.* **29**, 275.

Halpern, J.J.P., Holt, S.S.: 1992, *Nature* **357**, 222.

Halpern, J.P., Oke, J.B.: 1987, *Astrophys. J.* **312**, 91.

Hamann, F., Simon, M., Ridgway, S.T.: 1988, *Astrophys. J.* **326**, 859.

Hamann, F., Persson, S.E.: 1992a, *Astrophys. J. Suppl. Ser.* **82**, 247.

Hamann, F., Persson, S.E.: 1992b, *Astrophys. J.* **394**, 628.

Hameury, J.M., King, A.R., Lasota, J.P.: 1986, *Astron. Astrophys.* **162**, 71.

Harding, A.K.: 1981, *Astrophys. J.* **245**, 267.

Harding, A.K.: 1991, in *Frontier Objects in Astrophysics and Particle Physics*, F. Giovannelli & G. Mannocchi (eds.), Italian Physical Society, Editrice Compositori, Bologna, Italy **28**, 243.

Harding, A.K.: 1994, in *The Evolution of X-Ray Binaries*, S.S. Holt and C.S. Day (eds.), *AIP Conf. Proc.*, **308**, 429.

Harding, A.K., Stecker, F.W.: 1981, *Nature* **290**, 316.

Harris, S., Clegg, P., Hughes, J.: 1988, *Mon. Not. Roy. Astr. Soc.* **235**, 441.

Hartig, G.F., Davidsen, A.F.: 1978, *Bull. Amer. Astron. Soc.* **10**, 402.

Hartle, J.B., Thorne, K.S.: 1968, *Astrophys. J.* **153**, 807.

Hartman, R.C., Kniffen, D.A., Thompson, D.J., Fichtel, C.E., Ögelman, H.B., Tümer, T., *et al.*: 1979, *Astrophys. J.* **230**, 597.

Hartman, R.C., Bertsch, D.L., Fichtel, C.E., Hunter, S.D., Kanbach, G., Kniffen, D.A., *et al.*: 1992, in *The Compton Observatory Science Workshop*, C.R. Shrader, N. Gehrels & B. Dennis (eds.), p. 116.

Hartman, R.C., Bertsch. D.L., Bloom, S.D., Chen, A.W., Deines-Jones, P., Esposito, J.A., *et al.*: 1999, *Astrophys. J. Suppl. Ser.* **123**, 79.

Hartmann, L., Baliunas, S.L., Duncan, D.K., Noyes, R,V.: 1984, *Astrophys. J.* **279**, 778.

Hartmann, L., Kenyon, S.J.: 1987a, *Astrophys. J.* **312**, 243.

Hartmann, L., Kenyon, S.J.: 1987b, *Astrophys. J.* **322**, 393.

Hartquist, T., Havnes, O., Morfill, G.E.: 1993, *Fundamentals Cosm. Phys.* **15**, 107.

Hasinger, G.: 1994, in *Proc. of the IAU Symp. N. 161 on Astronomy from Wide Field Imaging*, Mac Gillivray (ed.).

Hasinger, G.: 1995, in *Seventeeth Texas Symposium on Relativistic Astrophysics and Cosmology*, H. Böhringer, G.E. Morfill & J.E. Trümper (eds.), *Annals of the New York Academy of Sciences* **759**, 200.

Hasinger, G.: 1996, *Astron. Astrophys. Suppl. Ser.* **120**, 607.

Hasinger, G., van der Klis, M.: 1989, *Astron. Astrophys.* **225**, 79.

Hasinger, G., Burg, R., Giacconi, R., Hartner, G., Schmidt, M., Trümper, J., *et al.*: 1993, *Astron. Astrophys.* **275**, 1.

Hasinger, G., Lehmann, I., Giacconi, R., Schmidt, M., Trümper, J., Zamorani, G.: 1999, in *Highlights in X-ray Astronomy: International Symposium in honour of Joachim Trümper's 65th birthday*, B. Aschenbach & M.J. Freyberg (eds.), Max-Planck-Institut für extraterrestrische Physik, Garching, Germany, *MPE report* **No. 272**, 199.

Hasinger, G., Miyaji, T., Schmidt, J.H.M.M.: 2000, *MPE-Report 1999* **273**, 83.

Hassall, B.J.M., Pringle, J.E., Schwarzenberg-Czerny, A., Wade, R.A., Whelan, J.A.J., Hill, P.W.: 1983, *Mon. Not. Roy Astr. Soc.* **203**, 865.

Hassall, B.J.M., Pringle, J.E., Verbunt, F.: 1985, *Mon. Not. Roy. Astr. Soc.* **216**, 353.

Hauser, M.G., Gillet, F.C., Low, F.J., Gautier, T.N., Beichman, C.A., Neugebauer, G., *et al.*: 1984, *Astrophys. J. Lett.* **278**, L15.

Hauser, M.G., Arendt, R.G., Kelsall, T., Dwek, E., Odegard, N., Weiland, J.L., *et al.*: 1998, *Astrophys. J.* **508**, 25.

Havnes, O., Morfill, G.E., Melandso, F.: 1992, *Icarus* **98**, 141.

Hawarden, T.G., Fairclough, J.H., Joseph, R.D., Leggett, S.K., Mountain, C.M.: 1986, in *Light on Dark Matter*, F.P. Israel (ed.), D. Reidel Publ. Co., Dordrecht, Holland, ASSL **124**, 455.

Hayakawa, S.: 1952, *Prog. Theor. Phys.* **8**, 571.

Hayashida, N., Honda, K., Honda, M., Imaizumi, S., Inoue, N., Kadota, K., *et al.*: 1994, *Phys. Rev. Lett.* **73**, 3491.

Hazard, C., Mitton, S.: 1979, *Active Galactic Nuclei*, Cambridge University Press, UK.

Heap, S.R.: 1979, in *Mass Loss and Evolution of O-Type Stars*, P.S. Conti & C.W.H. de Loore (eds.), D. Reidel Publ. Co., Dordrecht, Holland, p. 99.

Heise, J., Mewe, R., Brinkman, A.C., Gronenschild, E.H.B.M., den Boggende, A.J.F., Schrijver, J., *et al.*: 1978, *Astron. Astrophys. Lett.* **63**, L1.

Helfand, D.J.: 1998, in *The Relationship between Neutron Stars and Supernova Remnants*, R. Bandiera, E. Masini, F. Pacini, M. Salvati & L. Woltjer (eds.), *Mem. S.A.It.* **69 N. 4**, 791.

Helfand, D.J., Caillault, J.-P.: 1982, *Astrophys. J.* **253**, 760.

Helou, G.: 1986a, in *Star-Forming Dwarf Galaxies and Related Objects*, D. Kunth & T.X. Thuan (eds.), Ed. Front. Transl. by J. Thanh Van, Gif sur Yvette, France, p. 319.

Helou, G.: 1986b, *Astrophys. J. Lett.* **311**, L33.

Henrichs, H.F.: 1983, in *Accretion-Driven Stellar X-Ray Sources*, Cambridge University Press, p. 393.

Henry, R. C.: 1991, *Ann. Rev. Astron. Astrophys.* **29**, 89.

Henry, R.C.: 1999a, in *Multifrequency Behaviour of High Energy Cosmic Sources*, F. Giovannelli & L. Sabau-Graziati (eds.), *Mem. S.A.It.* **70**, 825.

Henry, R.C.: 1999b, *Astrophys. J. Lett.* **516**, L49.

Henry, R. C.: 1999c, in *The Low Surface Brightness Universe*, IAU Colloquium N. 171, J. I. Davies, C. Impey, & S. Phillips (eds.), *ASP Conf. Ser.* **170**, 357.

Henry, R.C.: 2002, in *Multifrequency Behaviour of High Energy Cosmic Sources*, F. Giovannelli & L. Sabau-Graziati (eds.), *Mem. S.A.It.* **73 N. 1**, 67.

Herbig, G.H.: 1962, *Ann. Rev. Astron. Astrophys.* **1**, 47.

Herbig, G.H.: 1970, *Mem. Soc. Roy. Sci. Liège Ser.* **5 9**, 13.

Herbig, G.H.: 1977, *Astrophys. J.* **214**, 747.

Herbig, G.H.: 1989, in *Low Mass Star Formation and Pre-Main Sequence Objects*, Bo Reipurth (ed.), ESO Conference and Workshop Proceedings **33**, 233.

Herman, J., Burger, J.H., Penninx, W.H.: 1986, *Astron. Astrophys.* **167**, 247.

Hermsen, W., Blömen, J.B.G.M.: 1983, in *Survey of the Southern Galaxy*, W.B. Burton & F.P. Israel (eds.), D. Reidel Publ. Co., Dordrecht, Holland, p. 65.

Hertz, P., Wood, K.S.: 1985, *Astrophys. J.* **290**, 171.

Herzberg, G.: 1981, *J. Spectrosc. Soc. Japan* **30**, 407.

Hess, V.F.: 1912, *Physik Zh.* **13**, 1084.

van den Heuvel, E.P.J.: 1976, in *Structure and Evolution of Close Binary Systems*, P. Eggleton (ed.), D. Reidel Publ. Co., Dordrecht, Holland, p. 35.
van den Heuvel, E.P.J., Rappaport, S.: 1987, in *Physics of Be Stars*, A. Slettebak & T.P. Snow (eds.), Cambridge and New York, Cambridge University Press, Proc. of the IAU Coll. N. 92, p. 291.
Hewish, A., Bell, S.J., Pilkington, J.D.H., Scott, P.F., Collins, R.A.: 1968, *Nature* **217**, 709.
Higdon, J.C., Lingenfelter, R.E.: 1976, *Astrophys. J. Lett.* **208**, L107.
Hildebrand, R.H., Spillar, E.J., Stiening, R.F.: 1981, *Astrophys. J.* **243**, 223.
Hillas, A.M.: 1984, in *High Energy Astrophysics*, 19th Rencontre de Moriond, J. Audouze, J. Tran Thanh Van (eds.), Editions Frontières, Gif-sur-Yvette, France, p. 11.
Hillas, A.M.: 1999, in *Frontier Objects in Astrophysics and Particle Physics*, F. Giovannelli & G. Mannocchi (eds.), Italian Physical Society, Editrice Compositori, Bologna, Italy **65**, 391.
Hillas, A.M., Johnson, A.P.: 1990, *Proc. 21st Intern. Cosmic Ray Conf. (Adelaide)* **4**, 19.
Hjellming, R.M.: 1996, in *Accretion Phenomena and Related Outflows*, D.T. Wickramasinghe, G.V. Bicknell & L. Ferrario (eds.), *ASP Conf. Ser.* **212**, 53.
Hjellming, R.M., Wade, C.M.: 1971a, *Astrophys. J. Lett.* **164**, L1.
Hjellming, R.M., Wade, C.M.: 1971b, *Astrophys. J. Lett.* **168**, L21.
Hjellming, R.M., Johnston, K.J.: 1988, *Astrophys. J.* **328**, 600.
Hjellming, R.M., Han, X., Roussel-Dupre, D.: 1990, *IAU Circ.* **N. 5112**.
Hjellming, R.M., Rupen, M.P., Hunstead, R.W., Campbell-Wilson, D., Mioduszewski, A.J., Gaensler, B.M., *et al.*: 2000, *Astrophys. J.* **544**, 977.
Hjorth, J., Fynbo, J., Dar, A., Courbiin, F., Moller, P.: 1999, *GCN Circ.* **N. 403**.
Hodapp, K.-W.: 1984, *Astron. Astrophys.* **141**, 255.
Holberg, J.B.: 1986, *Astrophys. J.* **311**, 969.
Holberg, J.B.: 1990, in *The Galactic and Extragalactic Background Radiation*, S. Bowyer & C. Leinert (eds.), Proc. IAU 139, Kluwer Academic Publ., p. 220.
Holland, S., Fynbo, J.P.U., Hjorth, J., Gorosabel, J., Pedersen, H., Andersen, M.I., *et al.*: 2001, *Astron. Astrophys.* **371**, 52.
Holt, S.S.: 1980, *Vistas in Astronomy* **24**, 301.
Holt, S.S., Boldt, E.A., Schwartz, D.A., Serlemitsos, P.J., Bleach, R.D.: 1971, *Astrophys. J. Lett.* **166**, L65.
Holt, S.S., Day, C.S. (eds.): 1993, *The Evolution of X-ray Binaries*, *AIP Conf. Proc.* **308**.
Honey, W.B., Bath, G.T., Charles, P.A., Whitehurst, R., Jonnes, D.H.P., Echevarria, J., *et al.*: 1989, *Mon. Not. Roy. Astr. Soc.* **236**, 727.
Horne, K.: 1994, in *Theory of Accretion Disks*, W.J. Duschl, J. Frank, F. Meyer, E. Meyer-Hofmeister & W.M. Tscharnuter (eds.), *NATO ASI Ser. C* **417**, 77.
Horne, K., Gomer, R.: 1980, *Astrophys. J.* **237**, 845.
Horne, K., Marsh, T.R., Cheng, F.H., Hubeny, I., Lanz, T.: 1994, *Astrophys. J.* **426**, 294.
Hoyle, F., Burbidge, G.: 1992, *Astrophys. J. Lett.* **399**, L9.
Hu, W., Peebles, P.J.E.: 2000 *Astrophys. J. Lett.* **528**, L61.
Huchra, J. P., Geller, M.J., Gallagher, J., Hunter, D., Hartmann, L., Fabbiano, G., *et al.*: 1983, *Astrophys. J.* **274**, 125.
van der Hucht, K.A.: 2001, *New Astron. Rev* **45**, 135.
van der Hucht, K.A., Conti, P.S., Lundstrom, I., Stenholm, B.: 1981, *Space Sci. Rev.* **28**, 227.
van der Hucht, K.A., Olnon, F.M.: 1985, *Astron. Astrophys. Lett.* **149**, L17.
Hudson, H.S., Bai, T., Gruber, D.E., Matteson, J.L., Nolan, P.L., Peterson, L.E.: 1980, *Astrophys. J. Lett.* **236**, L91.
Hulleman, F., van Kerkwijk, M.H., Kulkarni, S.R.: 2000, *Nature* **408**, 689.
Humason, M.L., Mayall, N.U., Sandage, A.: 1956, *Astron. J.* **61**, 97.
Hunter, D., Rice, W., Gallagher, J. Gillet, F.: 1987, in *Star Formation in Galaxies*, C. Persson (ed.), NASA, Washington, DC, US Govt. Print. Off., p. 253.

Hunter, S.D., Kinzer, R.L., Strong, A.W.: 1997, in *Proceedings Fourth Compton Symposium*, C.D. Dermer, M.S. Strickman & J.K. Kurfess (eds.), *AIP Conf. Proc.* **410**, 192.

Hurley, K.: 1994, *Nature* **372**, 652.

Hurley, K.: 2000a, in *Proc. 5th Huntsville GRB Symposium*, R.M. Kippen, R.S. Mallozzi & G.J. Fishman (eds.), *AIP Conf. Ser.* **526**, 763.

Hurley, K.: 2000b, in *The Fifth Compton Symposium*, M. L. McConnell & J.M. Ryan, *AIP Conf. Proc.* **510**, 515.

Hurley, K.: 2001, in *Gamma Ray Bursts in the Afterglow Era*, E. Costa, F. Frontera & J. Hjorth (eds.), Springer Verlag, Heidelberg, Germany, p. 3.

Hurley, K.: 2003, private communication.

Hurwitz, M., Bowyer, S.: 1988, *A Decade of UV Astronomy with the IUE Satellite* ESA **SP-281, Vol. 2**, 329.

Hurwitz, M., Bowyer, S.: 1995, *Adv. Space Res.* **16, N. 3**, 61.

van der Hucht, K.A., Olnon, F.M.: 1985, *Astron. Astrophys. Lett.* **149**, L17.

Hutchings, J.B.: 1980, *Astrophys. J.* **237**, 285.

Hutchings, J.B., Lequeux, J., Wolf, B: 1987, in Exploring the Universe with the IUE Satellite, Y. Kondo (ed.), D. Reidel Publ. Co., Dordrecht, Holland, p. 605.

Hutchinson, G.W.: 1952, *Phil. Mag.* **43**, 847.

Iben, I.Jr., Renzini, A.: 1983 *Ann. Rev. Astron. Astrophys.* **21**, 271.

Ikebe, Y.: 1995, *Ph.D. Thesis*, Department of Physics, University of Tokyo, Japan.

Ikebe, Y., Fukazawa, Y., Ishisaki, Y., Matsushita, K., Makishima, K., Kikuchi, K., *et al.*: 1996a, in *UV and X-ray Spectroscopy of Astrophysical and Laboratory Plasmas*, K. Yamashita & T. Watanabe (eds.), Universal Academy Press, Frontiers Science Series **15**, 371.

Ikebe, Y., Ezawa, H., Fukazawa, Y., Hirayama, M., Izhisaki, Y., Kikuchi, K., *et al.*: 1996b, *Nature* **379**, 427.

Ikebe, Y., Makishima, K., Fukazawa, Y., Tamura, T., Xu, H., Ohashi, T., *et al.*: 1999, in *Diffuse Thermal and Relativistic Plasma in Galaxy Clusters*, Böhringer, H., Feretti, L., Schuecker, P. (eds.), MPE-Report **271B**, 139.

Illarionov, A.F., Sunyaev, R.A.: 1975, *Astron. Astrophys.* **39**, 195.

Imhoff, C.L., Giamppa, M.S.: 1980, *Astrophys. J. Lett.* **239**, L115.

Imhoff, C.L., Appenzeller, Y.: 1987, in Exploring the Universe with the IUE Satellite, Y. Kondo (ed.), D. Reidel Publ. Co., Dordrecht, Holland, p. 295.

Infante, L., Garnavich, P.M., Stanek, K.Z., Wyrzykowski, L.: 2001, *GRB Coordinate Network* **1152**, 1.

Innes, D.E., Hartquist, T.W.: 1984, *Mon. Not. Roy. Astr. Soc.* **209**, 7.

Inoue, H., Koyama, K., Matsuoka, M., Ohashi, T., Tanaka, Y.: 1979, *Astrophys. J. Lett.* **227**, L85.

Inoue, H., Kii, T., Ogasaka, Y., Takahashi, T., Ueda, Y.: 1997, in *X-Ray Imaging and Spectroscopy of Cosmic Hot Plasmas*, F. Makino and K. Mitsuda (eds.), Universal Academy Press, Inc., *Frontier Science Series* **19**, 3.

Isern, J., Hernanz, M., García-Berro, E.: 1993, in *White Dwarfs: Advances in Observation and Theory*, M.A. Barstow (ed.), Kluwer Academic Publ., Dordrecht, Holland, *NATO ASI Ser.* **C403**, 139.

Isern, J., Hernanz, M., Abia, C., Jose', J.: 1997, in *Frontier Objects in Astrophysics and Particle Physics*, F. Giovannelli & G. Mannocchi (eds.), Italian Physical Society, Editrice Compositori, Bologna, Italy **57**, 113.

Isern, J., Giovannelli, F., Sabau, L., Bravo, E., Gomez-Gomar, J., Hernanz, M., *et al.*: 2001, *Astrophys. Space Sci.* **276**, 39.

Ishii, T. (for the K2K Collaboration): 2002, in *Le Rencontres de Physique de la Vallée d'Aoste: Results and Perspectives in Particle Physics*, M. Greco (ed.), *Frascati Physics Series* **XXVII**, 85.

Israel, G.L., Oosterbroek, T., Angelini, L., Campana, S., Mereghetti, S., Parmar, A.N., *et al.*: 1999, *Astron. Astrophys.* **346**, 929.

Israel, G.L., Oosterbroek, T., Stella, L., Campana, S. Mereghetti, S., Parmar, A.N.: 2001, *Astrophys. J. Lett.* **560**, L65.

Israel, G.L., Mereghetti, S., Stella, L.: 2002, in Soft Gamma Ray Repeaters, M. Feroci & S. Mereghetti (eds.), Mem. S.A.It. **73**, 465.

Jakobsen, P.: 1980, *Astron. Astrophys.* **81**, 66.

Jakobsen, P., Bowyer, S., Kimble, R., Jelinsky, P., Grewing, M., Krämer, G., *et al.*: 1984, *Astron. Astrophys.* **139**, 481.

Jakobsen, P., Albrecht, R., Barbieri, C., Blades, J.C., Boksenberg, A., Crane, P., *et al.*: Deharveng, J.M., Disney, M.J., 1991, *Astrophys. J. Lett.* **369**, L63.

Jakobsen, P., Macchetto, F., Panagia, N.: 1993, *Astrophys. J.* **403**, 736.

Jansen, F.A.: 1988, *Astrophys. J.* **331**, 949.

Jansen, F., Lumb, D., Altieri, B., Clavel, J., Ehle, M., Erd, C., *et al.*: 2001, *Astron. Astrophys. Lett.* **365**, L1.

Jaschek, M., Jaschek, C., Egret, D.: 1986, *Astron. Astrophys.* **158**, 325.

Jenkins, E.B.: 1978a, *Comments Astrophys.* **7**, 121.

Jenkins, E.B.: 1978b, *Astrophys. J.* **220**, 107.

Jenkins, E.B., Meloy, D.A.: 1974, *Astrophys. J. Lett.* **193**, L121.

Jenkins, E.B., Wallestein, G., Silk, J.: 1984, *Astrophys. J.* **278**, 649.

Jimenez, R., Band, D., Piran, T.: 2001, *Astrophys. J.* **561**, 171.

Johnson, W.N., Kinzer, R.L., Kurfess, J.D., Strickman, M.S., Purcell, W.R., Grabelssky, D.A., *et al.*: 1993, *Astrophys. J. Suppl. Ser.* **86**, 693.

Johnston, S., Walker, M.A., Bailes, M. (eds.): 1996, *Pulsars: Problems and progress*, IAU Coll. N. 160, Astron. Soc. Pacific Conf. Ser. **105**.

Jones, C., Forman, W.: 1978, *Astrophys. J.* **224**, 1.

Jones, C., Forman, W.: 1984, *Astrophys. J.* **276**, 38.

Jones, T.W., Ney, E.P., Stein, W.A.: 1981, *Astrophys. J.* **250**, 324.

Joss, P.C.: 1978, *Astrophys. J. Lett.* **255**, L123.

Joss, P.C., Rappaport, S.A.: 1984, *Ann. Rev. Astron. Astrophys.* **22**, 537.

Joubert, M., Masnou, J.L., Lequeux, J., Deharveng, J.M., Cruvellier, P.: 1983, *Astron. Astrophys.* **128**, 114.

Joy, A.H.: 1942, *Pub. Astron. Soc. Pacific* **54**, 15.

Joy, A.H.: 1945, *Astrophys. J.* **102**, 168.

Joy, A.H.: 1949, *Astrophys. J.* **110**, 424.

Joy, A.H.: 1956, *Astrophys. J.* **124**, 317.

Jura, M.A.: 1977, *Astrophys. J.* **212**, 634.

Jura, M.A.: 1986, *Astrophys. J.* **306**, 483.

Jura, M.A.: 1987, *Pub. Astron. Soc. Pacific* **99**,1123.

Jura, M., Kim, D.W., Knapp, G.R., Guhathakurta, P.: 1987, *Astrophys. J. Lett.* **312**, L11.

Kaastra, J.S., Ferrigno, C., Tamura, T., Paerels, F.B.S., Peterson, J.R., Mittaz, J.P.D.: 2001, *Astron. Astrophys. Lett.* **365**, L99.

Kaluzienski, L.J., Holt, S.S., Boldt, E.A., Serlemitsos, P.J.: 1976, *Astrophys. J. Lett.* **208**, L71.

Kamae, T., Oshugi, T., Thompson, D.J., Watanabe, K.: 2000, *Adv. Space Res.* **25**, 905.

Kappadath, S.C., Ryan, J., Bennet, K., Blömen, H., Forrest, D., Hermsen, W., *et al.*: 1996, *Astron. Astrophys. Suppl. Ser.* **120**, 619.

Karakuła, S.: 1997, in *Frontier Objects in Astrophysics and Particle Physics*, F. Giovannelli & G. Mannocchi (eds.), Italian Physical Society, Editrice Compositori, Bologna, Italy **57**, 355.

Karakuła, S., Tkaczyk, W., Giovannelli, F.: 1984, *Adv. Space Res.* **3 N. 10-1**, 335.

Karakuła, S., Tkaczyk, W.: 1985, in *Multifrequency Behaviour of Galactic Accreting Sources*, F. Giovannelli (ed.), Edizioni Scientifiche SIDEREA, Roma, p. 243.

Kaspi, V.M.: 1998, *Adv. Space Res.* **21**, 167.

Kaufmman Bernadó, M.M., Romero, G.E., Mirabel, I.F.: 2002, *Astron. Astrophys. Letter* **385**, L10.

Kawada, M., Bock, J.J., Hristov, V.V., Lange, A.E., Matsuhara, H., Matsumoto, T., Matsuura, S., Mauskopf, P.D., Richards, P.L., Tanaka, M.: 1994, *Astrophys. J. Lett.* **425**, L89.

Kawasaki, M., Sato, K.: 1986, *Phys. Lett.* **169B**, 280.

Kellerman, K.I.: 1993, *Nature* **361**, p. 134.

Kellerman, K.I., Shaffer, D.B., Clark, B.G., Geldzahler, B.J.: 1977, *Astrophys. J. Lett.* **214**, L61.

Kellog, E.M., Gursky, H., Tananbaum, H., Giacconi, R., Pounds, K.: 1972, *Astrophys. J. Lett.* **174**, L65.

Kellog, E.M., Murray, S.: 1974, *Astrophys. J. Lett.* **193**, L57.

Kellog, E.M., Baldwin, J.R., Koch, D.: 1975, *Astrophys. J.* **199**, 299.

Kenyon, S.J., Hartmann, L.: 1987, *Astrophys. J.* **323**, 714.

Kerr, F.J., Hindman, J.V., Gum, C.S.: 1959, *Aust. J. Phys.* **12**, 270.

Kerrick, A.D.: 1994, *IAU Circ.* **N. 5996**.

Kerrick, A.D., Akerlof, C.W., Biller, S.D., Buckley, J.H., Cawley, M.F., Chantell, M., *et al.*: 1995, *Astrophys. J. Lett.* **438**, L59.

Khokhlov, A., Melia, F.: 1996, *Astrophys. J. Lett.* **457**, L61.

Kidger, M.R., Beckman, J.E.: 1986, *Astron. Astrophys. Lett.* **164**, L25.

Kikuchi, K., Ohashi, T., Yamasaki, N., Ikebe, Y., Ishisaki, Y., Fukazawa, Y., *et al.*: 1996, in *UV and X-ray Spectroscopy of Astrophysical and Laboratory Plasmas*, K. Yamashita & T. Watanabe (eds.), Universal Academy Press, Frontiers Science Series **15**, 391.

Kim, K.T., Tribble, P.C., Kronberg, P.P.: 1991, *Astrophys. J.* **379**, 80.

Kiplinger, A.L.: 1979, *Astron. J.* **84**, 655.

Kirk, J.G.: 1997, in *Relativistic Jets in AGNs*, M. Ostrowski, M. Sikora, G. Madejski & M. Begelman (eds.), p. 145.

Klebesadel, R.W., Strong, I.B., Olson, R.A.: 1973, *Astrophys. J. Lett.* **182**, L85.

Kleinmann, S.G., Keel, W.C.: 1987, in *Star Formation in Galaxies*, C. Persson (ed.), NASA, Washington, DC, US Govt. Print. Off., p. 559.

van der Klis, M.: 1989, *Ann. Rev. Astron. Astrophys.* **27**, 517.

van der Klis, M.: 1995, in *Flares and Flashes*, J. Greiner, H.W. Duerbeck & R.E. Gershberg (eds.), *Lecture Notes in Phys.* **454**, 321.

van der Klis, M.: 1996, in *Compact Stars in Binaries*, J. van Paradijs, E.P.J. van den Heuvel & E. Kuulkers (eds.), Kluwer Academic Publ., Dordrecht, Holland, p. 301.

van der Klis, M.: 1997, in *Accretion Phenomena and Related Outflows*, D.T. Wickramasinghe, G.V. Bicknell & L. Ferrario (eds.), IAU Coll. N. 163, *ASP Conf. Ser.* **121**, 347.

van der Klis, M.: 2000, *Ann. Rev. Astron. Astrophys.* **38**, 717.

van der Klis, M.: 2001, *Astrophys. J.* **561**, 943.

van der Klis, M., Swank, J.H., Zhang, W., Jahoda, K., Morgan, E.H., Lewin, W.H.G., *et al.*: 1996, *Astrophys. J. Lett.* **469**, L1.

van der Klis, M., Wijnands, R.A.D., Horne, K., Chen, W.: 1997, *Astrophys. J. Lett.* **481**, L97.

Knapp, G.R.: 1990, in *The Interstellar Medium in Galaxies*, Thronson, H.A.Jr. & Shull, J.M. (eds.), Dordrecht, Netherlands, Kluwer Academic Publ., ASSL **161**, 3.

Kniffen, D.A.: 1993, *American Scientist* **81, N. 4**, 342.

Kniffen, D.A., Hartman, R.C., Thompson, D.J., Bignami, G.F., Fichtel, C.E., Tümer, T., *et al.*: 1974, *Nature* **251**, 397.

Kniffen, D.A., Bertsch, D.L., Dingus, B.L., Esposito, J.A., Fichtel, C.E., Hartman, R.C., *et al.*: 1996, *Astron. Astrophys. Suppl. Ser.* **120**, 615.

Knigge, C., Long, K.S., Blair, W.P., Wade, R.A.: 1997, *Astrophys. J.* **476**, 291.

Koch, R.H., Corcoran, M.F., Holenstein, B.D., McCluskey, G.E.: 1986, *Astrophys. J.* **306**, 618.

Kolb, E.W.: 1996, in *Results and Perspectives in Particle Physics*, M. Greco (ed.), Frascati Physics Series, INFN Laboratori Nazionali di Frascati **5**, 43.

Kohmura, Y., Fukazawa, Y., Ikebe, Y., Ishisaki, Y., Kamijo, S., Kaneda, H., *et al.*: 1993, in *EUV, X-Ray, and Gamma Ray Instrumentation for Astronomy IV*, Siegmund, O.H. (ed.), *SPIE* **2006**, 78.

Kondo, Y. (ed.): 1987, *Exploring the Universe with the IUE Satellite*, D. Reidel Publ. Co., Dordrecht, Holland.

Kondo, Y., Boggess, A., Maran, S.P.: 1989, *Ann. Rev. Astron. Astrophys.* **27**, 397.

Kornilov, V.G., Lipunov, V.M.: 1984, *Astron. Zhurnal* **61**, 686.

Koyama, K., Takano, S., Tawara, Y.,: 1991, *Nature* **350**, 135.

Koyama, K., Petre, R., Gotthelf, E.V., Hwang, U., Matsuura, M., Ozaki, M., *et al.*: 1995, *Nature* **378**, 255.

Koyama, K., Maeda, Y., Sonobe, T., Takeshima, T., Tanaka, Y., Yamauchi, S.: 1996, *Pub. Astron. Soc. Japan* **48**, 249.

Koyama, K., Kinugasa, K., Matsuzaki, K., Nishiuchi, M., Sugizaki, M., Torii, K., *et al.*: 1997, *Pub. Astron. Soc. Japan Lett.* **49**, L7.

Kozlovsky, B., Lingenfelter, R., Ramaty, R: 1987, *Astron. Astrophys.* **316**, 801.

Krabbe, A., Genzel, R., Drapatz, S., Rotaciuc, V.: 1991, *Astrophys. J. Lett.* **382**, L19.

Krabbe, A., Genzel, R., Eckart, A., Najarro, F., Lutz, D., Cameron, M., *et al.*: 1995, *Astrophys. J. Letter* **447**, L95.

Kraft, R.P.: 1967, *Astrophys. J.* **150**, 551.

Krennrich, F.: 1999, in *Frontier Objects in Astrophysics and Particle Physics*, F. Giovannelli, G. Mannocchi (eds.), Italian Physical Society, Editrice Compositori, Bologna, Italy **65**, 235.

Krichbaum, T.P., Witzel, A., Booth, R.S., Kus, A.J., Ronnang, B.O., Witzel, A., *et al.*: 1990, *Astron. Astrophys.* **237**, 3.

Kriss, G.A., Canizares, C.R., McClintock, J.E., Feigelson, E.D.: 1980, *Astrophys. J. Lett.* **235**, L61.

Kriss, G.A., Canizares, C.R., McClintock, J.E., Feigelson, E.D.: 1981, *Astrophys. J. Lett.* **245**, L51 (erratum for above).

Kriss, G.A., Cioffi, D.F., Canizares, C.R.: 1983, *Astrophys. J.* **272**, 439.

Krzeminski, W.: 1974a, *Mem. S.A.It.* **45**, 877.

Krzeminski, W.: 1974b, *Astrophys. J. Lett.* **192**, L135.

Ku, W., Chanan, G.: 1979, *Astrophys. J. Lett.* **323**, L59.

Kuhi, L.V., Cram, L.E.: 1989, in *FGK Stars and T Tauri Stars*, L.E. Cram & L.V. Kuhi (eds.), NASA **SP-502**, 99.

Kulikov, G.V., Khristiansen, G.B.: 1958, *JETP* **35**, 635.

Kulkarni, S.R., Narayan, R., Romani, R.W.: 1990, *Astrophys. J.* **356**, 174.

Kulkarni, S.R., Djorgovski, S.G., Ramaprakash, A.N., Goodrich, R., Bloom, J.S., Adelberger, K.L., *et al.*: 1998, *Nature* **393**, 35.

Kulkarni, S.R., Ramaprakash, A.N., Bloom, J., Djorgovski, S., Goodrich, R., Frail, D.: 1998, *The Astronomer's Telegram* **5**, 1.

Kulkarni, S.R., Berger, E., Bloom, J.S., Chaffee, F., Diercks, A., Djorgovski, S.G., *et al.*: 2001, in *Explosive Phenomena in Astrophysical Compact Objects*, Heon-Young Chang & Chang-Hwan Lee, Mannque Rho, & Insu Yi (eds.), *AIP Conf. Proc.* **556**, 240.

Kundt, W. (ed.): 1990, *Neutron Stars: Their Birth, Evolution, Radiation and Winds*, Kluwer Academic Publ., Dordrecht, The Netherlands.

Kundt, W.: 1996, in *Multifrequency Behaviour of High Energy Cosmic Sources*, F. Giovannelli & L. Sabau-Graziati (eds.), *Mem. S.A.It.* **67**, 439.

Kundt, W.: 1999, in *Frontier Objects in Astrophysics and Particle Physics*, F. Giovannelli & G. Mannocchi (eds.), Italian Physical Society, Editrice Compositori, Bologna, Italy **65**, 303.

Kundt, W.: 2001, in *Frontier Objects in Astrophysics and Particle Physics*, F. Giovannelli & G. Mannocchi (eds.), Italian Physical Society, Editrice Compositori, Bologna, Italy **73**, 301.

Kundt, W.: 2002, in *Multifrequency Behaviour of High Energy Cosmic Sources*, F. Giovannelli & L. Sabau-Graziati (eds.), *Mem. S.A.It.* **73**, 346.

Kunth, D., Sevre, F.: 1986, in *Star-Forming Dwarf Galaxies and Related Objects*, D. Kunth & T.X. Thuan (eds.), Ed. Front. Transl. by J. Thanh Van, Gif sur Yvette, France, p. 331.

Kunth, D., Weedman, D.: 1987, in Exploring the Universe with the IUE Satellite, Y. Kondo (ed.), D. Reidel Publ. Co., Dordrecht, Holland, p. 623.

Kurucz, R.L.: 1979, *Astrophys. J. Suppl. Ser.* **40**, 1.

Kuulkers, E., Wijnands, R., Belloni, T., Mendez, M., van der Klis, M., van Paradijs, J.: 1998a, *Astrophys. J.* **494**, 753.

Kuulkers, E., Wijnands, R., Belloni, T., Mendez, M., van der Klis, M., van Paradijs, J.: 1998b, in *The Active X-Ray Sky: Results from BeppoSAX and RXTE*, L. Scarsi, H. Bradt, P. Giommi, & F. Fiore (eds.), *Nucl. Phys. B, Proc. Suppl.* **69 Ns. 1-3**, 324.

Kuulkers, E., Fender, R.P., Spencer, R.E., Davis, R.J., Morison, I.: 1999, *Mon. Not. Roy. Astr. Soc.* **306**, 919.

Lacy, J.H., Townes, C.H., Geballe, T.R., Hollenbach, D.J.: 1980, *Astrophys. J.* **262**, 120.

Lada, C.J.: 1999, in *The Origin of Stars and Planetary Systems*, Charles J. Lada & N.D. Kylafis (eds.), Kluwer Academic Publ., p.143.

Lada, C.J., Wilking, B.A.: 1984, *Astrophys. J.* **287**, 610.

Lada, C.J., Lada, E.A.: 1991, in *The formation and Evolution of Stars Clusters*, K. Janes (ed.), *ASP Conf. Ser.* **13**, 3.

Lagache, G., Abergel, A., Boulanger, F., Désert, F.X., Puget, J.-L.: 1999, *Astron. Astrophys.* **344**, 322.

Lagage, P.O., Cesarsky, C.J.: 1983, *Astron. Astrophys.* **118**, 223.

Lamb, D.Q.: 2000, *astro-ph/0005028.*

Lamb, R.C.: 1996, *Space Sci. Rev.* **75**, 53.

Lamb, R.C..: 1997, in *Frontier Objects in Astrophysics and Particle Physics*, F. Giovannelli, G. Mannocchi (eds.), Italian Physical Society, Editrice Compositori, Bologna, Italy **57**, 247.

Lamers, H.J.G.L.M.: 1980, in *The Universe at Ultraviolet Wavelengths*, R.D. Chapman (ed.), NASA **CP-2171**, 93.

Lamers, H.J.G.L.M., Waters, L.B.F.M., Wesselius, P.R.: 1984, *Astron. Astrophys. Lett.* **134**, L17.

Lamzin, S.A.: 1989, *Soviet Astron.* **33**, 686.

Lamzin, S.A., Giovannelli, F., Rossi, C., Errico, L., Vittone, A.A., Bisnovatyi-Kogan, G.S., *et al.*: 1993, in *Stellar Jets and Bipolar Outflows*, L. Errico & A.A. Vittone (eds.), Kluwer Academy Publisher, Dordrecht, Holland, p. 115.

Lamzin, S.A., Bisnovatyi-Kogan, G.S., Errico, L., Giovannelli, F., Katysheva, N.A., Rossi, C., *et al.*: 1996, *Astron. Astrophys.* **306**, 877.

Lang, K.R.: 1992, *Astrophysical Data (Planet and Stars)*, (Berlin: Springer), pp. 350-61.

Lasota, J.-P.: 1995, in *Cape Workshop on Magnetic Cataclysmic Variables*, D.A.H. Buckley & B. Warner (eds.), *ASP Conf. Ser.* **85**, 68.

Lawrence, A., Walker, D., Rowan-Robinson, M., Leech, K.J., Penston, M.V.: 1986, *Mon. Not. Roy. Astr. Soc.* **219**, 687.

Lawrence, M.A., Reid, R.J.I., Watson, A.A.: 1991, *J. Phys. G.* **17**, 733.

Lea, S.M., De Young, D.S.: 1976, *Astrophys. J.* **210**, 647.

Leach, R.W., Ruffini, R.: 1973, *Astrophys. J. Lett.* **180**, L15.

Lebrun, F., Goldoni, P., Goldwurm, A., Laurent, P., Limousin, O., Paul, J.: 1999a, in *Proceedings Third Integral Workshop: The Extreme Universe*, A. Bazzano, G.G.C. Palumbo & C. Winkler (eds.), Astro. Lett. And Communications **38**, 457.

Lebrun, F., Ubertini, P., Di Cocco, G., Bassani, L., Bazzano, A., Bird, A.J., *et al.*: 1999b, in *Multi-frequency Behaviour of High Energy Cosmic Sources*, F. Giovannelli & L. Sabau-Graziati (eds.), *Mem. S.A.It.* **70**, 1299.

Leir, A.A., van den Bergh, S.: 1977, *Astrophys. J. Suppl.* **34**, 381.

Lena, P.: 1988, *Observational Astrophysics*, Springer–Verlag Berlin–Heidelberg, Germany.

Lepp, S., McCray, R.: 1983, *Astrophys. J.* **269**, 560.

Lequeux, J., Maucherat-Joubert, M., Deharveng, J.M., Kunth, D.: 1981, *Astron. Astrophys.* **103**, 305.
Leventhal, M., MacCallun, C.J., Stang, P.D.: 1978, *Astrophys. J. Lett.* **225**, L11.
Levin, S., Bensadoun, M., Bersanelli, M., De Amici, G., Kogut, A., Limon, *et al.*: 1992, *Astrophys. J.* **396**, 3.
Lewin, W.: 1980, in *Globular Clusters*, D. Hanes & B. Madore (eds.), Cambridge University Press, UK, Chapt. 18.
Lewin, W., Clark, G.: 1980, *Annals N.Y. Acad. Sci.* **336**, 451.
Lewin, W.H.G., van den Heuvel, E.P.J. (eds.): 1983, *Accretion Driven Stellar X-Ray Sources*, Cambridge University Press, UK.
Li, H., Miller, J.A., Colgate, S.A.: 1997, in *Relativistic Jets in AGNs*, M. Ostrowski, M. Sikora, G. Madejski & M. Begelman (eds.), p. 162.
Lichti, G.G., Balonek, T., Courvoisier, T.J.-L., Johnson, N., McConnell, M., McNamara, B., *et al.*: 1995, *Astron. Astrophys.* **298**, 711.
Lightman, A.P., White, T.R.: 1988, *Astrophys. J.* **335**, 57.
Lindegren, L.: 1995, *Astron. Astrophys.* **304**, 61.
Lindegren, L.: 1997, in *Hipparcos Venice '97*, B. Battrick (ed.), ESA Publication Division, ESTEC, Nordwijk, The Netherlands, ESA **SP-402**, 13.
Lindegren, L., Kovalevsky, J.: 1995, *Astron. Astrophys.* **304**, 189.
Linsky, J.L., Haisch, B.M.: 1979, *Astrophys. J. Lett.* **229**, L27.
Lipari, P.: 1997, in *Frontier Objects in Astrophysics and Particle Physics*, F. Giovannelli, G. Mannocchi (eds.), Italian Physical Society, Editrice Compositori, Bologna, Italy, **57**, 595.
Lipunov, V.M.: 1987, *Astrophys. Space Sci.* **132**, 1.
Lipunov, V.M.: 1989, *In the World of Binary Stars*, MIR Publishers, Moscow, USSR.
Lipunov, V.M.: 1995, in *Frontier Objects in Astrophysics and Particle Physics*, F. Giovannelli & G. Mannocchi (eds.), Italian Physical Society, Bologna, Italy **47**, 61.
Lipunov, V.M.: 1998, in *The Relationship Between Neutron Stars and Supernova Remnants*, R. Bandiera, E. Masini, F. Pacini, M. Salvati & L. Woltjer (eds.), *Mem. S.A.It.* **69 N. 4**, 1067.
Li Ti Pei, Mr., Wolfendale, A. W.: 1981, *Astron. Astrophys.* **103**, 19.
Liu, Q.Z., van Paradijs, J., van den Heuvel, E.P.J.: 2000, *Astron. Astrophys. Suppl. Ser.* **147**, 25.
Liu, Q.Z., van Paradijs, J., van den Heuvel, E.P.J.: 2001, *Astron. Astrophys.* **368**, 1021.
Lombardi, R., Giovannelli, F., Gaudenzi, S.: 1987, *Astrophys. Space Sci.* **130**, 275.
Long, K., White, R.L.: 1980, *Astrophys. J. Lett.* **239**, L65.
Long, K.S., Mauche, C.W., Raymond, J.C., Szkody, P., Mattei, J.A.: 1996, *Astrophys. J.* **469**, 841.
Longair, M.S.: 1981, *High Energy Astrophysics*, Oxford University Press, Oxford, UK, pp.16-58.
Longair, M.S.: 1991, *Adv. Space Res.* **11, N. 2**, 161.
de Loore, C., Giovannelli, F., van Dessel, E.L., Bartolini, C., Burger, M., Ferrari-Toniolo, M., *et al.*: 1984, *Astron. Astrophys.* **141**, 279.
Lorenz, E.: 1999, in *Frontier Objects in Astrophysics and Particle Physics*, F. Giovannelli, G. Mannocchi (eds.), Italian Physical Society, Editrice Compositori, Bologna, Italy **65**, 173.
Low, F.J., Beintema, D.A., Gautier, T.N., Gillet, F.C., Beichman, C.A., Neugebauer, G., *et al.*: 1984, *Astrophys. J. Lett.* **278**, L19.
Loewenstein, M., Mushotzky, R.F.: 1996, *Astrophys. J.* **466**, 695.
Lund,N., Westergaard, N.J., Budtz-Jørgensen, C.: 1999, in *Multifrequency Behaviour of High Energy Cosmic Sources*, Giovannelli, F. & Sabau-Graziati, L. (eds.), *Mem. S.A.It.* **70**, 1303.
Lynden-Bell, D.: 1969, *Nature* **233**, 690.
Lyndel-Bell, D., Rees, M.J.: 1971, *Mon. Not. Roy. Astr. Sooc.* **152**, 461.
MacMinn, D., Primack, J.R.: 1996, *Space Sci. Rev.* **75**, 413.
Macomb, D.J., Akerlof, C.W., Aller, H.D., Aller, M.F., Bertsch, D.L., Bruhweiler, F., *et al.*: 1995, *Astrophys. J. Lett.* **449**, L99.
Macomb, D.J., Akerlof, C.W., Aller, H.D., Aller, M.F., Bertsch, D.L., Bruhweiler, F., *et al.*: 1996, *Astrophys. J. Lett.* **459**, L111.

Madau, P.: 1998, in *The Young Universe: Galaxy Formation and Evolution at Intermediate and High Redshift*, S. D'Odorico, A. Fontana & E. Giallongo (eds.), *ASP Conf. Ser.* **146**, 289.
Madau, P., Ghisellini, G., Fabian, A.C.: 1994, *Mon. Not. Roy. Astr. Soc.* **270**, L17.
Maeda, Y., Koyama, K., Sakano, M., Takeshima, T., Yamauchi, S.: 1996, *Pub. Astron. Soc. Japan* **48**, 417.
Maeder, A.: 1983, *Astron. Astrophys.* **120**, 113.
Magazzú, A., Rebolo, R.: 1989, *Mem. S.A.It.* **60**, 105.
Mahoney, W.A., Ling, J.C., Wheaton, WM, A., Jacobson, A.S.: 1984, *Astrophys. J.* **286**, 578.
Makino, F., Ohashi, T. (eds.): 1994, *New Horizon of X-Ray Astronomy*, Frontier Science Series N. 12, Universal Academy Press, Tokyo, Japan.
Makino, F., Fink, H.H., Clavel, J.: 1992, in Frontiers of X-ray Astronomy, Y. Tanaka & K. Koyama (eds.), Universal Academic Press, Tokyo, Japan, p. 543.
Makino, F., Mitsuda, K. (eds.): 1997, *X-Ray Imaging and Spectroscopy of Cosmic Hot Plasmas*, Frontier Science Series N. 19, Universal Academy Press, Tokyo, Japan.
Makishima, K.: 1995, in *Dark Matter*, Holt, S.S. & Bennett, C.L. (eds.), *AIP Conf. Proc.* **336**, 172.
Makishima, K.: 1996, in *UV and X-Ray Spectroscopy of Astrophysical and Laboratory Plasmas*, K. Yamashita, T. Watanabe (eds.), Universal Academy Press, Inc., Tokyo, Japan, p. 167.
Malina, F.J.: 1946, *Army Ordnance* **31, N. 157**, 45.
Malkan, M.A.: 1983, *Astrophys. J.* **268**, 582.
Malkan, M.A., Sargent, W.L.W.: 1982, *Astrophys. J.* **254**, 22.
Malkan, M.A., Alloin, D., Shore, S.: 1987, in Exploring the Universe with the IUE Satellite, Y. Kondo (ed.), D. Reidel Publ. Co., Dordrecht, Holland, p. 655.
Malumuth, E.M., Kriss, G.A.: 1986, *Astrophys. J.* **308**, 10.
Manchanda, R.K., Polcaro, V.F., Norci, L., Giovannelli, F., Brinkmann, W., Radecke, H.D., *et al.*: 1996, *Astron. Astrophys.* **305**, 457.
Mandolesi, N., Calzolari, P., Cortiglioni, S., Morigi, G., Danese, L., De Zotti, G.: 1986, *Astrophys. J.* **310**, 561.
Mannheim, K.: 1993, *Phys. Rev. D* **48**, 2408.
Maran, S.P. (ed.): 1992, *The Astronomy and Astrophysics Encyclopedia*, Cambridge University Press, Cambridge, UK.
Maran, S.P., Aller, L.H., Gull, T.R., Stecher, T.P.: 1982, *Astrophys. J. Lett.* **253**, L43.
Maran, S.P., Kinney, A.N.: 1993, *Publ. Astron. Soc Pacific* **105**, 447.
Maraschi, L., Ghisellini, G., Celotti, A.: 1992, *Astrophys. J. Lett.* **397**, L5.
Marscher, A.P.: 1980, *Astrophys. J.* **235**, 386.
Marscher, A.P.: 1993, in *Sub-arcsecond Radio Astronomy*, R.J. Davis & R.S. Booth (eds.), Cambridge University Press, p. 297.
Marsden, P.L.: 1985, *The Observatory* **105**, 7.
Marshall, N., Warwick, R.S., Pounds, K.A.: 1981, *Mon. Not. Roy. Astr. Soc.* **194**, 987.
Martin, C.: 1986, *Ph.D. Thesis*, University of California, Berkeley, USA.
Martin, C., Bowyer, S.: 1989, *Astrophys. J.* **338**, 677.
Martin, C., Hurwitz, M., Bowyer, S.: 1991, *Astrophys. J.* **379**, 549.
Martinez-Pais, I.G., Giovannelli, F., Rossi, C.,Gaudenzi, S.: 1994, *Astron. Astrophys.* **291**, 455.
Martinez-Pais, I.G., Giovannelli, F., Gaudenzi, S,Rossi, C.: 1996, *Astron. Astrophys.* **308**, 833.
de Martino, D.: 1995a, in *Cataclysmic Variables*, A. Bianchini, M. della Valle & M. Orio (eds.), Kluwer Academic Publ. *ASSL* **205**, 397.
de Martino, D.: 1995b, in *Cape Workshop on Magnetic Cataclysmic Variables*, D.A.H. Buckley & B. Warner (eds.), *ASP Conf. Proc.* **85**, 238.
de Martino, D.: 1998, in *Ultraviolet Astrophysics Beyond the IUE Final Archive*, W. Wamstecker & R.Gonzalez-Riestra (eds.), ESA **SP-413**, 387.
de Martino, D.: 1999, in *UV Astronomy in Italy*, 1999, L. Buson & D. de Martino (eds.), Mem. Soc. Astron. It. **70**, 547.

de Martino, D., Waters, L.B.M.F., Giovannelli, F., Persi, P.: 1990a, in *Two-Topics in X-Ray Astronomy*, E. Rolfe (ed.), ESA **SP-296**, 519.

de Martino, D. Waters, L.B.F.M., Giovannelli, F., Bartolini, C., Guarnieri, A., Piccioni, A.: 1990b, in *Two-Topics in X-Ray Astronomy*, E. Rolfe (ed.), ESA **SP-296**, 521.

Masetti, N., Frontera, F., Stella, L., Orlandini, M., Parmar, A.N., Del Sordo, S., *et al.*: 2000, *Astron. Astrophys.* **363**, 188.

Maslennikov, K.L.: 1986, *Soviet Astron. Lett.* **12**, 191.

Masnou, J.L., Bennett, K., Bignami, G.F., Blömen, J.B.G.M., Buccheri, R., Caraveo, P.A., *et al.*: 1981, *Proc. 17th Intern. Cosmic Ray Conference (Paris)* **1**, 177.

Mason, K.O., Watson, M.G., White, N.E. (eds.): 1986, *The Physics of Accretion onto Compact Objects*, Lecture Notes in Physics **266**.

Massey, P.M., Hutchings, J.B., Bianchi, L.: 1985, *Astron. J.* **90**, 2239.

Massey, P., Conti, P.S.: 1983, *Astrophys. J.* **273**, 576.

Massey, P., Hunter, D.A.: 1998, *Astrophys. J.* **493**, 180.

Mather, J.C., Cheng, E.S., Epler Jr, R.E., Isaacman, R.B., Meyer, S.S., Shafer, R.A., *et al.*: 1990, *Astrophys. J. Lett.* **354**, L37.

Mather, J.C., Hauser, M.G., Bennet, C.L., Bogges, N.W., Cheng, E.S., Eplee Jr., R.E., *et al.*: 1991, *Adv. Space Res.* **11, N. 2**, 181.

Mather, J.C., Cheng, E.S., Cottingham, D.A., Eplee, R.E.Jr., Fixsen, D.J., Hewagama, T., *et al.*: 1994, *Astrophys. J.* **420**, 439.

Mathews, W.G., Baker, J.C.: 1971, *Astrophys. J.* **170**, 241.

Mathewson, D.S., Ford, V.L., Dopita, M.A., Tuohy, I.R., Long, K.S., Helfand, D.J.: 1983a, *Astrophys. J. Suppl. Ser.* **51**, 345.

Mathewson, D.S., Ford, V.L., Dopita, M.A., Tuohy, I.R., Long, K.S., Helfand, D.J.: 1983b, in *Supernova Remnants and Their X-Ray Emission*, D. Reidel Publ. Co., Dordrecht, Holland, p. 541.

Mathewson, D.S., Ford, V.L., Dopita, M.A., Tuohy, I.R., Mills, B.Y., Turtle, A.J.: 1984, *Astrophys. J. Suppl. Ser.* **55**, 189.

Matsumoto, H., Koyama, K., Awaki, H., Tomida, H., Tsuru, T., Mushotzky, R., *et al.*: 1996 *Publ. Astron. Soc. Japan* **48**, 201.

Matsumoto, T., Hayakawa, S., Matswo, H., Murakami, H., Sato, S., Lange, A.E., *et al.*: 1988a, *Astrophys. J. Lett.* **317**, L45.

Matsumoto, T., Akiba, M., Murakami, H.: 1988b, *Astrophys. J.* **332**, 575.

Matsuoka, M., Ikegami, T., Inoue, H., Koyama, K.: 1987, *Publ. Astr. Soc. Japan* **38**, 285.

Matsuoka, M., Piro, L., Yamauchi, M., Murakami, T.: 1990, *Astrophys. J.* **361**, 440.

Matsuoka, M., Cappi, M.: 1996, in *Multifrequency Behaviour of High Energy Cosmic Sources*, F. Giovannelli & L. Sabau-Graziati (eds.), *Mem. S.A.It.* **67**, 493.

Matsushita, K., Awaki, H., Fukazawa, Y., Makishima, K., Masai, K., Matsumoto, H., *et al.*: 1996, in *UV and X-ray Spectroscopy of Astrophysical and Laboratory Plasmas*, K. Yamashita & T. Watanabe (eds.), Universal Academy Press, Frontiers Science Series **15**, 335.

Matsuura, S., Kawada, M., Matsuhara, H., Matsumoto, T., Noda, M., Tanaka, M., *et al.*: 1994, *Publ. Astron. Soc. Pacific* **106**, 770.

Matt, G.: 1995, in *Frontier Objects in Astrophysics and Particle Physics*, F. Giovannelli & G. Mannocchi (eds.), Italian Physical Society, Editrice Compositori, Bologna, Italy **47**, 163.

Matt, G., Fabian, A.C.: 1994, *Mon Not. R. astron. Soc.* **267**, 187.

Matteucci, F., Giovannelli, F. (eds.): 2000, *The Evolution of the Milky Way: Stars versus Clusters*, ASSL, Kluwer Academic Publishers, Dordrecht, The Netherlands.

Matthews, H.E., Goss, W.M., Winnberg, A., Habing, H.J.: 1973, *Astron. Astrophys.* **29**, 309.

Matting, W.: 1958, *Astron. Nachr.* **284**, 109.

Matting, W.: 1959, *Astron. Nachr.* **285**, 1.

Mattox, J.R.: 1997, *GCN* **N. 4**.

Mattox, J.R., Hartman, R.C., Reimer, O.: 2001, *Astrophys. J. Suppl. Ser.* **135**, 155.

Mauche, C.W.: 1996, in *Cataclysmic Variables and Related Objects*, A.Evans & J.H. Wood (eds.), IAU Coll. N. 158, Kluwer Academic Publishers, p. 243.

Mauche, C.W.: 1998, in *Wild Stars in the Old West* S. Howell, E. Kuulkers & C. Woodward (eds.), *ASP Conf. Ser.* **137**, 113.

Mauche, C.W., Wade, R.A., Polidan, R.S., van der Woerd, H., Paerels, F.: 1991, *Astrophys. J.* **372**, 659.

Mauche, C.W., Raymond, J.C., Mattei, J.A.: 1995, *Astrophys. J.* **446**, 842.

Maucherat-Joubert, M., Deharveng, J.M., Cruvellier, P.: 1980, *Astron. Astrophys.* **88**, 323.

Mayer-Hasselwander, H.A., Kanbach, G., Bennet, K., Lichti, G.G., Bignami, G.F., Caraveo, P.A., *et al.*: 1982, *Astron. Astrophys.* **105**, 164.

Mazets, E.P., Golenetskii, S.V., Aptekar, R.L., Gurian, Iu.A., Ilinskii, V.N.: 1981, *Nature* **290**, 378.

Mazets, E.P., Golenetskii, S.V.: 1988, *Sov. Sci. Rev. E. Astrophys. Space Phys.* **6**, 283.

Mazurek, T.J., Wheeler, J.C.: 1980, *Fund. Cosmic Phys.* **5**, 193.

McCammon, D., Sanders, W.T.: 1990, *Ann. Rev. Astron. Astrophys.* **28**, 657.

McCammon, D., Burrows, D.N., Sanders, W.T., Kraushaar, W.L.: 1983, *Astrophys. J.* **269**, 107.

McCluskey, G.E., Sahade, J: 1987, in Exploring the Universe with the IUE Satellite, Y. Kondo (ed.), D. Reidel Publ. Co., Dordrecht, Holland, p. 427.

McCluskey, G.E., Kondo, Y., Olson, E.C.: 1988, *Astrophys. J.* **332**, 1019.

McConnell, M.L., Owens, A., Chupp, E.L., Dunphy, P.P., Forrest, D.J., Vestrand, W.T.: 1987, 20th Intern. Cosmic Ray Conf. (Moscow) **1**, 58.

McKee, C.F., Van Buren, D., Larareff, B.: 1984, *Astrophys. J. Lett.* **278**, L115.

McKellar, A.: 1940, *Pub. Soc. Astron. Soc. Pacific* **52**, 187.

McKellar, A.: 1941, *Pub. Soc. Astron. Soc. Pacific* **53**, 233.

McNamara, D.H., Madsen, J.B., Barnes, J., Ericksen, B.F.: 2000, *Pub. Soc. Astron. Soc. Pacific* **112**, 202.

Meekins, J.F., Gilbert, F., Chubb, T.A., Friedman, H., Henry, R.C.: 1971, *Nature* **231**, 107.

Melnick, J., Sargent, W.L.: 1977, *Astrophys. J.* **215**, 401.

Menard, F., Bastien, P.: 1992, *Astron. J.* **103**, 564.

Méndez, M., van der Klis, M., van Paradijs, J., Lewin, W.H.G., Vaughan, B.A., Kuulkers, E., *et al.*: 1998a, *Astrophys. J. Lett.* **494**, L65.

Méndez, M., van der Klis, M., Wijnands, R., Ford, E.C., van Paradijis, J., Vaughan, B.A.: 1998b, *Astrophys. J. Lett.* **505**, L23.

Méndez, M., van der Klis, M.: 1999, *Astrophys. J. Lett.* **517**, L51.

Mendoza, E.E.: 1966, *Astrophys. J.* **143**, 1010.

Mendoza, E.E.: 1968, *Astrophys. J.* **151**, 977.

Mereghetti, S.: 1998, in *The Relationship between Neutron Stars and Supernova Remnants*, R. Bandiera, E. Masini, F. Pacini, M. Salvati & L. Woltjer (eds.), *Mem. S.A.It.* **69 N. 4**, 819.

Mereghetti, S.: 2001a, in *The Neutron Star – Black Hole Connection*, C. Kouveliotou, J. Ventura & E. van den Heuvel (eds.), Kluwer Academic Publ., Dordrecht, Holland, *NATO ASI Ser.* **C567**, 351.

Mereghetti, S.: 2001b, in *Frontier Objects in Astrophysics and Particle Physics*, F. Giovannelli & G. Mannocchi (eds.), Italian Physical Society, Editrice Compositori, Bologna, Italy **73**, 239.

Mereghetti, S., Stella, L.: 1995, *Astrophys. J. Lett.* **442**, L17.

Meszaros, P.: 1999a, *astro-ph/ 9912546.*

Meszaros, P.: 1999b, in *Cosmic Explosions: Tenth Astrophysics Conference*, S.S. Holt & W.W. Zhang (eds.), *AIP Conf. Proc.* **522**, 213.

Meszaros, P.: 2002, *Ann. Rev. Astron. Astrophys.* **40**, 137.

Metzger, A.E., Trombka, J.I., Reedy, R.C., Arnold, J.R.: 1974, in *Proc. Lunar Sci. Conf. 5th*, Pergamon Press, New York, USA, p. 1067.

Metzger, M.R., Djorgovski, S.G., Kulkarni, S.R., Steidel, C.C., Adelberger, K.L., Frail, D.A., *et al.*: 1997, *Nature* **387**, 878.
Meyer, J.P.: 1981, *17th Intern. Cosmic Ray Conf. (Paris)* **12**, 265.
Meyer, J.P.: 1985, *Astrophys. J. Suppl. Ser.* **57**, 151.
Meyer, S.S., Cheng, E.S., Page, L.A.: 1991, *Astrophys. J. Lett.* **371**, L7.
Meynet, G.: 1994, *Astrophys. J. Suppl. Ser.* **92**, 441.
Meynet, G., Maeder, A., Schaller, G., Schaerer, D., & Charbonnel, C. 1994, *Astron. Astrophys. Suppl. Ser.* **103**, 97.
Mezger, P.G., Duschl, W.J., Zylka, R.: 1996, *Astron. Astrophys. Rev.* **7**, 289.
Mezger, P.G., Zylka, R., Salter, C.J., Wink, J.E., Chini, R., Kreysa, E., *et al.*: 1989, *Astron. Astrophys.* **209**, 337.
Middleditch, J., Nelson, J.: 1976, *Astrophys. J.* **208**, 567.
Mignard, F.: 1997, in *Hipparcos Venice '97*, B. Battrick (ed.), ESA Publication Division, ESTEC, Nordwijk, The Netherlands, ESA **SP-402**, 5.
Miley, G., Neugebauer, G., Soifer, B.T., Clegg, P.E., Harris, S., Rowan-Robinson, M., *et al.*: 1984, *Astrophys. J. Lett.* **278**, L79.
Miley, G.K., Neugebauer, G., Soifer, B.T.: 1985, *Astrophys. J. Lett.* **293**, L11.
Milgrom, M.: 1986, *Astrophys. J.* **302**, 617.
Milne, P.A., Kurfess, J.D., Kinzer, R.L., Leising, M.D., Dixon, D.D.: 2000, *The Fifth Compton Symposium*, M.L. McConnell & J.M. Ryan (eds.), *AIP Conf. Proc.* **510**, 21.
Mirabel, I.F., Rodriguez, L.F.: 1994, *Nature* **371**, 46.
Mirabel, I.F., Rodriguez, L.F.: 1998, *Nature* **392**, 673.
Mirabel, I.F., Rodriguez, L.F.: 1999, *Ann. Rev. Astron. Astrophys.* **37**, 409.
Miralda-Escudé, J., Gould, A.: 2000, *Astrophys. J.* **545**, 847.
Mitchell, R.J., Cuhlane, J.L., Davison, P.J., Ives, J.C.: 1976, *Mon. Not. Roy. Astr. Soc.* **176**, 29p.
Mitchell, R.J., Ives, J.C., Cuhlane, J.L.: 1977, *Mon. Not. Roy. Astr. Soc.* **178**, 25p.
Mitrofanov, I.G.: 1996, in *Multifrequency Behaviour of High Energy Cosmic Sources*, F. Giovannelli & L. Sabau-Graziati (eds.), *Mem. S.A.It.* **67**, 417.
Mitsuda, K., Inoue, H., Koyama, K., Makishima, K., Matsuoka, M., Ogawara, Y., Suzuki, K., *et al.*: 1984, *Pub. Astron. Soc. Japan* **36**, 741.
Mittaz, J.P.D., Kaastra, J.S., Tamura, T., Fabian, A.C., Mushotzky, R.F., Peterson, J.R., *et al.*: 2001, *Astron. Astrophys. Lett.* **365**, L93.
Miyaji, T., Hasinger, G., Schmidt, M.: 2000, *Astron. Astrophys.* **353**, 25.
Modisette, J.L., Kondo, Y.: 1980, *Astrophys. J.* **240**, 180.
Moffat, A.F.J., Marchenko, S.V.: 1996, *Astron. Astrophys. Lett.* **305**, L29.
von Montigny, C., Bertsch, D.L., Chiang, J., Dingus, B.L., Esposito, J.A., Fichtel, C.E., *et al.*: 1995, *Astrophys. J.* **440**, 525.
Montmerle, T.: 1992, in *Solar and Stellar Coronae*, R. Pallavicini (ed.), *Mem. S.A.It.* **63**, 663.
Montmerle, T.: 1997, in *Astronomical and Biochemical Origins and the Search for Life in the Universe*, C. Batalli Cosmovici, S. Bowyer & D. Werthimer (eds.), IAU Coll. N. 161, Editrice Compositori, Bologna, Italy, p. 267.
Montmerle, T., André, P.: 1989, in *Low Mass Star Formation and Pre-Main Sequence Objects*, Bo Reipurth (ed.), ESO Conference and Workshop Proceedings **33**, 407.
Moore, B, Davis, M.: 1994, *Mon. Not. Roy. Astr. Soc.* **207**, 209.
Morfill, G.E., Sholer, M.: 1987, in *Physical Processes in Interstellar Clouds*, G.E. Morfill & M. Scholer (eds.), NATO ASI Ser. **C210**, 564.
Morfill, G.E., Spruit, H., Levy, E.H.: 1993, in *Protostars and Planets III*, E.H. Levy & J.I. Lunine (eds.), University of Arizona Press, Tucson, AZ, USA, p. 939.
Morgan, E., Remillard, R., Greiner, J.: 1996, *IAU Circ.* **N. 6392**.
Morgan, W.W., Kayser, S., White, R.A.: 1975, *Astrophys. J.* **199**, 545.
Morisawa, K., Matsuoka, M., Takahara, F., Piro, L.: 1990, *Astron. Astrophys.* **236**, 299.

Morrison, R., McCammon, D.: 1983, *Astrophys. J.* **270**, 119.
Morselli, A.: 2002, in *Multifrequency Behaviour of High Energy Cosmic Sources*, F. Giovannelli & L. Sabau-Graziati (eds.), *Mem. S.A.It.* **73 N. 4**, 1199.
Moskalenko, I.V., Strong, A.W.: 1998, *Astrophys. J.* **493**, 694.
Mouchet, M., Bonnet-Bidaud, J.M., de Martino, D.: 1998, in *Ultraviolet Astrophysics, Beyond the IUE Final Archive*, W. Wamstecker & R. González-Riestra, (eds.), ESA Publication Division, ESTEC, Nordwijk, The Netherlands, ESA **SP-413**, 431.
Mufson, S.L., McCollough, M.L., Dickel, J.R., Petre, R., White, R., Chevalier, R.: 1986, *Astron. J.* **92**, 1349.
Mullan, D.J.: 1979, *Astrophys. J.* **234**, 588.
Müller, D.: 1993, in *Frontier Objects in Astrophysics and Particle Physics*, F. Giovannelli, G. Mannocchi (eds.), Italian Physical Society, Editrice Compositori, Bologna, Italy **40**, 391.
Muller, C.A., Westerhout, G., Schmidt, M.: 1957, *B.A.N.* **13**, 151, 201 and 247.
Mundt, R., Fried, J.W.: 1983, *Astrophys. J. Lett.* **274**, L83.
Murray, M.J., Dufton, P.L., Hibbert, A., York, D.G.: 1984, *Astrophys. J.* **282**, 481.
Murthy, J., Henry, R.C., Feldman, P.D., Tennyson, P.D.: 1989, *Astrophys. J.* **336**, 954.
Murthy, J., Hall, D., Earl, M., Henry, R. C., & Holberg, J. B.: 1999, *Astrophys. J.* **522**, 904.
Mushotzky, R.F.: 1980, in *X-Ray Astronomy*, R. Giacconi, G. Setti (eds.), D. Reidel Publ. Co., Dordrecht, Holland, p. 171.
Mushotzky, R.F.: 1984a, *Phys. Scripta* **T7**, 157.
Mushotzky, R.F.: 1984b, *Adv. Space Res.* **3**, 10.
Mushotzky, R.F.: 1992, *Review talk at Texas/PASCOS Meeting*, Berkeley, CA, (December 1992).
Mushotzky, R.F.: 1994, in *Clusters of Galaxies*, F. Durret, A. Mazure & J. Tran Thanh Van (eds.), Gif–sur–Yvette, France: Editions Frontieres, p.167.
Mushotzky, R.F.: 1996, in *Proc. Röntgenstrahlung from the Universe*, H.U. Zimmermann, J. Trümper & H. Yorke (eds.), *MPE Report* **263**, 545.
Mushotzky, R.F., Serlemitsos, P.J., Smith, B.W., Boldt, E.A., Holt, S.S.: 1978, *Astrophys. J.* **225**, 21.
Mushotzky, R.F., Loewenstein, M., Arnaud, K., Fukazawa, Y.: 1995, in *Dark Matter*, Holt, S.S. & Bennett, C.L. (eds.), *AIP Conf. Proc.* **336**, 231.
Mushotzky, R.F., Loewenstein, M., Arnaud, K.A., Tamura, T., Fukazawa, Y., Matsushita, K., *et al.*: 1996, *Astrophys. J.* **466**, 686.
Nagano, M., Teshima, M., Matsubara, Y., Dai, H.Y., Hara, T., Hayashida, N., *et al.*: 1992, *J. Phys. G.* **18**, 423.
Nagase, F.: 1989, *Publ. Astron. Soc. Japan* **41**, 1.
Nagase, F.: 1994, in *The Evolution of X-Ray Binaries*, S.S. Holt and C.S. Day (eds.), *AIP Conf. Proc.*, **308**, 567.
Nagase, F., Hayakawa, S., Kunieda, H., Makino, F., Masai, K., Tawara, Y., *et al.*: 1982, *Astrophys. J.* **263**, 814.
Nagle, D.E., Gaisser, T.K., Protheroe, R.J.: 1988, *Ann. Rev. Nucl. Part. Sci.* **38**, 609.
Najarro, F., Krabbe, A., Genzel, R., Lutz, D., Kudritzki, R.P., Hillier, D.J.: 1997, *Astron. Astrophys.* **325**, 700.
Nakano, T.: 1991, in *Young Star Clusters and Early Stellar Evolution*, F. Palla, P. Persi & H. Zinnecker (eds.), *Mem. S.A.It.* **62**, 841.
Nakano, T.: 1993, *NRO (Nobeyama Radio Observatory, Japan) Report* **N. 331**.
Nakar, E., Piran, T.: 2002, *Astrophys. J.* **579**, 699.
Nandra, K., Pounds, K.A., Fabian, A.C., Rees, M.J.: 1989, *Mon. Not. Roy. Astr. Soc.* **236**, 39p.
Nandra, K., Pounds, K. A., Stewart, G. C.: 1991, *Lecture Notes in Physics*, **385**, 177.
Narayan, R., Garcia, M.R., McClintock, J.E.:, 1997, *Astrophys. J. Lett.* **478**, L79.
Navarro, J.F., Frenk, C.S., White, S.D.M.: 1995a, *Mon. Not. Roy. Astr. Soc.* **275**, 56.
Navarro, J.F., Frenk, C.S., White, S.D.M.: 1995b, *Mon. Not. Roy. Astr. Soc.* **275**, 720.

Nel, H.I., de Jager, O.C.: 1994, in *The Second Compton Symposium*, C.E. Fichtel, N. Gehrels & J.P. Norris (eds.), *AIP Conf. Proc.* **304**, 91.
Neumann, D.M., Arnaud, M., Gastaud, R., Aghanim, N., Lumb, Dbriel, U.G., *et al.*: 2001, *Astron. Astrophys. Lett.* **365**, L74.
Nicastro, L., Cusumano, G., Antonelli, A., Amati, L., Frontera, F., Palazzi, E., *et al.*: 2001, in *Gamma Ray Bursts in the Afterglow Era*, E. Costa, F. Frontera & J. Hjorth (eds.), Springer Verlag, Heidelberg, Germany, p. 198.
Nishimura, J., Fujii, M., Taira, T., Aizu, E., Hiraiwz, H., Kobayashi, T., *et al.*: 1980, *Astrophys. J.* **238**, 394.
Nousek, J.A., Baluta, C.J., Corbet, R.H.D., Mukai, K., Osborne, J.P., Ishida, M.: 1994, *Astrophys. J. Lett.* **436**, L19.
Ødegaard, K.J.R.: 1996, in *Proc 33rd Liège Int. Astroph. Coll., Wolf–Rayet Stars in the Framework of Stellar Evolution*, (Liège: Univ. of Liège), p. 81.
Oemler, A.Jr.: 1974, *Astrophys. J.* **194**, 1.
Ögelman, H, van den Heuvel, E.P.J. (eds.): 1989, *Timing Neutron Stars*, Kluwer, Dordrecht, Holland.
Ohashi, T., Fukazawa, Y., Ikebe, Y., Ezawa, H., Tamura, T., Makishima, K.: 1994, in *New Horizon of X-Ray Astronomy*, Universal Academy Press, Tokyo, Japan, p. 273.
Ohashi,T., Kikuchi, K., Kaho, T., Yamasaki, N., Ikebe, Y., Fukazawa, Y., *et al.*: 1996, in *UV and X-ray Spectroscopy of Astrophysical and Laboratory Plasmas*, K. Yamashita & T. Watanabe (eds.), Universal Academy Press, Frontiers Science Series **15**, 151.
Ohta, K., Yamada, T., Nakanishi, K., Ogasaka, Y., Kii, T., Hayashida, K.: 1996, *Astrophys. J. Lett.* **458**, L57.
Oka, T., Hasegawa, T., Sato, F., Tsuboi, M., Miyazaki, A.: 1998, *Astrophys. J. Suppl. Ser.* **118**, 455.
Oke, J.B., Zimmerman, B.: 1979, *Astrophys. J. Lett.* **231**, L13.
Okeke, P.N., Rees, M.J.: 1980, *Astron. Astrophys.* **81**, 263.
Okuda, T., Sakashita, S.: 1977, *Astrophys. Space Sci.* **47**, 385.
Omont, A., Blommaert, J.: 1997, in *The Impact of Large Scale Near-IR Sky Surveys*, F. Garzón, N. Epchtein, A. Omont, B. Burton & P. Persi (eds.), Kluwer Academic Publishers, Dordrecht, The Netherlands, Series Astrophysics and Space Science Library Series **210**, 81.
Oosterbroek, T., van der Klis, M., Kuulkers, E., van Paradijs, J., Lewin, W.H.G.: 1995, *Astron. Astrophys.* **297**, 141.
Orlandini, M, dal Fiume, D.: 2001, *X-Ray Astronomy: Stellar Endpoints, AGN, and the Diffuse X-Ray Background*, N.E. White, G. Malaguti & G.C. Palumbo (eds.), *AIP Conf. Proc.* **599**, 283.
Orosz, J.A., Remillard, R.A., Bailyn, C.D., McClintock, J.E.: 1997, *Astrophys. J. Lett.* **478**, L83.
Orosz, J.A., Kuulkers, E., van der Klis, M., McClintock, J.E., Garcia, M.R., Callanan, P.J., *et al.*: 2001, *Astrophys. J.* **555**, 489.
Osaki, Y.: 1985, *Astron. Astrophys.* **144**, 369.
Osaki, Y.: 1996, *Pub. Astron. Soc. Pacific* **108**, 390.
Osborne, J.L., Wolfendale, A.W., Worral, D.M.: 1994, *J. Phys. G* **20**, 1089.
Ostriker, J.P.: 1978, in *The Large Scale Structure of the Universe*, M.S. Longair & J. Einasto (eds.), D. Reidel Publishing Co., *IAU Symposium* **79**, 357.
Ostriker, J.P., Gunn, J.E.: 1969, *Astrophys. J.* **1573**, 1395.
Ostrowski, M.: 1997, in *Relativistic Jets in AGNs*, M. Ostrowski, M. Sikora, G. Madejski & M. Begelman (eds.), p. 153.
Ostrowski, M.: 1999, in *Frontier Objects in Astrophysics and Particle Physics*, F. Giovannelli & G. Mannocchi (eds.), Italian Physical Society, Editrice Compositori, Bologna, Italy **65**, p. 313.
Ostrowski, M.: 2000, in *Plasma Turbulence and Energetic Particles in Astrophysics*, M. Ostrowski & R. Schlickeiser (eds.), Obserwatorium Astronomiczne, Uniwersytet Jagelloński, p. 256.
Özel, F., Psaltis, D., Kaspi, V.M.: 2001, *Astrophys. J.* **563**, 225.
Pacini, F.: 1967, *Nature* **216**, 567.
Paczynski, B.: 1998, *Astrophys. J. Lett.* **494**, L45.

Padovani, P.: 1997, in *From the Micro- to the Mega-Parsec*, A. Comastri, T. Venturi, M. Bellazzini (eds.), Mem. S.A.It. **68**, 47.

Padovani, P.: 1998, in *New Horizons from Multi-Wavelength Sky Survey*, B. McLean, D.A. Golombek, J.J.E. Hayes & H.E. Payne (eds.), *179th IAU Symposium*, Kluwer Academic Publ., p. 257.

Padovani, P.: 1999, in *Frontier Objects in Astrophysics and Particle Physics*, F. Giovannelli & G. Mannocchi (eds.), Italian Physical Society, Editrice Compositori, Bologna, Italy, **653**, 159.

Padovani, P.: 2002, in Blazar Astrophysics with BeppoSAx and Other Observatories, P. Giommi, E. Massaro & G. Palumbo (eds.), ASI Special Publication, printed at ESA-ESRIN, p. 101.

Paerels, F., Hur, M.Y., Mauche, C.W., Heise, J.: 1996, *Astrophys. J.* **464**, 884.

Pagel, B.E.J.: 1985, in *Production and Distribution of* C,N,O *Elements*, I.J. Danziger, F. Matteucci & K. Kjaer (eds.), ESO Conference and Workshop Proceedings, p.155.

Pallavicini, R.P., Golub, L., Rosner, R., Vaiana, G.S., Ayres, T., Linsky, J.L.: 1981, *Astrophys. J.* **248**, 279.

Pallavicini, R.P., Golub, L., Rosner, R., Vaiana, G.S.: 1982, in *Cool Stars, Stellar Systems, and the Sun*, M.S. Giampapa & L. Golub (eds.), SAO Rep. N. SP-392, Washington DC, USA, p. 77.

Paltani, S., Roques, J.-P., Goldoni, P., Paul, J., Churazov, E., Gilfanov, M., Sunyaev, R., Khavenson, N.: 1999, in *Proceedings Third Integral Workshop: The Extreme Universe*, A. Bazzano, G.G.C. Palumbo & C. Winkler (eds.), Astro. Lett. and Communications **38**, 449.

Panagia, N.: 1999a, in *Multifrequency Behaviour of High Energy Cosmic Sources*, F. Giovannelli & L. Sabau-Graziati (eds.), *Mem. S.A.It.* **70**, 831.

Panagia, N.: 1999b, in *Frontier Objects in Astrophysics and Particle Physics*, F. Giovannelli, G. Mannocchi (eds.), Italian Physical Society, Editrice Compositori, Bologna, Italy, **65**, 25.

Panagia, N.: 2002a, in *Multifrequency Behaviour of High Energy Cosmic Sources*, F. Giovannelli & L. Sabau-Graziati (eds.), *Mem. S.A.It.* **73, N. 1**, 88.

Panagia, N.: 2002b, in *Multifrequency Behaviour of High Energy Cosmic Sources*, F. Giovannelli & L. Sabau-Graziati (eds.), *Mem. S.A.It.* **73, N. 4**, 886.

Panagia, N., Gilmozzi, R., Macchetto, F., Adorf, H.-M., Kirshner, R.P.: 1991, *Astrophys. J. Lett.* **380**, L23.

Panagia, N., Weiler, K.W., Montes, M.J., Van Dyk, S.D., Sramek, R.A., Lacey, C.K.: 2001, in *Frontier Objects in Astrophysics and Particle Physics*, F. Giovannelli & G. Mannocchi (eds.), Italian Physical Society, Editrice Compositori, Bologna, Italy, **73**, 207.

Panaitescu, A., Kumar, P.: 2001, *Astrophys. J. Lett.* **560**, L49.

van Paradijs, J.: 1983, in *Accretion-Driven Stellar X-Ray Sources*, W.H.G. Lewin & E.P.J. van den Heuvel, Cambridge and New York, Cambridge University Press, p. 189.

van Paradijs, J.: 1996, *Astrophys. J. Lett.* **464**, L139.

van Paradijs, J.: 1998, in *The Many Faces of Neutron Stars*, R. Buccheri, J. van Paradijs & and M.A. Alpar (eds.), Kluwer Academic Publ., Dordrecht, Holland, p. 279.

van Paradijs, J., McClintock, J.E.: 1995, in *X-Ray Binaries*, W.H.G. Lewin, J. van Paradijs & E.P.J. van den Heuvel (eds.), Cambridge Univ. Press, p. 58.

van Paradijs, J., Taam, R.E., van den Heuvel, E.P.J.: 1995, *Astron. Astrophys. Lett.* **299**, L41.

van Paradijs, J., White, N.E.: 1995, *Astron. Astrophys. Lett.* **447**, L33.

van Paradijs, J., Groot, P.J., Galama, T., Kouveliotou, C., Strom, R.G., Telting, J., *et al.*: 1997, *Nature* **386**, 686.

Paredes, J.M., Martí, J., Ribó, M., Massi, M.: 2000, *Science* **288**, 2340.

Paresce, F., Jakobsen, P.: 1980, *Nature* **288**, 119.

Parmar, A.N., Stella, L., Giommi, P.: 1989, *Astron. Astrophys.* **222**, 96.

Parmar, A.N., Guainazzi, M., Oosterbroek, T., Orr, AA., Favata, F., Lamb, D., *et al.*: 1999, *Astron. Astrophys.* **345**, 611.

Parson, S.B.: 1981, Astrophys. J. **247**, 560.

Parson, S.B.: 1983, Astrophys. J. Suppl. Ser. **53**, 553.

Partridge, R.B.: 1987, in *Observational Cosmology*, IAU Symp. N. 124 (Beijing), A. Hewitt, G.R. Burbidge and L. Fang, (eds.), D. Reidel Publ. Co., Dordrecth, Holland, p. 31.
Partridge, R.B., Pebles, P.J.E.: 1967a, *Astrophys. J.* **147**, 868.
Partridge, R.B., Pebles, P.J.E.: 1967b, *Astrophys. J.* **148**, 377.
Patterer, R.J., Ramsey, L., Huenemoerder, D.P., Welty, A.D.: 1993, *Astron. J.* **105**, 1519.
Patterson, J.: 1979, *Astrophys. J.* **231**, 789.
Patterson, J.: 1981, *Astrophys. J. Suppl. Ser.* **45**, 517.
Patterson, J.: 1984, *Astrophys. J. Suppl. Ser.* **54**, 443.
Patterson, J., Robinson, E.L., Nather, R.E.: 1978, *Astrophys. J. Lett.* **226**, L37.
Patterson, J., Raymond, J.C.: 1985a, *Astrophys. J.* **292**, 535.
Patterson, J., Raymond, J.C.: 1985b, *Astrophys. J.* **292**, 550.
Paul, J.A., Bennett, K., Bignami, G. F., Caraveo, P., Buccheri, R., Scarsi, L., *et al.*: 1978, *Astron. Astrophys. Lett.* **63**, L31.
Paul, J., Ballet, J., Cantin, M., Cordier, B., Goldwurm, A., Lambert, A., *et al.*: 1991, *Adv. Space Res.* **11**, 289.
Pawsey, J.L.: 1951, *Nature* **168**, 358.
Peacock, J.A., Cole, S., Norberg, P., Baugh, C.M., Bland-Hawthorn, J., Bridges, T., *et al.*: 2001, *Nature* **410**,169.
Pedichini, F., Di Paola, A., Stella, L., Buonanno, R., Boattini, A., Gandolfi, G., *et al.*: 1997, *Astron. Astrophys. Lett.* **327**, L36.
Penzias, A.A., Wilson, R.W.: 1965, *Astrophys. J.* **142**, 419.
Perotti, F., Ventura, A.D., Sechi, G., Villa, G., Di Cocco, G, Backer, R.E., *et al.*: 1979, *Nature* **282**, 484.
Perryman, M.A.C.: 1997, in *Hipparcos Venice '97*, B. Battrick (ed.), ESA Publication Division, ESTEC, Nordwijk, The Netherlands, ESA **SP-402**, 1.
Perryman, M.A.C., Lindegren, L., Kovalevsky, J., Turon, C., Hoeg, E., Grenon, M., *et al.*: 1995, *Astron. Astrophys.* **304**, 69.
Perryman, M.A.C., Lindegren, L., Kovalevsky, J., Hoeg, E., Bastian, U., Bernacca, P.L., *et al.*: 1997, *Astron. Astrophys.* **323**, 49.
Persi, P.: 2003, in *The Bridge between the Big Bang and Biology: Magnetic Fields of Celestial Bodies: Their Link to Life*, F. Giovannelli (ed)., CNR, Special Volume, Roma, Italy, in press.
Persi, P., Ferrari-Toniolo, M., Spada, G., Conti, G., di Benedetto, P., Tanzi, E.G., *et al.*: 1979, *Mon. Not. Roy. Astr. Soc.* **187**, 293.
Peterson, L.E.: 1973, in *X-Ray and Gamma Ray Astronomy*, IAU Symposium N. 55, H. Bradt & R. Giacconi (eds.), D. Reidel Publ. Co., Dordrecht, Holland, p. 51.
Peterson, J.R., Paerels, F.B.S., Kaastra, J.S., Arnaud, M., Reiprich, T.H., Fabian, A.C., *et al.*: 2001, *Astron. Astrophys. Lett.* **365**, L104.
Peterson, L.E., Winckler, J.B.: 1958, *Phys. Rev. Lett.* **1**, 205.
Pian, E.: 2002, *astro-ph/0110051*.
Piccinotti, G., Mushotzky, R.F., Bolt, E.A., Holt, S.S., Marshall, F.E., Serlemitsos, P.J., *et al.*: 1981, *Astrophys. J.* **252**, 485.
Piccioni, A., Bartolini, C., Bernabei, S., Galleti, S., Giovannelli, F., Guarnieri, A., *et al.*: 2000, in *The Be Phenomenon in Early-Type Stars*, M.A. Smith, H.F. Henrichs & J. Fabregat (eds.), IAU Coll. N. 175, *Astron. Soc. Pacific Conf. Ser.* **214**, 569.
Pierce, M.J., Welch, D.L., McClure, R.D., van den Berg, S., Racine, R., Stetson, P.B.: 1994, *Nature* **371**, 385.
Pierre, M., Le Borgne, J.F., Soucail, G., Kneib, J.P.: 1996, *Astron. Astrophys.* **311**, 413.
Piran, T.: 1999a, in *Gamma Ray Bursts: The First Three Minutes*, J. Poutanen & R. Svensson (eds.), *ASP Conf. Ser.* **190**, 3.
Piran, T.: 1999b, *Phys. Rep.* **314**, 575.
Piran, T.: 2000, *Phys. Rep.* **333**, 529.

Piran, T., Kumar, P., Panaitescu, A., Piro, L.: 2001, *Astrophys. J. Lett.* **560**, L167.
Piro, L.: 1996, in *Multifrequency Behaviour of High Energy Cosmic Sources*, F. Giovannelli & L. Sabau-Graziati (eds.), *Mem. S.A.It.* **67**, 575.
Piro, L., Matsuoka, M., Yamauchi, M.: 1989, in *23rd ESLAB Symposium on Two Topics in X-Ray Astronomy*, J. Hunt & B. Battrick (eds.), ESA **SP-296**, 819.
Piro, L., Yamauchi, M., Matsuoka, M.: 1990, *Astrophys. J. Lett.* **360**, L35.
Piro, L., de Rosa, A., Dadina, M., Nicastro, F., Matt, G., Perola, G.C., *et al.*: 2000a, *Adv. Space Res.* **25**, 453.
Piro, L., Garmire, G., Garcia, M., Stratta, G., Costa, E., Feroci, M., *et al.*: 2000b, *Science* **290**, 955.
Piro, L., Soffitta, P., Antonelli, L.A., in't Zand, J., Heise, J., Nicastro, L., *et al.*: 2001, *GRB Coordinate Network* **1172**, 1.
Plionis, M. Georgantopoulos, I. (eds.): 2000, *Large Scale Structure in the X-Ray Universe*, Atlantisciences, Biarritz.
Plavec, M.: 1968, *Adv. Astron. Astrophys.* **6**, 202.
Pohl, M.: 1994, *Astron. Astrophys.* **287**, 453.
Pohl, M.: 1997, *Astron. Astrophys.* **317**, 441.
Polcaro, V.F., Rossi, C., Norci, L., Giovannelli, F.: 1989, *Acta Astron.* **39**, 323.
Polcaro, V.F., Giovannelli, F., Manchanda, R.K., Norci, L., Rossi, C.: 1990, *Proc 21st Intern. Cosmic Ray Conf. (Adelaide)* **1**, 245.
Polcaro, V.F., Giovannelli, F., Manchanda, R.K., Pollock, A., Norci, L., *et al.*: 1991a, in *Wolf–Rayet Stars and Interrelations with Other Massive Stars in Galaxies*, K.A. van der Hucht & B. Hidayat (eds.), Kluwer Academic Publ., Dordrecht, The Netherlands, p. 103.
Polcaro, V.F., Giovannelli, F., Manchanda, R.K., Norci, L., Rossi, C.: 1991b, *Astron. Astrophys.* **252**, 590.
Polcaro, V.F., Brinkman, W., Giovannelli, F., Manchanda, R.K., Manteiga, M., Norci, L., Rossi, C.: 1991c, *22nd Intern. Cosmic Ray Conf. (Dublin)*, **OG 3.4.4**, 188.
Polcaro, V.F., Villada, M., Giovannelli, F.: 1993, *Astron. Astrophys. Lett.* **273**, L49.
Polcaro, V.F., Manchanda, R.K., Norci, L., Brinkman, W., Kanbach, G., Mayer-Hasselwander, H., *et al.*: 1995, in *Proc. 24th Intern. Cosmic Ray Conf. (Roma)* **2**, 215.
Polidan, R.S., Holberg, J.B.: 1984, *Nature* **309**, 528.
Polidan, R.S., Holberg, J.B.: 1987, *Mon. Not. Roy. Astr. Soc.* **309**, 528.
Pollock, A.M.T.: 1987, *Astron. Astrophys.* **171**, 135.
Porcas, R.W.: 1985, in *Active Galactic Nuclei*, J.E. Dyson (ed.), Manchester University Press, UK, p. 20.
Pottash, S.R.: 1986, in *Light on Dark Matter*, F.P. Israel (ed.), D. Reidel Publ. Co., Dordrecht, Holland, p. 131.
Pottash, S.R., Baud, B., Beintema, D., Emerson, J., Habing, H.J., Harris, S., *et al.*: 1984, *Astron. Astrophys.* **138**, 10.
Pounds, K.A.: 1990, *Vistas in Astronomy* **33**, 83.
Pounds, K.A. Nandra, K., Stewart, G.C., George, I.M., Fabian, A.C.: 1990, *Nature* **344**, 132.
Prantzos, N., Hashimoto, M., Nomoto, K.: 1990, *Astron. Astrophys.* **234**, 211.
Prantzos, N., Diehl, R.: 1996, *Phys. Rep.* **267**, 1.
Prantzos, N., Tosi, M., von Steiger, R. (eds.): 1998, *Primordial Nuclei and their Galactic Evolution*, Kluwer Academic Publishers.
Preece, R.D., Briggs, M.S., Mallozzi, R.S., Pendleton, G.N., Paciesas, W.S., Band, D.L.: 2000, *Astrophys. J. Suppl. Ser.* **126**, 19.
Predehl, P., Trümper, J.: 1994, *Astron. Astrophys. Letter* **290**, L29.
Pretzl, K.: 1995, in *Frontier Objects in Astrophysics and Particle Physics*, F. Giovannelli & G. Mannocchi (eds.), Italian Physical Society, Editrice Compositori, Bologna, Italy **47**, 89.
Price, P.A., Berger, E., Reichart, D.E., Kulkarni, S.R., Yost, S.A., Subrahmanyan, R., *et al.*: 2002, *Astrophys. J. Lett.* **572**, L51.

Priedhorsky, W.C.: 1986, *Astrophys. J. Lett.* **306**, L97.
Priedhorsky, W.C., Terrell, J.: 1983, *Nature* **303**, 681.
Primak, J., Sadoulet, B., Seckel, D.: 1988, *Ann. Rev. Nucl. Part. Sci.* **38**, 751.
Prince, T.A., Ling, J.C., Mahoney, W.A., Riegler, G.R., Jacobson, A.S.: 1982, *Astrophys. J. Lett.* **255**, L81.
Pringle, J.E., Wade, R.A. (eds.): 1985, *Interacting Binary Stars*, Cambridge Univ. Press.
Prince, T.A., Bildsten, L., Chakrabarty, D., Wilson, R.B., Finger, M.H.: 1994, in *The Evolution of X-Ray Binaries*, S.S. Holt & C.S. Day (eds.), *AIP Conf. Proc.*, **308**, 235.
Protheroe, R.J.: 1986, *Pub. Astron. Soc. Australian* **6**, 280.
Puget, J.-L., Stecker, F.W.: 1975, *Proc. 14th Intern. Cosmic Ray Conf. (Mñich)* **2**, 734.
Puget, J.-L, Abergel, A., Bernard, J.-P., Boulanger, F., Burton, W.B., Desert, F.-X., *et al.*: 1996, *Astron. Astrophys. Lett.* **308**, L5.
Punch, M., Akerlof, C.W., Cawley, M.F., Chantell, M., Fegan, D.J., Fennell, S., *et al.*: 1992, *Nature* **358**, 477.
Purcell, W.R., Bouchet, L., Johnson, W.N., Jung, G., Kinzer, R.L., Kurfess, J.D., *et al.*: 1996, *Astron. Astrophys. Suppl. Ser.* **120**, 389.
Purcell, W.R., Cheng, L.-X., Dixon, D.D., Kinzer, R.L., Kurfess, J.D., Leventhal, M., *et al.*: 1997, *Astrophys. J.* **491**, 725.
Puxley, P.J., Hawarden, T.G., Mountain, C.M., Leggett, S.K.: 1987, in *Star Formation in Galaxies*, C. Persson (ed.), NASA, Washington, DC, US Govt. Print. Off., p. 619.
Quintana, H., Melnick, J.: 1982, *Astron. J.* **87**, 972.
Rachen, J.P., Biermann, P.L.: 1993, *Astron. Astrophys.* **272**, 161.
Radhakrishnan, V., Srinivasan, G.: 1982, *Curr. Sci.* **51**, 1096.
Raine, D.J.: 1988, *Vistas in Astron.* **32, pt. 4**, 321.
Ramana Murthy, P.V., Wolfendale, A.W.: 1986, *Gamma Ray Astronomy*, Cambridge Astrophysics Series, Cambridge University Press.
Ramaprakash, A.N., Kulkarni, S.R., Frail, D.A., Koresko, C., Kuchner, M., Goodrich, R., *et al.*: 1998, *Nature* **393**, 43.
Ramaty, R., Kozlovsky, B., Lingenfelter, R.E.: 1975, *Space Sci. Rev.* **18**, 341.
Ramaty, R., Kozlovsky, B., Lingenfelter, R.E.: 1979, *Astrophys. J. Suppl. Ser.* **40**, 487.
Ramaty, R., Lingenfelter, R.E.: 1979, *Nature* **278**, 127.
Ramaty, R., Lingenfelter, R.E.: 1982, *Ann. Rev. Nucl. Part. Sci.* **32**, 235.
Rappaport, S., Doxsey, R., Zaumen, W.: 1971, *Astrophys. J. Lett.* **168**, L43.
Rappaport, S., Zaumen, W., Doxsey, R.: 1971, *Astrophys. J. Lett.* **168**, L17.
Rappaport, S., Cash, W., Doxsey, R., McClintock, J., Moore, G.: 1974, *Astrophys. J. Lett.* **187**, L5.
Rappaport, S., Stothers, R., Joss, P.C.: 1980, *Astrophys. J.* **235**, 570.
Rappaport, S., Joss, P.C.: 1981, in *X-Ray Astronomy with the Einstein Satellite*, R. Giacconi (ed.), D. Reidel Publ. Co., Dordrecht, Holland, p. 123.
Rebolo, R.: 1996, in *Lecture Notes in Physics*, **470**, 98.
Reedy, R.C.: 1978, in *Proc. of the Lunar and Planetary Science Conference, 9th*, Pergamon Press, Inc., New York, Proceedings. Volume **3**, 2961.
Rees, M.J.: 1984, *Ann. Rev. Astron. Astrophys.* **22**, 471.
Rees, M.J.: 1985 in *19th Intern. Cosmic Ray Conf. (La Jolla)* **9**, 1.
Rees, M.J.: 1989, in *Big Bang, Active Galactic Nuclei and Supernovae*, University Academy Press, Inc. and Yamada Science Foundation, p. 121.
Rees, M.J.: 1998, in *Primordial Nuclei and their Galactic Evolution*, N. Prantzos, M. Tosi, & R. von Steiger (eds.), *Space Sci. Rev.* **84**, 43.
Rees, M.J.: 1999, *Astron. Astrophys. Suppl. Ser.* **138**, 491.
Reich, W.: 2002, in Proc. of the 270. WE-Heraeus Seminar on Neutron Stars, Pulsars, and Supernova Remnants, W. Becker, H. Lesch, & J. Trümper (eds.), MPE Report **278**, 1.
Reichart, D.E.: 1999, *Astrophys. J. Lett.* **521**, L111.

Reichart, D.E., Lamb, D.Q., Fenimore, E.E., Ramirez-Ruiz, E., Cline, T.L., Hurley, K.: 2001, *Astrophys. J.* **552**, 57.

Reig, P., Fabregat, J., Coe, M.J.: 1997, *Astron. Astrophys.* **322**, 193.

Ressel, M.T., Turner, M.S.: 1990, *Comments Astrophys.* **14**, 323.

Reuter, H.-P., Lesch, H.: 1996, *Astron. Astrophys. Letter* **310**, L5.

Rhoads, J.E.: 1997, *Astrophys. J. Lett.* **487**, L1.

Rhoads, C., Ruffini, R.: 1974, *Phys. Rev. Lett.* **32**, 324.

Richards, P.L.: 1988, in *Comets to Cosmology, Proc. Third IRAS Conference (London)*, A. Lawrence (ed.), Springer-Verlag, Berlin, FRG, p. 289.

Ricketts, M.J., King, A.R., Raine, D.J.: 1979, *Mon. Not. R. astr. Soc.* **186**, 233.

Riegler, G.R., Ling, J.C., Mahoney, W.A., Wheaton, W.A., Willet, J.B., Jacobson, A.S., *et al.*: 1981, *Astrophys. J. Lett.* **248**, L13.

Rieke, G.H., Lebofsky, M.J.: 1979, *Ann. Rev. Astron. Astrophys.* **17**, 477.

Ritter, H., Kolb, U.: 1998, *Astron. Astrophys. Suppl. Ser.* **129**, 83.

Rix, H.-W., Maoz, D., Turner, E.L., Fukugita, M.: 1994, *Astrophys. J.* **435**, 49.

Robinson, E.L.: 1976, *Ann. Rev. Astron. Astrophys.* **14**, 119.

Rocchia, R., Arnaud, M., Blondel, C., Cheron, C., Christy, J.C., Rothenflug, R., *et al.*: 1984, *Astron. Astrophys.* **130**, 53.

Roche, P.F., Aitken, D.K..: 1986, *Mon. Not. Roy. Astr. Soc.* **221**, 63.

Romero, G.E.: 2001, in *The Nature of Unidentified Galactic Gamma Ray Sources*, A. Carraminana, O. Reimer & D. Thompson (eds.), Kluwer Academic publishers, Dordrecht, Holland, p. 65.

Romero, G.E., Benaglia, P., Torres, D.F.: 1999, *Astron. Astrophys.* **348**, 868.

Romero, G.E., Kaufman Bernadó, M.M., Combi, J.A., Torres, D.F.: 2001, *Astron. Astrophys.* **376**, 599.

Roncadelli, M.: 2003, *astro-ph/0301537.*

Rood, H.J., Sastry, G.N.: 1971, *Publ. Astron. Soc. Pac.* **83**, 313.

Rosa, M.: 1980, *Astron. Astrophys. Lett.* **85**, L21.

Rosa, M., Joubert, M., Benvenuti, P.: 1984, *Astron. Astrophys. Suppl. Ser.* **57**, 361.

Rosenberg, F.D., Eyles, C.J., Skinner, G.K., Willmore, A.P.: 1975, N̂ature **256**, 628.

Rosner, R., Golub, L., Vaiana, G.: 1985, *Ann. Rev. Astron. Astrophys.* **23**, 413.

Rossi, B.B.: 1973, in *X-Ray and Gamma Ray Astronomy*, IAU Symposium N. 55, H. Bradt & R. Giacconi (eds.), D. Reidel Publ. Co., Dordrecht, Holland, p. 1.

Rothschild, R., Boldt, E., Holt, S., Serlemitsos, P.: 1974, *Bull. Amer. Astron. Soc.*, **6**, 277.

Röttgering, H., Snellen, I., Miley, G., de Jong, J.P., Hanish, B., Perley, R.: 1994, *Astrophys. J.* **436**, 654.

Rucinski, S.M.: 1985, *Astron. J.* **90**, 2321.

Ruffini, R.: 1975, *Neutron Stars, Black Holes and Binary X-Ray Sources*, H. Gursky & R. Ruffini (eds.), D. Reidel Publ. Co., Dordrecht, Holland, p. 59.

de Rujula, A., Glashow, S.L.: 1980, *Phys. Rev. Lett.* **45**, 942.

Rutten, R.G.M., Schrijver, C.J., Lemmens, A.F.P., Zwaan, C.: 1991, *Astron. Astrophys.* **252**, 203.

Sabau-Graziati, L.: 1990, *Ph.D. Thesis*, University of Barcelona, Spain.

Sacher, W., Schönfelder, V.: 1983, *Space Sci. Rev.* **36**, 249.

Sackett, P.D., Pogge, R.W.: 1995, in *Dark Matter*, Holt, S.S. & Bennett, C.L. (eds.), *AIP Conf. Proc.* **336**, 141.

Sadakane, K., Nishida, M.: 1986, *Publ. Astron. Soc. Pac.* bf 98, 685.

Safier, P.N.: 1993a, *Astrophys. J.* **408**, 115.

Safier, P.N.: 1993b, *Astrophys. J.* **408**, 148.

Saggion, A., Bastieri, D.: 2002, in *Multifrequency Behaviour of High Energy Cosmic Sources*, F. Giovannelli & L. Sabau-Graziati (eds.), *Mem. S.A.It.* **73 N. 4**, 812.

Sahu, K., Livio, M., Petro, L., Macchetto, F.D., van Paradijs, J., Kouveliotou, C., *et al.*: 1997, *Nature* **387**, 476.

Sahu, K.C., Vreeswijk, P., Bakos, G., Menzier, J.W., Bragaglia, A., Frontera, F., *et al.*: 2000, *Astrophys. J.* **540**, 74.

Sakima, Y., Tawara, Y., Yamashita, K.: 1994, *Autumn Meeting of Japanese Astronomical Society* (in Japanese).

Salamon, M.H., Stecker, F.W.: 1994, *Astrophys. J. Lett.* **430**, L21.

Salamon, M.H., Stecker, F.W., de Jager, O.C.: 1994, *Astrophys. J. Lett.* **423**, L1.

di Salvo, T., Stella, L., Robba, N.R., van der Klis, M., Burderi, L., Israel, G.L., *et al.*: 2000, *Astrophys. J. Lett.* **544**, L119.

Sams, B.J., Eckart, A., Sunyaev, R.: 1996, *Nature* **328**, 47.

Sánchez-Fernández, C.: 2003, *Ph. D. Thesis*, University 'Autonoma', Madrid, Spain.

Sanders, D.B., Mirabel, I.F.: 1985, *Astrophys. J. Lett.* **298**, L31.

Sanders, D. B., Scoville, N.Z., Young, J.S., Soifer, B.T., Schloerb, F.P., Rice, W.L., *et al.*: 1986, *Astrophys. J. Lett.* **305**, L45.

Sanders, D.B., Scoville, N.Z., Soifer, B.T., Young, J.S., Danielson, G.E.: 1987, *Astrophys. J. Lett.* **312**, L5.

Sanders, D.B., Soifer, B.T., Elias, J.H., Neugebauer, G., Matthews, K.: 1988: *Astrophys. J. Lett.* **328**, L35.

Sanders, D.B., Mirabel, I.F.: 1996, *Ann. Rev. Astron. Astrophys.* **34**, 749.

Sanduleak, N.: 1971, *Astrophys. J. Lett.* **164**, L71.

Sanford, P.W. (ed.): 1979, *Observations of Compact X-Ray Sources*, Royal Society, London, UK, Proceedings Series A – Mathematical and Physical Sciences **366, N. 1726**.

Santangelo, A., Segreto, A., Giarrusso, S., dal Fiume, D., Orlandini, M., Parmar, A.N., *et al.*: 1999, *Astrophys. J. Lett.* **523**, L85.

Saraceno, P., Benedettini, M., di Giorgio, A. M., Giannini, T., Nisini, B., Lorenzetti, D., *et al.*: 1999, in *The Physics and Chemistry of the Interstellar Medium*, V. Ossenkopf, J. Stutzki, & G. Winnewisse (eds.), GCA-Verlag Herdecke, p. 279.

Saraceno, P., Orfei, R.: 2001, in *The Bridge between the Big Bang and Biology: Stars, Planetary Systems, Atmospheres, Volcanoes: Their Link to Life*, F. Giovannelli (ed)., CNR President Bureau, Special Volume, Roma, Italy, p. 77.

Sarazin, C.L.: 1988a, *X-Ray Emission from Clusters of Galaxies*, Cambridge University Press, Cambridge, UK.

Sarazin, C.L.: 1988b, *Rev. of Modern Phys.* **58**, 1.

Sarazin, C.L.: 1996, in *UV and X-Ray Spectroscopy of Astrophysical and Laboratory Plasmas*, K. Yamashita, T. Watanabe (eds.), Universal Academy Press, Inc., Tokyo, Japan, p. 9.

Sari, R., Piran, T.: 1997, *Astrophys. J.* **485**, 270.

Sauvage, M., Blommaert, J., Boulanger, F., Cesarsky, C.J., Cesarsky, D.A., Desert, F.X., *et al.*: 1996, *Astron. Astrophys. Lett.* **315**, L89.

Savage, B.D., Drake, J.F., Budich, W., Bohlin, R.C.: 1977, *Astrophys. J.* **216**, 291.

Savage, B.D., Massa, D.: 1987, *Astrophys. J.* **314**, 380.

Savage, A., Clowes, R.G., MacGillivray, H.T., Wolstencroft, R.D., Leggett, S.K., Puxley, P.J.: 1987, in *Star Formation in Galaxies*, C. Persson (ed.), NASA, Washington, DC, US Govt. Print. Off., p. 537.

Savage, B.D., Sembach, K.R.: 1996, *Ann. Rev. Astron. Astrophys.* **34**, 279.

Scarsi, L., Bennett, K., Bignami, G. F., Boella, G., Buccheri, R., Hermsen, W., *et al.*: 1977, in *ESA Recent Advances in Gamma Ray Astronomy*, p. 3.

Schaaf, R., Pietsch, W., Biermann, P. L., Kronberg, P. P., Schmutzler, T.: 1989, *Astrophys. J.* **336**, 722.

Schaefer, B.E., Palmer, D., Dingus, B.L., Schneid, E.J., Schönfelder, V., Ryan, J., *et al.*: 1998, *Astrophys. J.* **492**, 696.

Schaudel, D., Becker, W., Voges, W., Aschenbach, B., Reich, W., Weisskopf, M.: 2002, in *Neutron Stars in Supernova Remnants*, P.O. Slane & B.M. Gaensler (eds.), *ASP Conf. Ser.* **271**, 391.

Scheepmaker, A., Ricker, G., Brecher, K., Ryckman, S., Ballantine, J., Doty, J., *et al.*: 1976, *Astrophys. J. Lett.* **205**, L65.

Schindler, S.: 1996, in *UV and X-Ray Spectroscopy of Astrophysical and Laboratory Plasmas*, K. Yamashita, T. Watanabe (eds.), Universal Academy Press, Inc., Tokyo, Japan, p. 157.

Schindler, S.: 2001, in *Frontier Objects in Astrophysics and Particle Physics*, F. Giovannelli & G. Mannocchi (eds.), Italian Physical Society, Editrice Compositori, Bologna, Italy **73**, 33.

Schindler, S.: 2002a, in *Multifrequency Behaviour of High Energy Cosmic Sources*, F. Giovannelli & L. Sabau-Graziati (eds.), *Mem. S.A.It.* **73**, 125.

Schindler, S.: 2002b, in *Multifrequency Behaviour of High Energy Cosmic Sources*, F. Giovannelli & L. Sabau-Graziati (eds.), *Mem. S.A.It.* **73**, 965.

Schönfelder, V.: 1983, *Space Sci. Rev.* **36**, 249.

Schönfelder, V.: 1995, *Adv. Space Res.* **15, N. 5**, 1.

Schönfelder, V., Bennet, K., Collmar, W., Connors, A., Deerenberg, A., Dihel, R., *et al.*: 1992, in *Data Analysis in Astronomy – IV*, p. 185.

Schönfelder, V., Weidenspointner, G.M.: 1999, *MPE Report* **273**, 85.

Schramm, D.N.: 1990, in *Astrophysical Ages and Dating Methods*, E. Vangioni-Flam *et al.* (eds.), Editions Frontiere, Gif-sur-Yvette, France, p. 365.

Schramm, D.N.: 1993, in *Results and Perspectives in Particle Physics*, M. Greco (ed.), Edition Frontières, Gif-sur-Yvette Cedex, France, **C60**, 13.

Schramm, D.N.: 1998, in *Primordial Nuclei and their Galactic Evolution*, N. Prantzos, M. Tosi, & R. von Steiger (eds.), *Space Sci. Rev.* **84**, 3.

Schreier, E., Gursky, H., Kellogg, E., Tananbaum, H., Giacconi, R.: 1971, *Astrophys. J. Lett.* **170**, L21.

Schreier, E., Levinson, R., Gursky, H., Kellogg, E., Tananbaum, H., Giacconi, R.: 1972a, *Astrophys. J. Lett.* **172**, L79.

Schreier, E., Levinson, R., Gursky, H., Kellogg, E., Tananbaum, H., Giacconi, R.: 1972b, *Astrophys. J. Lett.* **173**, L151 (erratum for above).

Schubnell, M.S., Akerlof, C.W., Biller, S., Buckley, J., Carter-Lewis, D.A., Cawley, M.F., *et al.*: 1996, *Astrophys. J.* **460**, 644.

Schwartz, D.A.; 1979, in *X-Ray Astronomy*, W.A. Baity, L.E. Peterson (eds.), Pergamon Press, Oxford, UK, p. 453.

Schwartz, D.A., Davis, M., Doxsey, R.E., Griffiths, R.E., Huchra, J., Johnston, M.D., *et al.*: 1980a, *Astrophys. J. Lett.* **238**, L53.

Schwartz, D.A., Schwartz, J., Tucker, W.: 1980b, *Astrophys. J. Lett.* **238**, L59.

Schwarz, H.E., Aspin, C., Corradi, R.L.M., Reipurth, B.: 1997, *Astron. Astrophys.* **319**, 267.

Schwartz, P.R., Simon, T., Zuckermann, B., Howell, R.R.: 1984, *Astrophys. J. Lett.* **280**, L23.

Schwartz, P.R., Simon, T., Campbell, R.: 1986, *Astrophys. J.* **303**, 233.

Schwarzenberg-Czerny, A.A., Ward, M., Hanes, D.A., Jones, D.H.P., Pringle, J.E., Verbunt, F., *et al.*: 1985, *Mon. Not. Roy. Astr. Soc.* **212**, 645.

Sciama, D.W.,: 1997, *Mon. Not. Roy. Astr. Soc.* **289**, 945.

Sciama, D.W.: 1998, it Astron. Astrophys. **335**, 12.

Scoville, N.Z.: 1992, in *The Astronomy and Astrophysics Encyclopedia*, Maran, S.P (ed.), Cambridge University Press, UK, p. 373.

Scoville, N.Z., Young, J.: 1983, *Astrophys. J.* **265**, 148.

Scoville, N.Z., Sanders, D.: 1986, in *Interstellat Processes*, H. Thronson & D. Hollenbach (eds.), D. reidel Publ. Co., Dordrecht, Holland, p. 178.

Scoville, N.Z., Good, J.C.: 1987, in *Star Formation in Galaxies*, C.J. Lonsdale (ed.), NASA PRINT OFF, Washington DC, USA, p. 3.

Sedov, L.: 1959, *Similarity and Dimensional Methods in Mechanics*, Academy Press, New York, USA.

Sekiguchi, K.: 1987, in *Star Formation in Galaxies*, C. Persson (ed.), NASA, Washington, DC, US Govt. Print. Off., p. 507.

Serabyn, E., Morris, M.: 1994, *Astrophys. J. Lett.* **424**, L91.

Serlemitsos, P.J., Smith, B.W., Boldt, E.A., Holt, S.S., Swank, J.H.: 1977, *Astrophys. J. Lett.* **211**, L63.

Serlemitsos, P.J., Loewenstein, M., Mushotzky, R.F., Marshall, F.E., Petre, R.: 1993, *Astrophys, J.* **413**, 518.

Setti, G., Woltjer, L.: 1994, *Astrophys. J. Suppl. Ser.* **92**, 629.

Seyfert, C.K.: 1943, *Astrophys, J.* **97**, 28.

Shafer, R.A., Fabian, A.C.: 1983, in *Early Evolution of the Universe and Its Present Structure*, G.O. Abell, G. Chincarini (eds.), D. Reidel Pub. Co., Dordrecht, Holland, p. 333.

Shakura, N., Sunyaev, R.A.: 1973, *Astron. Astrophys.* **24**, 337.

Shapiro, P.R., Field, G.B.: 1976, *Astrophys, J.* **205**, 762.

Sheffer, E.K., Kopaeva, I.F., Averintsev, M.B., Bisnovatyi-Kogan, G.S., Golynskaya, I.M., Gurin, L.S., *et al.*: 1992, *Soviet Astron* **36(1)**, 41.

Sherman, R.D., Silk, J.: 1979, *Astrophys. J.* **238**, 794.

Shibazaki, N., Lamb, F.K.: 1987, *Astrophys. J.* **318**, 767.

Shlosman, I., Vitello, P.A.J.: 1993, *Astrophys. J.* **409**, 372.

Shlosman, I., Vitello, P.A.J., Mauche, C.W.: 1996, *Astrophys. J.* **461**, 377.

Shu, F.H., Adams, F.C., Lizano, S: 1987, *Ann. Rev. Astron. Astrophys.* **25**, 23.

Shull, J.M., Beckwith, S.: 1982, *Ann. Rev. Astron. Astrophys.* **20**, 163.

Shull, J.M., Van Steenberg, M.E.: 1985, *Astrophys. J.* **294**, 599.

Shull, J.M., Slavin, J.D.: 1994, *Astrophys. J.* **427**, 784.

Shulman, S., Fritz, G., Meekins, J., Friedman, H.: 1971, *Astrophys. J. Lett.* **168**, L49.

Schulz, N. S., Hasinger, G., Trüemper, J.: 1989, *Astron. Astrophys.* **225**, 48.

Sidoli, L.: 2000, *Ph. D. Thesis*, University of Milan, Italy.

Sidoli, L.: 2001, in *Frontier Objects in Astrophysics and Particle Physics*, F. Giovannelli & G. Mannocchi (eds.), Italian Physical Society, Editrice Compositori, Bologna, Italy **73**, 193.

Sidoli, L., Mereghetti, S., Israel, G.L., Chiappetti, L., Treves, A., Orlandini, M.: 1999, *Astrophys. J.* **525**, 215.

Sidoli, L., Parmar, A.N., Oosterbroek, T., Stella, L., Verbunt, F., Masetti, N., *et al.*: 2001, *Astron. Astrophys.* **368**, 451.

Silberberg, R., Tsao, C.H., Shapiro, M.M., Biermann, P.L.: 1991, in *Cosmic Rays, Supernovae and the Interstellar Medium*, M.M. Shapiro, R. Silberberg, J.P. Wefel (eds.), *NATO ASI Ser.*, Kluwer Academic Publishers, Dordrecht, Holland **337**, 97.

Silk, J.: 1989, in *Particle Astrophysics: Forefront Experimental Issues*, E.B. Norman (ed.), World Scientific Pub. Co. Pte Ltd, Singapore, p. 229.

Silk, J., Stebbins, A.: 1983, *Astrophys. J.* **269**, 1.

Simon, T., Herbig, G., Boesgaard, A.M.: 1985, *Astrophys. J.* **293**, 551.

Sion, E.M.: 1986, *Pub. Astron. Soc. Pacific* **98**, 821.

Sion, E.M.: 1991, *Astron. J.* **102**, 295.

Sion, E.M., Leckemby, H.J., Szkody, P.: 1990, *Astrophys. J. Lett.* **364**, L41.

Sion, E.M., Long, K.S., Szkody, P., Huang, M.: 1994, *Astrophys. J. Lett.* **430**, L53.

Sion, E.M., Cheng, F.H., Sparks, W.M., Szkody, P., Huang, M., Hubeny, I.: 1997, *Astrophys. J. Lett.* **480**, L17.

Skinner, G.K., Bedford, D. K., Elsner, R.F., Leahy, D., Weisskopf, M.C., Grindlay, J.: 1982, *Nature* **297**, 568.

Skinner, G.K., Kennea, J.A.: 1999, in *Multifrequency Behaviour of High Energy Cosmic Sources*, F. Giovannelli & L. Sabau-Graziati (eds.), *Mem. S.A.It.* **70**, 1057.

Skumanich, A.: 1972, *Astrophys. J.* **171**, 565.

Skuppin, R., Grewing, M., Bianchi, L., de Boer, K.S.: 1987, *Astron. Astrophys.* **177**, 228.

Smak, J.: 1985a, in *Multifrequency Behaviour of Galactic Accreting Sources*, F. Giovannelli (ed.), SIDEREA, Roma, Italy, p. 17.

Smak, J.: 1985b, in *Multifrequency Behaviour of Galactic Accreting Sources*, F. Giovannelli (ed.), SIDEREA, Roma, Italy, p. 3.

Smak, J.: 2002, in *Multifrequency Behaviour of High Energy Cosmic Sources*, F. Giovannelli & L. Sabau-Graziati (eds.), *Mem. S.A.It.* **73 N. 1**, 206.

Smialkowski, A., Wolfendale, A.W., Zhang, L.: 1997, in *Frontier Objects in Astrophysics and Particle Physics*, F. Giovannelli & G. Mannocchi (eds.), Italian Physical Society, Editrice Compositori, Bologna, Italy **57**, 269.

Smith, A., Davelaar, J., Peacock, A., Taylor, B.G., Morini, M., Robba, N.R.: 1988, *Astrophys. J.* **325**, 288.

Smith, B., Mushotzky, R., Serlemitsos, P.J.: 1979, *Astrophys. J.* **227**, 37.

Smith, L.J., Willis, A.J.: 1983, *Astron. Astrophys. Suppl. Ser.* **54**, 229.

Smoot, G.F., De Amici, G., Friedman, S.D., Witebsky, C., Sironi, G., Bonelli, G., *et al.*: 1985, *Astrophys. J. Lett.* **291**, L23.

Smoot, G.F., Bennett, C.L., Kogut, A., Aymon, J., Backus, C., De Amici, G., *et al.*: 1991, *Astrophys. J. Lett.* **371**, L1.

Smoot, G.F., Bennett, C.L., Kogut, A., Wright, E.L., Aymon, J., Bogges, N.W., *et al.*: 1992, *Astrophys. J. Lett.* **396**, L1.

Snowden, S.L., Cox, D.P., McCammon, D., Sanders, W.T.: 1990, *Astrophys. J.* **354**, 211.

Sodorski, T.J., Dwek, E., Hauser, M.G., Kerr, F.J.: 1987, in *Star Formation in Galaxies*, C.J. Lonsdale (ed.), NASA PRINT OFF, Washington DC, USA, p. 99.

Soifer, B.T., Houck, J.R., Harwit, M.: 1971, *Astrophys. J. Lett.* **168**, L73.

Soifer, B.T., Sanders, D.B., Neugebauer, G., Danielson, G.E., Lonsdale, C.J., Madore, B.F., *et al.*: 1986a, *Astrophys. J. Lett.* **303**, L41.

Soifer, B.T., Rice, W.L., Mould, J.R., Gillett, F.C., Rowan-Robinson, M., Habing, H.J.: 1986b, *Astrophys. J.* **304**, 651.

Soifer, B.T., Neugebauer, G., Houck, J.R.: 1987, *Ann. Rev. Astron. Astrophys.* **25**, 187.

Soifer, B.T., Neugebauer, G.: 1991, *Astron. J.* **101**, 354.

Soifer, B.T., Sanders, D.B., Madore, B.F., Neugebauer, G., Danielson, G.E., Elias, J.H., *et al.*: 1987, *Astrophys. J.* **320**, 238.

Sokolov, V.V.: 2001a, in *Gamma Ray Bursts in the Afterglow Era*, E. Costa, F. Frontera & J. Hjorth (eds.), Springer Verlag, Heidelberg, Germany, p. 136.

Sokolov, V.V., Fatkhullin, T.A., Castro-Tirado, A.J., Fruchter, A.S., Komarova, V.N., Kasimova, E.R., *et al.*: 2001b, *Astron. Astrophys.* **372**, 438.

Solinger, A.B., Tucker, W.H.: 1972, *Astrophys. J. Lett.* **175**, L107.

Solinger, A., Rappaport, S., Buff, J.: 1975, *Astrophys. J.* **201**, 381.

Soltan, A., Henry, J.P.: 1983, *Astrophys. J.* **271**, 442.

Sonett, C.P., Morfill, G.E., Jokipii, J.R.: 1987, *Nature* **330**, 458.

Songaila, A., Cowie, L.L., Hogan, C.J., Rugers, M.: 1994, *Nature* **368**, 599.

Sonneborn, G.: 1988, in *A Decade of UV Astronomy with the IUE Satellite*, E. Rolfe (ed.), ESA **SP-281 N. 1**, 111.

Sonneborn, G., Altner, B., Kirshner, R.P.: 1987, *Astrophyys. J. Lett.* **323**, L35.

Spencer, J.H., Schwartz, P.R.: 1974, *Astrophyys. J. Lett.* **188**, L105.

Spinoglio, L., Persi, P., Ferrari-Toniolo, M., Giovannelli, F., Bassani, L., di Cocco, G., *et al.*: 1985, *Astron. Astrophys.* **153**, 55.

Spinoglio, L., Giannini, T., Nisini, B., van den Ancker, M.E., Caux, E., Di Giorgio, A.M., *et al.*: 2000, *Astron. Astrophys.* **353**, 1055.

Spinrad, H., Stone, R.P.S.: 1978, *Astrophys. J.* **226**, 609.

Spitzer, L.: 1956, *Astrophys. J.* **124**, 20.

Spitzer, L., Drake, J.F., Jenkins, E.B., Morton, D.C., Rogerson, J.B., York, D.G.: 1973, *Astrophys. J. Lett.* **181**, L116.
Spitzer, Jr., L., Fitzpatrick, E.L.: 1992, *Astrophys. J. Lett.* **391**, L41.
Sreekumar, P., Bertsch, D.L., Dingus, B.L., Fichtel, C.E., Hartman, R.C., Hunter, *et al.*: 1992, *Astrophys. J. Lett.* **400**, L67.
Sreekumar, P., Bertsch, D.L., Dingus, B.L., Fichtel, C.E., Hartman, R.C., Hunter, *et al.*: 1993, *Phys. Rev. Lett.* **70**, 127.
Sreekumar, P., Bertsch, D.L., Dingus, B.L., Esposito, J. A., Fichtel, C.E., Hartman, R.C., *et al.*: 1998, *Astrophys. J.* **494**, 523.
Srinivasan, R., Boyle, P.J., Buckley, J.H., Carter-Lewis, D.A., Catanese, M., Cawley, M.F., *et al.*: 1997, *Astrophys. J.* **489**, 170.
Stahler, W.S.: 1986, *Pub. Astron. Soc. Pacific* **98**, 1081.
Stahler, S.W., Shu, F.H., Taam, R.E.: 1980, *Astrophys. J* **241**, 637.
Stamenov, J.N., Ushev, S.Z., Nikolsky, S.I., Yakovlev, V.I.: 1983, *Proc. 18th Intern. Cosmic Ray Conf. (Bangalore)* 6, 54.
Stanek, K.Z., Garnavich, P.M., Kaluzny, J., Pych, W., Thompson, I.: 1999, *Astrophys. J. Lett.* **522**, L39.
Stanev, T., Biermann, P.L., Lloyd-Evans, J., Rachen, J.P., Watson, A.A.: 1995, *Phys. Rev Lett.* **75**, 3056.
Stanev, T.: 2001, in *Frontier Objects in Astrophysics and Particle Physics*, F. Giovannelli & G. Mannocchi (eds.), Italian Physical Society, Editrice Compositori, Bologna, Italy **73**, 397.
Staude, J., Appenzeller, I.: 2000, *Ann. Rep. 1998 of Max-Planck-Institute für Astronomie*, Heidelberg-Königstuhl, Germany, pp. 20-21.
Staude, J., Rix, H.-W. (eds.): 2001a, *Ann. Rep. 1999 of Max-Planck-Institute für Astronomie*, Heidelberg-Königstuhl, Germany, pp. 16-17.
Staude, J., Rix, H.-W. (eds.): 2001b, *Ann. Rep. 2000 of Max-Planck-Institute für Astronomie*, Heidelberg-Königstuhl, Germany, pp. 23-24.
Stebbins, A., Silk, J.: 1986, *Astrophys. J.* **300**, 1.
Stecker, F.W.: 1970, *Astrophys. Space Sci.* **6**, 377.
Stecker, F.W., de Jager, O.C., Salamon, M.H.: 1992, *Astrophys. J. Lett.* **390**, L49.
Stecker, F.W., Salamon, M.H., Malkan, M.A.: 1993, *Astrophys. J. Lett.* **410**, L71.
Steigman, G., Hata, N., Felten, J.E.: 1998, in Primordial Nuclei and their Galactic Evolution, N. Prantzos, M. Tosi, & R. von Steiger (eds.), *Space Sci. Rev.* **84**, 31.
Stella, L.: 2002, in *Multifrequency Behaviour of High Energy Cosmic Sources*, F. Giovannelli & L. Sabau-Graziati (eds.), *Mem. S.A.It.* **73 N. 1**, 288.
Stella, L., White, N.E., Priedhorsky, W.: 1987, *Astrophys. J. Lett.* **312**, L17.
Stella, L., Mereghetti, S., Israel, G.L.: 1996, in *Highlights of European Astrophysics*, M. Rodonó & S. Catalano (eds.), *Mem. S.A.It.* **67**, 1053.
Stepanian, A.A.: 1981, *17th Intern. Cosmic Ray Conf. (Paris)* **1**, 50.
Stephens, S.A., Badhwar, G.D.: 1981, *Astrophys. Space Sci.* **76**, 213.
Stephenson, C.B.: 1966, *Astron. J.* **71**, 477.
Stephenson, F.R., Clark, D.H.: 1978, *Applications of Early Astronomical Records*, Hilger Ltd., Bristol, U.K..
Stern, B., Poutanen, J., Svensson, R.: 1999, *Astrophys. J.* **510**, 312.
Stern, R.A.: 1983, *Adv. Space Res.* **2**, 39.
Stover, R.J.: 1981, *Astrophys. J.* **248**, 684.
Stover, R.J., Robinson, E.L., Nather, R.E., Montemayor, T.J.: 1980, *Astrophys. J.* **240**, 597.
van Straaten, S., Ford, E.C., van der Klis, M., Méndez, M., Kaaret, P.: 2000, *Astrophys. J.* **520**, 1049.
Stringfellow, G.S.: 1991, in *Young Star Clusters and Early Stellar Evolution*, F. Palla, P. Persi & H. Zinnecker (eds.), *Mem. S.A.It.* **62**, 847.

Strohmayer, T.E., Zhang, W., Swank, J.H., Smale, A., Titarchuk, L., Day, C., *et al.*: 1996, *Astrophys. J. Lett.* **469**, L9.

Strom, S.E.: 1972, *Publ. Astron. Soc. Pacific* **84**, 745.

Strong, A.W.: 1995, *Space Sci. Rev.* **76**, 205.

Strong, A.W., Wdowczyk, J., Wolfendale, A.W.: 1974, *J. Phys. A* **7**, 1767.

Strong, A.W., Wolfendale, A.W., Worral, D.M.: 1976, *Mon. Not. Roy. Astr. Soc.* **175**, 23p.

Strong, A.W., Wolfendale, A.W., Bennet, K., Wills, R.D.: 1978, *Mon. Not. Roy. Astr. Soc.* **182**, 751.

Strong, A.W., Bennet, K., Blömen, H., Diehl, R., Hermsen, W., Purcell, W., *et al.*: 1996, *Astron. Astrophys. Suppl. Ser.* **120**, 381.

Strong, A.W., Moskalenko, I.V.: 1997, in *Proceedings of the Fourth Compton Symposium*, C.D. Dermer, M.S. Strickman & J.K. Kurfess (eds.), *AIP Conf. Proc.* **410**, 1162.

Strong, A.W., Moskalenko, I.V.: 1998, *Astrophys. J.* **509**, 212.

Struble, M.F., Rood, H.J.: 1982, *Astron. J.* **87**, 7.

Strukov, I.A., Bryukhanov, A.A., Skulachev, D.P., Sazhin, M.V.: 1992, *Sov. Astron. Lett.* **18, 3**, 153.

Strukov, I., Skulachev, D., Brjukhanov, A.: 1995, in *Frontier Objects in Astrophysics and Particle Physics*, F. Giovannelli & G. Mannocchi (eds.), Italian Physical Society, Editrice Compositori, Bologna, Italy **47**, 139.

Sturrock, P.A.: 1971, *Astrophys. J.* **164**, 529.

Sturrock, P.A., Baker, K.B.: 1979, *Astrophys. J.* **234**, 612.

Sunyaev, R.A., Titarchuk, L.G.: 1980, *Astron. Astrophys.* **86**, 121.

Sunyaev, R.A., Arefev, V.A., Borozdin, K.N., Gilfanov, M.R., Efremov, V.V., Kaniovskii, A.S., *et al.*: 1991, *Soviet Astron. Lett.* **17, N. 6**, 409.

Sunyaev, R., Revnivtsev, M.: 2000, *Astron. Astrophys.* **358**, 617.

Suzuki, Y.: 1998, *Space Sci. Rev.* **85**, 91.

Swanenburg, B.N., Hermsen, W., Bennett, K., Bignami, G.F., Caraveo, P., Kanbach, G., *et al.*: 1978, *Nature* **275**, 298.

Swanenburg, B.N., Bennet, K., Bignami, G.F., Buccheri, R., Caraveo, P., Hermsen, W., *et al.*: 1981, *Astrophys. J. Lett.* **243**, L69.

Swank, J.H., Kouveliotou, C., Hurley, K. (eds.): 1998, *X-Ray Timing and Cosmic Gamma Ray Bursts*, Elsevier Science.

Swing, P., Rosenfeld, L.: 1937, *Astrophys. J.* **86**, 483.

Swordy, S.P., L'Heureux, J., Meyer, P., Müller, D.: 1993, *Astrophys. J.* **403**, 658.

Szkody, P.: 1998, in *Ultraviolet Astrophysics, Beyond the IUE Final Archive*, W. Wamstecker & R. González-Riestra, (eds.), ESA Publication Division, ESTEC, Nordwijk, The Netherlands, ESA **SP-413**, 381.

Takahashi, T., Kataoka, J., Kii, T., Kubo, H., Makino, F., Makishima, K., *et al.*: 1996, in *Multifrequency Behaviour of High Energy Cosmic Sources*, F. Giovannelli & L. Sabau-Graziati (eds.), *Mem. S.A.It.* **67 Ns. 1-2**, 533.

Takahashi, T., Kubo, H., Madejski, G., Tashiro, M., Makino, F.: 1997, in *All-sky X-ray Observations in the Next Decade*, M. Matsuoka & N. Kawai (eds.), The Institute of Physical and Chemical Research (RIKEN), Wako, Saitama, Japan, p. 213.

Tammann, G.A., Labhardt, L., Federspiel, M., Sandage, A., Saha, A., Macchetto, F.D., *et al.*: 1996, in *Science with Hubble Space Telescope – II*, P. Benvenuti, F.D. Macchetto and E.J. Schreier (eds.), Space Telescope Science Institute, p. 9.

Tamura, T., Fuzakawa, Y., Makishima, K.: 1996, in *UV and X-ray Spectroscopy of Astrophysical and Laboratory Plasmas*, K. Yamashita & T. Watanabe (eds.), Universal Academy Press, Frontiers Science Series **15**, 387.

Tamura, T., Kaastra, J.S., Peterson, J.R., Paerels, F.B.S., Mittaz, J.P.D., Trudolyubov, S.P., *et al.*: 2001, *Astron. Astrophys. Lett.* **365**, L92.

Tanaka, Y., Inoue, H., Holt, S.S.: 1994, *Publ. Astron. Soc. Japan Lett.* **46, N. 3**, L37.

Tanaka, Y., Lewin, W.H.G.: 1995, in *X-Ray Binaries*, W. Lewin, J. van Paradijis, & E. van den Heuvel (eds.), Cambridge University Press, p. 126.

Tananbaum, H.D.: 1973, in *X-Ray and Gamma Ray Astronomy*, IAU Symposium N. 55, H. Bradt & R. Giacconi (eds.), D. Reidel Publ. Co., Dordrecht, Holland, p. 9.

Tananbaum, H.D., Gursky, H., Kellogg, E.M., Levinson, R., Schreier, E., Giacconi, R.: 1972, *Astrophys. J. Lett.* **174**, L143.

Tanimori, T., Tsukagoshi, T., Kifune, T., Edwards, P.G., Fujimoto, M., Hara, T., *et al.*: 1994, *Astrophys. J. Lett.* **429**, L61.

Tanzi, E.G., Tarenghi, M., Treves, A., Howarth, I.D., Willis, A.J., Wilson, R.: 1980, *Astron. Astrophys.* **83**, 270.

Tashiro, M.: 1992, *Ph. D. Thesis*, University of Tokyo, Japan.

Tauris, T.M., van den Heuvel, E.P.J.: 2003, *astro-ph/0303456*.

Tavani, M.: 1996, *Phys. Rev. Lett.* **76**, 3478.

Taylor, G.B., Allen, S.W., Fabian, A.C.: 1999, in *Diffuse Thermal and Relativistic Plasma in Galaxy Clusters*, Böhringer, H., Feretti, L., Schuecker, P. (eds.), *MPE-Report* **271B**, p. 77.

Tennyson, P.D., Henry, R.C., Feldman, P.D., Hartig, G.F.: 1988, *Astrophys. J.* **330**, 435.

Terebey, S., van Buren, D., Padgett, D.L., Hancock, T., Brundage, M.: 1998, *Astrophys. J. Lett.* **507**, L71.

Terquem, C., Bertout, C.: 1993, *Astron. Astrophys.* **274**, 291.

Thompson, C.: 2002, in *Soft Gamma Ray Repeaters*, M. Feroci & S. Mereghetti (eds.), *Mem. S.A.It.* **73**, 477.

Thompson, C., Duncan, R.C.: 1995, *Mon. Not. Roy. Astr. Soc.* **275**, 255.

Thompson, C., Duncan,. R.C.: 1996, *Astrophys. J.* **473**, 322.

Thompson, C., Lyutikov, M., Kulkarni, S.R.: 2002, *Astrophys. J.* **574**, 332.

Thompson, D.J.: 2000, *Adv. Space Res.* **25**, 659.

Thompson, D.J., Bertsch, D.L., Dingus, B.L., Esposito, J.A., Etienne, A., Fichteli, C.E., *et al.*: Friedlander, D.P., Hartman, R.C., 1995, *Astrophys. J. Suppl. Ser.* **101**, p. 259.

Thompson, D.J., Bertsch, D.L., Dingus, B.L., Esposito, J.A., Etienne, A., Fichtel, C.E., *et al.*: 1996, *Astrophys. J. Suppl. Ser.* **107**, 227.

Thompson, D.J., Harding, A.K., Hermsen, W., Ulmer, M.P.: 1997, *Proc. 4th Compton Symp.*, C.D. Dermer, M.S. Strickman & J.D. Kurfess (eds.), *AIP Conf. Proc.* **410**, 39.

Thronson, H.A.Jr., Harper, D.A., Bally, J., Dragovan, M., Mozurkewich, D., Greenhouse, M.A., *et al.*: 1986, *Astron. J.* **91**, 1350.

Thronson, H.A., Bally, J.: 1987, in *Star Formation in Galaxies*, C. Persson (ed.), NASA, Washington, DC, US Govt. Print. Off., p. 267.

Tiengo, A., Göler, E., Staubert, R., Mereghetti, S.: 2002, *Astron. Astrophys.* **383**, 182.

Tinsley, B.: 1972a, *Astron. Astrophys.* **20**, 383.

Tinsley, B.: 1972b, *Astrophys. J. Lett.* **173**, L93.

Tinsley, B.M.: 1980, *Fund. Cosmic Phys.* **5**, 287.

Tkachenko, A., Terekhov, O.V., Sunyaev, R.A., Burenin, R.A., Barat, C., Dezalay, *et al.*: 2000, *Astron. Astrophys. Lett.* **358**, L41.

Tkaczyk, W., Wdowczyk, J., Wolfendale, A.W.: 1975, *J. Phys. A* **8**, 1518.

Tkaczyk, W., Karakuła, S., Giovannelli, F.: 1984, in *X-Ray Astronomy '84*, M. Oda & R. Giacconi (eds.), ISAS, Tokyo, Japan, p. 499.

Toller, G.N.: 1983, *Astrophys. J. Lett.* **266**, L79.

Topka, K., Avni, Y., Golub, L., Gorenstein, P., Harnden, F.R.Jr., Rosner, R., *et al.*: 1982, *Astrophys. J.* **259**, 677.

Trammell, S.R., Goodrich, R.W., Dinerstein, H.L.: 1995, *Astrophys. J.* **453**, 761.

Trimble, V.: 1992, *Pub. Astr. Soc. Pacific* **104**, 1.

Trimble, V.: 1993, *Pub. Astr. Soc. Pacific* **105**, 1.

Trimble, V., Leonard, P.J.T.: 1994, *Pub. Astr. Soc. Pacific* **106**, 1.

Trimble, V., Leonard, P.J.T.: 1995, *Pub. Astr. Soc. Pacific* **107**, 1.

Trimble, V., Leonard, P.J.T.: 1996, *Pub. Astr. Soc. Pacific* **108**, 8.

Trimble, V., McFadden, L.A.: 1997, *Pub. Astr. Soc. Pacific* **109**, 78.

Trimble, V., McFadden, L.A.: 1998, *Pub. Astr. Soc. Pacific* **110**, 223.

Trimble, V., Aschwanden, M.J.: 1999, *Pub. Astr. Soc. Pacific* **111**, 385.

Trimble, V., Aschwanden, M.J.: 2000, *Pub. Astr. Soc. Pacific* **112**, 434.

Trimble, V., Aschwanden, M.J.: 2001, *Pub. Astr. Soc. Pacific* **113**, 1025.

Trimble, V., Aschwanden, M.J.: 2002, *Pub. Astr. Soc. Pacific* **114**, 475.

Trinchieri, G., Fabbiano, G.: 1985, *Astrophys. J.* **296**, 447.

Trombka, J.I., Dyer, C.S., Evans, L.G., Bielefeld, M.J., Seltzer, S.M., Metzger, A.E.: 1977, *Astrophys. J.* **212**, 925.

Trümper, J.: 1990, in *Observatories in Earth Orbit and Beyond*, Y. Kondo (ed.), Kluwer Academic Publ., Dordrecht, Boston, p. 61.

Trümper, J., Pietsch, W., Reppin, C., Voges, W., Staubert, R., Kendziorra, E.: 1978, *Astrophys. J. Lett.* **219**, L105.

Trümper, J., Becker, W.: 1998, *Adv. Space Res.* **21**, 203.

Trushkin, S.: 2002, *SNR catalog*, (http://cats.sao.com)

Tsunemi, H., Miyata, E., Kinugasa, K., Hiraga, J.: 2000, *Adv. Space Res.* **25**, 539.

Tsuru, T., Koyama, K., Hughes, J.P., Arimoto, N., Kii, T., Hattori, M.: 1996, in *UV and X-Ray Spectroscopy of Astrophysical and Laboratory Plasmas*, K. Yamashita, T. Watanabe (eds.), Universal Academy Press, Inc., Tokyo, Japan, p. 375.

Turner, D.G., Forbes, D.: 1982, *Pub. Astron. Soc. Pacific* **94**, 789.

Turner, M.J.L., Thomas, H.D., Patchett, B.E., Reading, D.H., Makishima, K., Ohashi, T., *et al.*: 1989, *Publ. Astr. Soc. Pacific* **41**, 345.

Turner, T.J., Pounds, K.A.: 1989, *Mon. Not. Roy. Astr. Soc.* **240**, 833.

Ueda, Y., Takahashi, T., Inoue, H., Tsuru, T., Sakano, M., Ishisaki, Y., *et al.*: 1999, *Astrophys. J.* **518**, 656.

Ueda, Y., Akiyama, M., Hangere, G., Inoue, H., Ishisaki, Y., Makishima, K., *et al.*: 2000, *Adv. Space Res.* **25**, 839.

Ueda, Y., Yamaoka, K., Sánchez-Fernández, C., Dhawan, V., Chaty, S., Grove, J.E., *et al.*: 2002, *Astrophys. J.* **571**, 918.

Ulla, A., Garzón, F., De Graauw, T., López, R.: 1999, in *Multifrequency Behaviour of High Energy Cosmic Sources*, F. Giovannelli & L. Sabau-Graziati (eds.), *Mem. S.A.It.* **70**, 1201.

Ulla, A., Giovannelli, F., Martinez-Pais, I.G.: 1999, in *Multifrequency Behaviour of High Energy Cosmic Sources*, F. Giovannelli & L. Sabau-Graziati (eds.), *Mem. S.A.It.* **70**, 971.

Ulrich, M.H., Altamore, A., Perola, G.C., Boksenberg, A., Penston, M.V., Snijders, M.A.J., *et al.*: 1985, *Nature* **313**, 747.

Urry, C.M., Padovani, P.: 1995, *Pub. Astron. Soc. Pacific* **107**, 803.

Vacanti G., Cawley M.F., Colombo E., Fegan D.J., Hillas A.M., Kwok P.W., *et al.*: 1991, *Astrophys. J.* **377**, 467.

Vagnetti, F., Cavaliere, A., Giallongo, E.: 1991, *Astrophys. J.* **368**, 366.

Vagnetti, F., Spera, R.: 1994, *Astrophys. J.* **436**, 611.

Vaiana, G.S., Cassinelli, J.P., Fabbiano, G., Giacconi, R., Golub, R., Gorenstein, P., *et al.*: 1981, *Astrophys. J.* **245**, 163.

Van Steenberg, M.E., Shull, J.M.: 1988, *Astrophys. J.* **330**, 942.

van der Veen, W.E.C.J.: 1987, in *The Late Stages of Stellar Evolution*, S. Kwok & S.R. Pottash (eds.), D. Reidel Publ. Co., Dordrecht, Holland, p. 213.

Vedrenne, G., Jean, P., Kandel, B., Albernhe, F., Borrel, V., Mandrou, P., *et al.*: 1999, in *Multifrequency Behaviour of High Energy Cosmic Sources*, Giovannelli & L. Sabau-Graziati (eds.), *Mem. S.A.It.* **70**, 1289.

Verbunt, F.: 1987, *Astron. Astrophys. Suppl. Ser.* **71**, 339.

Verbunt, F.: 2001, *Astron. Astrophys.* **368**, 137.
Verbunt, F., Hassall, B.J.M., Pringle, J.E., Warner, B., Marang, F.: 1987, *Mon. Not. Roy. Astr. Soc.* **225**, 113.
Verbunt, F., Belloni, T., Johnston, H.M., van der Klis, M., Lewin, W.H.G.: 1994, *Astron. Astrophys.* **285**, 903.
Verbunt, F., Bunk, W., Hasinger, G., Johnston, H.M.: 1995, *Astron. Astrophys.* **300**, 732.
Verbunt, F., Bunk, W.H., Ritter, H., Pfeffermann, E.: 1997, *Astron. Astrophys.* **327**, 602.
Verbunt, F., Hasinger, G.: 1998, *Astron. Astrophys.* **336**, 895.
Verschuur, G.L.: 1980, *Fund. Cosmic Phys.* **5**, 113.
Verschuur, G.L.: 1989, *Interstellar Matters: Essays on Curiosity and Astronomical Discovery*, Springer-Verlag, New York Inc.
Viala, Y.P., Letzelter, C., Eidelsberg, M., Rostas, F.: 1988, *Astron. Astrophys.* **193**, 265.
Vietri, M.: 1998, in *Le Rencontres de Physique de la Vallée d'Aoste: Results and Perspectives in Particle Physics*, M. Greco (ed.), Frascati Physics Series **Vol. XII**, 5.
Vietri, M., Stella, L.: 1998, *Astrophys. J. Lett.* **507**, L45.
Vigroux, L., Mirabel, F., Altieri, B., Boulanger, F., Cesarsky, C., Cesarsky, D., *et al.*: 1996, *Astron. Astrophys. Lett.* **315**, L93.
Vihlu, O., Poutanen, J., Nikula, P., Nevalainen, J.: 2001, *Astrophys. J. Lett.* **553**, L51.
Villada, M., Giovannelli, F., Polcaro, V.F.: 1992, *Astron. Astrophys. Lett.* **259**, L1.
Villada, M., Rossi, C., Polcaro, V.F., Giovannelli, F.: 1999, *Astron. Astrophys.* **344**, 277.
Villard, P.: 1900, *Compt. Rend. Acad. Sci. Paris* **130**, 1010.
Vitello, P., Shlosman, I.: 1993, *Astrophys. J.* **410**, 815.
Vladilo, G.: 1986, *Ph. D. Thesis*, ISAS, Trieste, Italy.
Vladilo, G., Beckman, J.E., Crivellari, L., Franco, M.L., Molaro, P.: 1985, *Astron. Astrophys.* **144**, 81.
Völk, H.J., Biermann, P.L.: 1988, *Astrophys. J. Lett.* **333**, L65.
Vreeswijk, P.M., Galama, T.J., Rol, E., Pollard, K., Menzies, J., Sackett, P., *et al.*: 1999a, *GCN Circ.* **N. 310**.
Vreeswijk, P.M., Galama, T.J., Rol, E., Stappers, B., Palazzi, E., Pian, *et al.*: 1999b, *GCN Circ.* **N. 324**.
de Vries, H.W., Heithausen, A., Thaddeus, P.: 1987, *Astrophys. J.* **319**, 723.
Wagoner, R.V.: 1992, in *The Astronomy and Astrophysics Encyclopedia*, Maran, S.P (ed.), Cambridge University Press, UK, p. 161.
Wagoner, R.V., Fowler, W.A., Hoyle, F.: 1967, *Astrophys. J.* **148**, 3.
Walker, H.J.: 1985, *Astron. Astrophys.* **152**, 58.
Walker, H.J.: 1986, in *Hydrogen Deficient Stars*, IAU Coll. N. 87, D. Reidel Publ. Co., Dordrecht, Holland, p. 407.
Walker, M.F.: 1956, *Astrophys. J. Suppl. Ser.* **2**, 365.
Walker, T.: 1998, in *Primordial Nuclei and their Galactic Evolution*, N. Prantzos, M. Tosi, & R. von Steiger (eds.), *Space Sci. Rev.* **84**, 55.
Walter, F.M., Cash, W., Charles, P.A., Bowyer, C.S.: 1980, *Astrophys. J.* **245**, 671.
Walter, F.M., Bowyer, C.S.: 1981, *Astrophys. J.* **245**, 671.
Walter, F.M., Brown, A., Mathiew, R.D., Myers, P.C., Vrba, F.J.: 1988, *Astron. J.* **96**, 297.
Walterbos, R.A.M., Schwering, P.B.W.: 1986, in *Light on Dark Matter*, F.P. Israel (ed.), D. Reidel Publ. Co., Dordrecht, Holland, ASSL **124**, 401.
Wamstecker, W., González-Riestra, R. (eds.): 1998, *Ultraviolet Astrophysics Beyond the IUE Final Archive*, ESA Publication Division, ESTEC, Nordwijk, The Netherlands, ESA **SP-413**.
Ward, M.J.: 1988, *Mon. Not. Roy. Astr. Soc.* **231**, 1P.
Ward, M.J.: 1995, in *Science with the VLT*, J.R. Walsh & I.J. Danziger (eds.), Springer-Verlag Berlin, Heidelberg, New York, p. 304.
Warner, B.: 1976, *The Observatory* **96**, 49.

Warner, B.: 1995, in *Cataclysmic Variable Stars* Cambridge Astrophysics Series **28**, Cambridge University Press.
Warner, B., Wickramasinghe, D.T.: 1991, *Mon. Not. Roy. Astr. Soc.* **248**, 370.
Warwick, R.S., Pye, J.P., Fabian, A.C.: 1980, *Mon. Not. Roy. Astr. Soc.* **190**, 243.
Warwick, R.S., Norton, A.J., Turner, M.J.L., Watson, M.G., Willingale, R.: 1988, *Mon. Not. Roy. Astr. Soc.* **232**, 551.
Waters, L.B.F.M.: 1986a, *Astron. Astrophys. Lett.* **159**, L1.
Waters, L.B.F.M.: 1986b, *Astron. Astrophys.* **162**, 121.
Waters, L.B.F.M.: 1986c, in *Light on Dark Matter*, F.P. Israel (ed.), D. Reidel Publ. Co., Dordrecht, Holland, p. 83.
Waters, L.B.F.M.: 1987, *Ph.D. Thesis*, University of Utrecht, Holland.
Waters, L.B.F.M., Wesselius, P.R.: 1986, *Astron. Astrophys.* **155**, 104.
Watson, A.A.: 1984, *Adv. Space Res.* **4**, 35.
Watson, M.G., Willingale, R., Grindlay, J.E., Hertz, P.: 1981, *Astrophys. J.* **250**, 142.
Watson, M.G., King, A.R., Heise, J.: 1985, *Space Sci. Rev.* **40**, 127.
Watson, R.A., Gutierrez de la Cruz, C.M., Davies, R.D., Lasenby, A.N., Rebolo, R., Beckman, J.E., *et al.*: 1992, *Nature* **357**, 660.
Wdowczyk, J., Tkaczyk, W., Wolfendale, A.W.: 1972, *J. Phys. A.* **5**, 1419.
Wdowczyk, J., Wolfendale, A.W.: 1975, *Nature* 258, 217.
Wdowczyk, J., Wolfendale, A.W.: 1983, *Nature* **305**, 609.
Wdowczyk, J., Wolfendale, A.W.: 1984, *J. Phys. A.* **10**, 1453.
Wdowczyk, J., Wolfendale, A.W.: 1987, *J. Phys. G: Nucl. Phys.* **13**, 411.
Webster, B.L., Murdin, P.: 1972, *Nature* **235**, 37.
Weekes, T.C.: 1988, *Phys. Rep.* **160**, 1.
Weekes, T.C.: 1992, *Space Sci. Rev.* **59**, 315.
Weekes, T.C.: 1996, *Space Sci. Rev.* **75**, 1.
Weekes, T.C.: 1997, in *Frontier Objects in Astrophysics and Particle Physics*, F. Giovannelli & G. Mannocchi (eds.), Italian Physical Society, Editrice Compositori, Bologna, Italy **57**, 231.
Wefel, J.: 1988, in *Genesis and Propagation of Cosmic Rays*, Shapiro, M.M. & Wefel, J.P. (eds.), Dordrecht, D. Reidel Publishing Co., *NATO ASI Ser.* **C220**, 1.
Weidenspointner, G.: 1999, *The Origin of the Cosmic Gamma Ray Background in the Comptel Energy Range*, Dissertation, Max-Planck-Institut für Extraterrestrische Physik, Garching.
Weiland, J.L., Arendt, R.G., Berriman, G.B., Dwek, E., Freudenreich, H.T., Hauser, M.G., *et al.*: 1994, *Astrophys. J. Lett.* **425**, L81.
Weiler, K.W., Panagia, N.: 1978, *Astron. Astrophys.* **70**, 419.
Weiler, L.W., Sramek, R.A.: 1988, *Ann. Rev. Astron. Astrophys.* **26**, 295.
Weiner, B.J., Williams, T.B.: 1996, *Astron. J.* **111**, 1156.
Weintraub, D.A., Masson, C.R., Zuckerman, B.: 1987, *Astrophys. J.* **320**, 336.
Weintraub, D., Sandell, D., Duncan, W.D.: 1989, *Astrophys. J. Lett.* **340**, L69.
Weisskopf, M.C.: 1990, in *Observatories in Earth Orbit and Beyond*, Y. Kondo (ed.), Kluwer Academic Publ., Dordrecht, Boston, p. 71.
Weisskopf, M.C., Hester, J.J., Tennant, A.F., Elsner, R.F., Schulz, N.S., Marshall, H.L., *et al.*: 2000, *Astrophys. J. Lett.* **536**, L81.
Weller, C.S.: 1983, *Astrophys. J.* **268**, 899.
Weymann, R.: 1967, *Astrophys. J.* **147**, 887.
Wheaton, W.A., Doty, J.P., Primini, F.A., Cooke, B.A., Dobson, C.A., Goldman, A., *et al.*: 1979, *Nature* **282**, 240.
White, G.L., Ricketts, M.J.: 1979, *Mon. Not. Roy. Astr. Soc.* **187**, 757.
White, M., Scott, D., Silk, J.: 1994, *Ann. Rev. Astron. Astrophys.* **32**, 319.
White, N.E.: 1989, *Astron. Astrophys. Rev.* **1**, 85.

White, N. E., Kaluzienski, J.L., Swank, J.H.: 1984, in *High Energy Transients in Astrophysics*, Santa Cruz, p. 31.
White, N.E., Stella, L., Parmar, A.N.: 1988, *Astrophys. J.* **324**, 363.
White, R.A., Burns, J.O.: 1980, *Astron. J.* **85**, 117.
White, R.E. III, Chevalier, R.A.: 1983, *Astrophys. J.* **275**, 69.
White, S.D.M., Navarro J.F., Evrard, A.E., Frenk, C.S.: 1993, *Nature* **366**, 429.
White, S.M., Pallavicini, R., Kundu, M.R.: 1992a, *Astron. Astrophys.* **259**, 149.
White, S.M., Pallavicini, R., Kundu, M.R.: 1992b, *Astron. Astrophys.* **257**, 557.
Whitt, D.C.B., Bode, M.F., Longmore, A.J., Admason, A.J., McFadzean, A.D., Aitken, D.K., *et al.*: 1988, *Mon. Not. R. astr. Soc.* **233**, 321.
Wijers, R.A.M.J., Vreeswijk, P.M., Galama, T.J., Rol, E., van Paradijs, J., Kouveliotou, C., *et al.*: 1999, *Astrophys. J. Lett.* **523**, L33.
Wijnands, R.: 1999, *Ph.D. thesis*, Univ. Amsterdam, Holland.
Wijnands, R.: 2001, *Adv. Space Res.* **28**, 469.
Wijnands, R., van der Klis, M.: 1998, *Astrophys. J. Lett.* **507**, L63.
Wilking, B.A., Lada, C.J., Young, E.T.: 1989, *Astrophys. J.* **340**, 823.
Wilkinson, D.T.: 1987, in *13th Texas Symp. on Relativistic Astrophysics*, M.P. Ulmer (ed.), World Scientific, Singapore, p. 209.
Williamson, F.O., Sanders, W.T., Kraushaar, W.L., McCammon, D., Borken, R., Bunner, A.N.: 1974, *Astrophys. J. Lett.* **193**, L133.
Wills, B.J., Netzer, H., Wills, D.:, 1985, *Astrophys. J.* **288**, 94.
Wilson, C.T.R.: 1900, *Proc. Cambridge Phil. Soc.* **11**, 32.
Wilson, O.C., Vainu Bappu, M.K.: 1957, *Astrophys. J.* **125**, 661.
Wilson, R.: 1970, in *Ultraviolet Stellar Spectra and Related ground based Observations*, L. Houziaux & H.E. Butler (eds.), IAU Symp. N. 36, D. Reidel Pub. Co., Dordrecht, Holland, p.147.
Wilson, R.E.: 1972, *Astrophys. J. Lett.* **174**, L27.
Wilson, R.W., Jefferts, K.B., Penzias, A.A.: 1970, *Astrophys. J. Lett.* **161**, L43.
Wilson, T.L., Matteucci, F.: 1992, *Astron. Astrophys. Rev.* **4**, 1.
Wilson, T.L., Rood, R.T.: 1994, *Ann. Rev. Astron. Astrophys.* **32**, 191.
Withworth, A.: 1979, *Mon. Not. Roy. Astr. Soc.* **186**, 59.
Witt, A.N., Stecher, T.P., Boroson, T.A., Bohlin, R.C.: 1989, *Astrophys. J. Lett.* **336**, L21.
Wittman, D.M., Tyson, J.A., Kirkman, D., Dell'Antonio, I., Bernstein, G.: 2000, *Nature* **405**, 143.
van Woerden, H., Schwarz, U.J., Hulsbosch, A.N.M.: 1985, in *The Milky Way Galaxy*, H. van Woerden, R.J. Allen & W.B. Burton (eds.), *IAU Symposium, N. 106*, D. Reidel Publ. Co., Dordrecht, Holland, p. 387.
Wolf, M.: 1906, *Astron. Nachr.* **170**, 211.
Wolfendale, A.W.: 1995, in *Frontier Objects in Astrophysics and Particle Physics*, F. Giovannelli & G. Mannocchi (eds.), Italian Physical Society, Editrice Compositori, Bologna, Italy **47**, 463.
Wolfendale, A.W., Zan, M.: 1997, in *Frontier Objects in Astrophysics and Particle Physics*, F. Giovannelli & G. Mannocchi (eds.), Italian Physical Society, Editrice Compositori, Bologna, Italy **57**, 325.
Woltjer, L.: 1972, *Ann. Rev. Astron. Astrophys.* **10**, 129.
Wood, J.H., Horne, K., Berriman, G., Wade, R.A.: 1989, *Astrophys. J.* **341**, 974.
Wood, K.S., Meekins, J.F., Yentis, D.J., Smathers, H.W., Menutt, D.P., Bleach, R.D., *et al.*: 1984, *Astrophys. J. Suppl. Ser.* **56**, 507.
Woods, J.A., Verbunt, F., Cameron, A.C., Drew, J.E., Piters, A.: 1992, *Mon. Not. Roy. Astr. Soc.* **225**, 237.
Woosley, S.E.: 1993, *Astrophys. J.* **405**, 273.
Woosley, S.E.: 2001, in *Gamma Ray Bursts in the Afterglow Era*, E. Costa, F. Frontera & J. Hjorth (eds.), Springer Verlag, Heidelberg, Germany, p. 257.
Wright, E.L.: 1991, *Astrophys. J.* **375**, 608.

Wright, E.L., Smoot, G.F., Kogut, A., Hinshaw, G., Tenorio, L., Lineweaver, C., *et al.*: 1994, *Astrophys. J.* **420**, 1.

Wu, C.-C.: 1977, *Astrophys. J. Lett.* **217**, L117.

Wynn-Williams, G.: 1992, *The Fullness of Space*, Cambridge University Press, p. 100.

Yadav, J.S.: 2001, *Astrophys. J.* **548**, 876.

Yamashita, K., Watanabe, T.(eds.): 1996, *UV and X-Ray Spectroscopy of Astrophysical and Laboratory Plasmas*, Universal Academy Press, Inc., Tokyo, Japan.

Yentis, D.J., Cruddace, R.G., Gursky, H., Stuart, B.V., Wallin, J.F., MacGillivray, H.T., *et al.*: 1992, in *Digitized Optical Sky Surveys*, H.T. MacGillivray, E.B. Thomson (eds.), Kluwer Academic Publishers, Dordrecht, Holland, p. 67.

York, D.G.: 1983, *Astrophys. J.* **264**, 172.

York, D.G., Spitzer, L., Bohlin, R.C., Hill, J., Jenkins, E.B., Savage, B.D., *et al.*: 1983 *Astrophys. J. Lett.* **266**, L55.

York, D.G., Green, R.F., Bechtold, J., Chaffee, F.H.Jr.: 1984, *Astrophys. J. Lett.* **280**, L1.

Young, J.S., Scoville, N.: 1982, *Astrophys. J. Lett.* **260**, L11.

Yusef-Zadeh, F., Morris, M.: 1987, *Astrophys. J.* **320**, 545.

Yusef-Zadeh, F., Roberts, D.A., Goss, W.M., Frail, D.A., Green, A.J.: 1996, *Astrophys. J. Lett.* **466**, L25.

Yusef-Zadeh, F., Purcell, W., Gotthelf, E.: 1997, in *Proceedings Fourth Compton Symposium*, C.D. Dermer, M.S. Strickman and J.D. Kurfess (eds.), *AIP Conf. Proc.* **410**, p. 1027.

Zappala, R.R.: 1972, *Astrophys. J.* **172**, 57.

Zatsepin, V.I.: 1995, *J. Phys. G.* **21**, L31.

Zdziarski, A.A., Krolik, J.H.: 1993, *Astropys. J. Lett.* **409**, L33.

Zdziarski, A.A., Zycki, P.T., Krolik, J.H.: 1993, *Astropys. J. Lett.* **414**, L81.

Zdziarski, A.A., Grove, J.E., Poutanen, J., Rao, A.R., Vadawale, S.V.: 2001, *Astropys. J. Lett.* **554**, L45.

Zel'dovich, Ya.,B., Sunyaev, R.A.: 1969, *Astrophys. Space Sci.* **4**, 301.

Zhang, S.N., Cui W., Chen W.: 1997, *Astrophys. J. Lett.* **482**, L155.

Zhang, W., Morgan E.H., Jahoda K., Swank J.H., Strohmayer T.E., Jernigan G., *et al.*: 1996, *Astropys. J. Lett.* **469**, L29.

Zhang, W., Smale, A.P., Strohmayer, T.E., Swank, J.H.: 1998, *Astropys. J. Lett.* **500**, L171.

Ziółkowski, J.: 1997a, in *Relativistic Jets in AGNs*, M. Ostrowski, M. Sikora, G. Madejski & M. Begelman (eds.), p. 38.

Ziółkowski, J.: 1997b, in *Frontier Objects in Astrophysics and Particle Physics*, F. Giovannelli & G. Mannocchi (eds.), Italian Physical Society, Editrice Compositori, Bologna, Italy **57**, 301.

Ziółkowski, J.: 1999, in *Frontier Objects in Astrophysics and Particle Physics*, F. Giovannelli & G. Mannocchi (eds.), Italian Physical Society, Editrice Compositori, Bologna, Italy **65**, 215.

Ziółkowski, J.: 2002, in *Multifrequency Behaviour of High Energy Cosmic Sources*, F. Giovannelli & L. Sabau-Graziati (eds.), *Mem. S.A.It.* **73 N. 1**, 300.

Zombeck, M.V.: 1990, *Handbook of Space Astronomy and Astrophysics*, Cambridge University Press, p. 212.

Zuckerman, B.: 1986, in *Light on Dark Matter*, F.P. Israel (ed.), D. Reidel Publ. Co., Dordrecht, Holland, p. 93.

Zuckerman, B., Dyck, H.M.: 1986a, *Astrophys. J.* **304**, 394.

Zuckerman, B., Dyck, H.M.: 1986b, *Astrophys. J.* **311**, 345.

MIX
Papier aus verantwortungsvollen Quellen
Paper from responsible sources
FSC® C105338

If you have any concerns about our products,
you can contact us on
**ProductSafety@springernature.com**

In case Publisher is established outside the EU,
the EU authorized representative is:
**Springer Nature Customer Service Center GmbH**
**Europaplatz 3, 69115 Heidelberg, Germany**

Printed by Libri Plureos GmbH
in Hamburg, Germany